Biology:
Exploring Life

SECOND EDITION

Biology: Exploring Life

Gil Brum
California State Polytechnic University, Pomona

Larry McKane
California State Polytechnic University, Pomona

Gerry Karp
Formerly of the University of Florida, Gainesville

JOHN WILEY & SONS, INC.

New York / Chichester / Brisbane / Toronto / Singapore

Acquisitions Editor   *Sally Cheney*
Developmental Editor   *Rachel Nelson*
Marketing Manager   *Catherine Faduska*
Associate Marketing Manager   *Deb Benson*
Senior Production Editor   *Katharine Rubin*
Text Designer/"Steps" Illustration Art Direction   *Karin Gerdes Kincheloe*
Manufacturing Manager   *Andrea Price*
Photo Department Director   *Stella Kupferberg*
Senior Illustration Coordinator   *Edward Starr*
"Steps to Discovery" Art Illustrator   *Carlyn Iverson*
Cover Design   *Meryl Levavi*
Text Illustrations   *Network Graphics/Blaize Zito Associates, Inc.*
Photo Editor/Cover Photography Direction   *Charles Hamilton*
Photo Researchers   *Hilary Newman, Pat Cadley, Lana Berkovitz*
Cover Photo   *James H. Carmichael, Jr./The Image Bank*

This book was set in New Caledonia by Progressive Typographers and printed and bound by Von Hoffmann Press. The cover was printed by Lehigh. Color separations by Progressive Typographers and Color Associates, Inc.

Recognizing the importance of preserving what has been written, it is a policy of John Wiley & Sons, Inc. to have books of enduring value published in the United States printed on acid-free paper, and we exert our best efforts to that end.

The paper in this book was manufactured by a mill whose forest management programs include sustained yield harvesting of its timberlands. Sustained yield harvesting principles ensure that the number of trees cut each year does not exceed the amount of new growth.

Library of Congress Cataloging in Publication Data:
Brum, Gilbert D.
    Biology : exploring life/Gil Brum, Larry McKane, Gerry Karp.—2nd ed.
        p.   cm.
    Includes bibliographical references and index.
    ISBN 0-471-54408-6 (cloth)
    1. Biology.   I. McKane, Larry.   II. Karp, Gerald.   III. Title.
    QH308.2.B78   1993
    574—dc20   93-23383
                                                                      CIP

Unit I ISBN 0-471-01827-9 (pbk)
Unit II ISBN 0-471-01831-7 (pbk)
Unit III ISBN 0-471-01830-9 (pbk)
Unit IV ISBN 0-471-01829-5 (pbk)
Unit V ISBN 0-471-01828-7 (pbk)
Unit VI ISBN 0-471-01832-5 (pbk)

Printed in the United States of America

10   9   8   7   6   5   4   3   2   1

For the Student, we hope this book helps you discover the thrill of exploring life and helps you recognize the important role biology plays in your everyday life.

To Margaret, Jan, and Patsy, who kept loving us even when we were at our most unlovable.

To our children, Jennifer, Julia, Christopher, and Matthew, whose fascination with exploring life inspires us all. And especially to Jenny–we all wish you were here to share the excitement of this special time in life.

Preface to the Instructor

Biology: Exploring Life, Second Edition is devoted to the process of investigation and discovery. The challenge and thrill of understanding how nature works ignites biologists' quests for knowledge and instills a desire to share their insights and discoveries. The satisfactions of knowing that the principles of nature can be understood and sharing this knowledge are why we teach. These are also the reasons why we created this book.

Capturing and holding student interest challenges even the best of teachers. To help meet this challenge, we have endeavored to create a book that makes biology relevant and appealing, that reveals biology as a dynamic process of exploration and discovery, and that emphasizes the widening influence of biologists in shaping and protecting our world and in helping secure our futures. We direct the reader's attention toward principles and concepts to dispel the misconception of many undergraduates that biology is nothing more than a very long list of facts and jargon. Facts and principles form the core of the course, but we have attempted to show the *significance* of each fact and principle and to reveal the important role biology plays in modern society.

From our own experiences in the introductory biology classroom, we have discovered that

- emphasizing principles, applications, and scientific exploration invigorates the teaching and learning process of biology and helps students make the significant connections needed for full understanding and appreciation of the importance of biology; and

- students learn more if a book is devoted to telling the story of biology rather than a recitation of facts and details.

Guided by these insights, we have tried to create a process-oriented book that still retains the facts, structures, and terminology needed for a fundamental understanding of biology. With these goals in mind, we have interwoven into the text

1. an emphasis on the ways that science works,
2. the underlying adventure of exploration,
3. five fundamental biological themes, and
4. balanced attention to the human perspective.

This book should challenge your students to think critically, to formulate their own hypotheses as possible explanations to unanswered questions, and to apply the approaches learned in the study of biology to understanding (and perhaps helping to solve) the serious problems that affect every person, indeed every organism, on this planet.

THE DEVELOPMENT STORY

The second edition of *Biology: Exploring Life* builds effectively on the strengths of the First Edition by Gil Brum and Larry McKane. For this edition, we added a third author, Gerry Karp, a cell and molecular biologist. Our complementary areas of expertise (genetics, zoology, botany, ecology, microbiology, and cell and molecular biology) as well as awards for teaching and writing have helped us form a balanced team. Together, we exhaustively revised and refined each chapter until all three of us, each with our different likes and dislikes, sincerely believed in the result. What evolved from this process was a satisfying synergism and a close friendship.

THE APPROACH

The elements of this new approach are described in the upcoming section "To the Student: A User's Guide." These pedagogical features are embedded in a book that is written in an informal, accessible style that invites the reader to explore the process of biology. In addition, we have tried to keep the narrative focused on *processes,* rather than on static facts, while creating an underlying foundation that helps students make the connections needed to tie together the information into a greater understanding than that which comes from memorizing facts alone. One way to help students make these connections is to relate the fundamentals of biology to humans, revealing the human perspective in each biological principle, from biochemicals to ecosystems. With each such insight, students take a substantial step toward becoming the informed citizens that make up responsible voting public.

We hope that, through this textbook, we can become partners with the instructor and the student. The biology teacher's greatest asset is the basic desire of students to understand themselves and the world around them. Unfortunately, many students have grown detached from this natural curiosity. Our overriding objective in creating this book was to arouse the students' fascination with exploring life, building knowledge and insight that will enable them to make real-life judgments as modern biology takes on greater significance in everyday life.

THE ART PROGRAM

The diligence and refinement that went into creating the text of *Biology: Exploring Life,* Second Edition characterizes the art program as well. Each photo was picked specifically for its relevance to the topic at hand and for its aesthetic and instructive value in illustrating the narrative concepts. The illustrations were carefully crafted under the guidance of the authors for accuracy and utility as well as aesthetics. The value of illustrations cannot be overlooked in a discipline as filled with images and processes as biology. Through the use of cell icons, labeled illustrations of pathways and processes, and detailed legends, the student is taken through the world of biology, from its microscopic chemical components to the macroscopic organisms and the environments that they inhabit.

SUPPLEMENTARY MATERIALS

In our continuing effort to meet all of your individual needs, Wiley is pleased to offer the various topics covered in this text in customized paperback "splits." For more details, please contact your local Wiley sales representative. We have also developed an integrated supplements package that helps the instructor bring the study of biology to life in the classroom and that will maximize the students' use and understanding of the text.

The *Instructor's Manual,* developed by Michael Leboffe and Gary Wisehart of San Diego City College, contains lecture outlines, transparency references, suggested lecture activities, sample concept maps, section concept map masters (to be used as overhead transparencies), and answers to study guide questions.

Gary Wisehart and Mark Mandell developed the test bank, which consists of four types of questions: fill-in questions, matching questions, multiple-choice questions, and critical thinking questions. A computerized test bank is also available.

A comprehensive visual ancillary package includes four-color transparencies (200 figures from the text), *Process of Science* transparency overlays that break down various biological processes into progressive steps, a video library consisting of tapes from Coronet MTI, and the *Bio Sci* videodisk series from Videodiscovery, covering topics in biochemistry, botany, vertebrate biology, reproduction, ecology, animal behavior, and genetics. Suggestions for integrating the videodisk material in your classroom discussions are available in the instructor's manual.

A comprehensive study guide and lab manual are also available and are described in more detail in the User's Guide section of the preface.

Acknowledgments

*I*t was a delight to work with so many creative individuals whose inspiration, artistry, and vital steam guided this complex project to completion. We wish we were able to acknowledge each of them here, for not only did they meet nearly impossible deadlines, but each willingly poured their heart and soul into this text. The book you now hold in your hands is in large part a tribute to their talent and dedication.

There is one individual whose unique talent, quick intellect, charm, and knowledge not only helped to make this book a reality, but who herself made an enormous contribution to the content and pedagogical strength of this book. We are proud to call Sally Cheney, our biology editor, a colleague. Her powerful belief in this textbook's new approaches to teaching biology helped instill enthusiasm and confidence in everyone who worked on it. Indeed, Sally is truly a force of positive change in college textbook publishing–she has an uncommon ability to think both like a biologist and an editor; she knows what biologists want and need in their classes and is dedicated to delivering it; she recognizes that the future of biology education is more than just publishing another look-alike text; and she is knowledgeable and persuasive enough to convince publishers to stick their necks out a little further for the good of educational advancement. Without Sally, this text would have fallen short of our goal. With Sally, it became even more than we envisioned.

Another individual also helped make this a truly special book, as well as made the many long hours of work so delightful. Stella Kupferberg, we treasure your friendship, applaud your exceptional talent, and salute your high standards. Stella also provided us with two other important assets, Charles Hamilton and Hilary Newman. Stella and Charles tirelessly applied their skill, and artistry to get us images of incomparable effectiveness and beauty, and Hilary's diligent handling helped to insure there were no oversights.

Our thanks to Rachel Nelson for her meticulous editing, for maintaining consistency between sometimes dissimilar writing styles of three authors, and for keeping track of an incalculable number of publishing and biological details; to Katharine Rubin for expertly and gently guiding this project through the myriad levels of production, and for putting up with three such demanding authors; to Karin Kincheloe for a stunningly beautiful design; to Ishaya Monokoff and Ed Starr for orchestrating a brilliant art program; to Network Graphics, especially John Smith and John Hargraves, who executed our illustrations with beauty and style without diluting their conceptual strength or pedagogy, and to Carlyn Iverson, whose artistic talent helped us visually distill our "Steps to Discovery" episodes into images that bring the process of science to life.

We would also like to thank Cathy Faduska and Alida Setford, their creative flair helped us to tell the story behind this book, as well as helped us convey what we tried to accomplish. And to Herb Brown, thank you for your initial confidence and continued support. A very special thank you to Deb Benson, our marketing manager. What a joy to work with you, Deb, your energy, enthusiasm, confidence, and pleasant personality bolstered even our spirits.

We wish to acknowledge Diana Lipscomb of George Washington University for her invaluable contributions to the evolution chapters, and Judy Goodenough of the University of Massachusetts, Amherst, for contributing an outstanding chapter on Animal Behavior.

To the reviewers and instructors who used the First Edition, your insightful feedback helped us forge the foundation for this new edition. To the reviewers, and workshop and conference participants for the Second Edition, thank you for your careful guidance and for caring so much about your students.

Dennis Anderson, *Oklahoma City Community College*
Sarah Barlow, *Middle Tennessee State University*
Robert Beckman, *North Carolina State University*
Timothy Bell, *Chicago State University*
David F. Blaydes, *West Virginia University*
Richard Bliss, *Yuba College*
Richard Boohar, *University of Nebraska, Lincoln*
Clyde Bottrell, *Tarrant County Junior College*
J. D. Brammer, *North Dakota State University*
Peggy Branstrator, *Indiana University, East*
Allyn Bregman, *SUNY, New Paltz*
Daniel Brooks, *University of Toronto*

Gary Brusca, *Humboldt State University*
Jack Bruk, *California State University, Fullerton*
Marvin Cantor, *California State University, Northridge*
Richard Cheney, *Christopher Newport College*
Larry Cohen, *California State University, San Marcos*
David Cotter, *Georgia College*
Robert Creek, *Eastern Kentucky University*
Ken Curry, *University of Southern Mississippi*
Judy Davis, *Eastern Michigan University*
Loren Denny, *southwest Missouri State University*
Captain Donald Diesel, *U. S. Air Force Academy*
Tom Dickinson, *University College of the Cariboo*

Mike Donovan, *Southern Utah State College*
Robert Ebert, *Palomar College*
Thomas Emmel, *University of Florida*
Joseph Faryniarz, *Mattatuck Community College*
Alan Feduccia, *University of North Carolina, Chapel Hill*
Eugene Ferri, *Bucks County Community College*
Victor Fet, *Loyola University, New Orleans*
David Fox, *Loyola University, New Orleans*
Mary Forrest, *Okanagan University College*
Michael Gains, *University of Kansas*
S. K. Gangwere, *Wayne State University*
Dennis George, *Johnson County Community College*
Bill Glider, *University of Nebraska*
Paul Goldstein, *University of North Carolina, Charlotte*
Judy Goodenough, *University of Massachusetts, Amherst*
Nathaniel Grant, *Southern Carolina State College*
Mel Green, *University of California, San Diego*
Dana Griffin, *Florida State University*
Barbara L. Haas, *Loyola University of Chicago*
Richard Haas, *California State University, Fresno*
Fredrick Hagerman, *Ohio State University*
Tom Haresign, *Long Island University, Southampton*
Jane Noble-Harvey, *University of Delaware*
W. R. Hawkins, *Mt. San Antonio College*
Vernon Hedricks, *Brevard Community College*
Paul Hertz, *Barnard College*
Howard Hetzle, *Illinois State University*
Ronald K. Hodgson, *Central Michigan University*
W. G. Hopkins, *University of Western Ontario*
Thomas Hutto, *West Virginia State College*
Duane Jeffrey, *Brigham Young University*
John Jenkins, *Swarthmore College*
Claudia Jones, *University of Pittsburgh*
R. David Jones, *Adelphi University*
J. Michael Jones, *Culver Stockton College*
Gene Kalland, *California State University, Dominiquez Hills*
Arnold Karpoff, *University of Louisville*
Judith Kelly, *Henry Ford Community College*
Richard Kelly, *SUNY, Albany*
Richard Kelly, *University of Western Florida*
Dale Kennedy, *Kansas State University*
Mirium Kittrell, *Kingsboro Community College*
John Kmeltz, *Kean College New Jersey*
Robert Krasner, *Providence College*
Susan Landesman, *Evergreen State College*
Anton Lawson, *Arizona State University*
Lawrence Levine, *Wayne State University*
Jerri Lindsey, *Tarrant County Junior College*
Diana Lipscomb, *George Washington University*
James Luken, *Northern Kentucky University*

Ted Maguder, *University of Hartford*
Jon Maki, *Eastern Kentucky University*
Charles Mallery, *University of Miami*
William McEowen, *Mesa Community College*
Roger Milkman, *University of Iowa*
Helen Miller, *Oklahoma State University*
Elizabeth Moore, *Glassboro State College*
Janice Moore, *Colorado State University*
Eston Morrison, *Tarleton State University*
John Mutchmor, *Iowa State University*
Douglas W. Ogle, *Virginia Highlands Community College*
Joel Ostroff, *Brevard Community College*
James Lewis Payne, *Virginia Commonwealth University*
Gary Peterson, *South Dakota State University*
MaryAnn Phillippi, *Southern Illinois University, Carbondale*
R. Douglas Powers, *Boston College*
Robert Raikow, *University of Pittsburgh*
Charles Ralph, *Colorado State University*
Aryan Roest, *California State Polytechnic Univ., San Luis Obispo*
Robert Romans, *Bowling Green State University*
Raymond Rose, *Beaver College*
Richard G. Rose, *West Valley College*
Donald G. Ruch, *Transylvania University*
A. G. Scarbrough, *Towsow State University*
Gail Schiffer, *Kennesaw State University*
John Schmidt, *Ohio State University*
John R. Schrock, *Emporia State University*
Marilyn Shopper, *Johnson County Community College*
John Smarrelli, *Loyola University of Chicago*
Deborah Smith, *Meredith College*
Guy Steucek, *Millersville University*
Ralph Sulerud, *Augsburg College*
Tom Terry, *University of Connecticut*
James Thorp, *Cornell University*
W. M. Thwaites, *San Diego State University*
Michael Torelli, *University of California, Davis*
Michael Treshow, *University of Utah*
Terry Trobec, *Oakton Community College*
Len Troncale, *California State Polytechnic University, Pomona*
Richard Van Norman, *University of Utah*
David Vanicek, *California State University, Sacramento*
Terry F. Werner, *Harris-Stowe State College*
David Whitenberg, *Southwest Texas State University*
P. Kelly Williams, *University of Dayton*
Robert Winget, *Brigham Young University*
Steven Wolf, *University of Missouri, Kansas City*
Harry Womack, *Salisbury State University*
William Yurkiewicz, *Millersville University*

Gil Brum
Larry McKane
Gerry Karp

Brief Table of Contents

Contents

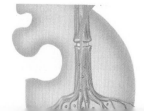

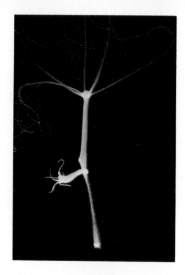

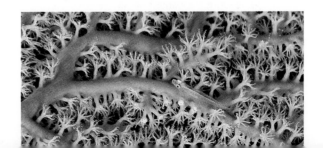

To The Student:
A User's Guide

*B*iology is a journey of exploration and discovery, of struggle and breakthrough. It is enlivened by the thrill of understanding not only what living things do but also how they work. We have tried to create such an experience for you.

Excellence in writing, visual images, and broad biological coverage form the core of a modern biology textbook. But as important as these three factors are in making difficult concepts and facts clear and meaningful, none of them reveals the excitement of biology—the adventure that unearths what we know about life. To help relate the true nature of this adventure, we have developed several distinctive features for this book, features that strengthen its biological core, that will engage and hold your attention, that reveal the human side of biology, that enable every reader to understand how science works, that stimulate critical thinking, and that will create the informed citizenship we all hope will make a positive difference in the future of our planet.

Steps to Discovery

The process of science enriches all parts of this book. We believe that students, like biologists, themselves, are intrigued by scientific puzzles. Every chapter is introduced by a "Steps to Discovery" narrative, the story of an investigation that led to a scientific breakthrough in an area of biology which relates to that chapter's topic. The "Steps to Discovery" narratives portray biologists as they really are: human beings, with motivations, misfortunes, and mishaps, much like everyone experiences. We hope these narratives help you better appreciate biological investigation, realizing that it is understandable and within your grasp.

Throughout the narrative of these pieces, the writing is enlivened with scientific work that has provided knowledge and understanding of life. This approach is meant not just to pay tribute to scientific giants and Nobel prize winners, but once again to help you realize that science does not grow by itself. Facts do not magically materialize. They are the products of rational ideas, insight, determination, and, sometimes, a little luck. Each of the "Steps to Discovery" narratives includes a painting that is meant primarily as an aesthetic accompaniment to the adventure described in the essay and to help you form a mental picture of the subject.

STEPS TO DISCOVERY
A Factor Promoting the Growth of Nerves

Rita Levi-Montalcini received her medical degree from the University of Turin in Italy in 1936, the same year that Benito Mussolini began his anti-Semitic campaign. By 1939, as a Jew, Levi-Montalcini had been barred from carrying out research and practicing medicine, yet she continued to do both secretly. As a student, Levi-Montalcini had been fascinated with the structure and function of the nervous system. Unable to return to the university, she set up a simple laboratory in her small bedroom in her family's home. As World War II raged throughout Europe, and the Allies systematically bombed Italy, Levi-Montalcini studied chick embryos in her bedroom, discovering new information about the growth of nerve cells from the spinal cord into the nearby limbs. In her autobiography *In Praise of Imperfection*, she writes: "Every time the alarm sounded, I would carry down to the precarious safety of the cellars the Zeiss binocular microscope and my most precious silver-stained embryonic sections." In September 1943, German troops arrived in Turin to support the Italian Fascists. Levi-Montalcini and her family fled southward to Florence, where they remained in hiding for the remainder of the w[...]

After the war ended, Levi-Montalcini continued research at the University of Turin. In 1946, she acce[...] an invitation from Viktor Hamburger, a leading expe[...] the development of the chick nervous system, to co[...] Washington University in St. Louis to work with him f[...] semester; she remained at Washington University [...] years.

A chick embryo and one of its nerve cells helped scientists discover nerve growth factor (NGF).

One of Levi-Montalcini's first projects was the reexamination of a previous experiment of Elmer Bueker, a former student of Hamburger's. Bueker had removed a limb from a chick embryo, replaced it with a fragment of a mouse connective tissue tumor, and found that nerve fibers grew into this mass of implanted tumor cells. When Levi-Montalcini repeated the experiment she made an unexpected discovery: One part of the nervous system of these experimental chick embryos—the sympathetic nervous system—had grown five to six times larger than had its counterpart in a normal chick embryo. (The sympathetic nervous system helps control the activity of internal organs, such as the heart and digestive tract.) Close examination revealed that the small piece of tumor tissue that had been grafted onto the embryo had caused sympathetic nerve fibers to grow "wildly" into all of the chick's internal organs, even causing some of the blood vessels to become obstructed by the invasive fibers. Levi-Montalcini hypothesized that the tumor was releasing some soluble substance that induced the remarkable growth of this part of the nervous system. Her hypothesis was soon confirmed by further experiments. She called the active substance **nerve growth factor (NGF)**.

The next step was to determine the chemical nature of NGF, a task that was more readily performed by growing the tumor cells in a culture dish rather than an embryo. But Hamburger's laboratory at Washington University did not have the facilities for such work. To continue the project, Levi-Montalcini boarded a plane, with a pair of tumor-bearing mice in the pocket of her overcoat, and flew to Brazil, where she had a friend who operated a tissue culture laboratory. When she placed sympathetic nervous tissue in the proximity of the tumor cells in a culture dish, the nervous tissue sprouted a halo of nerve fibers that grew toward the tumor cells. When the tissue was cultured in the absence of NGF, no such growth occurred.

For the next 2 years, Levi-Montalcini's lab was devoted to characterizing the substance in the tumor cells that possessed the ability to cause nerve outgrowth. The work was carried out primarily by a young biochemist, Stanley Cohen, who had joined the lab. One of the favored approaches to studying the nature of a biological molecule is to determine its sensitivity to enzymes. In order to determine if nerve growth factor was a protein or a nucleic acid, Cohen treated the active material with a small amount of snake venom, which contains a highly active enzyme that degrades nucleic acid. It was then that chance stepped in.

Cohen expected that treatment with the venom w[...] ther destroy the activity of the tumor cell fraction [...] was a nucleic acid) or leave it unaffected (if NG[...] protein). To Cohen's surprise, treatment with the [...] *increased* the nerve-growth promoting activity of the [...] rial. In fact, treatment of sympathetic nerve tissue w[...] venom alone (in the absence of the tumor extract) ind[...] the growth of a halo of nerve fibers! Cohen soon disco[...] why: The snake venom possessed the same nerve g[...] factor as did the tumor cells, but at much higher conce[...] tion. Cohen soon demonstrated that NGF was a prot[...]

Levi-Montalcini and Cohen reasoned that since s[...] venom was derived from a *modified* salivary gland, [...] other salivary glands might prove to be even better sou[...] of the protein. This hypothesis proved to be correct. Wh[...] Levi-Montalcini and Cohen tested the salivary glands fr[...] male mice, they discovered the richest source of NGF ye[...] source 10,000 times more active than the tumor cells a[...] ten times more active than snake venom.

A crucial question remained: Did NGF play a role [...] the normal development of the embryo, or was its ability t[...] stimulate nerve growth just an accidental property of the [...] molecule? To answer this question, Levi-Montalcini and [...] Cohen injected embryos with an antibody against NGF, [...] which they hoped would inactivate NGF molecules wher[...] ever they were present in the embryonic tissue. The em[...] bryos developed normally, with one major exception: They [...] virtually lacked a sympathetic nervous system. The re[...] searchers concluded that NGF must be important during [...] normal development of the nervous system; otherwise, in[...] activation of NGF could not have had such a dramatic [...] effect.

By the early 1970s, the amino acid sequence of NGF [...] had been determined, and the protein is now being synthe[...] sized by recombinant DNA technology. During the past [...] decade, Fred Gage, of the University of California, has [...] found that NGF is able to revitalize aged or damaged nerve [...] cells in rats. Based on these studies, NGF is currently being [...] tested as a possible treatment of Alzheimer's disease. For [...] their pioneering work, Rita Levi-Montalcini and Stanley [...] Cohen shared the 1987 Nobel Prize in Physiology and [...] Medicine.

*M*any students are overwhelmed by the diversity of living organisms and the multitude of seemingly unrelated facts that they are forced to learn in an introductory biology course. Most aspects of biology, however, can be thought of as examples of a small number of recurrent themes. Using the thematic approach, the details and principles of biology can be assembled into a body of knowledge that makes sense, and is not just a collection of disconnected facts. Facts become ideas, and details become parts of concepts as you make connections between seemingly unrelated areas of biology, forging a deeper understanding.

All areas of biology are bound together by evolution, the central theme in the study of life. Every organism is the product of evolution, which has generated the diversity of biological features that distinguish organisms from one another and the similarities that all organisms share. From this basic evolutionary theme emerge several other themes that recur throughout the book:

- **Relationship between Form and Function**
- **Biological Order, Regulation, and Homeostatis**
- **Acquiring and Using Energy**
- **Unity Within Diversity**
- **Evolution and Adaptation**

We have highlighted the prevalent recurrence of each theme throughout the text with an icon, shown above. The icons can be used to activate higher thought processes by inviting you to explore how the fact or concept being discussed fits the indicated theme.

FORM AND FUNCTION
The length and shape of this tiger lily and that of the hummingbird's beak match, enabling the hummingbird to gather nutritious nectar from the flower base, while the flower deposits or collects pollen for reproduction.

ACQUIRING AND USING ENERGY
All organisms acquire and use energy whether by directly trapping the energy in sunlight or by harvesting energy stored in the bodies of plants or animals.

ORDER, REGULATION AND HOMEOSTASIS
Whether active or dormant, enveloped by snow or blistering heat, all organisms must maintain order and regulate internal conditions to remain alive.

UNITY WITHIN DIVERSITY
Despite the remarkable variation in different kinds of organisms on earth, all organisms are composed of one or more cells.

EVOLUTION AND ADAPTATION
Evolution has produced the astounding variety of life on earth. Like the lizard hidden on the bark of this tree, each kind of organism possesses adaptations that enable it to survive and reproduce in its particular habitat.

Reexamining the Themes

*E*ach chapter concludes with a "Reexamining the Themes" section, which revisits the themes and how they emerge within the context of the chapter's concepts and principles. This section will help you realize that the same themes are evident at all levels of biological organization, whether you are studying the molecular and cellular aspects of biology or the global characteristics of biology.

---

84 • PART 2 / *Chemical and Cellular Foundations of Life*

When two organisms have the same protein, the difference in amino acid sequence of that protein can be correlated with the evolutionary relatedness of the organisms. The amino acid sequence of hemoglobin, for example, is much more similar between humans and monkeys—organisms that are closely related—than between humans and turtles, who are only distantly related. In fact, the evolutionary tree that emerges when comparing the structure of specific proteins from various animals very closely matches that previously constructed from fossil evidence.

The fact that the amino acid sequences of proteins change as organisms diverge from one another reflects an underlying change in their genetic information. Even though a DNA molecule from a mushroom, a redwood tree, and a cow may appear superficially identical, the sequences of nucleotides that make up the various DNA molecules are very different. These differences reflect evolutionary changes resulting from natural selection (Chapter 34).

Virtually all differences among living organisms can be traced to evolutionary changes in the structure of their various macromolecules, originating from changes in the nucleotide sequences of their DNA. (See CTQ #7.)

REEXAMINING THE THEMES

Relationship between Form and Function

The structure of a macromolecule correlates with a particular function. The unbranched, extended nature of the cellulose molecule endows it with resistance to pulling forces, an important property of plant cell walls. The hydrophobic character of lipids underlies many of their biological roles, explaining, for example, how waxes are able to provide plants with a waterproof covering. Protein function is correlated with protein shape. Just as a key is shaped to open a specific lock, a protein is shaped for a particular molecular interaction. For example, the shape of each polypeptide chain of hemoglobin enables a molecule of oxygen to fit perfectly into its binding site. A single alteration in the amino acid sequence of a hemoglobin chain can drastically reduce the molecule's oxygen-carrying capacity.

Biological Order, Regulation, and Homeostasis

Both blood sugar levels and body weight in humans are controlled by complex homeostatic mechanisms. The level of glucose in your blood is regulated by factors acting on the liver, which stimulate either glycogen breakdown (which increases blood sugar) or glycogen formation (which decreases blood sugar). Your body weight is, at least partly, determined by factors emanating from fat cells which either increase metabolic rate (which tends to decrease body weight) or slow down metabolic rate (which tends to increase body weight).

Acquiring and Utilizing Energy

The chemical energy that fuels biological activities is stored primarily in two types of macromolecules: polysaccharides and fats. Polysaccharides, including starch in plants and glycogen in animals, function primarily in the short-term storage of chemical energy. These polysaccharides can be rapidly broken down to sugars, such as glucose, which are readily metabolized to release energy. Gram-for-gram, fats contain even more energy than polysaccharides and function primarily as a long-term storage of chemical energy.

Unity within Diversity

All organisms, from bacteria to humans, are composed of the same four families of macromolecules, illustrating the unity of life. The precise nature of these macromolecules and the ways they are organized into higher structures differ from organism to organism, thereby building diversity. Plants, for example, polymerize glucose into starch and cellulose, while animals polymerize glucose into glycogen. Similarly, many proteins (such as hemoglobin) are present in a variety of organisms, but the precise amino acid sequence of the protein varies from one species to the next.

Evolution and Adaptation

Evolution becomes very apparent at the molecular level when we compare the structure of macromolecules among diverse organisms. Analysis of the amino acid sequences of proteins and the nucleotide sequences of nucleic acids reveals a gradual change over time in the structure of macromolecules. Organisms that are closely related have proteins and nucleic acids whose sequences are similar than are those of distantly related organisms. To a large degree, the differences observed among diverse organisms derives from the evolutionary differences in nucleic acid and protein sequences.

---

242 • PART 3 / *The Genetic Basis of Life*

The segregation of alleles and their independent assortment during meiosis increase genotype diversity by promoting new combinations of genes. But the shuffling of existing genes alone does not explain the presence of such a vast diversity of life. If all organisms descended from a common ancestor, with its relatively small complement of genes, where did all the genes present in today's millions of species come from? The answer is mutation.

Most mutant alleles are detrimental; that is, they are more likely to disrupt a well-ordered, smoothly functioning organism than to increase the organism's fitness. For example, a mutation might change a gene so that it produces an inactive enzyme needed for a critical life function. Occasionally, however, one of these stable genetic changes creates an advantageous characteristic that fitness of the offspring. In this way, mutati raw material for evolution and the diversifie earth.

One of the requirements for genes is stabili remain basically the same from generation to the fitness of organisms would rapidly deter same time, there must be some capacity f change; otherwise, there would be no potent tion. Alterations in genes do occur, albeit rare changes (mutations) represent the raw mater tion. (See CTQ #7.)

REEXAMINING THE THEMES

Biological Order, Regulation, and Homeostasis

Mendel discovered that the transmission of genetic factors followed a predictable pattern, indicating that the processes responsible for the formation of gametes, including the segregation of alleles, must occur in a highly ordered manner. This orderly pattern can be traced to the process of meiosis and the precision with which homologous chromosomes are separated during the first meiotic division. Mendel's discovery of independent assortment can also be connected with the first meiotic division, when each pair of homologous chromosomes becomes aligned at the metaphase plate in a manner that is independent of other pairs of homologues.

of certain genetic diseases, such as cystic fibrosis. A the mechanism by which genes are transmitted is u the genes themselves are highly diverse from one o to the next. It is this genetic difference among spec forms the very basis of biological diversity

Unity within Diversity

All eukaryotic, sexually reproducing organisms follow the same "rules" for transmitting inherited traits. Although Mendel chose to work with peas, he could have come to the same conclusions had he studied fruit flies or mice or had he scrutinized a family's medical records on the transmission

Evolution and Adaptation

Mendel's findings provided a critical link in our k edge of the mechanism of evolution. A key tenet i theory of evolution is that favorable genetic variation crease the likelihood that an individual will survive t productive age and that its offspring will exhibit these s favorable characteristics. Mendel's demonstration units of inheritance pass from parents to offspring with being blended revealed the means by which advantage traits could be preserved in a species over many gene tions. The subsequent discovery of genetic change by m tation revealed how new genes appeared in a populatic thus providing the raw material for evolution.

SYNOPSIS

Gregor Mendel discovered the pattern by which inherited traits are transmitted from parents to offspring. Mendel discovered that inherited traits were controlled by pairs of factors (genes). The two factors for a given trait in an individual could be identical (homozygous) or different (heterozygous). In heterozygotes, one of the gene variants (alleles) may be dominant over the other, recessive allele. Because of dominance, the appearance (phenotype) of the heterozygote (genotype of *Aa*) is identical to that of the homozygote with two dominant alleles

The Human Perspective

Students will naturally find many ways in which the material presented in any biology course relates to them. But it is not always obvious how you can use biological information for better living or how it might influence your life. Your ability to see yourself in the course boosts interest and heightens the usefulness of the information. This translates into greater retention and understanding.

To accomplish this desirable outcome, the entire book has been constructed with you—the student—in mind. Perhaps the most notable feature of this approach is a series of boxed essays called "The Human Perspective" that directly reveals the human relevance of the biological topic being discussed at that point in the text. You will soon realize that human life, including your own, is an integral part of biology.

PART 2 / Chemical and Cellular Foundations of Life

◁ THE HUMAN PERSPECTIVE ▷
Obesity and the Hungry Fat Cell

FIGURE 1
Actor Robert DeNiro in (left) a scene from the movie *Raging Bull* and (right) a recent photograph.

It has become increasingly clear in recent years that people who are exceedingly overweight—that is, obese—are at increased risk of serious health problems, including heart disease and cancer. By most definitions, a person is obese if he or she is about 20 percent above "normal" or desirable body weight. Approximately 35 percent of adults in the United States are considered obese by this definition, twice as many as at the turn of the century. Among young adults, high blood pressure is five times more prevalent and diabetes three times more prevalent in a group of obese people than in a group of people who are at normal weight. Given these statistics, together with the social stigma facing the obese, there would seem to be strong motivation for maintaining a "normal" body weight. Why, then, are so many of us so overweight? And, why is it so hard to lose unwanted pounds and yet so easy to gain them back? The answers go beyond our fondness for high-calorie foods.

Excess body fat is stored in fat cells (*adipocytes*) located largely beneath the skin. These cells can change their volume more than a hundredfold, depending on the amount of fat they contain. As a person gains body fat, his or her fat cells become larger and larger, accounting for the bulging, sagging body shape. If the person becomes sufficiently overweight, and their fat cells approach their maximum fat-carrying capacity, chemical messages are sent through the blood, causing formation of new fat cells that are "hungry" to begin accumulating their own fat. Once a fat cell is formed, it may expand or contract in volume, but it appears to remain in the body for the rest of the person's life.

○ Although the subject remains controversial, current research findings suggest that body weight is one of the properties subject to physiologic regulation in humans. Apparently, each person has a particular weight that his or her body's regulatory machinery acts to maintain. This particular value—whether 40 kilograms (80 pounds) or 200 kilograms (400 pounds)—is referred to as the person's **set-point**.

People maintain their body weight at a relatively constant value by balancing energy intake (in the form of food calories) with energy expenditure (in the form of calories burned by metabolic activities or excreted). Obese individuals are thought to have a higher set-point than do persons of normal weight. In many cases, the set-point value appears to have a strong genetic component. For instance, studies reveal there is no correlation between the body mass of adoptees and their adoptive parents, but there is a clear relationship between adoptees and their biological parents, with whom they have not lived.

The existence of a body-weight set-point is most evident when the body weight of a person is "forced" to deviate from the regulated value. Individuals of normal body weight who are fed large amounts of high-calorie foods under experimental conditions tend to gain increasing amounts of weight. If these people cease their energy-rich diets, however, they return quite rapidly to their previous levels, at which point further weight loss stops. This is illustrated by actor Robert DeNiro, who reportedly gained about 50 pounds for the filming of the movie "Raging Bull" (Figure 1), and then lost the weight prior to his next acting role. Conversely, a person who is put on a strict, low-calorie diet will begin to lose weight. The drop in body weight soon triggers a decrease in the person's resting metabolic rate; that is, the amount of calories burned when the person is not engaged in physical activity. The drop in metabolic rate is the body's compensatory measure for the decreased food intake. In other words, it is the body's attempt to halt further weight loss. This effect is particularly pronounced among obese people who diet and lose large amounts of weight: Their pulse rate and blood pressure drop markedly, their fat cells shrink to "ghosts" of their former selves, and they tend to be continually hungry. If these obese individuals go back to eating a *normal* diet, they tend to regain the lost weight rapidly. The drive of these formerly obese persons to increase their food intake is probably a response to chemical signals emanating from the fat cells as they shrink below their previous size.

---

630 • PART 5 / Form and Function of Animal Life

◁ THE HUMAN PERSPECTIVE ▷
Dying for a Cigarette?

... average, smoking cigarettes will cut ...imately 6 to 8 years off your life, ...han 5 minutes for every cigarette ... Cigarette smoking is the greatest ... preventable death in the United ...ccording to a 1991 report by the ...for Disease Control (CDC), ...0,000 Americans die each year ...ng-related causes. Smoking ac...87 percent of all lung-cancer ...smokers are more susceptible ...he esophagus, larynx, mouth, ... bladder than are nonsmok...reased incidence of lung ...mong smokers compared to ...hown in Figure 1a, and the ...d by quitting is shown in ...ffects of smoking on lung ...Figure 2. Atherosclero... and peptic ulcers also ...greater frequency than ...rs. For example, long-...5 times more likely to ...terial disease than are ...sema (a condition ...tion of lung tissue, ...culty in breathing) ...nmation of the air-...r prevalent among

...ger other people. ...sponsible for the ...nocent bystand-...re the same air ... passive (invol-...own); second-... seriously ill ...ers have dou-...ry infections ...osed to to-...g married ...us; 20 per-...ong non-...utable to inhaling other people's tobacco smoke. Another "innocent bystander" is a fetus developing in the uterus of a woman who smokes. Smoking increases the incidence of miscarriage and stillbirth and decreases the birthweight of the infant. Once born, these babies suffer twice as many respiratory infections as do babies of nonsmoking mothers.

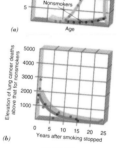

(a)

(b)

Why is smoking so bad for your health? The smoke emitted from a burning cigarette contains more than 2,000 identifiable substances, many of which are either irritants or carcinogens. These compounds include carbon monoxide, sulfur dioxide, formaldehyde, nitrosamines, toluene, ammonia, and radioactive isotopes. Autopsies of respiratory tissues from smokers (and from nonsmokers who have lived for long periods with smokers) show widespread cellular changes, including the presence of precancerous cells (cells that may become malignant, given time) and a marked reduction in the number of cilia that play a vital role in the removal of bacteria and debris from the airways.

Of all the compounds found in tobacco (including smokeless varieties), the most important is nicotine, not because it is carcinogenic, but because it is so addictive. Nicotine is addictive because it acts like a neurotransmitter by binding to certain acetylcholine receptors (page 477), stimulating postsynaptic neurons. The physiological effects of this stimulation include the release of epinephrine, an increase in blood sugar, an elevated heart rate, and the constriction of blood vessels, causing elevated blood pressure. A smoker's nervous system becomes "accustomed" to the presence of nicotine and decreases the output of the natural neurotransmitter. As a result, when a person tries to stop smoking, the sudden absence of nicotine, together with the decreased level of the natural transmitter, decreases stimulation of postsynaptic neurons, which creates a craving for a cigarette—a "nicotine fit." Ex-smokers may be so conditioned to the act of smoking that the craving for cigarettes can continue long after the physiological addiction disappears.

Biolines

*T*he "Biolines" are boxed essays that highlight fascinating facts, applications, and real-life lessons, enlivening the mainstream of biological information. Many are remarkable stories that reveal nature to be as surprising and interesting as any novelist could imagine.

◁ BIOLINE ▷
DNA Fingerprints and Criminal Law

On February 5, 1987, a woman and her 2-year-old daughter were found stabbed to death in their apartment in the New York City borough of the Bronx. Following a tip, the police questioned a resident of a neighboring building. A small bloodstain was found on the suspect's watch, which was sent to a laboratory for DNA fingerprint analysis. The DNA from the white blood cells in the stain was amplified using the PCR technique and was digested with a restriction enzyme. The restriction fragments were then separated by electrophoresis, and a pattern of labeled fragments was identified with a radioactive probe. The banding pattern produced by the DNA from the suspect's watch was found to be a perfect match to the pattern produced by DNA taken from one of the victims. The results were provided to the opposing attorneys, and a pretrial hearing was called in 1989 to discuss the validity of the DNA evidence.

During the hearing, a number of expert witnesses for the prosecution explained the basis of the DNA analysis. According to these experts, with the exception of identical twins, have the same nucleotide sequence in their DNA. Moreover, differences in DNA sequence can be detected by comparing the lengths of the fragments produced by restriction-enzyme digestion of different DNA samples. The patterns produce a DNA fingerprint" (Figure 1) that is as to an individual as is a set of conventional fingerprints lifted from a glass. In DNA fingerprints had already been in more than 200 criminal cases in the States and had been hailed as the important development in forensic (the application of medical facts

FIGURE 1
Alec Jeffreys of the University of Leicester, England, examining a DNA fingerprint. Jeffreys was primarily responsible for developing the DNA fingerprint technique and was the scientist who confirmed the death of Josef Mengele.

to legal problems) in decades. The widespread use of DNA fingerprinting evidence in court had been based on its general acceptability in the scientific community. According to a report from the company performing the DNA analysis, the likelihood that the same banding patterns could be obtained by chance from two *different* individuals in the community was only one in 100 million.

What made this case (known as the Castro case, after the defendant) memorable and distinct from its predecessors was that the defense also called on expert witnesses to scrutinize the data and to present

their opinions. While these experts firmed the capability of DNA fingerprinting to identify an individual out of a population, they found serious techn flaws in the analysis of the DNA samp used by the prosecution. In an unprecedented occurrence, the experts who earlier testified *for the prosecution* agre that the DNA analysis in this case unreliable and should not be used as e dence! The problem was not with the tec nique itself but in the way it had carried out in this particular case. Conse quently, the judge threw out the evidence

In the wake of the Castro case, the us of DNA fingerprinting to decide guilt o innocence has been seriously questioned Several panels and agencies are working to formulate guidelines for the licensing of forensic DNA laboratories and the certifi cation of their employees. In 1992, a panel of the National Academy of Sciences re leased a report endorsing the general reli ability of the technique but called for the institution of strict standards *to be set by* scientists.

Meanwhile, another issue regarding DNA fingerprinting has been raised and hotly debated. Two geneticists, Richard Lewontin of Harvard University and Daniel Hartl of Washington University, coauthored a paper published in December 1991, suggesting that scientists do not have enough data on genetic variation within different racial or ethnic groups to calculate the odds that two individuals — a suspect and a perpetrator of the crime — are one and the same on the basis of an identical DNA fingerprint. The matter remains an issue of great concern in both the scientific and legal communities and has yet to be resolved.

◁ BIOLINE ▷
The Fish That Changes Sex

In vertebrates, gender is generally a biologically inflexible commitment: An individual develops into either a male or a female as dictated by the sex chromosomes acquired from one's parents. Yet, even among vertebrates, there are organisms that can reverse their sexual commitment. The Australian cleaner fish (Figure 1), a small animal that sets up "cleaning stations" to which larger fishes come for parasite removal, can change its gender in response to environmental demands. Most male cleaner fish travel alone rather than with a school. Except for a single male, schools of cleaner fish are comprised entirely of females. Although it might seem logical to conclude that maleness engenders solo travel, it is actually the other way around: Being alone fosters maleness. A cleaner fish that develops away from a school *becomes* a male, whereas the same fish developing in a school would have become a female.

FIGURE 1
The small Australian wrasse (cleaner fish) is seen on a much larger grouper.

But what of the one male in the school—the one with the harem? He may have developed as a solo fish and then found a school in need of his spermatogenic services. But there is another way a school may acquire a male. If the male in a school dies (or is removed experimentally), one of the females, the one at the top of a behavioral hierarchy that exists in each school, becomes uncharacteristically aggressive and takes over the behavioral role of the missing male. She begins to develop male gonads, and within a few weeks, the female becomes a reproductively competent male, indistinguishable from other males. Furthermore, the sex change is reversible. If a fully developed male enters the school during the sexual transition, the almost-male fish developmentally backpedals, once again assuming the biological and behavioral role of a female.

⚠ Not all organisms follow the mammalian pattern of sex determination. In some animals, most notably birds, the opposite pattern is found: The female's cells have an X and a Y chromosome, while the male's cells have two Xs. An exception to this rule of a strict relation between sex and chromosomes is discussed in the Bioline: The Fish That Changes Sex. Although some plants possess sex chromosomes and gender distinctions between individuals, most have only autosomes; consequently, each individual produces both male and female parts.

SEX LINKAGE

For fruit flies and humans alike, there are hundreds of genes on the X chromosome that have no counterpart on the smaller Y chromosome. Most of these genes have nothing to do with determining gender, but their effect on phenotype usually *depends on* gender. For example, in females, a recessive allele on one X chromosome will be masked (and not expressed) if a dominant counterpart resides on the other X chromosome. In males, it only takes one recessive allele on the single X chromosome to determine the individual's phenotype since there is no corresponding allele on the Y chromosome. Inherited characteristics determined by genes that reside on the X chromosome are called **X-linked characteristics.**

So far, some 200 human X-linked characteristics have been described, many of which produce disorders that are found almost exclusively in men. These include a type of heart-valve defect (*mitral stenosis*), a particular form of mental retardation, several optical and hearing impairments, muscular dystrophy, and red-green colorblindness (Figure 13-8).

One X-linked recessive disorder has altered the course of history. The disease is **hemophilia**, or "bleeder's disease," a genetic disorder characterized by the inability to produce a clotting factor needed to halt blood flow quickly following an injury. Nearly all hemophiliacs are males. Although females can inherit two recessive alleles for hemophilia, this occurrence is extremely rare. In general, women who have acquired the rare defective allele are heterozygous **carriers** for the disease. The phenotype of a carrier

*S*everal ethical issues are discussed in the Bioethics essays which add provocative pauses throughout the text. Biological Science does not operate in a vacuum but has profound consequences on the general community. Because biologists study life, the science is peppered with ethical considerations. The moral issues discussed in these essays are neither simple nor easy to resolve, and we do not claim to have any certain answers. Our goal is to encourage you to consider the bioethical issues that you will face now and in the future.

Coordinating the Organism: The Role of the Nervous System / CHAPTER 23 • 489

◁ B I O E T H I C S ▷
Blurring the Line between Life and Death
By ARTHUR CAPLAN
Division of the Center for Biomedical
Ethics at the University of Minnesota

Theresa Ann Campo Pearson didn't have a very long life. When she died in 1992, she was only 10 days old. Despite her short life, she became the center of a very strange, sad, and wrenching ethical controversy. Theresa died because her brain had failed to form. She had anencephaly, a condition in which only the brainstem, located at the top of the spinal cord, is present. Her parents wanted to donate Theresa's organs; the courts said no. Some people found it strange that Theresa's parents, Laura Campo and Justin Pearson, did not get their way. Why not allow donation, when every day in North America a baby dies because there is no heart, lung, or liver available for transplantation?

Anencephaly is best described as completely "unabling," not disabling. Children born with anencephaly cannot think, feel, sense, or be aware of the world. Many are stillborn; the majority of the rest die within days of birth. A mere handful live for a few weeks. Theresa's parents

knew all this. But rather than abort the pregnancy, they chose to have their baby. In fact, the baby was born by Caesarean section, at least partly in the hope that it would be born alive, thereby making organ donation possible. When Theresa died at Broward General Medical Center in Fort Lauderdale, Florida, however, no organs were taken. Two Florida courts ruled that the baby could not be used as a source of organs unless she was brain-dead, and Theresa Ann Campo was never pronounced brain-dead.

Brain death refers to a situation in which the brain has irreversibly lost all function and activity. Babies born with anencephaly have some brain function in their brainstem so, while they cannot think or feel, they are alive. According to Florida law—and the law in more than 40 other states—only those individuals declared brain-dead can donate organs. The courts of Florida had no other option but to deny the request for organ donation.

One obvious solution is to change the law so that states could decide that organs can be removed upon parental consent from either those who are born brain-dead or from babies who are born with anencephaly. Another solution is to rewrite the definition of death to say that death occurs either when the brain has totally ceased to function or if a baby is born anencephalic. Do you feel that either of these changes should be made? Some may argue that medicine will fudge the line between life and death in order to get organs for transplant. Do you agree with this concern? How do you think redefining death will affect a person's decision to check off the donation box on the back of a driver's license? Do you think people may worry that if they are known to be potential donors they won't be aggressively treated at the hospital? In your opinion, would changing the definition of death to include anencephaly be beneficial or deleterious?

Like the brain, the spinal cord is composed of white matter (myelinated axons) and gray matter (dendrites and cell bodies). However, the arrangement of these types of matter is reversed in the spinal cord, compared to their arrangement in the brain: The spinal cord's white matter surrounds the gray matter (Figure 23-16).

The human central nervous system is the most complex and highly evolved assembly of matter. Among its functions are the processing of sensory information collected from both the external and internal environment; the regulation of internal physiological activities; the coordination of complex motor activities; and the endowment of such intangible "mental" qualities as emotions, creativity, language, and the ability to think, learn, and remember. (See CTQ #6.)

ARCHITECTURE OF THE PERIPHERAL NERVOUS SYSTEM

The peripheral nervous system provides the neurological bridge between the central nervous system and the various parts of the body. The peripheral nervous system is made up of paired nerves that extend into the periphery from the CNS at various levels along the body. Each nerve is composed of a large bundle of myelinated axons surrounded by a connective tissue sheath. Twelve pairs of **cranial nerves** emerge from the central stalk of the human brain, and 31 pairs of **spinal nerves** extend from the spinal cord out between the vertebrae of humans (Figure 23-16). For the most part, the cranial nerves *innervate* (supply nerves to) tissues and organs of the head and neck, whereas the spinal nerves innervate the chest, abdomen, and limbs.

Additional Pedagogical Features

We have worked to assure that each chapter in this book is an effective teaching and learning instrument. In addition to the pedagogical features discussed above, we have included some additional tried-and-proven-effective tools.

KEY POINTS

Key points follow each major section and offer a condensation of the relevant facts and details as well as the concepts discussed. You can use these key points to reaffirm your understanding of the previous reading or to alert you to misunderstood material before moving on to the next topic. Each key point is tied to a Critical Thinking Question found at the end of the chapter; together, they encourage you to analyze the information, taking it beyond mere memorization.

Plant Tissues and Organs / CHAPTER 18 • 361

➠ Many plants replenish old and dying cells with vigorous new cells. But since each plant cell has a surrounding cell wall (Chapter 7) old plant cells do not just wither and disappear when they die. As a result, the longer a plant lives, the more complex its anatomy becomes. **Annuals** are plants that live for 1 year or less, such as corn and marigolds. Because they live for such a brief period, these plants do not completely replace old cells. As a result, annuals are anatomically less complex than are **biennials** — plants that live for 2 years — and **perennials** — herbs, shrubs, and trees that live longer than 2 years. Biennials (carrots, Queen Anne's lace) and perennials (rosebushes, apple trees) are able to live longer than annuals because they produce new cells to replace those that cease functioning or die, providing a continual supply of young, vigorous cells.

▣ In this chapter, we will focus on the body construction of flowering plants, the most evolutionarily advanced, and structurally complex of any group in the plant kingdom. All flowering plants are **vascular plants;** that is, they contain specialized cells that circulate water, minerals, and food (organic molecules) throughout the plant. Botanists divide flowering plants into two main groups: **dicotyledons,** or dicots (*di* = two, *cotyledon* = embryonic seed leaf), and **monocotyledons,** or monocots (*mono* = one). Table 18-1 illustrates the many differences that distinguish dicots from monocots and will be used as a reference throughout the chapter.

SHOOTS AND ROOTS

The flowering plant body is a study in contradictions. A typical plant grows through the soil and the air simultaneously, two very different habitats with very different conditions. As a result, the two main parts of the plant differ dramatically in form (anatomy) and function (physiology): The underground **root system** anchors the plant in the soil and absorbs water and nutrients, while the aerial **shoot system** absorbs sunlight and gathers carbon dioxide for photosynthesis (Figure 18-2). The shoot system also produces stems, leaves, flowers, and fruits. Interconnected vascular tissues transport materials between the aerial shoot system and the underground root system. These connections allow water and minerals absorbed by the root to be conducted to shoot tissues, and for food produced by the shoot to be transported to root tissues. We will discuss the various components of these two systems in more detail later in the chapter.

Over 90 percent of all plant species are flowering plants. Flowering plants are the most recently evolved plant group, having undergone rapid evolution during the past 1 million to 2 million years as environmental conditions on land became more variable. (See CTQ # 2.)

TABLE 18-1

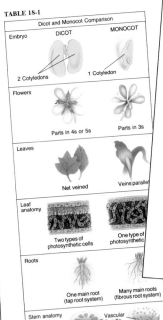

Dicot and Monocot Comparison

	DICOT	MONOCOT
Embryo	2 Cotyledons	1 Cotyledon
Flowers	Parts in 4s or 5s	Parts in 3s
Leaves	Net veined	Veins parallel
Leaf anatomy	Two types of photosynthetic cells	One type of photosynthetic
Roots	One main root (tap root system)	Many main roots (fibrous root system)
Stem anatomy	Vascular bundles in rings	Scattered vascular bundles
Root anatomy	Xylem in center	Pith in center
Secondary growth	Yes	No

---

the corresponding polypeptide. The cumulative effect of gradual changes in polypeptides over evolutionary time has been the generation of life's diversity.

Evolution and Adaptation

➠ Evolutionary change from generation to generation depends on genetic variability. Much of this variability arises from reshuffling maternal and paternal genes during meiosis, but somewhere along the way *new* genetic information must be introduced into the population. New genetic information arises from mutations in existing Some of these mutations arise during replication; occur as the result of unrepaired damage as the DNA "sitting" in a cell. Mutations that occur in an indivi germ cells can be considered the raw material on natural selection operates; whereas harmful mutation duce offspring with a reduced fitness, beneficial muta produce offspring with an increased fitness.

SYNOPSIS

Experiments in the 1940s and 1950s established conclusively that DNA is the genetic material. These experiments included the demonstration that transforming bacteria from one genetic strain to another; that bacteriophages injected their DNA into a host cell during infection; and that the injected DNA was transmitted to the bacteriophage progeny.

DNA is a double helix. DNA is a helical molecule consisting of two chains of nucleotides running in opposite directions, with their backbones facing inward like rungs on a ladder. Adenine-containing nucleotides on one strand always pair with thymine-containing nucleotides on the other strand, likewise for guanine- and cytosine-containing nucleotides. As a result, the two strands of a DNA molecule are complementary to one another. Genetic information is encoded in the specific linear sequence of nucleotides that make up the strands.

DNA replication is semiconservative. During replication, the double helix separates, and each strand serves as a template for the formation of a new, complementary strand. Nucleotide assembly is carried out by the enzyme DNA polymerase, which moves along the two strands in opposite directions. As a result, one of the strands is synthesized continuously, while the other is synthesized in segments that are covalently joined. Accuracy is maintained by a proofreading mechanism present within the polymerase.

Information flows in a cell from DNA to RNA to protein. Each gene consists of a linear sequence of nucleotides that determines the linear sequence of amino

acids in a polypeptide. This is accomplished in two ma steps: transcription and translation.

During transcription, the information spelled out the gene's nucleotide sequence is encoded in a mole cule of messenger RNA (mRNA). The mRNA contain a series of codons. Each codon consists of three nucleotides Of the 64 possible codons, 61 specify an amino acid, and the other 3 stop the process of protein synthesis.

During translation, the sequence of codons in the mRNA is used as the basis for the assembly of a chain of specific amino acids. Translating mRNA messages occurs on ribosomes and requires tRNAs, which serve as decoders. Each tRNA is folded into a cloverleaf structure with an anticodon at one end — which binds to a complementary codon in the mRNA — and a specific amino acid at the other end — which becomes incorporated into the growing polypeptide chain. Amino acids are added to their appropriate tRNAs by a set of enzymes. The sequential interaction of charged tRNAs with the mRNA results in the assembly of a chain of amino acids in the precise order dictated by the DNA.

Mutation is a change in the genetic message. Gene mutations may occur as a single nucleotide substitution, which leads to the insertion of an amino acid different from that originally encoded. In contrast, the addition of one or two nucleotides throws off the reading frame of the ribosome as it moves along the mRNA, leading to the incorporation of incorrect amino acids "downstream" from the point of mutation. Exposure to mutagens increases the rate of mutation.

SYNOPSIS

The synopsis section offers a convenient summary of the chapter material in a readable narrative form. The material is summarized in concise paragraphs that detail the main points of the material, offering a useful review tool to help reinforce recall and understanding of the chapter's information.

REVIEW QUESTIONS

Along with the synopsis, the Review Questions provide a convenient study tool for testing your knowledge of the facts and processes presented in the chapter.

---

224 • PART 2 / *Chemical and Cellular Foundations of Life*

Key Terms

zygote (p. 214)
meiosis (p. 214)
life cycle (p. 214)
germ cell (p. 214)
somatic cell (p. 214)
meiosis I (p. 216)

reduction division (p. 216)
synapsis (p. 216)
tetrad (p. 216)
crossing over (p. 216)
genetic recombination (p. 216)
synaptonemal complex (p. 218)

maternal chromosome (p. 219)
paternal chromosome (p. 219)
independent assortment (p. 219)
meiosis II (p. 219)

Review Questions

1. Match the activity with the phase of meiosis in which it occurs.

   a. synapsis
   b. crossing over
   c. kinetochores split
   d. independent assortment
   e. homologous chromosomes
      separate
   f. cytokinesis

   1. prophase I
   2. metaphase I
   3. anaphase I
   4. telophase I
   5. prophase II
   6. anaphase II
   7. telophase II

2. How do crossing over and independent assortment increase the genetic variability of a species?

3. Why is meiosis I (and not meiosis II) referred to as the reduction division?

4. Suppose that one human sperm contains *x* amount of DNA. How much DNA would a cell just entering meiosis contain? A cell entering meiosis II? A cell just completing meiosis II? Which of these three cells would have a haploid number of chromosomes? A diploid number of chromosomes?

Critical Thinking Questions

1. Why are disorders, such as Down syndrome, that arise from abnormal chromosome numbers, characterized by a number of seemingly unrelated abnormalities?

2. A gardener's favorite plant had white flowers and long seed pods. To add some variety to her garden, she transplants some plants of the same type, but with pink flowers and short seed pods from her neighbor's garden. To her surprise, in a few generations, she grows plants with white flowers and short seed pods and plants with pink flowers and long seed pods, as well as the original combinations. What are two ways in which these new combinations could have arisen?

3. Set up the meiosis template in the diagram below on a large sheet of paper. Then use pieces of colored yarn or pipe cleaners to simulate chromosomes and make a model of the phases of meiosis. (*See template on opposite page*)

4. Would you expect two genes on the same chromosome, such as yellow flowers and short stems, always to be exchanged during crossing over? How might they remain together *in spite of* crossing over?

5. Suppose paternal chromosomes always lined up on the same side of the metaphase plate of cells in meiosis I. How would this affect genetic variability of offspring? Would they all be identical? Why or why not?

Additional Readings

Chandley, A. C. 1988. Meiosis in man. *Trends in Gen.* 4:79–83. (Intermediate)
Hsu, T. C. 1979. *Human and mammalian cytogenetics.* New York: Springer-Verlag. (Intermediate)
John, B. 1990. *Meiosis.* New York: Cambridge University Press. (Advanced)

Moens, P. B. 1987. *Meiosis.* Orlando: Academic. (Advanced)
Patterson, D. 1987. The causes of Down syndrome. *Sci. Amer.* Feb:52–60. (Intermediate-Advanced)
White, M. J. D. 1973. *The chromosomes.* Halsted. (Advanced)

STIMULATING CRITICAL THINKING

Each chapter contains as part of its end material a diverse mix of Critical Thinking Questions. These questions ask you to apply your knowledge and understanding of the facts and concepts to hypothetical situations in order to solve problems, form hypotheses, and hammer out alternative points of view. Such exercises provide you with more effective thinking skills for competing and living in today's complex world.

ADDITIONAL READINGS

Supplementary readings relevant to the Chapter's topics are provided at the end of every chapter. These readings are ranked by level of difficulty (introductory, intermediate, or advanced) so that you can tailor your supplemental readings to your level of interest and experience.

Careers in Biology

*T*he appendices of this edition include "Careers in Biology," a frequently overlooked aspect of our discipline. Although many of you may be taking biology as a requirement for another major (or may have yet to declare a major), some of you are already biology majors and may become interested enough to investigate the career opportunities in life sciences. This appendix helps students discover how an interest in biology can grow into a livelihood. It also helps the instructor advise students who are considering biology as a life endeavor.

APPENDIX

◄ **D** ►

Careers in Biology

Although many of you are enrolled in biology as a requirement for another major, some of you will become interested enough to investigate the career opportunities in life sciences. This interest in biology can grow into a satisfying livelihood. Here are some facts to consider:

- Biology is a field that offers a very wide range of possible science careers

- Biology offers high job security since many aspects of it deal with the most vital human needs: health and food

- Each year in the United States, nearly 40,000 people obtain bachelor's degrees in biology. But the number of newly created and vacated positions for biologists is increasing at a rate that exceeds the number of new graduates. Many of these jobs will be in the newer areas of biotechnology and bioservices.

Biologists not only enjoy job satisfaction, their work often changes the future for the better. Careers in medical biology help combat diseases and promote health. Biologists have been instrumental in preserving the earth's life-supporting capacity. Biotechnologists are engineering organisms that promise dramatic breakthroughs in medicine, food production, pest management, and environmental protection. Even the economic vitality of modern society will be increasingly linked to biology.

Biology also combines well with other fields of expertise. There is an increasing demand for people with backgrounds or majors in biology complexed with such areas as business, art, law, or engineering. Such a distinct blend of expertise gives a person a special advantage.

The average starting salary for all biologists with a Bachelor's degree is $22,000. A recent survey of California State University graduates in biology revealed that most were earning salaries between $20,000 and $50,000. But as important as salary is, most biologists stress job satisfaction, job security, work with sophisticated tools and scientific equipment, travel opportunities (either to the field or to scientific conferences), and opportunities to be creative in their job as the reasons they are happy in their career.

Here is a list of just a few of the careers for people with degrees in biology. For more resources, such as lists of current openings, career guides, and job banks, write to Biology Career Information, John Wiley and Sons, 605 Third Avenue, New York, NY 10158.

A SAMPLER OF JOBS THAT GRADUATES HAVE SECURED IN THE FIELD OF BIOLOGY°

Agricultural Biologist	Bioanalytical Chemist	Brain Function	Environmental Center
Agricultural Economist	Biochemical/Endocrine	Researcher	Director
Agricultural Extension	Toxicologist	Cancer Biologist	Environmental Engineer
Officer	Biochemical Engineer	Cardiovascular Biologist	Environmental Geographer
Agronomist	Pharmacology Distributor	Cardiovascular/Computer	Environmental Law Specialist
Amino-acid Analyst	Pharmacology Technician	Specialist	Farmer
Analytical Biochemist	Biochemist	Chemical Ecologist	Fetal Physiologist
Anatomist	Biogeochemist	Chromatographer	Flavorist
Animal Behavior	Biogeographer	Clinical Pharmacologist	Food Processing Technologist
Specialist	Biological Engineer	Coagulation Biochemist	Food Production Manager
Anticancer Drug Research	Biologist	Cognitive Neuroscientist	Food Quality Control
Technician	Biomedical	Computer Scientist	Inspector
Antiviral Therapist	Communication Biologist	Dental Assistant	Flower Grower
Arid Soils Technician	Biometerologist	Ecological Biochemist	Forest Ecologist
Audio-neurobiologist	Biophysicist	Electrophysiology/	Forest Economist
Author, Magazines & Books	Biotechnologist	Cardiovascular Technician	Forest Engineer
Behavioral Biologist	Blood Analyst	Energy Regulation Officer	Forest Geneticist
Bioanalyst	Botanist	Environmental Biochemist	Forest Manager

Study Guide

Written by Gary Wisehart and Michael Leboffe of San Diego City College, the *Study Guide* has been designed with innovative pedagogical features to maximize your understanding and retention of the facts and concepts presented in the text. Each chapter in the *Study Guide* contains the following elements.

Concepts Maps

In Chapter 1 of the *Study Guide*, the beginning of a concept map stating the five themes is introduced. In each subsequent chapter, the concept map is expanded to incorporate topics covered in each chapter as well as the interconnections between chapters and the five themes. "Connector" phrases are used to link the concepts and themes, and the text icons representing the themes are incorporated into the concept maps.

Go Figure!

In each chapter, questions are posed regarding the figures in the text. Students can explore their understanding of the figures and are asked to think critically about the figures based on their understanding of the surrounding text and their own experiences.

Self-Tests

Each chapter includes a set of matching and multiple-choice questions. Answers to the Study Guide questions are provided.

Concept Map Construction

The student is asked to create concept maps for a group of terms, using appropriate connector phrases and adding terms as necessary.

Laboratory Manual

*B*iology: Exploring Life, Second Edition is supplemented by a comprehensive *Laboratory Manual* containing approximately 60 lab exercises chosen by the text authors from the National Association of Biology Teachers. These labs have been thoroughly class-tested and have been assembled from various scientific publications. They include such topics as

- Chaparral and Fire Ecology: Role of Fire in Seed Germination *(The American Biology Teacher)*
- A Model for Teaching Mitosis and Meiosis *(American Biology Teacher)*

- Laboratory Study of Climbing Behavior in the Salt Marsh Snail *(Oceanography for Landlocked Classrooms)*
- Down and Dirty DNA Extraction *(A Sourcebook of Biotechnology Activities)*
- Bioethics: The Ice-Minus Case *(A Sourcebook of Biotechnology Activities)*
- Using Dandelion Flower Stalks for Gravitropic Studies *(The American Biology Teacher)*
- pH and Rate of Enzymatic Reactions *(The American Biology Teacher)*

CHAPTER
◄ 22 ►

An Introduction to Animal Form and Function

STEPS TO DISCOVERY
The Concept of Homeostasis

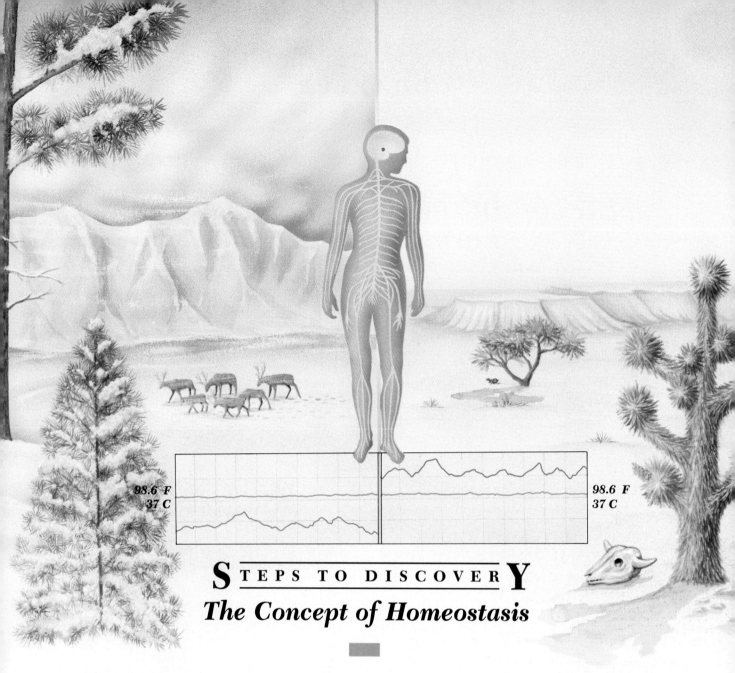

98.6 F
37 C

98.6 F
37 C

S TEPS TO DISCOVER Y
The Concept of Homeostasis

*I*n 1843, at the age of 30, Claude Bernard moved from a small French town, where he had been a pharmacist and an aspiring playwright, to Paris, where he planned to pursue his literary career. Bernard's plans changed, however, when he enrolled in medical school, and he went on to become one of the leading physiologists of the nineteenth century. During his prestigious career, Bernard studied body temperature, stomach juices, the role of nerves in controlling the diameter of blood vessels, and the function of the liver, but he focused particular attention on the fluids that bathe

the body's internal tissues. Bernard carefully monitored the temperature, acidity, and salt and sugar concentrations within the blood under various conditions and found that the body's fluids resisted change, providing cells with a stable, ordered environment. In his own translated words: "It is the fixity of the internal environment which is the condition of free and independent life. . . . All the vital mechanisms, however varied they may be, have only one object, that of preserving constant the conditions of life in the internal environment."

Even in extreme environments, humans maintain a constant body temperature. The regulatory center is the hypothalamus

Bernard's work provided the foundation for a diverse array of studies of the mechanisms by which the body maintains its internal constancy. For example, in the 1880s, Charles Richet in France and Isaac Ott in the United States independently determined that one part of the brain, the hypothalamus, played a key role in maintaining body temperature. If this part of the brain of laboratory animals was damaged, these animals lost their ability to hold their body temperature constant in the face of increasing or decreasing environmental temperatures. In 1912, Henry Barbour of Yale University developed a technique to selectively warm or cool this part of the brain. Barbour accomplished this feat by implanting fine silver tubes into the hypothalamus, and then circulating water of particular temperatures through the tubes. Cooling the hypothalamus below normal body temperature caused the animals to increase heat production, just as if they had been outside on a cold winter night. Warming the hypothalamus evoked the opposite response; the animals reacted by losing heat, just as if they had been exposed to a hot desert sun. The results of these studies suggested that one of the functions of the hypothalamus was to determine if the body's temperature was rising or falling and then to elicit an appropriate response that would return the temperature to its normal value.

One of the most influential physiologists of the twentieth century was Walter Cannon of Harvard University. Working in a field hospital in France during the fierce trench warfare of the last few months of World War I, Cannon was struck by the body's ability to withstand the terrible trauma resulting from severe wounds and to restore the orderly conditions necessary for survival. After the war, Cannon returned to Harvard and turned his attention to studying the mechanisms by which the body "fights" to maintain the internal environment. In 1926, he coined the term "homeostasis" (*homeo* = sameness, *stasis* = standing still; balance). Cannon did not mean to suggest that the body was stagnant or incapable of change but, rather, that it had the capacity to respond in dynamic ways to situations that threatened to disrupt its internal stability.

Cannon summarized his views on homeostasis in a book entitled *The Wisdom of the Body*. In his book, Cannon noted that humans could be exposed *for short periods* of time to dry heat above 115°C (239°F), without raising their body temperature, or to high altitudes with greatly reduced oxygen, such as when flying in an airplane or climbing a mountain, without showing serious effects of oxygen deprivation. He also noted that the body could resist disturbances that arise from within. For example, a person running for 20 minutes produces enough heat to "cause some of the albuminous substances of the body to become stiff, like a hard-boiled egg" and enough lactic acid (the acid of sour milk) to "neutralize all the alkali contained in the blood." Yet, the body becomes neither overheated nor poisoned.

The concept of homeostasis identified by Bernard and Cannon pervades virtually every aspect of the physiologic sciences. As we consider the function of major organs, such as the lungs, heart, kidney, and liver, we will see how each of these organs helps maintain a stable, ordered internal environment that promotes our health and well-being. We will also examine some of the dangers that lurk should these homeostatic controls fail.

of the brain.

Go without water on a hot summer day and you build up a thirst that has to be quenched. Spend a cold winter night in a bed with inadequate covers and you sleep huddled into a ball, trying to keep warm. Stay up all night studying for an exam and you find yourself so tired you can barely stay awake. All of these common events in our lives illustrate cases where our bodies are "telling" us that we need to take some action to maintain the stability of our internal environment. Most homeostatic activities take place without the need for these types of behavioral responses. For example, without conscious effort, our bodies "automatically" prevent themselves from becoming too salty or dilute, too hot or too cold; our bodies prevent waste products from accumulating to dangerous levels and stop oxygen or ATP concentrations from dropping below required amounts. All organisms have homeostatic mechanisms that enable them to respond to changes that threaten to upset their internal environment. In animals, these responses are carried out by teams of cells, tissues, organs, and organ systems.

▼   ▼   ▼

HOMEOSTATIC MECHANISMS

As you are reading this chapter or listening to your professor's lecture in class, you are probably in an environment where the temperature is actively maintained at a relatively constant value. Maintaining a constant room temperature requires more than just a heater or an air conditioner. The system requires

1. a *sensor* that continually monitors the temperature in the room, sending information about the air temperature to

2. a *control unit, or thermostat*, which is set at a particular temperature. If the information from the sensor indicates that the temperature in the room is either cooler or warmer than that for which the thermostat is set, a signal is transmitted to

3. an *effector*—the furnace or air conditioner that generates hot or cold air. When the air in the room is restored to the temperature set by the thermostat, the effector is turned off, until the temperature again deviates from the thermostat's set-point.

The operation of this temperature-control system is an example of a **negative-feedback mechanism.** The term "feedback" indicates that a change in a property is "fed back" to a control unit, inducing a subsequent change. Feedback is "negative" when a change in the property, such as air temperature, *reverses* further changes. For example, a rise in room temperature triggers a response to lower the temperature, while a drop in room temperature triggers the opposite response.

↻ Maintaining a constant *body* temperature requires an analogous set of biological components (Figure 22-1) and provides one of the best examples of how homeostatic mechanisms maintain the constancy of the internal environment. Your body remains at a temperature of 37°C (98.6°F) as a result of

1. *sensors* (called *temperature receptors*) located both at the body surface and deep within the body's interior; sensors monitor the temperature of both the external and internal environments. Information from the temperature receptors is passed to

2. a *control unit*, or *integrator*, (known as the body's "thermostat") located within the hypothalamus of the brain, which is normally "set" at 37°C. When the thermostat receives information that the body temperature is *deviating* from 37°C, signals are sent to the body's

3. *effectors*, which consist of widely scattered muscles and glands. If, for example, the temperature receptors detect a rise in body temperature, the brain initiates a response that causes skin glands to release sweat, cooling the body back down to 37°C (Chapter 28).

Negative-feedback mechanisms operate by reversing change. In contrast, **positive-feedback mechanisms** operate when a change in the body feeds back to *increase* the magnitude of the change. Blood clotting is an example of positive feedback. Your blood contains proteins that have the potential to form a blood clot that can plug a wound in a blood vessel which can be a life-saving response. The clotting of the blood at the site of an injured vessel begins on a very small scale. First, a small amount of clotted protein is deposited. As the clot forms, it induces a response that triggers the deposition of additional clotted protein, which, in turn, triggers even more deposition. As a result of this positive-feedback mechanism, what began as a small response is rapidly amplified so that vessel damage can be controlled quickly, halting any further loss of blood and maintaining internal stability.

We have described several examples of homeostasis and the role of negative- and positive-feedback mechanisms, but we have yet to discuss the parts of the body responsible for these physiologic activities. Look back for a moment at Figure 1-4 and the discussion of the hierarchy of biological organization. We saw in earlier chapters how atoms, molecules, and organelles are organized into increasingly complex biological structures, including cells, which are the fundamental units of life. Let us now continue farther up the ladder of biological organization and see how the cells of an animal carry out their activities in concert

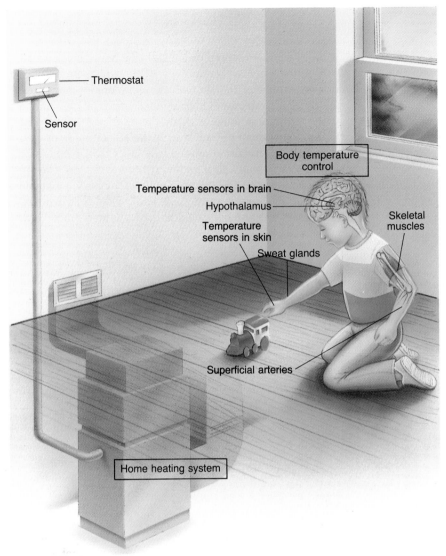

FIGURE 22-1

Controlling temperature levels by negative feedback. The temperature within the room and the body temperature of the child are maintained at a relatively constant value by an analogous set of elements that include temperature sensors, thermostats, and effectors. In the regulation of body temperature, the temperature sensors are located in the skin and brain; the thermostat is in the hypothalamus; and the effectors are the sweat glands, which produce sweat, and the skeletal muscles, which cause shivering.

with one another to meet the many needs of the entire animal.

Maintaining a stable internal environment requires a diverse array of sensors, integrators, and response mechanisms that detect and reverse any tendency toward physiologic disorder. (See CTQ #2.)

UNITY AND DIVERSITY AMONG ANIMAL TISSUES

The human body contains several hundred recognizably different types of cells, each of which works with other types of cells to accomplish a common function. The simplest of these teams of cells is a **tissue**—an organized group of cells with a similar structure and a common function. Tissues are the "fabrics" from which complex animals (and plants) are constructed. The use of the word "fabric" to describe tissues is not just a recent invention of biology writers; the term is derived from *tissu*, an Old English word meaning a "fine cloth." Like intricately woven fabric, tissues are the products of finely organized biological designs and not mere aggregates of cells randomly packed together in a unit. These living fabrics form the foundation of the multicellular organism.

◢ Despite the tremendous diversity that exists in the types of cells present within a single animal (and even greater diversity found among different animals), there is a striking overall unity of function. All of the diverse cells found among animals can be classified as part of one of four fundamental types of tissues of which all organs and organ systems are composed. Each of your organs, for example, is composed entirely of epithelial, connective, muscle, and/or nervous tissue.

EPITHELIAL TISSUE

Epithelial tissue (Figure 22-2) is present as an **epithelium**—a sheet of tightly adhering cells that lines the spaces of the body, such as the outer edge of an organ or the inner lining of a blood vessel or duct. The surface of the entire body is covered by an epithelial layer, which constitutes the outer layer (*epidermis*) of the skin. Some epithelia, such as that of the skin or the lining of the mouth, are primarily protective. Other epithelia, such as the lining of the intestine or lungs, regulate the movement of materials from one side of the cell layer to the other. Epithelial cells are often specialized as glandular cells that manufacture and *secrete* materials into the extracellular space. The epithelial lining of the trachea (Figure 22-2b), for example, contains mucous-secreting cells and ciliated cells that work together to move debris out of the respiratory tract. Other

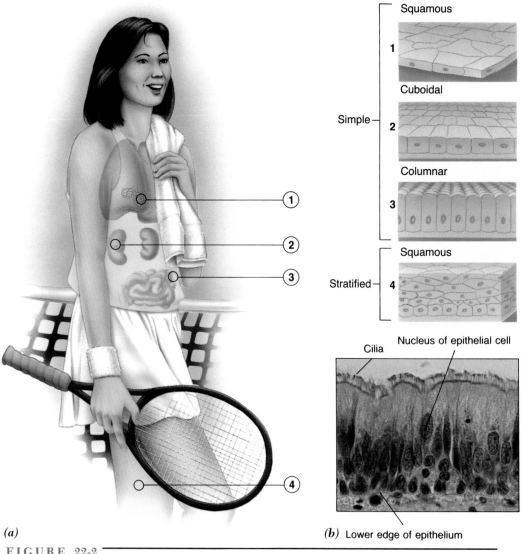

(a)

(b) Lower edge of epithelium

FIGURE 22-2

Epithelial tissue. *(a)* Epithelial tissues are categorized as either *simple* or *stratified* depending on whether they consist of one or more than one layer of cells. Epithelia are also categorized as either squamous (flattened), cuboidal, or columnar depending on the shapes of the cells. Lung tissue is an example of a simple squamous epithelium; kidney tubules contain a simple cuboidal epithelium; the inner lining of the small intestine is a simple columnar epithelium; and the outer part of the skin is a stratified epithelium. *(b)* A stained section showing the ciliated epithelum that lines the inner surface of the trachea (windpipe). Even though the nuclei are found at different levels, this epithelium is only one cell layer thick; it is called a pseudostratified epithelium.

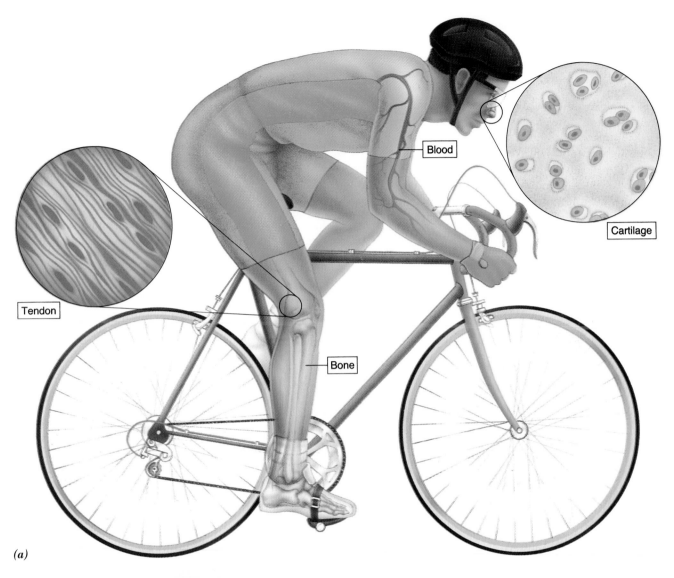

(a)

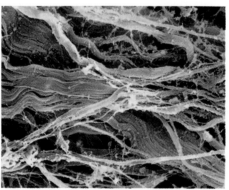

(b)

FIGURE 22-3

Connective tissue. *(a)* Connective tissues are quite diverse in structure and function and include bone, cartilage, blood, and tendons. All of these tissues are categorized by extensive extracellular materials. *(b)* Bundles of extracellular collagen fibers give connective tissues their strength.

secretory epithelia release hormones, oils, sweat, milk, and digestive enzymes. The structure and functions of epithelia will be discussed in greater detail in later chapters on the skin, respiratory tract, digestive tract, and kidney.

CONNECTIVE TISSUE

The body is held together by **connective tissues** (Figure 22-3), which consist of a loosely organized array of cells surrounded by a nonliving extracellular matrix. The properties of connective tissue are due largely to the extracellular matrix, which contains proteins and polysaccharides secreted by the "entrapped" cells. For example, collagen, the most common component of the extracellular matrix (and the most abundant protein in the human body) is an inelastic molecule that provides resistance to pulling forces. Fibers of collagen (Figure 22-3*b*) provide tendons and ligaments with the strength to connect muscles to bones and bones to one another. Looser arrangements of collagen protect delicate structures, such as the eyeball, much like packing material prevents breakage of fragile glassware. Connective tissues may also contain elastic fibers that provide some tissues, such as skin and vocal cords, with the capacity to stretch and recoil. Other examples of connective tissues include skeletal elements composed of cartilage and bone; sheets (*mesenteries*) that support the visceral (internal) organs; the transparent outer layer (*cornea*) of the eyeball; fat (*adipose*) deposits; and the deeper layer (*dermis*) of the skin. Blood is also a type of connective tissue; blood cells are surrounded by an extracellular matrix with a fluid consistency. The structure and functions of connective tissues will be discussed in greater detail in the chapters on the skin, skeleton and circulatory system.

MUSCLE TISSUE

Muscle tissue (Figure 22-4) consists of muscle cells—cells that contain elongated protein filaments that slide over one another, causing the muscle cells to shorten (contract). When muscle tissue contracts, it generates the forces needed for motion—either to move the animal (or its parts) or to propel substances within the body. Muscle tissue is present in large masses capable of moving the largest bones of the body and is the major element of the heart. Muscle tissue can also be found scattered more subtly throughout the body's internal organs. Among other consequences, when such muscle tissue contracts, blood vessels constrict; food moves through the digestive tract; and urine is emptied from the bladder. The various types of muscle tissue will be described in Chapter 26.

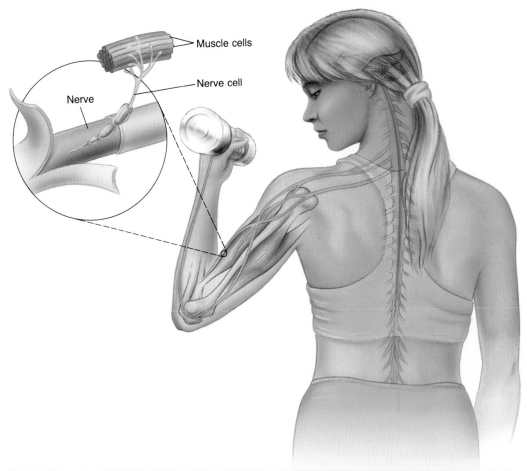

FIGURE 22-4

Muscle and nerve tissues. The structure of one type of muscle tissue is exemplified by a large skeletal muscle of the arm. Other types of muscle tissue are found in the heart and scattered less noticeably throughout the internal organs of the body. Nerve tissue is concentrated within the brain and spinal cord, but also extends from these central neural organs into most of the distant regions of the body.

NERVE TISSUE

Nerve tissue (Figure 22-4) consists of highly elongated nerve cells—cells that are specialized for long-distance communication within an animal. Communication is accomplished when a signal is sent along the membrane of a nerve cell to its terminal end and is relayed to another cell—a nerve cell, muscle cell, or gland cell—thereby evoking a response. Nerve tissue is responsible for coordinating such diverse bodily activities as breathing, sweating, sexual responses, and defecation, as well as providing some animals with memory, emotions, and consciousness. The structure and function of nervous tissues will be discussed in Chapter 23.

Physiologic activities are accomplished by groups of similar cells working together as tissues. Four broad types of tissues are distinguished, but numerous subtypes can be identified. (See CTQ #3.)

UNITY AND DIVERSITY AMONG ORGAN SYSTEMS

Animals as diverse as sponges, whales, earthworms, and anteaters have common needs: All animals must obtain oxygen, digest food, eliminate wastes, protect themselves from microbes and predators, and so forth. Each of these basic needs are met by **organs**—structures that are composed of different tissues working together to perform a particular function. A major function of the kidney, for example, is the removal of metabolic waste products from the blood. The elimination of waste products from the body requires additional organs, including the bladder and a series of tubular pathways (the ureter and urethra) that lead to the body surface. All of the various organs that work together to perform a common task, such as the elimination of waste products, make up an **organ system.**

Common needs often lead to similar, but independent, evolutionary solutions. Even though an earthworm and a whale are so distantly related that their most recent common ancestor *may* have been a single-celled protist, these two animals possess organ systems that have similar overall functions. In the following chapters, we will examine each type of organ system found in animals, focusing on human systems. Despite similarities in overall function, such as uptake of oxygen or elimination of waste products, the organ systems of different animals may be constructed very differently.

Keep in mind that each animal's environment is a critical factor in determining the types of homeostatic mechanisms that will be adaptive for that animal. For example, sea animals must cope with very salty surroundings and, consequently, must possess mechanisms to prevent the loss of water by osmosis. In contrast, animals inhabiting a freshwater stream live in a highly dilute environment that has very little available salt. As a result, freshwater animals must possess adaptations to eliminate water that floods their body by osmosis. In addition, freshwater animals must cope with marked fluctuations in water temperature, oxygen availability, and the risk of the evaporation of their home. Terrestrial (land) animals face even harsher conditions. They must cope with rapid changes in climate, a shortage of external water, a propensity to lose water to the surrounding air, and the lack of a medium to help support their body weight. Each of these different environments presents different physiologic challenges and selects for different types of homeostatic responses.

TYPES OF ORGAN SYSTEMS

Figure 22-5 provides an introduction to the types of organ systems we will encounter in the following chapters. We will summarize them briefly here.

- *Digestive systems* break down *(digest)* the macromolecules present in food. The breakdown products are absorbed across the lining of the digestive tract, providing the body with small-molecular-weight nutrients that supply energy and building materials.

- *Excretory/osmoregulatory (urinary) systems* accomplish two interrelated functions: They rid the body of waste products, the unusable byproducts of metabolism, and they maintain balanced internal concentrations of salt and water. Waste products include potentially toxic nitrogen-containing compounds, most notably ammonia, which must be either eliminated directly or modified to a less toxic compound, such as urea, and then eliminated. Similarly, many biological processes, ranging from enzyme activity to nerve-impulse conduction, are very sensitive to the concentration of dissolved salts, which must be strictly regulated.

- *Respiratory systems* absorb the oxygen animals need to oxidize the organic molecules that fuel their biological activities. The respiratory system may also expel carbon dioxide, a waste product of metabolism. Animals that live in water and absorb the oxygen dissolved in their aquatic medium have different types of respiratory organs than do animals that breathe air.

- *Circulatory systems* transport materials inside an animal's body. Food absorbed by digestive systems and oxygen absorbed by respiratory systems must be distributed to those locations in the body where they are needed; wastes generated by cell metabolism must be carried to a site of elimination. Most animals, particularly those of larger size, distribute materials via a system of branched vessels to all parts of the body.

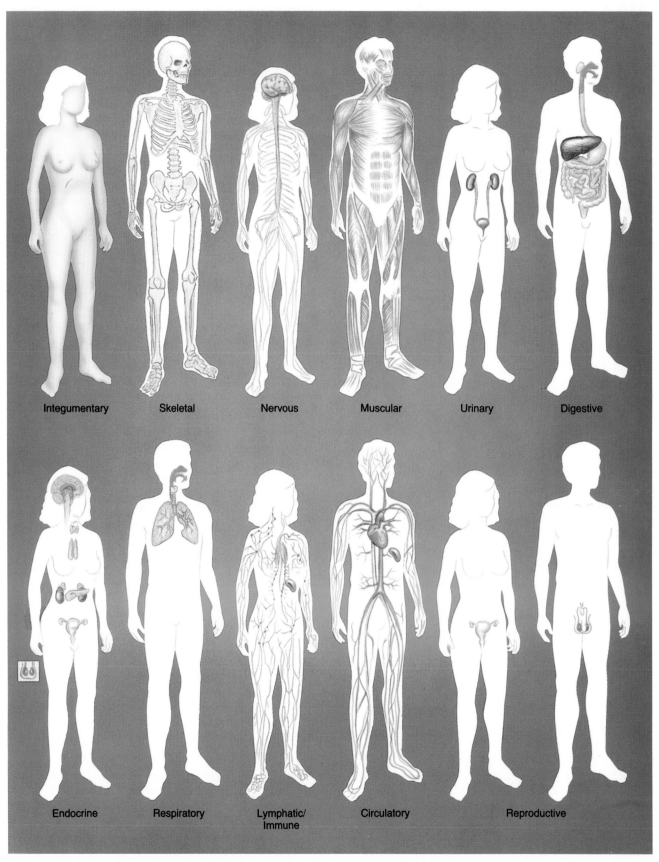

Integumentary   Skeletal   Nervous   Muscular   Urinary   Digestive

Endocrine   Respiratory   Lymphatic/ Immune   Circulatory   Reproductive

FIGURE 22-5

The types of organs systems found in the human body.

- *Immune systems* protect animals from foreign substances and invading microorganisms. The body's immunological defenses consist of various cells that recognize and ingest foreign materials as well as soluble blood proteins, called *antibodies,* which specifically bind to foreign materials and inactivate them.

- *Integumentary systems* cover the surfaces of animals and provide a barrier between the animal's body and the external environment. These systems protect the animal from dehydration, provide physical support, prevent invasion by foreign microorganisms, and help regulate internal temperature.

- *Skeletal and muscular systems* work together to provide both support and movement. Most animals possess rigid skeletal structures that provide support and maintain body shape. In addition, skeletal structures are often involved in *movement* (a shift in position of a part of the body) and *locomotion* (a shift of the entire animal from one place to another). Movement is often accomplished as rigid skeletal elements are pulled in one direction or another by the contraction of attached muscles. Locomotor structures, such as wings, legs, or flagella, typically project from the body as *appendages* (Figure 22-6) that act upon the external environment.

- *Reproductive systems* produce gametes (sperm or eggs) for fertilization and the subsequent development of new individuals of a species. All organisms have a limited life span. Individual bacteria live for minutes, a Galapagos tortoise for hundreds of years, and a giant redwood tree for thousands of years. Yet, these *species* have survived for millions of years—the products of continual reproduction. Reproduction also supplies the means by which variability among individuals in an animal population is generated and is the basis for biological evolution.

The Nervous And Endocrine Systems: Regulating Bodily functions

⟳ The multitude of physiologic processes that proceed simultaneously within an animal must be continually monitored and regulated. If, for example, oxygen levels in the body's tissues drop, mechanisms are triggered that increase the uptake of oxygen from the environment. Or, if an animal is confronted by a potential predator, a protective response is triggered that increases the animal's chance for survival. Two types of systems—the nervous and endocrine systems—regulate and coordinate bodily functions.

- *Nervous systems* are networks of nerve cells (*neurons*) that receive information concerning changes in the external and internal environment, integrate the information, and send out directives to the body's muscles and glands to respond in an appropriate manner. Information moves along the pathways of the nervous system in the form of impulses traveling along the plasma membranes of the nerve cells. These impulses provide a mechanism by which all parts of the body can rapidly communicate with one another. Information enters the nervous system via a series of *sensory* structures that detect changes in the external and internal environment.

- *Endocrine systems* also regulate and coordinate many of an animal's internal activities. Endocrine systems consist of a disconnected network of glands that release chemical messengers (hormones) into the blood. Hormones circulate through the bloodstream and ultimately interact with their particular target cells, triggering specific responses. While responses triggered by the nervous system tend to occur rapidly, those triggered by the endocrine system occur more slowly and often include metabolic changes. Endocrine responses include changes in the level of glucose in the blood, metamorphosis of a caterpillar into a butterfly, and sexual maturation (Figure 22-7).

THE COST OF RUNNING THE BODY'S BUSINESS

☀ The operation of each physiologic system requires the expenditure of energy. An animal takes in chemical energy that is stored in food. Energy-containing macromolecules enter the digestive system, where they are disassembled; the energy-rich products are then absorbed into the body, where they are made available to all of the body's cells. The biggest consumers of energy are the muscular and nervous systems. Even when you are at rest, muscular contraction is at work, pumping blood through your vessels, food through your digestive tract, and air into your lungs. Maintaining an elevated body temperature can be the most costly of all activities, consuming approximately 90 percent of the energy you expend when you are at rest. If your physical activity should increase, muscle activity takes a much larger share of your energy supply.

As a group, animals have a common set of physiologic needs, which are met by organ systems that fall into distinct categories, according to function. The structure of the organs, their mechanism of action, and their organization into systems is highly varied among diverse animals. (See CTQ #4.)

THE EVOLUTION OF ORGAN SYSTEMS

Much of what we know about the evolutionary relationship among animals is derived from studies of fossil remains left behind by individuals living millions of years ago. Fossils almost invariably consist of the hardened skeletal parts of

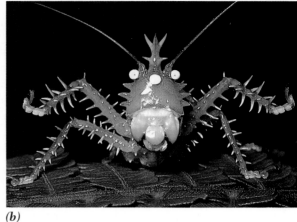

(a)   *(b)*

FIGURE 22-6

A gallery of animal appendages. *(a)* A giant Pacific octopus with a shark caught in the suction grip of its tentacles. *(b)* An inhabitant of the South American rain forest, this katydid exhibits a number of appendages including legs, antenna, and even its mouthparts which are derived from modified embryonic appendages. *(c)* The wings of this African cape gannet are appendages used for flight. *(d)* The appendages of a chimpanzee are similar in structure and function to our own arms and legs.

ancient animals; they reveal very little direct information about any of the other systems introduced in this chapter. Consequently, our knowledge about the evolution of most organ systems is based largely on comparative studies of living animals.

▐▶ This approach always raises the question of origin. Are the similarities in an organ between two distantly related living animals due to the fact that the organ evolved from a similar organ present in a common ancestor (in which case, the organs are said to be *homologous*) or from independent courses of evolution? Two organs with similar functions often resemble each other, even though they are not de-

rived from a common ancestral organ, because they have evolved in response to similar types of selective pressures. In such cases, the organs are said to be *analogous*. For example, the excretory organs of both a human and a lobster consist of microscopic tubules. While the tubules may look superficially similar in both organisms, they have evolved independently to meet a similar physiologic need. This phenomenon is an example of "convergent evolution" (Chapter 33).

When biologists compare the organ systems of invertebrates—animals lacking backbones—they often find a progression in anatomic complexity from simpler

(a)   *(b)*

FIGURE 22-7

The result of hormones. This emperor fish is transformed from a juvenile *(a)* to a sexually mature adult *(b)* as the result of hormones secreted by the fish's gonads.

(c)   *(d)*

animals, such as sea anemones and flatworms, to more complex animals, such as beetles and octopuses. For example, the nervous system of a sea anemone consists of a diffuse net of nerve cells, whereas that of an octopus includes a large, complex brain that coordinates the animal's intricate behavior.

A comparison of most organ systems in vertebrates—animals that possess backbones—provides some of the clearest insight into how natural selection can lead to the modification of structures to meet different physiologic challenges. In some cases, a particular structure has undergone a transformation from one form and function to another. For example, the bones situated just behind your

eardrum are derived from bones that formed part of the jaws of your vertebrate ancestors. The evolutionary movement of vertebrates from the water onto the land required certain changes in the way sound vibrations were transmitted from the environment to the sensory receptors in the ear. Fortuitously, one of the bones used to support the jaws of fishes was no longer needed as a jaw brace in the early amphibians. Instead, this bone became "pressed into service" as an ear bone (Figure 22-8*a*). The other two bones in the middle ear of mammals are derived from bones that were previously part of the jaws of our amphibian ancestors (Figure 22-8*b*). We can see from this description that, during the course of vertebrate evolution, these bones under-

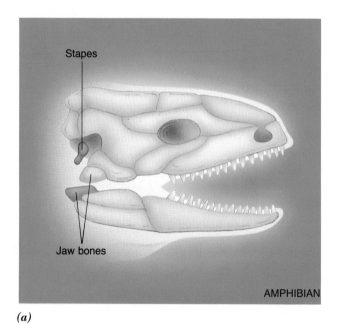
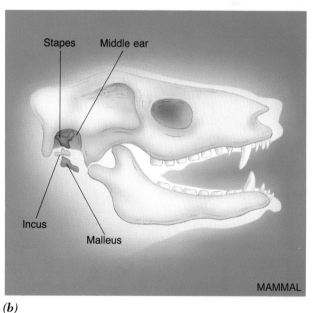

(a)   *(b)*

FIGURE 22-8

The evolution of the bones of the ear. *(a)* The middle ear of amphibians contains a single bone, the stapes, which can be traced to a bone in the skull of ancestral fishes. *(b)* Mammals have three bones in the middle ear—the incus, malleus, and stapes (often called the anvil, hammer, and stirrup, respectively). The incus and malleus have evolved from bones that are present in the upper and lower jaws of our amphibian ancestors.

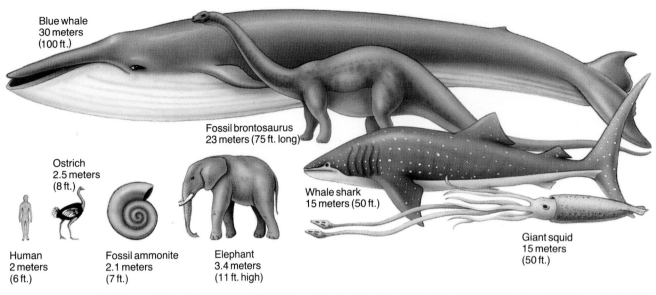

Blue whale
30 meters
(100 ft.)

Fossil brontosaurus
23 meters (75 ft. long)

Ostrich
2.5 meters
(8 ft.)

Whale shark
15 meters (50 ft.)

Giant squid
15 meters
(50 ft.)

Human
2 meters
(6 ft.)

Fossil ammonite
2.1 meters
(7 ft.)

Elephant
3.4 meters
(11 ft. high)

FIGURE 22-9
Animals are found in many different sizes.

went a dramatic change in position, shape, and function. What were once food-gathering structures evolved into transmitters of sound waves.

The characteristics of each organ system within an animal can be understood by considering the animal's ancestry and the environmental challenges with which the system must cope. (See CTQ #6.)

BODY SIZE, SURFACE AREA, AND VOLUME

Multicellular animals span a range of size more than five orders of magnitude (100,000 times), from microscopic, aquatic rotifers (100 micrometers long) to the blue whale (30 meters long). As biologists measured physiologic activities in animals of widely different dimensions (Figure 22-9), it became apparent that the levels of these activities did not increase in a proportional manner. For example, when an elephant and a mouse are compared *per gram of body weight*, the elephant utilizes less oxygen, eats less food, and has a greater skeletal mass than does the mouse. These lack of proportionalities are known as **scaling effects.**

Many scaling effects in biology can be explained on the basis of unequal changes in surface area and volume as animals increase or decrease in size. Surface area, which might be measured in units from *square* micrometers to *square* meters, is the area that covers the outside of the body. Volumes of these same animals would then be mea-

sured in units ranging from *cubic* micrometers to *cubic* meters. (We have italicized the words "square" and "cubic" to emphasize that the surface area of an animal is a function of the square of the animal's length, width, and height, while the volume is a function of the cube of these dimensions.) As the dimensions of an object—whether it be a simple sphere or a complex animal—increase, the surface area of the object increases to less of a degree than does its volume (Figure 22-10). In other words, the larger the animal, the smaller the **surface area/volume ratio (SA/V).** This relationship has very important physiologic consequences.

An animal's surface is the boundary between itself and its external environment. All materials and energy (food, water, respiratory gases, waste products, heat) that pass between the animal and its environment must cross the body surface, which includes internal surfaces, such as the digestive tract and the lungs, as well as the external skin. The rate at which this exchange can occur is directly proportional to the surface area available to be crossed. In contrast, the mass, or weight, of an animal is proportional to its volume; the greater the volume, the more food and oxygen the animal needs, and the more waste materials it produces.

Imagine, for a moment, that an animal, let's say a frog, absorbs most of its oxygen across its body surface. As the frog grows larger, its *need* for oxygen will increase in proportion to its volume, while its *ability to absorb* oxygen will only increase in proportion to its surface area. As a result, the frog's need for oxygen will outstrip its ability to provide this vital substance. As we will see later on, larger animals have evolved special adaptations to counteract this scaling problem. For example, the respiratory organs of large ani-

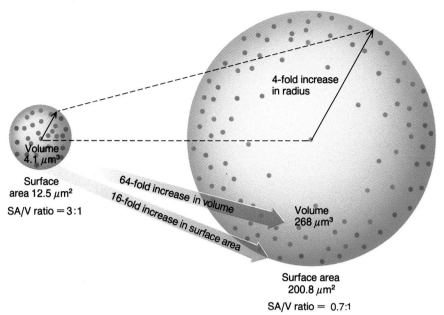

Volume
4.1 μm^3

Surface
area 12.5 μm^2

SA/V ratio = 3:1

4-fold increase
in radius

64-fold increase in volume

16-fold increase in surface area

Volume
268 μm^3

Surface area
200.8 μm^2

SA/V ratio = 0.7:1

FIGURE 22-10

Relationship between surface area and volume. Quadrupling the size of a sphere increases its surface area 16 times and its volume 64 times. Suppose the spheres were single-celled organisms—floating protozoa, for example. The plasma membrane of the larger cell would have to provide nutrients and oxygen (shown as blue dots) for a mass of living protoplasm that was 64 times greater than the protoplasm of the smaller cell, but through a surface only 16 times as large. Without special mechanisms, the large cell would not be able to maintain the same concentration of nutrients and oxygen as would the smaller cell.

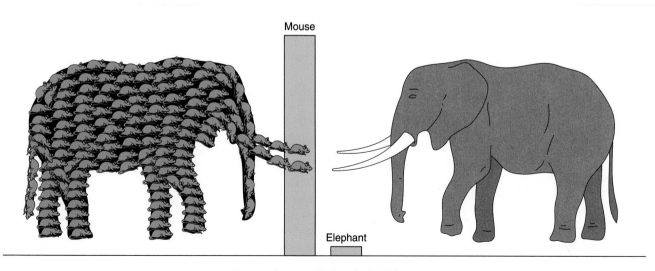

Mouse

Elephant

Oxygen Consumption/gm Body Weight

FIGURE 22-11

The mouse and the elephant. An average mouse weighs about 25 grams, compared to an average elephant that weighs nearly 4 million grams. Because of their difference in size, the elephant has a much smaller surface area/volume. This difference is reflected in the metabolic rates (volume of oxygen consumed per gram of body weight per hour) of these animals—as indicated by the height of the bars, the metabolic rate of the mouse is 24 times greater than that of the elephant.

mals contain extensive folds, or partitions, that greatly increase the area across which oxygen can be absorbed. Similarly, the lining of a large animal's digestive tract contains folds and projections that greatly increase the surface area available for absorbing nutrients. Reducing surface area may also be adaptive. For example, people tend to curl up or huddle together with friends when cold, in an attempt to decrease the surface area across which heat is lost.

But why does an elephant eat less food and utilize less oxygen *per gram of body weight* than does a mouse (Figure 22-11)? At least part of the explanation can be traced to the fact that elephants have a much smaller surface area/volume ratio than do mice. Most of the energy that a mouse or elephant obtains from its food is utilized to maintain its high body temperature. The loss of body heat, however, depends on the body's surface area; on a cool night, a small mouse will lose its body heat much more rapidly than will a large elephant. Consequently, the mouse must take in *relatively* more calories in its food and produce relatively more heat (as measured by its oxygen consumption) than does the elephant.

The level of various physiologic activities exhibited by different animals is often understood when their relative surface area/volume ratio is considered. (See CTQ #7.)

REEXAMINING THE THEMES

Relationship between Form and Function

As we will see throughout this section of the book, the structures of various tissues, organs, and organ systems are closely correlated with their functions. For example, an epithelium may separate compartments of different solute concentration because this tissue consists of a continuous sheet of tightly adhering cells. Similarly, connective tissue can function to protect and support parts of the body because it contains extracellular materials that have strength, hardness, elasticity, and/or resiliency.

Biological Order, Regulation, and Homeostasis

The basic function of most organ systems is to maintain the constancy of the internal environment. Each system has a particular role to play in this overall "drama," from maintaining a stable, elevated temperature, to keeping the salt and water level of body fluids at a relatively constant concentration, to eliminating wastes, to protecting the body from foreign invasion. Two systems—the nervous and endocrine systems—coordinate the activities of the other organ systems. The nervous and endocrine systems must collect information about conditions within the body and send out appropriate messages to the various effectors, whose responses maintain the ordered state.

Acquiring and Using Energy

All of the various organ systems require energy to fuel their activities. Virtually every activity—from contracting a muscle to thinking about the answer to a question on an exam—requires energy. Chemical energy is derived from the food we eat, which must be disassembled by our digestive system. The products of digestion are then absorbed through the lining of our digestive tract and carried by the circulatory system to the sites where the energy is needed.

Unity within Diversity

Animals exhibit remarkably diverse sizes, shapes, and internal body plans, yet they display an equally remarkable physiologic unity. Virtually all animals have the same types of organ systems, even if these systems have evolved independently. The presence of similar organ systems reflects the fact that all animals have similar needs that are basic to maintaining life. Similarly, even though animals possess a diverse array of cells, these cells are organized into a small number of similar types of tissues.

Evolution and Adaptation

Organ systems are adapted to the particular challenges presented by an animal's environment. For example, fishes living in the sea possess an osmoregulatory system that maintains water balance in the face of water loss by osmosis, while their close relatives living in a freshwater stream possess an osmoregulatory system that maintains water balance in the face of water gain by osmosis. Each body part of an animal has evolved from a structure that was present in an ancestor and was then modified by natural selection. In some cases, a structure can undergo dramatic changes in form and function, as organisms adapt to new conditions. For example, the movement of vertebrates from water to land was accompanied by a shift in certain bones from the jaws to the middle ear.

SYNOPSIS

Multicellular animals must conduct three critical functions in order to maintain a homeostatic state. The body must (1) receive information from sensors that detect changes; (2) pass the information to a control unit, typically located in the brain, which is "set" to maintain a certain value; and (3) send out signals to effectors (muscles and glands) to initiate a response that restores conditions to that which is set by the control unit. Maintaining a property at a constant level requires negative feedback. In certain cases, homeostasis requires that a change in the body is amplified, rather than reversed. This can be accomplished by positive feedback, as illustrated by the rapid formation of a blood clot, which prevents further loss of blood.

Multicellular animals have similar types of organ systems that meet common needs. These include a digestive system, which disassembles food matter and provides the body with nutrients; an excretory/osmoregulatory system, which eliminates metabolic waste products and maintains a balanced concentration of salt and water; a respiratory system, which absorbs oxygen and often expels carbon dioxide; a circulatory system, which transports materials from place to place in the body; an immune system, which protects against foreign substances and invading microorganisms; an integumentary system, which covers the animal; skeletal and muscular systems, which facilitate support, movement, and locomotion; and a reproductive system, which produces sperm and eggs.

The nervous and endocrine systems control bodily activities. These two systems coordinate the activities of the other organ systems by collecting information about conditions in both the internal and external environment and providing signals to effectors that carry out a particular response.

Although diverse in form and function, cells are organized into four basic types of tissues. Epithelial tissues consist of sheets of cells that act as linings. Their functions include protection, exchange, and secretion. Connective tissues consist of cells surrounded by a nonliving extracellular matrix. As skeletal materials, connective tissues provide support and facilitate movement; as ligaments and tendons, they connect parts of the body and resist stretching; as the cornea, they provide a transparent layer for vision; and as the blood, they distribute materials from place to place. Muscle tissue provides the force for movement of materials within the body and movement of attached skeletal elements. Nerve tissue forms a communication network that transmits information used in coordinating bodily activities.

Animals with markedly different surface area/volume ratios face different physiologic challenges. As an animal (or series of animals) increases in size, its surface area increases as a function of the square of its linear dimensions, while its volume increases as a function of the cube of those dimensions. Consequently, surface area/volume ratio decreases as the size of an animal increases. Exchange of substances between the animal and its environment are determined by the body's surface area. Larger animals require special adaptations of their respiratory and digestive tracts in order to provide sufficient surface area to facilitate the absorption of the large amounts of oxygen and nutrients needed to sustain life. Smaller, warm-blooded animals tend to gain and lose heat more rapidly than do larger ones.

Key Terms

negative-feedback mechanism (p. 450)
positive-feedback mechanism (p. 450)
tissue (p. 451)
epithelial tissue (p. 452)

epithelium (p. 452)
connective tissue (p. 453)
muscle tissue (p. 454)
nerve tissue (p. 455)

organ (p. 455)
organ system (p. 455)
scaling effect (p. 460)
surface area/volume ratio (p. 460)

Review Questions

1. Describe the basic components your body employs to maintain homeostasis. Describe these components in connection with the maintenance of a constant body temperature.

2. Compare and contrast analogous and homologous organs; negative- and positive-feedback mechanisms; tissues and organs; connective tissue and epithelial tissue; and the functions of the digestive system and the excretory system.

3. Contrast the basic mechanism of operation of the nervous and endocrine systems in regulating bodily functions.

4. Consider an animal whose shape is a perfect cube. When it hatched, the animal measured 1 centimeter along each side; when it was fully grown, it measured 10 centimeters along each side. What are the differences in surface area and volume between the animal's hatched and fully grown state? How does its surface area/volume ratio differ at these two stages of life? What physiologic changes would you expect to find as the animal grows?

Critical Thinking Questions

1. In the Steps to Discovery vignette, several activities were mentioned that have the potential to disrupt the internal stability of our bodies but are kept from doing so by homeostatic mechanisms. Can you think of any other activities in which you engage that have this potential? Explain your answer.

2. When the concentration of carbon dioxide in your blood increases, the pH decreases. This stimulates the respiratory center in your brain. You then breathe faster and deeper, expel more carbon dioxide from your lungs, and the pH increases. In this manner, your body maintains a relatively constant blood pH. Identify the following parts of this system: sensor, control unit, effectors. Is this an example of negative or positive feedback? Explain.

3. You are cutting an apple with a sharp knife and accidentally cut your finger deeply. Which of the four types of tissues did the knife pass through? Support your answer.

4. How does the organization of flowering plants compare with that of higher animals, such as the human? What is there in the life styles of flowering plants and higher animals that could account for differences? What tissues in plants are analogous to the four animal tissue types? Do flowering plants have organ systems?

5. Select any five of the organ systems surveyed in this chapter and discuss the effects on your body if one of these systems should suddenly stop operating.

6. Discuss the different types of challenges that would face an animal living on land compared to one living in the ocean and the adaptations required in the excretory/osmoregulatory, skeletal/muscular, and respiratory systems of both animals to meet those challenges.

7. According to Jonathan Swift, the author of *Gulliver's Travels*, once Gulliver landed in Lilliputia and became accepted by the tiny emperor of that land, it was deemed that he should be provided with a daily allowance of meat and drink sufficient for the support of 1,728 Lilliputians. This value was determined by mathematicians, who calculated Gulliver's volume as equal to 1,728 times their own. Was this the appropriate diet for Gulliver? Why or why not?

Additional Readings

Benison, S., A. C. Barger, and E. Wolfe. 1987. *Walter B. Cannon: The life and times of a young scientist.* Cambridge, MA: Harvard University Press. (Introductory)

The following are intermediate–advanced level histology and physiology texts that describe all of the organ systems covered in this section.

Eckert, R. 1988. *Animal physiology: Mechanisms and adaptations.* 3rd ed. New York: W. H. Freeman.

Fawcett, D. 1986. *Bloom and Fawcett: A textbook of histology,* 11th ed. Philadelphia: Saunders.

Fox, S. I. 1984. *Human physiology.* Dubuque, IA: W. C. Brown.

Guyton, A. C. 1992. *Human physiology and mechanisms of disease.* Philadelphia: Saunders.

Vander, A. J., J. H. Sherman, and D. S. Luciano. 1990. *Human physiology: The mechanisms of body function,* 5th ed. New York: McGraw-Hill.

Weiss, L., Ed. 1988. *Cell and tissue biology: A textbook of histology,* 6th ed. Baltimore: Williams & Wilkins.

Coordinating the Organism: The Role of the Nervous System

STEPS TO DISCOVERY
A Factor Promoting the Growth of Nerves

A Factor Promoting the Growth of Nerves

Rita Levi-Montalcini received her medical degree from the University of Turin in Italy in 1936, the same year that Benito Mussolini began his anti-Semitic campaign. By 1939, as a Jew, Levi-Montalcini had been barred from carrying out research and practicing medicine, yet she continued to do both secretly. As a student, Levi-Montalcini had been fascinated with the structure and function of the nervous system. Unable to return to the university, she set up a simple laboratory in her small bedroom in her family's home. As World War II raged throughout Europe, and the Allies systematically bombed Italy, Levi-Montalcini studied chick embryos in her bedroom, discovering new information about the growth of nerve cells from the spinal cord into the nearby limbs. In her autobiography *In Praise of*

Imperfection, she writes: "Every time the alarm sounded, I would carry down to the precarious safety of the cellars the Zeiss binocular microscope and my most precious silver-stained embryonic sections." In September 1943, German troops arrived in Turin to support the Italian Fascists. Levi-Montalcini and her family fled southward to Florence, where they remained in hiding for the remainder of the war.

After the war ended, Levi-Montalcini continued her research at the University of Turin. In 1946, she accepted an invitation from Viktor Hamburger, a leading expert on the development of the chick nervous system, to come to Washington University in St. Louis to work with him for one semester; she remained at Washington University for 30 years.

A chick embryo and one of its nerve cells helped scientists discover nerve growth factor (NGF).

One of Levi-Montalcini's first projects was the reexamination of a previous experiment of Elmer Bueker, a former student of Hamburger's. Bueker had removed a limb from a chick embryo, replaced it with a fragment of a mouse connective tissue tumor, and found that nerve fibers grew into this mass of implanted tumor cells. When Levi-Montalcini repeated the experiment she made an unexpected discovery: One part of the nervous system of these experimental chick embryos—the sympathetic nervous system—had grown five to six times larger than had its counterpart in a normal chick embryo. (The sympathetic nervous system helps control the activity of internal organs, such as the heart and digestive tract.) Close examination revealed that the small piece of tumor tissue that had been grafted onto the embryo had caused sympathetic nerve fibers to grow "wildly" into all of the chick's internal organs, even causing some of the blood vessels to become obstructed by the invasive fibers. Levi-Montalcini hypothesized that the tumor was releasing some soluble substance that induced the remarkable growth of this part of the nervous system. Her hypothesis was soon confirmed by further experiments. She called the active substance **nerve growth factor (NGF).**

The next step was to determine the chemical nature of NGF, a task that was more readily performed by growing the tumor cells in a culture dish rather than an embryo. But Hamburger's laboratory at Washington University did not have the facilities for such work. To continue the project, Levi-Montalcini boarded a plane, with a pair of tumor-bearing mice in the pocket of her overcoat, and flew to Brazil, where she had a friend who operated a tissue culture laboratory. When she placed sympathetic nervous tissue in the proximity of the tumor cells in a culture dish, the nervous tissue sprouted a halo of nerve fibers that grew toward the tumor cells. When the tissue was cultured in the absence of NGF, no such growth occurred.

For the next 2 years, Levi-Montalcini's lab was devoted to characterizing the substance in the tumor cells that possessed the ability to cause nerve outgrowth. The work was carried out primarily by a young biochemist, Stanley Cohen, who had joined the lab. One of the favored approaches to studying the nature of a biological molecule is to determine its sensitivity to enzymes. In order to determine if nerve growth factor was a protein or a nucleic acid, Cohen treated the active material with a small amount of snake venom, which contains a highly active enzyme that degrades nucleic acid. It was then that chance stepped in.

Cohen expected that treatment with the venom would either destroy the activity of the tumor cell fraction (if NGF was a nucleic acid) or leave it unaffected (if NGF was a protein). To Cohen's surprise, treatment with the venom *increased* the nerve-growth promoting activity of the material. In fact, treatment of sympathetic nerve tissue with the venom alone (in the absence of the tumor extract) induced the growth of a halo of nerve fibers! Cohen soon discovered why: The snake venom possessed the same nerve growth factor as did the tumor cells, but at much higher concentration. Cohen soon demonstrated that NGF was a protein.

Levi-Montalcini and Cohen reasoned that since snake venom was derived from a *modified* salivary gland, then other salivary glands might prove to be even better sources of the protein. This hypothesis proved to be correct. When Levi-Montalcini and Cohen tested the salivary glands from male mice, they discovered the richest source of NGF yet, a source 10,000 times more active than the tumor cells and ten times more active than snake venom.

A crucial question remained: Did NGF play a role in the normal development of the embryo, or was its ability to stimulate nerve growth just an accidental property of the molecule? To answer this question, Levi-Montalcini and Cohen injected embryos with an antibody against NGF, which they hoped would inactivate NGF molecules wherever they were present in the embryonic tissues. The embryos developed normally, with one major exception: They virtually lacked a sympathetic nervous system. The researchers concluded that NGF must be important during normal development of the nervous system; otherwise, inactivation of NGF could not have had such a dramatic effect.

By the early 1970s, the amino acid sequence of NGF had been determined, and the protein is now being synthesized by recombinant DNA technology. During the past decade, Fred Gage, of the University of California, has found that NGF is able to revitalize aged or damaged nerve cells in rats. Based on these studies, NGF is currently being tested as a possible treatment of Alzheimer's disease. For their pioneering work, Rita Levi-Montalcini and Stanley Cohen shared the 1987 Nobel Prize in Physiology and Medicine.

*I*magine that you are sitting at your desk reading a book when, out of the corner of your eye, you catch a glimpse of a dark furry-looking object resting on your right forearm. If you are like most people, within a second you will have flicked the object off your arm with the back of your left hand. You will then get out of your chair, bend over, and try to get a better look at the insect or spider you presume to have wounded. When you find the object, you identify it as a clump of thread that must have come loose from the sweater you're wearing. You smile, realize your heart is beating a little faster, sit back down, and return to your book.

The events we have just described—the glimpse of an object, the "instantaneous" determination of a threat, the quick muscular response by your hand, the curiosity as to the nature of the object, the increase in your heart rate, the humor you find in the events that have occurred, and the desire to return to your book—are all a direct result of activities taking place in your nervous system. So too is the visualization of these events in your "mind's eye."

This one brief sequence of events points out many of the functions of the nervous system. Most importantly, the nervous system communicates information from one part of the body to another and, in so doing, regulates the body's activities and maintains homeostasis. Your very survival depends on continuous neural activity. For example, without orders from the nervous system, you would be unable to contract the muscles that draw air into your lungs; you would be unable to activate your sweat glands to release fluids needed to lower your body temperature; and you would be unable to chew, swallow, or send food along your digestive tract.

The nervous system also controls your basic drives (such as hunger, thirst, and sexual desires) and emotional responses (including anger and fear). Consciousness itself is derived from neural activity. Despite the recent advances in cell and molecular biology, we are still very far from understanding the underlying neural basis of thought, learning, memory, perception, and behavior. One thing is certain: The operation of the nervous system will remain a fertile ground for investigation for many years to come.

▼ ▼ ▼

NEURONS AND THEIR TARGETS

Each nerve cell, or **neuron,** is specialized for conducting messages, in the form of moving *impulses,* from one part of the body to another. Messages can be sent along these cellular "transmission lines" at speeds of over 100 meters per second (225 miles per hour). The effect of the impulse depends on two properties: the nature of the neuron, and the type of target cell that responds to the neuron.

- *Nature of the neuron.* Some neurons, called **excitatory neurons,** stimulate their target cells into activity; others, called **inhibitory neurons,** oppose a response, encouraging target cells to remain at rest.

- *Nature of the target cell.* Only three basic types of cells can respond *directly* to an arriving impulse. They are

1. muscle cells, which respond to excitatory stimulation by contracting, thereby exerting force;
2. gland cells, which respond to excitation by secreting a substance;
3. other nerve cells, which may generate impulses of their own, thereby relaying the message to another target cell.

NEURONS: FORM AND FUNCTION

Describing the form of a "typical" nerve cell is like trying to describe a "typical" human personality. Your body contains more than 100 billion neurons, and no two are exactly alike. Nonetheless, all neurons are composed of the same basic parts (Figure 23-1), which allow them to collect, conduct, and transmit impulses.

The form of a neuron is readily correlated with its function. The nucleus of the neuron is located within an expanded region, called the **cell body,** which is the metabolic center of the cell and the site where most of its material contents are manufactured. Extending from the cell bodies of most neurons are a number of miniscule extensions, called **dendrites,** which receive *incoming* information from external sources, typically other neurons. Also emerging from the cell body is a single, more prominent extension, the **axon,** which conducts *outgoing* impulses away from the cell body and toward the target cell. Impulses are generally initiated in the region where the cell body merges into the axon. While some axons may be only a few micrometers in length, others extend for considerable distance. The neurons that carry impulses from a giraffe's spinal cord to its legs, for example, may extend 3 meters, placing them among the longest cells in the animal world.

Most axons split near their ends into smaller processes, each ending in a **synaptic knob**—a specialized site where impulses are transmitted from neuron to target cell. Some cells in your brain may end in thousands of synaptic knobs, allowing these brain cells to communicate with thousands of potential targets.

Neuroglia: The Supporting Cast

Only about 10 percent of the cells in the human nervous system are neurons; the rest are **neuroglial cells.**

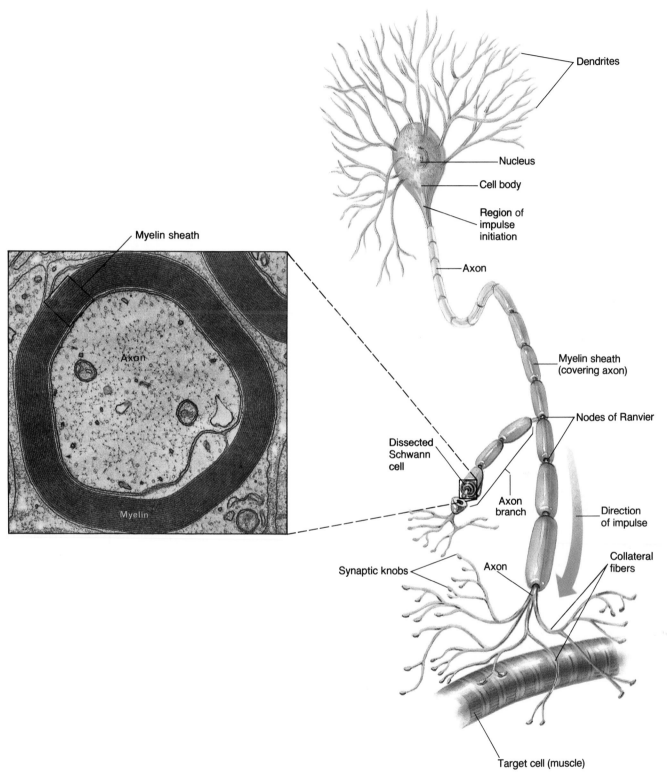

FIGURE 23-1

Anatomy of a neuron.   Information enters the nerve cell through a branching network of dendrites. The incoming signals impinge on the cell body, which contains the nucleus and the cell's synthetic machinery. The cell body merges with the elongated axon, which often branches into numerous smaller processes, each of which ends in a synaptic knob. Most vertebrate axons are wrapped in a myelin sheath composed of Schwann cells. The inset shows an electron micrograph of a cross section through an axon surrounded by a myelin sheath.

These "accessory cells" include **Schwann cells,** which consist almost wholly of plasma membrane, with very little cytoplasm (see inset, Figure 23-1). These cells wrap themselves around axons, forming a layered jacket of cell membranes, called a **myelin sheath.** Since cell membranes consist predominantly of lipids, which are poor conductors of electricity, the myelin sheath functions as living "electrical tape," insulating the axon against electrical interference from its neighbor. The insulation is not complete, however; tiny, naked gaps remain exposed between adjacent Schwann cells. These uninsulated gaps, called the **nodes of Ranvier,** help speed nerve impulses along an axon by a "leap-frog" mechanism that enables the impulses to skip from gap to gap. This phenomenon is discussed in more detail later in the chapter.

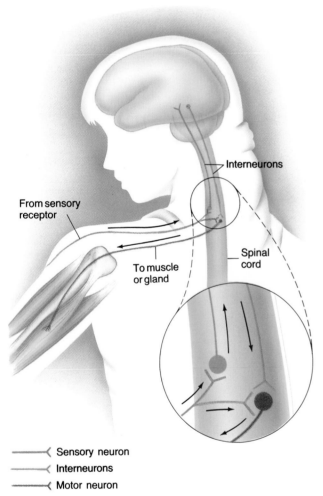

— Sensory neuron
— Interneurons
— Motor neuron

FIGURE 23-2

Three types of neurons. Sensory neurons carry impulses from the periphery (sensory receptors) to the CNS (as indicated by the spinal cord). The various pathways that run up and down the human brain and spinal cord are composed of large numbers of interneurons, which are located entirely within the brain or spinal cord. Motor neurons carry impulses from the CNS to effector cells in the periphery.

Classification of Neurons According to Function

Nerve cells can be grouped into three classes—sensory neurons, interneurons, and motor neurons—depending on the direction impulses are carried and the type of cells to which they are functionally linked (Figure 23-2). **Sensory neurons** carry information about changes in the external and internal environments *toward* the central nervous system (the brain and spinal cord). Once they enter the central nervous system (CNS), impulses from sensory neurons are transmitted to **interneurons,** which transmit impulses from one part of the CNS to another. Interneurons route incoming or outgoing impulses, integrating the millions of messages constantly racing through the CNS. Outgoing impulses are carried by **motor neurons,** which stretch from the CNS to the body's muscles or glands whose contraction or secretion may be stimulated or inhibited.

> **Neurons are highly specialized, cellular communication systems, equipped for receiving, conducting, and transmitting information. Neurons regulate biological function by stimulating or inhibiting actions in target cells. (See CTQ #2.)**

GENERATING AND CONDUCTING NEURAL IMPULSES

Propagation of an impulse along a neuron is often compared to the conduction of a pulse of electricity along a wire. But this analogy fails to take into account basic differences: Electricity is a flow of electrons along a wire, but a nerve impulse involves neither electrons nor a flow of charged particles along an axon. Rather, the nerve impulse occurs as the result of the movement of ions *across* the plasma membrane, rather than along its length. Why do ions move across the plasma membrane? And how can ionic movement across the membrane at one point lead to an impulse that speeds along the entire length of a neuron to an awaiting target cell? The answers to these questions will be revealed as we compare a "resting" neuron to one that has been triggered to conduct an impulse.

THE MEMBRANE POTENTIAL OF A NEURON "AT REST"

The concentrations of specific ions on the two sides of the plasma membrane of a resting neuron—one that is not conducting an impulse—are very different (Figure 23-3). The concentration of potassium ions (K^+) is approximately 30 times higher inside the cell than outside, while the concentration of sodium (Na^+) and chloride (Cl^-) ions are approximately 10 to 15 times higher outside the cell than inside. (These *ionic gradients* are established by the sodium–potassium pump, discussed on page 142.) You might expect that as a result of their concentration gradients, potassium ions would diffuse out of the cell and sodium ions would diffuse inward. But the ability of ions to move across a membrane is not automatic; rather, it depends on the permeability of the cell membrane. Recall from Chapter 7 that ions move across the plasma membrane through specific channels. Nerve cell membranes have two types of channels: *leak channels*, which are always open, and *gated channels*, which can be either open or closed. In the resting state, potassium ions diffuse out of a cell through potassium leak channels. In contrast, since the nerve cell lacks sodium leak channels, the plasma membrane is virtually impermeable to sodium ions.

Potassium ions are positively charged. The movement of positive charges out of the cell leaves the inside of the membrane more negatively charged than the outside. This separation of positive and negative charge is called a **potential difference,** or **voltage.** The voltage across a cell membrane can be measured by inserting microscopic electrodes into a nerve cell. The **resting potential**—the potential difference when the cell is at rest—measures ap-

proximately − 70 millivolts. The negative value, due largely to the outward diffusion of potassium ions, indicates the negativity of the inside of the cell, relative to the outside. In the resting state, the membrane is said to be *polarized*.

ACTION POTENTIALS: TRIGGERED BY A REDUCTION IN MEMBRANE POTENTIAL

Physiologists first learned about membrane potentials in the 1930s from studies on the giant axons of the squid. These axons, which are approximately 1 millimeter in diameter, carry impulses at high speeds, enabling the squid to escape rapidly from predators. If the membrane of a resting squid axon is stimulated by poking it with a fine needle or jolting it with a very small electric current, the axon responds by opening the gates of some of its sodium channels, allowing a number of sodium ions to move into the cell. This movement of positive charges into the cell reduces the

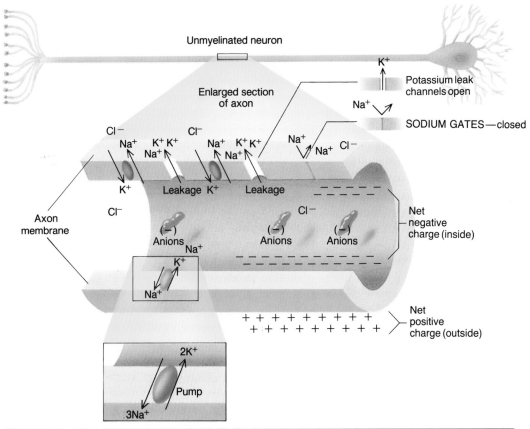

FIGURE 23-3

Formation of a membrane potential in an unmyelinated, "resting" neuron. An ATP-driven pump maintains a steep ionic gradient across the membrane by pumping sodium ions (Na⁺) out of the cell in exchange for potassium ions (K⁺), which are pumped inward (the lower square shows the details of the pump). The membrane of the resting neuron is permeable to potassium and impermeable to sodium (insets at upper right). As a result, positively charged potassium ions diffuse out of the cell through potassium leak channels, resulting in a separation of charge (a *membrane potential*), with the inside of the cell negative relative to the outside. The negativity of the inside of the cell is maintained largely by negatively charged anions, including protein molecules, bicarbonate ions, and phosphate groups.

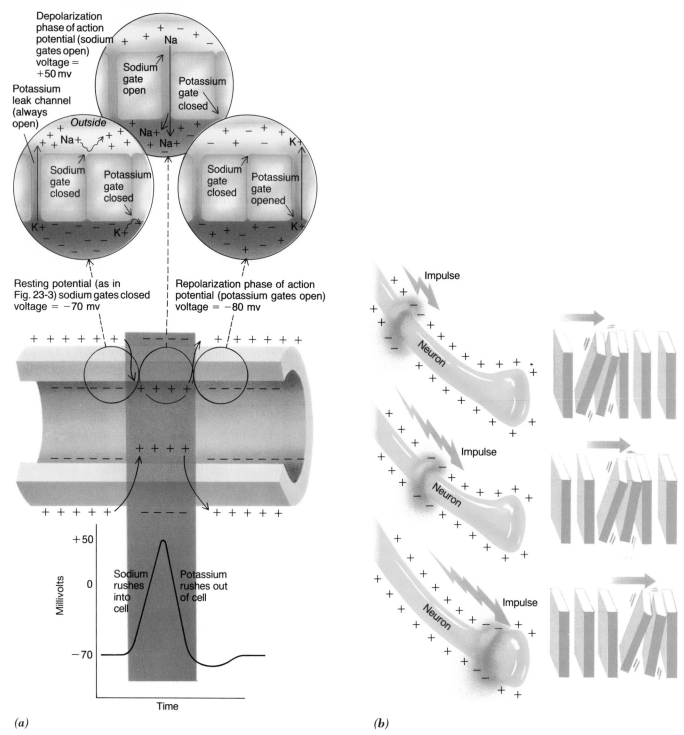

Depolarization phase of action potential (sodium gates open) voltage = +50 mv

Sodium gate open

Potassium gate closed

Na

Potassium leak channel (always open)

Outside

Na+

Sodium gate closed

Potassium gate closed

K+

K+

Na+

Na+

Sodium gate closed

Potassium gate opened

K+

K+

Resting potential (as in Fig. 23-3) sodium gates closed voltage = −70 mv

Repolarization phase of action potential (potassium gates open) voltage = −80 mv

+ + + + + + − − − − + + + + + +

− − − + + + − −

+ + +

+ + + + + − − − − + + + + +

+50

0

−70

Millivolts

Sodium rushes into cell

Potassium rushes out of cell

Time

(a)

Impulse

Neuron

Impulse

Neuron

Impulse

Neuron

(b)

FIGURE 23-4

Formation and propagation of an action potential. *(a)* When depolarization of the membrane exceeds the threshold value, the membrane's sodium gates open and allow positively charged sodium ions to move into the cell. The influx of sodium causes a fleeting reversal in the polarity of the membrane potential, typically from the resting value of − 70 millivolts to + 50 millivolts. This charge reversal is known as an action potential. Within a brief fraction of a second, the sodium gates close and the potassium gates open, allowing potassium to diffuse across the membrane and reestablish an even more negative potential at that location (− 80 millivolts) than that of the resting potential. Almost as soon as they open, the potassium gates close, leaving the potassium leak channels as the primary path of ion movement across the membrane, and reestablishing the resting potential. *(b)* **Propagating an action potential as an impulse.** The disruption in polarity of the membrane that accompanies the action potential at one site triggers the opening of the sodium gates at an adjacent site, initiating an action potential farther along the axon. This process repeats itself from point to point, causing an impulse to race along the axon to the end of the neuron, much like falling dominoes, but with one important difference: Each "domino" along the neuron rights itself immediately after falling.

membrane voltage, making it less negative. Since the reduction in membrane voltage causes a decrease in the polarity between the two sides of the membrane it is called a *depolarization*.

If the stimulus causes the membrane to depolarize by only a few millivolts, say from − 70 to − 50 millivolts, the membrane rapidly returns to its resting potential as soon as the stimulus has ceased (Figure 23-3; 23-4*a* left circle). If the stimulus is great enough, however, the membrane becomes depolarized beyond a certain point, called the **threshold.** When this happens, a new series of events is launched. The change in voltage causes the gates on the sodium channels to swing open, and sodium ions flood into the cell (Figure 23-4*a*, middle circle). As a result of the inflow of sodium ions, the membrane potential briefly reverses itself (Figure 23-4*a*, lower plot), becoming a positive potential of about + 50 millivolts. Then, as the membrane potential reaches its peak positive value, the sodium gates close again, and the gated potassium channels open (Figure 23-4*a*, right circle). As a result, potassium ions flood out of the cell, and the membrane potential swings back to a negative value (Figure 23-4*a*, lower plot). The large negative potential causes the gated potassium channels to close, leaving only the potassium leak channels open. As a result, the membrane is *repolarized* and returns to its resting state. These collective changes in membrane potential are called an **action potential;** the entire sequence occurs in a few milliseconds (thousandths of a second).

The movements of ions across the plasma membrane of nerve cells forms the basis for neural communication. Certain *local* (topically applied) anesthetics, such as Xylocaine and Novocaine, the anesthetic used primarily by dentists, act by closing the gates of ion channels in the membranes of nerve cells. As long as these ion channels remain closed, the affected neurons are unable to generate action potentials and cannot inform the brain of the painful insults being experienced by your gums or teeth. The next time you are in a dentist's chair listening to the sound of the drill, think of the millions of plugged ion channels in the sensory neurons leading from the roots of your teeth to your brain.

Propagation of Action Potentials as an Impulse

The action potential that occurs at one point along the neuron is the "spark" that creates a neural impulse. Like falling dominoes, the propagation of action potentials along the entire length of the axon is the result of a chain reaction (Figure 23-4*b*). An action potential at one site of a membrane induces a depolarization at the adjacent site of the membrane farther along the neuron, initiating an action potential at that site. This new action potential stimulates the next region of the membrane, and the chain reaction continues, producing a wave of action potentials that travels along the entire length of the excited neuron. Once a wave of action potentials is triggered, the wave passes down the entire length of the neuron without any loss of intensity, arriving at its target cell with the same strength it had at its point of origin.

Speed Is of the Essence

When you accidentally stumble, the only thing that keeps your face from hitting the floor is your ability to thrust your foot or hands forward fast enough to arrest your forward plunge—faster than gravity can plaster you against the ground. The speed of your response is even more remarkable when you realize that neural impulses have to travel to the central nervous system (notably, the spinal column) so they can be routed to the arm and leg muscles that prevent you from losing your race against gravity. Think of how often the speed of your response has saved you from personal injury—jumping out of the path of an oncoming car, maneuvering your car when you are cut off by another vehicle, withdrawing your hand from a hot object fast enough to prevent serious injury. The speed of all these actions depends on how fast a nerve impulse reaches a muscle and stimulates it to contract. The speed an impulse travels depends on (1) the diameter of the axon, and (2) whether or not the axon is jacketed in a myelin sheath.

In general, the larger a neuron's diameter, the faster the neuron conducts impulses. This is the adaptive advantage of giant axons in squids and other invertebrates; the larger diameter of these animals' axons increases the speed at which the animals can escape danger. Vertebrate evolution improved on this utility with the adaptation of the myelin sheath. Composed almost entirely of lipid-containing membranes, the myelin sheath is ideally suited to preventing the passage of ions across the plasma membrane. As a result, action potentials can only occur in the unwrapped nodes between the Schwann cells (the nodes of Ranvier). An action potential at one node is strong enough to trigger another action potential at the next node. Consequently, a nerve impulse skips along myelinated neurons from node to node, rather than taking the slower continuous route along the membrane. This "hopping" type of propagation is called **saltatory conduction,** after the Latin "saltare," meaning "to leap." Impulses are able to travel along a myelinated axon at speeds up to 120 meters per second, which is nearly 20 times faster than the speed of impulses of an unmyelinated neuron of the same diameter.

The importance of myelination is dramatically illustrated by multiple sclerosis, a disease that results from the gradual deterioration of the myelin sheath that surrounds axons in various parts of the nervous system. The disease usually begins in young adulthood; victims will experience weakness in their hands, difficulty in walking, and/or problems with their vision. The disease is characterized by progressive muscular dysfunction, often culminating in permanent paralysis.

Because of the ionic gradients generated across its plasma membrane, a resting neuron is always primed and ready. A slight depolarizing stimulus may trigger a self-propagating wave of activity that sweeps down the length of the axon, undiminished in intensity. This wave constitutes a nerve impulse—the mechanism of communication within the nervous system. (See CTQ #3.)

NEUROTRANSMISSION: JUMPING THE SYNAPTIC CLEFT

Neurons are linked with their target cells at specialized junctions called **synapses.** Careful examination of a synapse reveals that the two cells do not make direct contact but are separated from each other by a narrow gap of about 20 to 40 nanometers. This gap is called the **synaptic cleft** (Figure 23-5). Somehow the presynaptic neuron's impulse must "jump" across this cleft in order to affect the postsynaptic target cell.

↻ The first indication that neural messages are carried across the synaptic cleft by chemicals came from an ingenious experiment conducted by Otto Loewi in 1921 (Figure 23-6). The design of the experiment (for which Loewi was awarded the Nobel Prize) came to the scientist in a dream. The heart rate of a vertebrate is regulated by the balance of input from two opposing (antagonistic) nerves, each consisting of a large number of neurons. Loewi isolated a frog's heart together with both nerves. When he stimulated the inhibitory *(vagus)* nerve, a chemical was released from the heart preparation into a salt solution, which was allowed to drain into a second, isolated heart. The rate of the second heart slowed dramatically, as though its own inhibitory nerve had been activated. The substance responsible for inhibiting the frog's heart (and the human heart) was later identified as *acetylcholine,* the first *neurotransmitter* to be discovered.

NEUROTRANSMITTERS: CHEMICALS THAT CARRY THE MESSAGE

Acetylcholine is only one of a number of different chemicals (Table 23-1) that act as **neurotransmitters**—molecules released from neurons, stimulating or inhibiting target cells. These substances, stored in numerous membrane-bound packets called **synaptic vesicles** found inside the synaptic knob, have no influence as long as they remain

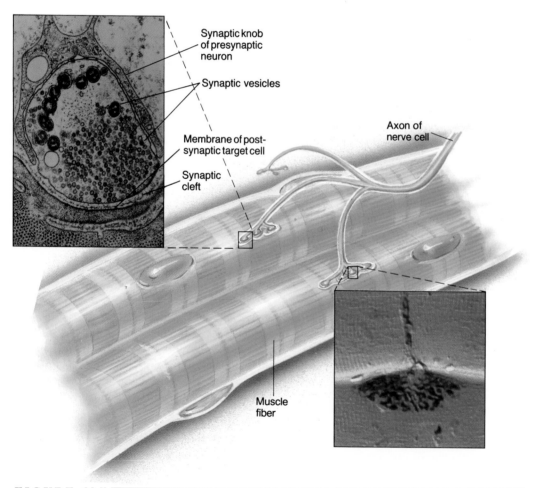

Synaptic knob of presynaptic neuron

Synaptic vesicles

Membrane of post-synaptic target cell

Synaptic cleft

Axon of nerve cell

Muscle fiber

F I G U R E 23-5

Synaptic junction between a neuron and an effector cell. Each synaptic knob abuts the target cell membrane very closely; the synaptic vesicles within the knob are indicated in the micrograph in the upper-left box.

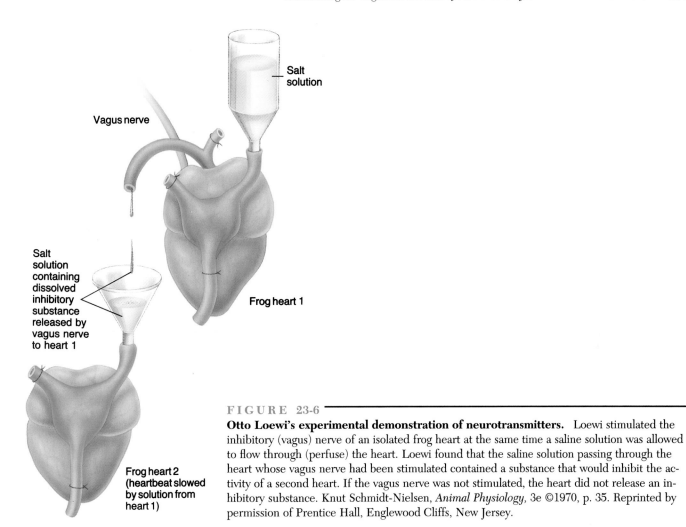

FIGURE 23-6

Otto Loewi's experimental demonstration of neurotransmitters. Loewi stimulated the inhibitory (vagus) nerve of an isolated frog heart at the same time a saline solution was allowed to flow through (perfuse) the heart. Loewi found that the saline solution passing through the heart whose vagus nerve had been stimulated contained a substance that would inhibit the activity of a second heart. If the vagus nerve was not stimulated, the heart did not release an inhibitory substance. Knut Schmidt-Nielsen, *Animal Physiology,* 3e ©1970, p. 35. Reprinted by permission of Prentice Hall, Englewood Cliffs, New Jersey.

TABLE 23-1

SOME NEUROTRANSMITTERS AND THEIR EFFECTS

Neurotransmitter	*+ or −*[a]	*Most Common Target Cells*	*Predominant Effect*
Acetylcholine	+ −	Voluntary muscles Heart muscle	Stimulates muscle contraction. Increases threshold of contraction.
Glycine	−	Motor neurons to voluntary muscles	Raises threshold of excitation, checking uncontrolled muscle contraction.
Dopamine	−	Neurons that produce acetylcholine	Prevents overactivity of neurons that activate muscles. (Deficiencies result in uncontrolled muscle contractions of Parkinson's disease.)
Norepinephrine (noradrenaline)	+	Neurons of central nervous system responsible for arousal, attention, and mood; involuntary muscles (e.g., heart); glands	Increases alertness and attention; heightens readiness for muscular activity.
GABA	−	Motor neurons to voluntary muscles	Prevents uncontrolled muscle contraction.
Serotonin	−	Neurons in the brain that maintain wakefulness	Induces sleep; may modulate mood.

[a] "+" = excitatory; "−" = inhibitory.

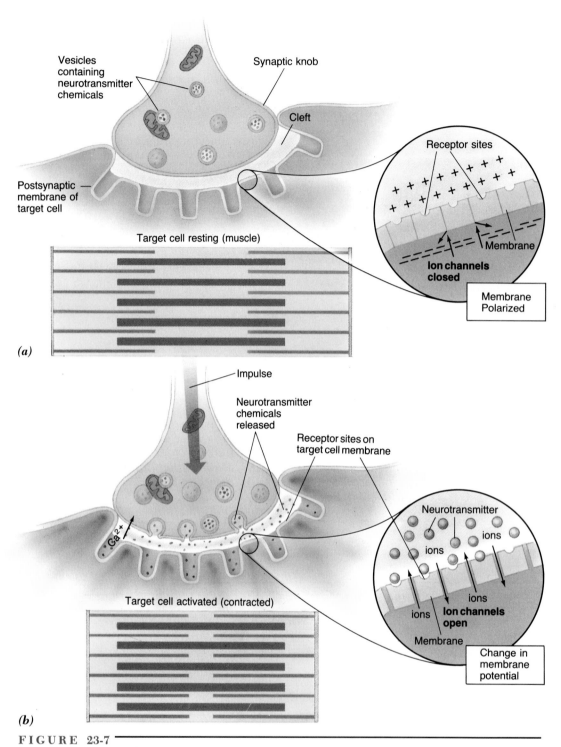

Vesicles containing neurotransmitter chemicals

Synaptic knob

Cleft

Postsynaptic membrane of target cell

Target cell resting (muscle)

Receptor sites

+ + + + + + + + +
+ + + + + + + +

Membrane

Ion channels closed

Membrane Polarized

(a)

Impulse

Neurotransmitter chemicals released

Receptor sites on target cell membrane

Ca²⁺

Target cell activated (contracted)

Neurotransmitter

ions

ions

ions

ions

ions

Ion channels open

Membrane

Change in membrane potential

(b)

FIGURE 23-7

Synaptic transmission between a neuron and target cell (in this case a muscle fiber). *(a)* Neuron at rest. The synaptic cleft contains no neurotransmitter, the receptor sites remain empty, and the target remains unaffected by the neuron. *(b)* When the impulse reaches the tip of a neuron, calcium ions enter the synaptic knob, neurotransmitter molecules are released and bind to receptor proteins on the postsynaptic neuron, sodium ion channels swing open, and the muscle fiber contracts.

packaged in the neuron (Figure 23-7*a*). When an impulse reaches the end of a presynaptic neuron, however, it triggers an opening of calcium ion channels in that part of the membrane, allowing calcium ions to flow *into* the synaptic knobs of the cell (Figure 23-7*b*). Calcium ions are a potent inducer of exocytosis, the discharge of materials out of a cell (page 99). As calcium ions flow into the cell, the membranes of a number of synaptic vesicles fuse with the plasma membrane, and the vesicles spill their contents of neurotransmitter molecules into the synaptic cleft. The discharged neurotransmitter molecules then diffuse across the narrow gap and bind specifically to *receptor proteins* in the membrane of the postsynaptic target cell. The interaction between a neurotransmitter and its specific receptor can have one of two opposite effects depending on the target cell:

1. it can decrease the membrane potential (*depolarize* the membrane), which will *excite* the target cell, making it more likely to respond, as in Figure 23-7, or
2. it can increase the membrane potential (*hyperpolarize* the membrane) which will *inhibit* the target cell, making it less likely to respond.

If you are running to class, for example, neurons stimulate the muscles in your legs to contract by releasing acetylcholine, which *decreases* the membrane potential of the voluntary muscle cells in your leg, making it easier to excite the cells. A different neurotransmitter, *norepinephrine,* is released by nerves that stimulate heart contraction.

↻ If the synaptic story ended here, the target cell would be unable to recover from the chemical message. Neurochemical excitation at the synapse would lock an activated target into a perpetually excited state. This is not the case, however. Target cells are prevented from maintaining a perpetual state of excitation in at least two ways: by enzymes that destroy neurotransmitter molecules almost as soon as they react with their receptors, and by enzymes that transport neurotransmitter molecules back to the neuron that orginally released them—a process called *reuptake.* Because of the destruction and/or reuptake of neurotransmitter molecules, the effect of each impulse lasts no more than a few milliseconds; order is maintained.

Neurobiologists have discovered over 30 different chemicals that act as neurotransmitters, some of which are described in Table 23-1. Most of these neurotransmitters act within the brain alone and, as we will discuss later in the chapter, some neurotransmitters, such as dopamine, can have dramatic effects on our emotional state.

Unfortunately, neurons do not always function properly. Things can go dreadfully wrong at the synapse. When the synaptic cleft is occupied by chemicals that interfere with neurotransmitters, for example, the resulting disorder can lead to paralysis or even death (see Bioline: "Deadly Meddling at the Synapse").

SYNAPSES: SITES OF INFORMATION INTEGRATION

Synaptic transmission does not operate on the basis of "one impulse in, one impulse out." In fact, a single impulse transmitted by a single neuron rarely initiates a response in a target cell because it fails to exceed the target's threshold. Activating a target cell requires a number of excitatory signals which, added together (*summated*), exceed the threshold. Summation can result from (1) simultaneous arrival of multiple stimuli from several adjacent neurons (Figure 23.8) or (2) a virtual "nonstop" barrage of impulses from just one neuron. If every neuron that spontaneously "goes off" were to generate a response from its target, the resulting chaos would disrupt the body's homeostasis. It would be similar to a car alarm's sensitivity being set so high that every little breeze would activate the siren.

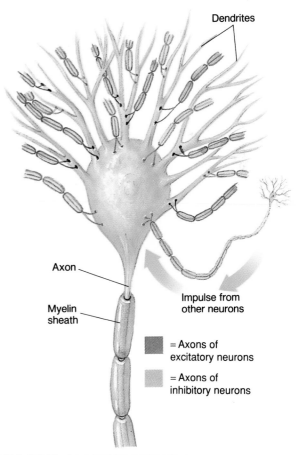

Dendrites

Axon

Myelin sheath

Impulse from other neurons

■ = Axons of excitatory neurons

□ = Axons of inhibitory neurons

FIGURE 23-8
Multiple synapses on a single neuron can run into the thousands. Whether or not the neuron is activated and relays the message depends on whether or not the combined stimulation of the excitatory and inhibitory neurons exceeds the membrane threshold of the target neuron.

◁ B I O L I N E ▷
Deadly Meddling at the Synapse

Nerve poisons! Few substances have acquired such notorious reputations as this group of chemicals, and none deserves the reputation more. Tiny quantities of these substances can sabotage the transmission of impulses across synaptic junctions, interfering with essential motor activity. A number of mechanisms lead to these neurological disasters; the following neurotoxins exemplify just a few.

CURARE

For centuries, hunters in South America have enhanced their predatory skills by dipping their weapons in extracts prepared from tropical plants. For example, a small dart from a blowgun becomes a missile of death when coated with curare, a powerful nerve poison contained in one such botanical extract. Curare blocks the acetylcholine receptor sites on the membrane of muscle cells. As a result, muscles are paralyzed in a state of relaxation and fail to contract in response to neural commands, even though the synaptic cleft is saturated with acetylcholine. Death from suffocation occurs quickly once the muscles used for breathing stop functioning. Curare is now used under clinically-controlled conditions (under the name *Tubocurarine*) as a mus-

cle relaxant to prevent muscle damage that can occur from overcontraction as the result of tetanus or during electroconvulsive shock therapy.

BOTULISM

The bacterium *Clostridium botulinum* releases one of the most potent toxins known; a few ounces is enough to kill every person on earth. A person who eats food containing this toxin (usually a result of improperly canned food) is attacked by a poison that spreads through the bloodstream from the intestine to all the neurons in the body. The toxin prevents the release of acetylcholine from motor neurons. Its effect is therefore similar to curare, only 50 times more potent. The symptoms of botulism include double vision, loss of coordination, and eventual fatal paralysis.

ORGANOPHOSPHATES AND NERVE GAS

Some neurotoxins block the enzymatic destruction of acetylcholine so that the neurotransmitter remains in the synapse, preventing the muscles from returning to their relaxed state once stimulated. Victims suffer sustained paralysis; their muscles re-

main in a state of permanent contraction —just the opposite of the cause of death by curare. Nerve gas exerts its lethal effect in this way, as do organophosphate pesticides, which can be as deadly to humans as they are to the insects for which they are intended.

TETANUS

Another type of paralysis occurs if a wound becomes infected by a common soil bacterium. This disease is known as *tetanus*. As the bacteria grow in the wound, they release a powerful neurotoxin that seeps into the bloodstream and blocks inhibitory synapses on motor neurons throughout the body. The removal of inhibitory influences on these neurons creates a hyperexcitable state; consequently, even low levels of excitatory impulses set off unchecked muscle spasms throughout the body. Although the jaw muscles are the first to be affected (which is why tetanus is commonly known as "lockjaw"), the toxin eventually strikes all of the body's voluntary muscles. Muscular spasms in the back bend the spine into an exaggerated arch, which can snap the backbone. The victim eventually loses the ability to breathe and dies of asphyxiation.

Such hypersensitization leading to accidental activation of target cells is further prevented by the presence of inhibitory neurons and their neurotransmitters, which increase the threshold; that is, they decrease the response sensitivity of the target. Most target cells are influenced by impulses from hundreds of different excitatory and inhibitory neurons (Figure 23-8). If the combined input from all the incoming synapses is enough to trigger a response, the cell performs its activities. Until then, the target cell remains at rest.

◑ Taken as a group, synapses are more than just connecting sites between adjacent neurons; they are key determi-

nants in the orderly routing of impulses throughout the nervous system. The billions of synapses that exist in a complex mammalian nervous system act like gates stationed along the various pathways; some pieces of information are allowed to pass from one neuron to another, while other pieces are held back or rerouted in some other direction. Such synaptic integration allows us to focus on the book we are reading or the music we are listening to, while simultaneously ignoring all of the distracting background noise with which we are constantly bombarded. We will return to the importance of synapses when we discuss the basis of learning and memory later in the chapter.

Now that we have described the form and function of neurons and the way they transmit information to other cells, we will examine how nerve cells are organized into more complex neural structures.

Neurons accomplish their regulatory functions by passing information—in the form of chemical transmitter substances—to other cells, across synaptic clefts. The interaction between neurotransmitters and membrane receptors then initiates or inhibits a response in the target cell. Synapses determine the paths by which information travels through the nervous system. (See CTQ #4.)

THE NERVOUS SYSTEM

The nervous system of a vertebrate is divided into two major divisions: the central nervous system and the peripheral nervous system. The **central nervous system (CNS)** consists of two major parts: the **brain,** which is the center of neural integration, and the **spinal cord,** which contains billions of neurons that run to and from the brain and also mediates many of the body's reflex responses. All neurons, or parts of neurons, situated outside the central nervous system are part of the **peripheral nervous system,** "peri" meaning "around the edge," as in *peri*meter. The peripheral nervous system connects the various organs and tissues of the body with the brain and spinal cord. The neurons of the peripheral nervous system are grouped into **nerves**—"living cables" composed of large numbers of individual neurons bundled together in parallel alignment, together with their supporting cells (Figure 23-9). All incoming and outgoing impulses are routed through the CNS, which functions as a centralized "command and control center." The simplest example of neural centralization is the reflex arc.

THE REFLEX ARC: SHORTENING REACTION TIME

The fastest motor reactions to stimuli are reflex responses. A **reflex** is an involuntary response to a stimulus—a response that occurs "automatically" and requires no conscious deliberation or awareness of the stimulus. Reflex responses occur so rapidly because the impulse travels the shortest route possible: from the site of the stimulus, through the central nervous system, to the responding effector. The chain of neurons that mediate a reflex make up a **reflex arc.** One of the simplest reflex arcs—that which is responsible for your foot jumping forward when the doctor strikes the area below your knee with a rubber hammer—is illustrated in Figure 23-10*a*.

A reflex arc begins with a **sensory receptor**—a cell that responds to a change in its environment. In many cases, the receptor is the tip of a sensory neuron that is specialized to respond to a stimulus. For example, the *stretch receptor* responsible for the knee-jerk response is a sensory neuron whose end is wrapped around a muscle fiber embedded within a muscle of the thigh. When the tendon of the knee is tapped, the attached muscle is stretched, activating the receptor, which generates a neural impulse. Some of the terminal processes of the sensory neuron end in the spinal cord (Figure 23-10*a*) in synapses with motor neurons leading directly back to the thigh muscle. Impulses traveling back along these motor neurons cause the muscle to contract, and the leg extends forward.

◎ The *stretch reflex* just described didn't evolve to help doctors evaluate the state of your nervous system. Rather, the reflex is an adaptive response that helps you maintain your posture and balance. The same stretch reflex that keeps you upright also works for the mountain goats pictured in Figure 23-10b.

Your day is full of adaptive reflex responses. For example, an overambitious sip of steaming coffee triggers a reflex activation of the muscles in your tongue, jaw, and mouth.

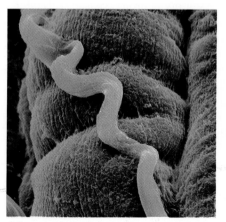
(a)

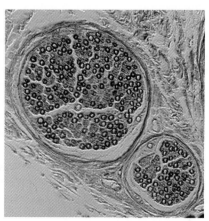
(b)

FIGURE 23-9

Nerves. *(a)* A nerve winding across the surface of muscle fibers. *(b)* In this cross section, the individual neurons in the nerve are seen bound together in a sheath of connective tissue.

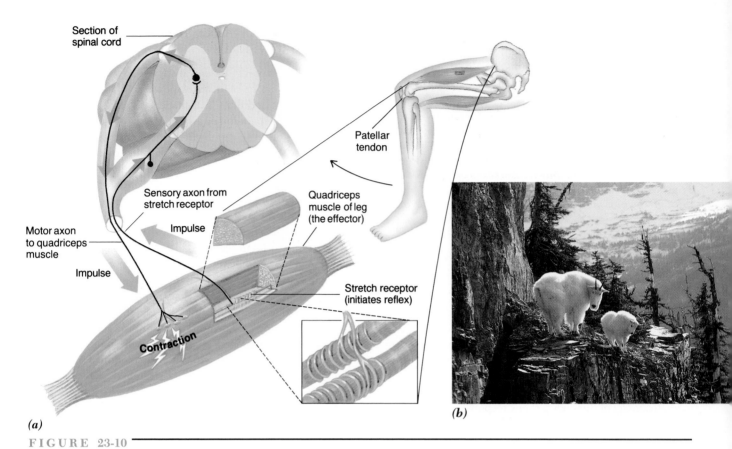

FIGURE 23-10

The reflex arc that mediates the knee-jerk response. *(a)* A reflex arc includes a receptor, which generates a nerve impulse after it is activated by some change in the environment; a sensory neuron, which carries the impulse to the CNS; a motor neuron, which sends a command signal back to the periphery; and an effector (either a muscle or gland), which provides the response. In the case of the knee-jerk response, the receptor is actually the end of a sensory neuron, which is wrapped around a special muscle fiber situated within the quadriceps muscle of the thigh. When the muscle fiber is stretched, it activates the sensory neuron, sending impulses to the spinal cord. In this stretch reflex, impulses are transmitted directly from the sensory neuron to the motor neuron, causing the quadriceps muscle to contract and extend the leg forward. Unlike this stretch reflex, most reflexes include one or more interneurons in the arc. *(b)* These mountain goats depend on stretch reflexes to maintain their footing on steep trails.

The tongue automatically jumps to the back of the mouth, which may open and discharge the liquid before your brain can consider the reaction of the other people at the table. Other reflexes include regulation of blood pressure, control of pupil size in response to changes in light intensity, and withdrawal of the hand when it encounters a sharp or hot object. Reflexes are involuntary responses that occur very rapidly in a *stereotyped* manner; that is, the reflex is the same every time the same simple stimulus is encountered.

MORE COMPLEX CIRCUITS

Not all of the sensory neurons from the stretch receptors in the thigh muscle terminate at a motor neuron which returns to the muscle. Many sensory neurons synapse with interneurons of the spinal cord. These interneurons carry impulses upward into the brain, where we perceive the stimulus (such as the hammer striking our knee) and send commands to our muscles. Our ability to stand upright, for example, depends on sensory input reaching the brain from many different muscles. Similarly, the motor neurons that contract the thigh muscle are not activated solely by a simple reflex arc. Rather, a single motor neuron may be covered with synaptic knobs from thousands of different interneurons. As a result, motor neurons leading to the thigh muscle can be activated by many unrelated activities, such as willingly kicking a football or automatically jumping out of the way of an oncoming truck.

Each neuron is a link in a chain, joined to other links by synapses. Neurons are organized into functional circuits that allow information to be routed in meaningful ways through the billions of cells that make up a complex nervous system. The simplest circuits consist of reflex arcs, whereby sensory information can direct motor activity without the participation of higher neural centers. (See CTQ #5.)

ARCHITECTURE OF THE HUMAN CENTRAL NERVOUS SYSTEM

The central nervous system performs the most complex neural functions. It collects information about the internal and external environment, "sorts" through the information, relays impulses of its own along different pathways to various parts of the brain and spinal cord, and then acts on the information by sending command messages to peripheral effectors. The human CNS is the most complex, highly organized structure found on earth. We will now take a closer look at the two main components of the human CNS: the brain and the spinal cord.

THE BRAIN

Although the brain constitutes only about 2.5 percent of your body weight, it consumes 25 percent of your body's oxygen supplies while generating the ATP needed to fuel its activities. If not replenished, the brain's oxygen content would be exhausted in about 10 seconds. Within only a few minutes, the damage to the brain would be irreversible.

The human brain (Figure 23-11) is a mass of nearly 1.5 kilograms (3 pounds) of gelatinlike tissue. It contains a darker outer region—**gray matter**—in which the cell bodies and dendrites of the brain's neurons reside. Gray matter is rich in neuron-to-neuron synapses, places where

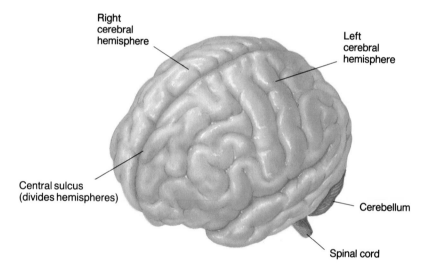

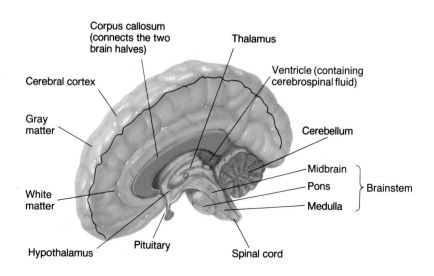

FIGURE 23-11

Two views of the brain. The intact human brain reveals mostly cerebrum, which is divided into a right and left cerebral hemisphere. Bisecting the brain reveals some of the brain's structural complexity. The medulla, pons, and midbrain constitute the brainstem, which controls visceral functions, such as breathing and cardiovascular activity. The thalamus is primarily a way station for sensory information that passes to the cerebrum, while the hypothalamus is one of the brain's homeostatic control centers. The right and left halves of the cerebral cortex are connected by a thick mass of nerve fibers that form the corpus callosum. The entire brain is covered by the meninges and bony cranium (not shown).

neural associations are made. The inner region of the brain consists largely of **white matter** composed of myelinated axons; its whitish cast is provided by the light-colored myelin sheaths, which insulate the neurons. The tissue of the brain surrounds a series of distinct but interconnected chambers, called **ventricles,** which are filled with a protein-rich liquid, the **cerebrospinal fluid,** which cushions and nourishes the brain. This fluid also surrounds the delicate brain (and spinal cord), cushioning it against injury. The brain and its surrounding fluid is encased in a complex, protective, watertight sheath, called the *meninges,* which, in turn, is enclosed within a protective, bony case, the *cranium.*

The brain can be divided in various ways, depending on different criteria. For our purposes, we will restrict the discussion to three functional groups: (1) the cerebrum, (2) the cerebellum, and (3) a series of interrelated networks that form the brainstem, limbic system, and reticular formation.

The Cerebrum

The cerebrum is the most prominent part of the human brain (Figure 23-11). Its two halves, called **cerebral hemispheres,** are generally associated with "higher" brain func-

tions, such as speech and rational thought. Actually, these functions are attributes of the **cerebral cortex,** the outer, highly wrinkled layer of the cerebrum (*cortex* = rind). Every cubic inch of this thin layer of gray matter contains 10,000 miles of interconnecting neurons. The convolutions (wrinkles) in the cerebrum increase the surface area of the cortex without enlarging the space required to house it. Convolutions in the cerebral cortex are presumed to be an evolutionary sign of higher cerebral capabilities, such as intellect; artistic and creative abilities; and a greater capacity for language, learning, and memory, compared to those of the smooth-brained, lower vertebrates, such as amphibians and reptiles.

Each cerebral hemisphere is composed of four lobes —temporal, frontal, occipital, and parietal—each of which has a unique set of functions (Figure 23-12). The two cerebral hemispheres are connected by the *corpus callosum*—a thick cable made up of hundreds of millions of neurons— which allows the left and right hemispheres to communicate with each other. As we will discuss in the Human Perspective box in Chapter 24, most of the sensory information from the right side of the body is transmitted through the corpus callosum to the left cerebral hemisphere, and vice versa.

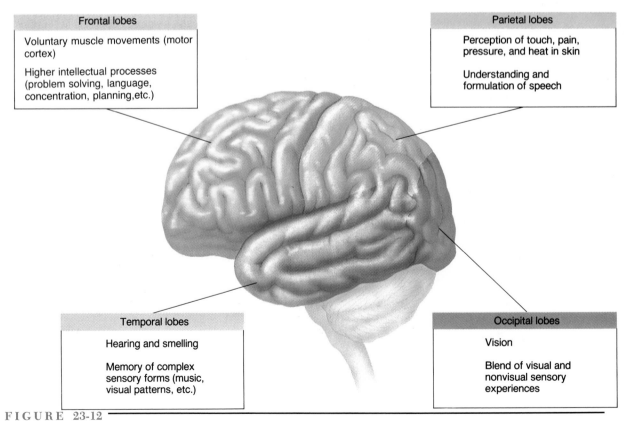

Frontal lobes

Voluntary muscle movements (motor cortex)

Higher intellectual processes (problem solving, language, concentration, planning, etc.)

Parietal lobes

Perception of touch, pain, pressure, and heat in skin

Understanding and formulation of speech

Temporal lobes

Hearing and smelling

Memory of complex sensory forms (music, visual patterns, etc.)

Occipital lobes

Vision

Blend of visual and nonvisual sensory experiences

FIGURE 23-12

The four major lobes of the cerebrum and some of their roles in physiology and behavior.

Memory   Imagine, for a moment, that you are standing in front of the entrance to your former high school. You walk through the doors, through the halls, and into the room where you took one of your more memorable classes. Your former teacher is standing in the front of the class, and you can hear the sound of her voice. A minute earlier, none of this was on your mind, and now you have "directed" your brain to recall a series of specific, ordered images about which you can consciously reminisce. How is this possible? Scientists cannot yet answer this question, but they can provide some interesting insights. The types of images you have just brought to "mind" can also be evoked by electrical stimulation of various parts of the cerebral cortex. When this procedure is performed, the subject may see an image of something they thought they had long forgotten, or hear a voice, or smell an aroma. The information is "stored" in the cerebral cortex, probably in the pattern of neural circuits formed by the billions of neurons that make up this portion of the brain.

Scientists have also learned about memory by studying individuals who have suffered specific brain injuries that have affected their memory. During the 1950s, a young man (known as H. M.) came to the attention of Brenda Milner of the Montreal Neurological Institute. A portion of the man's brain had been removed during surgery in an attempt to stop severe seizures. The operation had unexpected and tragic results. H. M. was able to remember events from his childhood and adolescence, but he was unable to memorize any new information. For example, if H. M. were given a number to remember, he could do so only as long as he focused all of his attention on it. As soon as he lost his attention, he would forget the number, as well as any recollection of ever being told about it. Over and over, H. M. would greet as strangers the same doctors and researchers he worked with on a daily basis. Perhaps most tragic, H. M. was not able to grasp the reason for his problem because he could not remember what he was told.

The case of H. M. revealed for the first time the dual nature of human memory. H. M. could remember events from the past, similar to your having remembered your high-school classroom, because his *long-term memory* remained basically intact. In contrast, *short-term memory* allows us to retain a piece of information only for a matter of seconds or minutes. Short-term memory can be considered "working memory" since we use it to carry out our daily activities. Remembering a phone number from a telephone book until we get a chance to write it down is an example of short-term memory. As we go about our activities, certain pieces of information from our short-term memory become processed subconsciously in some unknown way so that they become part of our long-term memory. Short- and long-term memory apparently involve different types of neural processes and different parts of the brain.

Studies on H. M.—and others with similar conditions —have indicated that short-term memory is processed in the temporal lobes of the cerebrum and the hippocampus, a part of the limbic system discussed below. These are the same regions of the brain that degenerate in individuals with Alzheimer's disease (see The Human Perspective: Alzheimer's Disease: A Status Report). Like H. M., Alzheimer's victims often vividly remember events from the distant past but are unable to remember what has happened to them a few minutes earlier. Long-term memory appears to be more diffusely located in the cerebrum, as evidenced by the fact that stimulation to many different parts of the cerebral cortex will evoke memories of past events.

Learning   is a change in behavior that results from prior experience. Learning occurs as the nervous system processes sensory stimuli and uses the information to make changes in the nervous system that lead to new types of responses. Virtually all animals are capable of modifying their behavior as the result of prior experience. Most of what we know about the cellular mechanisms responsible for learning have come from studies of simpler animals, particularly the sea slug, *Aplysia,* a marine mollusk whose nervous system consists of approximately 10,000 unusually large neurons. These studies have been carried out largely by Eric Kandel and his colleagues at Columbia University.

As a sea slug moves across the sea bottom or the surface of a rock, water passes through its siphon and over the gills. If the siphon is gently stimulated, the animal responds by retracting its gills. This *gill-withdrawal reflex* is a protective mechanism. But if the stimulation is due to wave action, or contact with a rock, there would be no reason for gill withdrawal. In fact, repeated gentle contact with the siphon soon leads to a cessation of the gill withdrawal response; the behavior has been *habituated.* Habituation is a simple form of learning which is "remembered" for several days.

The gill-withdrawal reflex in *Aplysia* is mediated by a simple reflex arc. Studies have shown that, following habituation, action potentials continue to be generated in the sensory neuron leading out of the siphon, but these impulses are not transmitted to the target motor neuron leading to the gills. This is because successive impulses in the sensory neuron trigger the release of a decreasing quantity of neurotransmitter, which, in turn, evokes a smaller stimulus to the motor neuron. You might conclude that the sensory neuron is simply running out of transmitter substance, but this is not the case. Recall that neurotransmitter release is mediated by an influx of calcium ions at the tip of the neuron. The underlying cause of habituation of the gill-withdrawal reflex is a regulated decrease in the number of calcium channels that open in response to successive action potentials reaching the tip of the sensory neuron. From this finding, we can conclude that, in *Aplysia,* the memory of recent events is stored in a short-term change in individual neurons; if one stops touching the siphon for a period of time, the sensory neuron reverts to its original state in which the maximum number of calcium channels open upon stimulation.

The studies on *Aplysia* indicate that learning is accompanied by structural modifications of synapses. Studies on

◁ THE HUMAN PERSPECTIVE ▷
Alzheimer's Disease: A Status Report

First it robs victims of their memory of recent events, followed by a loss of simple reasoning and the ability to feed or clean oneself. Finally, it takes the victim's life. Alzheimer's disease strikes about 5 percent of people age 65 or older, and possibly as many as 50 percent of those 85 or older. This insidious disease destroys nerve cells in the brain, particularly those that release acetylcholine. The hippocampus, a region of the brain important for memory, is affected most severely. The severe effects of Alzheimer's disease appear in brain scans of afflicted persons (Figure 1).

The brains of Alzheimer's patients show two microscopic abnormalities not found in normal brains. The cytoplasm of degenerating neurons contains strange tangled fibrils, and the space outside the neurons contains dense plaques. The tangled fibrils consist of disorganized cytoskeletal proteins, while the plaques consist of fragments of a membrane protein called beta-amyloid. Two questions that are the subject of intense current interest are whether or not the tangles or the plaques cause Alzheimer's disease, and why either of these abnormal formations develop.

Many neurobiologists believe that individuals who suffer from Alzheimer's disease have sustained genetic damage that has caused their nerve cells to produce abnormal gene products. Evidence for this belief comes from a rare genetic disorder in which victims develop all of the charac-

teristics of Alzheimer's at a much earlier age than the disease normally appears. This form, termed *familial Alzheimer's disease (FAD)*, is characterized by the same tangled fibrils and plaques seen in individuals with *sporadic* (noninherited) cases. More than one gene may be responsible for FAD. One of these genes was isolated in 1991 and was shown to code for the pre-

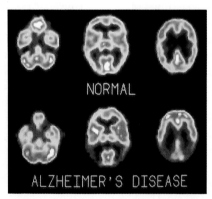

FIGURE 1

Characteristic signs of Alzheimer's disease. These PET (positron-emission tomography) scans, which depict levels of metabolic activity within living tissue, reveal marked differences between the brain of a normal person and one suffering from Alzehimer's disease.

cursor of the beta-amyloid protein found in the plaques that characterize the disease. In studies of two separate families, individuals afflicted with FAD were found to have alterations in this particular gene which changed the amino acid sequence of the protein. This finding strengthens the belief that the amyloid plaques are a cause of the disease rather than simply an effect. Researchers have now successfully introduced this mutant gene into mice in an attempt to induce an Alzheimer's-like disease in laboratory animals and to establish an animal model for possible treatments. This is particularly important since current treatments have had little success. As we discussed in the opening pages of the chapter, NGF may eventually prove a useful treatment.

Researchers are also attempting to develop diagnostic procedures that would detect Alzheimer's disease at a very early stage, even before symptoms develop. A number of biotechnology companies are working on diagnostic procedures based on substances that might appear in the blood of people with the disease. Now that a gene for FAD has been isolated, the possibility also exists for developing genetic screening tests. The same ethical question arises in this case as in the case of Huntington's disease (page 350): Who among us would want to learn that we will ultimately be stricken with a horrible, degenerative disease for which no treatment is available?

more complex animals have corroborated this view, indicating that as new tasks are learned, physical changes (such as the loss or gain of dendrites or the phosphorylation of membrane proteins) may occur at key synapses, which weaken or strengthen the connections between the neurons involved in the response. In humans, these synaptic changes occur primarily in the cerebrum.

Language Communication between members of a species occurs throughout the animal world. Fireflies communicate by flashes of light; gypsy moths communicate by the release of airborne chemicals; and birds and dolphins communicate by vocalization. Nowhere does the complexity of language approach that used by humans, however. Children learn to communicate verbally at a very early age, long before they are capable of learning complex motor activities. Studies with infants as early as a few months old have convinced most researchers that humans are born with an innate, genetically determined ability to learn languages without formal instruction. For most people, especially right-handed individuals, this ability is localized in the left frontal lobe of the cerebrum. (In 10 to 20 percent of people, the center is localized in either the right frontal lobe or in both frontal lobes.) The localization of the language center in the left frontal lobe can be dramatically revealed by anesthetizing this part of the brain, leaving the person awake and alert but unable to speak.

The importance of the left frontal lobe in language abilities has been confirmed in studies of deaf children who learn to communicate via sign language at an early age. Sign language is a form of communication which is performed by motor activities of the hands. You might expect that, like other motor activities, control over signing would emanate from the motor areas of the brain, but researchers have found that it is actually the language center that controls the motions used in sign communication.

Motor Activities Within the cerebral cortex, some regions are specialized for receiving sensory input, and other regions are specialized for initiating motor output. The locations and functions of the sensory portions of the cortex will be discussed in the next chapter, together with the sensory organs themselves. For now, we will concentrate on the **motor cortex** (Figure 23-13)—the paired strips of gray matter found in the frontal lobe of the brain, where most of the impulses that command our muscles emanate. Stimulation of various parts of the motor cortex results in contraction of specific groups of muscle fibers. Those parts of the body which perform the most intricate and delicate movements, including the fingers, lips, tongue, and vocal cords, have the greatest representation in the motor cortex (Figure 23-13). As a result, stimulation of just a few cells may produce reactions ranging from a slight movement of a finger to a major movement of the torso.

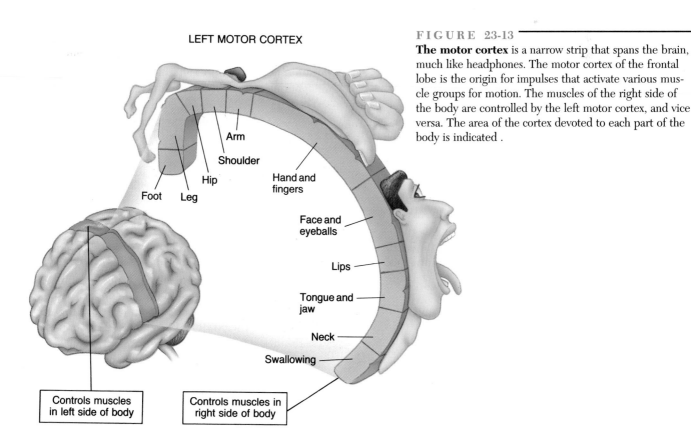

LEFT MOTOR CORTEX

Arm
Shoulder
Hip
Hand and fingers
Foot   Leg
Face and eyeballs
Lips
Tongue and jaw
Neck
Swallowing

Controls muscles in left side of body

Controls muscles in right side of body

FIGURE 23-13

The motor cortex is a narrow strip that spans the brain, much like headphones. The motor cortex of the frontal lobe is the origin for impulses that activate various muscle groups for motion. The muscles of the right side of the body are controlled by the left motor cortex, and vice versa. The area of the cortex devoted to each part of the body is indicated .

The Cerebellum

The cerebellum is a bulbous structure (Figure 23-11) that receives information from receptors located in the muscles, joints, and tendons, as well as from the eyes and ears. The amount of information converging on the cerebellum is revealed by examining the huge number of synaptic contacts made by some of these cells; some individual cerebellar neurons receive input from as many as 80,000 other neurons!

⟳ Vast amounts of sensory information can be processed within a fraction of a second. Messages are then sent to the motor cortex of the cerebrum, where they are used in directing such complex motor activities as playing a musical instrument, writing, or shooting a basketball (Figure 23-14). For this reason, a person who sustains cerebellar damage will have difficulty performing smooth, coordinated movements; motor activities lose their subconscious basis, and the person may have to think about each movement that would otherwise be performed automatically.

The Brainstem, Reticular Formation, and Limbic System

The brainstem forms the central stalk of the brain (Figure 23-11) and is responsible for regulating most of the involuntary, visceral activities of the body, such as breathing and swallowing, as well as maintaining heart rate and blood pressure. Consequently, permanent damage to the brainstem usually leads to coma or death. Whereas the cerebral cortex is the most recently evolved part of the vertebrate brain, the *brainstem* is probably the oldest part; it makes up the bulk of the brain of lower vertebrates, such as fish and amphibians.

Most of the sensory information streaming into the cerebral cortex is routed through the *thalamus*—the part of the brain located just beneath the cerebrum (Figure 23-11). The thalamus also coordinates outgoing motor signals. Associated with the thalamus is the *reticular formation,* which is composed of several interconnected sites in the core of the brainstem that selectively arouse conscious activity, producing a state of wakeful alertness. The reticular formation can be activated by any number of internal factors or external stimuli—the sound of an approaching horn, a flashlight shining in your eyes, or the middle-of-the-night realization that you forgot to study for tomorrow morning's biology exam. When the reticular system fails to maintain arousal, the brain falls asleep and stays asleep until the reticular formation is activated by sensory input from, say, an alarm clock.

The reticular formation is no larger than your little finger. Nonetheless, it does much more than simply keep you awake and alert (no small task itself). The reticular formation helps you cope with the millions of impulses that assault the brain every waking second (Figure 23-15). It screens out the trivial signals, while allowing vital or unusual signals to pass through and alert the mind.

FIGURE 23-14
Michael Jordan about to score two.

⟳ Beneath the thalamus (Figure 23-11) is the **hypothalamus** (*hypo* = below), a portion of the brain that is only about the size of the tip of your thumb. Despite its small size, the hypothalamus is critical for maintaining homeostasis. The hypothalamus regulates body temperature, keeping it around 37°C (98.6°F). It also helps control blood pressure, heart rate, and the body's urges, such as hunger, thirst, and sex drive. The hypothalamus also controls the nearby *pituitary gland* (Figure 23-11), which is a key component of the endocrine system (discussed in Chapter 25).

The hypothalamus and the thalamus comprise part of an interconnected group of structures buried in the core of the brain. This complex is called the **limbic system.** As discussed above, one part of the limbic system, the hippocampus, is an important center for short-term memory; without it, you would have already forgotten how this sentence began. The limbic system is also associated with pleasure and joy, pain and fury, and an ability to balance emotions. Stimulation of different parts of the limbic system with electrodes can induce immediate anger or euphoria, sexual arousal, or deep relaxation.

The importance of the limbic system in regulating our emotions is dramatically illustrated by "Julie," an expatient of Vernon Mark, a neurosurgeon at Harvard Medical School. Since childhood, Julie had suffered from epileptic seizures, which initiated outbursts of violent behavior. In one episode, at age 18, Julie was overcome by a seizure in a movie theatre. She then went to the restroom, where she stabbed a woman who accidentally bumped against her. Mark identified the *amygdala,* a part of the limbic system, as the site in the brain that was being electrically disturbed during Julie's seizures. When this site was stimulated with an electrode, the patient became unresponsive for a few seconds, then reacted violently by smashing her fists against the wall or swinging an object she held in her hand. Mark removed the abnormal brain tissue, and Julie never again demonstrated an uncontrolled rage. The importance of the limbic system in controlling our emotions is also revealed by mood-altering drugs.

The Effect of Cocaine on the Limbic System When certain parts of the limbic system are stimulated—either artificially (by electrodes) or naturally (by incoming impulses)—a strong feeling of pleasure is elicited. This feeling results from the release of the neurotransmitter dopamine, which binds to certain neurons in the "pleasure centers" of the limbic system. Normally, the effects of dopamine are short-lived because the neurotransmitter is rapidly removed from the synaptic cleft by the quick reuptake into the neuron that released it. However, cocaine, a compound extracted from the leaves of the South American coca plant (*Erythroxylum coca*) and inhaled by most users, interferes with the reuptake of dopamine. The sustained presence of dopamine in the synaptic clefts of the limbic center produces a feeling of euphoria—a "high" that lasts for several minutes. The high is followed by a "crash," in which the person feels depressed, irritable, and anxious. Since the fastest way to relieve the unpleasant effects of the crash is to inhale more cocaine, the drug can rapidly become addictive.

In addition to causing mood elevations, cocaine also elevates a person's heart rate, blood pressure, body temperature, and blood sugar. The drug has also been known to trigger seizures and heart failure, the latter of which accounts for nearly 1,000 deaths per year in the United States alone.

Opium Derivatives and Endogenous Opiates Morphine and heroin are structurally related molecules, called *opiates*, derived from the opium poppy. Like cocaine, opiates interfere with the transmission of impulses by neurons of the limbic system. Unlike cocaine, opiates act by binding to specific receptors situated in the postsynaptic membrane of neurons found in several areas of the brain. Among their targets, opiates bind to neurons in the pain pathways of the limbic system, blocking the perception of pain.

The discovery of opiate receptors in the early 1970s raised a key question: Why should neurons of the brain possess receptors that specifically bind substances derived from an opium poppy? One likely explanation was that the brain produced one or more substances that were similar in structure to opiates and that bound to the same receptors. After an intense search carried out by a number of laboratories, two classes of peptides were discovered. These peptides, called *endorphins* (a contraction for "endogenous morphinelike substances") and *enkephalins* (from the Greek meaning "in the head"), normally bind to receptors located on neurons that are part of pain-transmitting pathways. Like morphine or heroin, endorphins and enkephalins interfere with the delivery of information through this pathway to the brain, thereby blocking the sensation of pain. Even strenuous exercise or painful workouts may lead to the release of endorphins and enkephalins, causing a feeling of euphoria. This phenomenon is commonly referred to as "runner's high." Other psychoactive drugs have also been found to act by binding to receptors for neurotransmitters in the brain. For example, nicotine binds to acetylcholine receptors; lysergic acid diethylamide (LSD) binds to serotonin receptors; and mescaline (extracted from the peyote cactus) binds to norepinephrine receptors.

FIGURE 23-15

The limbic system allows this student to focus on her book and to ignore the surrounding noise.

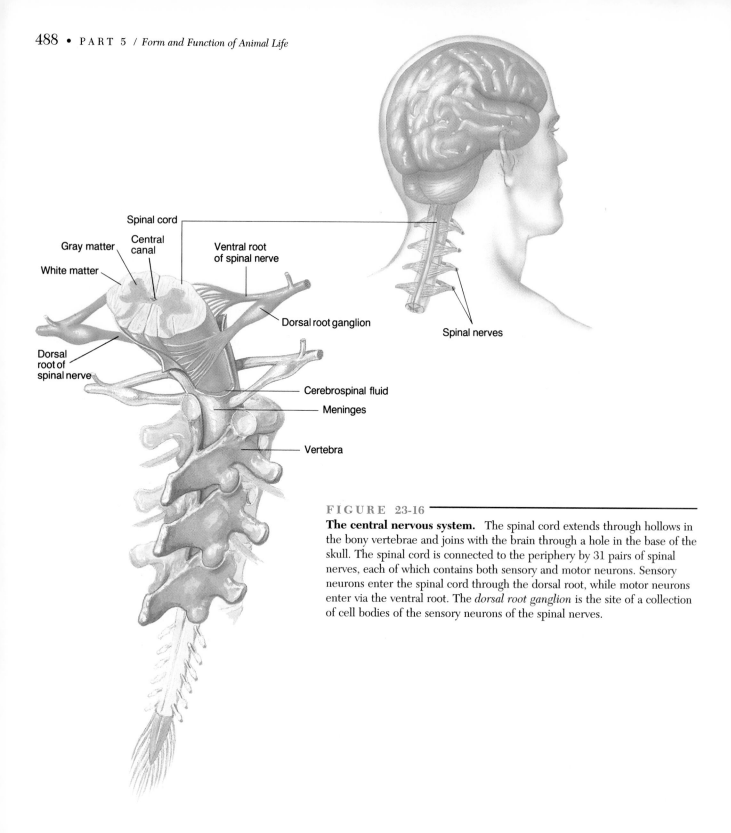

Spinal cord

Central canal

Gray matter

White matter

Ventral root of spinal nerve

Dorsal root ganglion

Dorsal root of spinal nerve

Cerebrospinal fluid

Meninges

Vertebra

Spinal nerves

FIGURE 23-16

The central nervous system. The spinal cord extends through hollows in the bony vertebrae and joins with the brain through a hole in the base of the skull. The spinal cord is connected to the periphery by 31 pairs of spinal nerves, each of which contains both sensory and motor neurons. Sensory neurons enter the spinal cord through the dorsal root, while motor neurons enter via the ventral root. The *dorsal root ganglion* is the site of a collection of cell bodies of the sensory neurons of the spinal nerves.

THE SPINAL CORD

The brainstem merges with the second major component of the central nervous system—the *spinal cord*—a thick-walled, tubular mass of neural tissue that extends from the top of the neck to the lower back. The diameter of the spinal cord is similar to that of your little finger. Like the brain, the spinal cord is surrounded by the meninges, which, in turn, is surrounded by the hollow bony vertebrae that make up the "backbone" (Figure 23-16). A narrow *central canal* that is filled with cerebrospinal fluid runs the length of the spinal cord and opens into the ventricles of the brain. Numerous medical procedures are performed following injection of anesthetics into the central canal, a process that "deadens" a limited portion of the body.

◁ BIOETHICS ▷

Blurring the Line between Life and Death

By ARTHUR CAPLAN
**Division of the Center for Biomedical
Ethics at the University of Minnesota**

Theresa Ann Campo Pearson didn't have a very long life. When she died in 1992, she was only 10 days old. Despite her short life, she became the center of a very strange, sad, and wrenching ethical controversy. Theresa died because her brain had failed to form. She had anencephaly, a condition in which only the brainstem, located at the top of the spinal cord, is present. Her parents wanted to donate Theresa's organs; the courts said no. Some people found it strange that Theresa's parents, Laura Campo and Justin Pearson, did not get their way. Why not allow donation, when every day in North America a baby dies because there is no heart, lung, or liver available for transplantation?

Anencephaly is best described as completely "unabling," not disabling. Children born with anencephaly cannot think, feel, sense, or be aware of the world. Many are stillborn; the majority of the rest die within days of birth. A mere handful live for a few weeks. Theresa's parents

knew all this. But rather than abort the pregnancy, they chose to have their baby. In fact, the baby was born by Caesarean section, at least partly in the hope that it would be born alive, thereby making organ donation possible. When Theresa died at Broward General Medical Center in Fort Lauderdale, Florida, however, no organs were taken. Two Florida courts ruled that the baby could not be used as a source of organs unless she was brain-dead, and Theresa Ann Campo was never pronounced brain-dead.

Brain death refers to a situation in which the brain has irreversibly lost all function and activity. Babies born with anencephaly have some brain function in their brainstem so, while they cannot think or feel, they are alive. According to Florida law—and the law in more than 40 other states—only those individuals declared brain-dead can donate organs. The courts of Florida had no other option but to deny the request for organ donation.

One obvious solution is to change the law so that states could decide that organs can be removed upon parental consent from either those who are born brain-dead or from babies who are born with anencephaly. Another solution is to rewrite the definition of death to say that death occurs either when the brain has totally ceased to function or if a baby is born anencephalic. Do you feel that either of these changes should be made? Some may argue that medicine will fudge the line between life and death in order to get organs for transplant. Do you agree with this concern? How do you think redefining death will affect a person's decision to check off the donation box on the back of a driver's license? Do you think people may worry that if they are known to be potential donors they won't be aggressively treated at the hospital? In your opinion, would changing the definition of death to include anencephaly be beneficial or deleterious?

Like the brain, the spinal cord is composed of white matter (myelinated axons) and gray matter (dendrites and cell bodies). However, the arrangement of these types of matter is reversed in the spinal cord, compared to their arrangement in the brain: The spinal cord's white matter surrounds the gray matter (Figure 23-16).

The human central nervous system is the most complex and highly evolved assembly of matter. Among its functions are the processing of sensory information collected from both the external and internal environment; the regulation of internal physiological activities; the coordination of complex motor activities; and the endowment of such intangible "mental" qualities as emotions, creativity, language, and the ability to think, learn, and remember. (See CTQ #6.)

ARCHITECTURE OF THE PERIPHERAL NERVOUS SYSTEM

The peripheral nervous system provides the neurological bridge between the central nervous system and the various parts of the body. The peripheral nervous system is made up of paired nerves that extend into the periphery from the CNS at various levels along the body. Each nerve is composed of a large bundle of myelinated axons surrounded by a connective tissue sheath. Twelve pairs of **cranial nerves** emerge from the central stalk of the human brain, and 31 pairs of **spinal nerves** extend from the spinal cord out between the vertebrae of humans (Figure 23-16). For the most part, the cranial nerves *innervate* (supply nerves to) tissues and organs of the head and neck, whereas the spinal nerves innervate the chest, abdomen, and limbs.

All of the spinal nerves contain both sensory and motor neurons, which separate from one another a short distance after leaving the spinal cord, forming two distinct roots (Figure 23-16). Sensory impulses enter the spinal cord through *dorsal roots*, while motor impulses leave the spinal cord through *ventral roots*.

The sensory neurons that form part of the cranial and spinal nerves provide the CNS with the sensory input needed for conscious awareness and the ability to respond to changes in the body and in the external environment. The motor neurons present in these nerves carry messages to the body's muscles and glands. These outbound motor neurons constitute two distinct divisions of the peripheral nervous system: the somatic and the autonomic divisions.

THE SOMATIC AND AUTONOMIC DIVISIONS OF THE PERIPHERAL NERVOUS SYSTEM

The **somatic division** of the peripheral nervous system carries messages to the skin and to those muscles that move the skeleton, such as the major muscles of the head, trunk, and limbs. These movements generally follow voluntary orders from the brain, although stereotyped reflex movements, such as the knee-jerk reflex, are also mediated by the somatic division.

The **autonomic division** controls the involuntary, homeostatic activities of the body's internal organs and blood vessels. Autonomic motor neurons regulate heart rate, blood-vessel diameter, respiratory rate, glandular secretions, excretory functions, digestion, sexual responses, and so forth. Although this system's operations are largely involuntary, many autonomic functions can be influenced by conscious control. Biofeedback training, for example, teaches how to modify certain types of autonomic activity to help reduce blood pressure or to lower physiological stress, responses that are classically considered beyond voluntary control.

The Sympathetic and Parasympathetic Divisions of the Autonomic Nervous System

The autonomic division can be further divided into two parts: the **sympathetic system** and the **parasympathetic system** (Figure 23-17). Most organs of the body receive neurons from both of these systems, which evoke opposite (*antagonistic*) responses. Your heart rate, for example, is precisely regulated by the balance that is maintained between continuous stimulatory influences from sympathetic neurons and inhibitory influences from parasympathetic neurons.

A survey of the specific responses evoked by the autonomic division (Figure 23-17) reveals the different adaptive functions of the sympathetic and parasympathetic systems. When effector organs are stimulated by the sympathetic system, an adaptive response is triggered which better enables the body to cope with stressful situations, as might occur if you were suddenly confronted with a person coming toward you with a weapon. In such situations, the heart

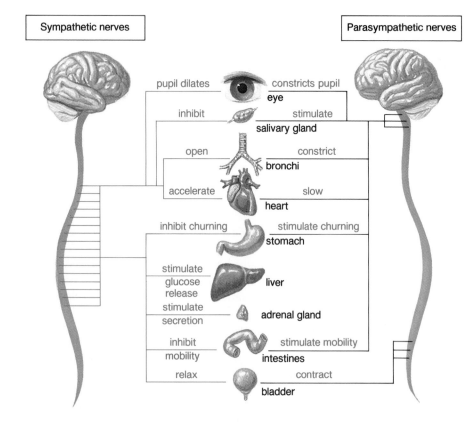

Sympathetic nerves

Parasympathetic nerves

pupil dilates — eye — constricts pupil

inhibit — salivary gland — stimulate

open — bronchi — constrict

accelerate — heart — slow

inhibit churning — stomach — stimulate churning

stimulate glucose release — liver

stimulate secretion — adrenal gland

inhibit mobility — intestines — stimulate mobility

relax — bladder — contract

FIGURE 23-17
The two divisions of the autonomic nervous system control most of the same internal organs but under opposite circumstances. The effects of the parasympathetic system predominate during normal, relaxed activity. During times of stress, the sympathetic system predominates, eliciting physiological changes that increase the level of performance to better adapt to the stressful situation. Parasympathetic neurons release acetylcholine at synapses with their target organs, while most sympathetic neurons release norepinephrine.

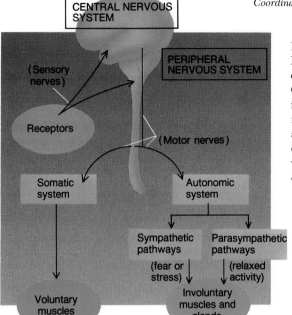

FIGURE 23-18

Interplay between the central and peripheral nervous systems. Receptors relay information about the internal and external environment to the CNS by way of sensory (or *afferent*) neurons. The CNS responds to this input by sending impulses to target cells through motor (or *efferent*) neurons. Motor nerves are part of two systems. (1) the somatic system, which acts on voluntary muscles; and (2) the autonomic system, which either excites or inhibits involuntary muscles and glands. The autonomic system is divided into the sympathetic and parasympathetic pathways, which typically elicit opposite responses in target organs.

rate increases, and blood is shunted away from the periphery toward the lungs for added oxygen uptake and toward the skeletal muscles, whose energy demands soar during periods of activity. The liver is stimulated to release additional glucose into the blood, while the activity of the digestive organs is reduced. The level of metabolism is increased, causing a rise in body temperature, which triggers the release of fluids by the sweat glands. Taken together, these changes constitute the "fight-or-flight" response, whereby the body is physiologically prepared either to confront the danger or to beat a fast retreat. Sympathetic neurons also stimulate the adrenal gland, which releases hormones that further prepare the body to meet the crisis at hand. This teamwork is one example of the interrelatedness of the nervous and endocrine systems (Chapter 25).

In contrast, the parasympathetic system handles the mundane, "housekeeping" functions of the body, such as digestion and excretion (Figure 23-17). For example, emptying the urinary bladder is a reflex that is mediated by the parasympathetic system and can be overridden by voluntary control. A person who has suffered a spinal-cord injury loses this voluntary control and thus cannot prevent voiding urine whenever the stretch receptors in the bladder wall launch the bladder-contraction reflex. The functional relationship between the central and peripheral nervous system is summarized in Figure 23-18.

The central nervous system receives information about conditions in the external environment and governs the activities of the body's organs by receiving and sending impulses over the neurons that make up the peripheral nervous system. (See CTQ #7.)

THE EVOLUTION OF COMPLEX NERVOUS SYSTEMS FROM SINGLE-CELLED ROOTS

Many people tend to think that a large and sophisticated brain is the ultimate achievement of evolution and that humans therefore reside at this "evolutionary pinnacle." Yet, the nervous systems of simpler animals (some of which lack even the earliest traces of a brain) are just as effective in enabling these organisms to survive and reproduce as is the large brain of humans. Our "advanced" brains are merely one adaptive strategy. Some animals, notably sponges, survive without a single neuron, as do all plants, fungi, protists, and monerans. That is not to say that these organisms lack neuron-like communication. Even unicellular organisms respond to the presence of external stimuli. For example, a chemical detected by a bacterial cell activates or inhibits flagella at the opposite end of the cell. In this way, the cell moves toward desirable chemicals, such as nutrients, and away from harmful substances, such as toxins. Impulses are relayed from "sensors" at one end of the cell to the flagella at the other end of the bacterial cell by a wave of depolarization similar to that responsible for neural impulse transmission. From these simple beginnings, more complicated nervous systems evolved.

Figure 23-19 illustrates some of the stages in the evolution of complex nervous systems. Even though there is remarkable diversity in the design and complexity of these various nervous systems, they are all built from the same type of nerve cells that utilize the same basic mechanism for generating, conducting, and transmitting nerve impulses. The simplest multicellular nervous systems are found in cnidarians: hydras, jellyfish, and sea anemones. These animals possess a *nerve net*—a diffuse system of individual nerve cells that gives these animals the ability to move around slowly, to seize and paralyze prey with their stinging tentacles, and to shove the incapacitated victim into the mouth (Figure 23-20*a*).

Compare the anemone's nerve net with the primitive nervous system of the flatworm in Figure 23-19. Unlike the anemone, which is a sedentary animal, the flatworm slowly creeps over a surface, its head end leading the way. This type of directed movement is correlated with the evolutionary beginnings of **cephalization**—the concentration of nerve cells near the anterior end of the animal, forming a primitive "brain." In most animals, the brain is located near the area of greatest concentration of sense organs, which is situated in the head. As evolution generated more and more

complex animal forms, this tendency toward cephalization became more pronounced, and the remainder of the central nervous system became more complex as well.

The brains of flatworms and earthworms provide so little control that the worm can lose its head and still continue to live while regenerating a replacement. The brains of insects, lobsters, and spiders play a more prominent and indispensable role in integrating the responses of various parts of a complex body. The most complex brain of any invertebrate belongs to the octopus. An octopus can distinguish among relatively similar objects and can accomplish such complex tasks as removing a stopper from a jar (Figure 23-20*b*). Some parts of the octopus's brain are concerned with memory and decision-making activities. If, for example, an octopus's higher brain centers are damaged, the animal may still be able to attack a crab placed in its aquar-

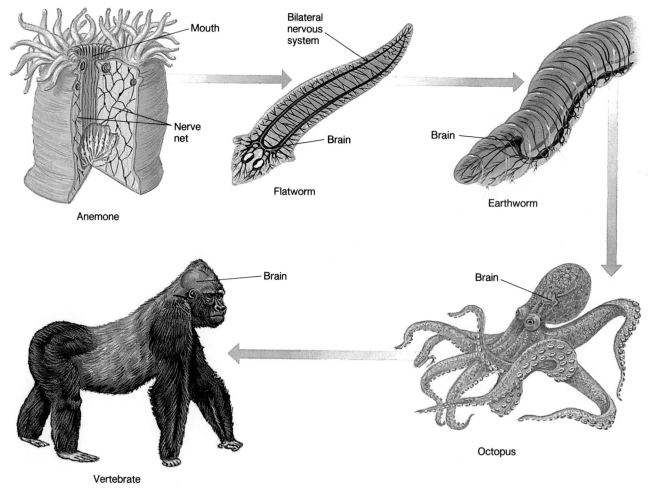

FIGURE 23-19

Evolution of the nervous system. (The shaded arrows indicate increasing neural complexity, *not* an evolutionary pathway.) The simplest nervous systems are found in cnidarians, such as the sea anemone, which possess diffuse nerve nets but lack a central system. The trend toward cephalization (the development of control centers in the head) first appears in flatworms and becomes successively more pronounced in the earthworm, octopus, and vertebrate.

(a)   *(b)*

FIGURE 23-20

Animals with nervous systems of contrasting complexity. *(a)* The anemone's predatory skills rely on a simple system of sensory neurons that launch a feeding response when they detect prey. In this instance the prey is a sea star that has made contact with the anemone's tentacles. *(b)* An octopus has a highly complex nervous system that allows the animal to figure out how to dine on a crab trapped inside a stoppered jar.

ium, but if the crab were to hide under a rock, the octopus might "forget" that food is nearby. Without its normal brain function, the octopus illustrates a case of "out of sight, out of mind."

⯈ It is likely that sea anemones, flatworms, earthworms, and octopuses are part of a single evolutionary pathway and that the increased complexity of their nervous systems occurred by stepwise advances of the same basic plan. Even though vertebrates evolved along a different pathway, their nervous systems show unmistakable similarities to those of complex invertebrates. The nervous system of an octopus and a human, for example, both consist of a peripheral portion, where impulses initiate and terminate, and a cen-

tral portion, where impulses are received, sorted, and then relayed to the appropriate destination, illustrating once again, that similar physiological needs often lead to similar evolutionary solutions.

All organisms possess an inherent ability to respond to stimuli; this ability formed the basis for the evolution of primitive nervous systems. The evolution of more active, anatomically complex animals was accompanied by the evolution of more complex nervous systems that are capable of mediating more complex physiological and behavioral responses. (See CTQ #8.)

REEXAMINING THE THEMES

Relationship between Form and Function

Both the single nerve cell and the entire nervous system are structurally adapted for their tasks of collecting and disseminating information. This can be illustrated by comparing the living components of the nervous system to that of a telephone network. A telephone system contains a

collecting device at one end (a mouthpiece); a long, thin, transmission line to carry coded information over a distance; and a speaker at the other end to deliver information that can be interpreted by the target listener. Similarly, nerve cells contain a network of dendrites that collect information from various sources, a cablelike axon that propagates the information in coded form, and synaptic knobs that transmit the information to a variety of targets. Like a telephone network, a nervous system contains a central communications center, where incoming messages from scattered sites are sorted, and outgoing messages are routed to selected target sites.

Biological Order, Regulation, and Homeostasis

The nervous system is more than just a communications network; it is the control center for regulating bodily function. The nervous system monitors the environment, integrates sensory input, and sends out command signals to initiate appropriate responses when changes in the internal or external environment are detected. If your body becomes too warm, nerve cells stimulate your sweat glands to secrete a cooling layer of water; if the concentration of oxygen in your blood becomes too low, nerve cells stimulate your breathing muscles to increase their rate of contraction; if you are threatened by a hungry predator, your nervous system will send out directives either to outsmart the beast or to beat a fast retreat.

Unity within Diversity

While the architecture and complexity of nervous systems are highly diverse, all nervous systems are built from very similar types of nerve cells that conduct and transmit impulses by the same basic mechanisms. The complexity of the human nervous system (as well as that of other "higher" animals) can be traced to the intricate circuitry of our system rather than to an increased complexity of the units that make up the system. Similarly, the nervous systems of virtually all animals, from flatworms to mammals, consist of a peripheral portion in which impulses are initiated and terminated, and a central portion in which impulses are received and sorted, and then relayed to the appropriate destinations.

Evolution and Adaptation

The basic mechanism responsible for neural activity can be found in the simplest unicellular organisms—bacteria—where communication from one part of the cell to another is accomplished by waves of depolarization that pass along the plasma membrane. With the evolution of multicellular animals, communication became more complex; specialized cells evolved that stretched considerable distances across the body. In its simplest form (as exemplified by hydras and sea anemones), the nervous system consists of a simple, diffuse net, with no evidence of centralization. With the evolution of more complex animals came a central concentration of nerve tissue, usually near the anterior end of the animal, close to the sense organs of the head.

SYNOPSIS

The nervous system communicates information from one part of the body to another and, in doing so, regulates the body's activities and maintains homeostasis. Nerve cells (or neurons) are specialized in that they have dendrites, for collecting information; a cell body that houses the nucleus and most of the cellular machinery; an axon, for conducting information in the form of impulses; and terminal synaptic knobs, for transmitting information to a target cell (a muscle cell, gland cell, or another neuron).

Neural impulses result from sequential changes in the distribution of ions across the plasma membrane. Molecular pumps in the plasma membrane of a neuron maintain steep gradients of sodium (Na^+) and potassium (K^+) across the membrane. In the resting neuron, potassium ions leak outward across the membrane creating a resting potential. Should the neuron receive a stimulus that causes the potential to drop (depolarize) past the threshold, an action potential is automatically triggered, whereby gated sodium channels open, allowing an inward movement of sodium ions and a reversal in the potential. This change triggers the closing of the sodium channels and the opening of the gated potassium channels, causing the potential to return to a negative value. When the gated potassium channels close, the membrane returns to its resting state. Once triggered, a wave of action potentials passes down the length of the membrane without losing intensity. In vertebrates, the speed of neural impulses is greatly increased by wrapping the axons in a lipid-containing myelin sheath. The sheath is punctuated along its length by nodes, and the impulse travels along the axon by skipping from node to node.

An impulse is transmitted from a neuron to its target cell by the release of a chemical-transmitter substance that diffuses across the narrow synaptic cleft between the two cells. Neurotransmitter substances are stored in vesicles within the synaptic knobs of a neuron. When an impulse arrives at the synaptic knob, calcium ions flow into the neuron, triggering the fusion of the membrane of a portion of the vesicles with the plasma membrane and releasing the neurotransmitter into the synaptic cleft. When the neurotransmitter binds to a specific receptor on the target cell membrane, it may either excite (depolarize) or inhibit (hyperpolarize) the target cell. In most cases, one excitatory transmission is not sufficient to stimulate a target cell to respond. Several excitatory impulses may become additive if they arrive at the target cell membrane close enough together in either time or space. Synapses act as gates that route information in the proper directions through the nervous system. Moreover, synapses can change over time, as occurs during learning and adapting to new situations.

The nervous system of vertebrates consists of a central nervous system (brain and spinal cord) and a peripheral nervous system (cranial and spinal nerves). Sensory neurons carry information about changes in the environment from the periphery toward the CNS, while motor neurons carry command signals to the body's muscles and glands. Interneurons convey impulses within the CNS.

The fastest motor reactions to stimuli are mediated by reflex arcs, which consist of at least one sensory and one motor neuron. This type of simple circuit provides the shortest possible route for the flow of information through the CNS. Information also passes to and from higher centers in the CNS, which mediate more complex responses.

The human brain consists of several distinct parts, including the cerebrum, the cerebellum, and the brainstem. The cerebrum is covered by the cerebral cortex, a layer of gray matter that houses the sites of higher brain functions, such as speech and rational thought. The cerebellum, in conjunction with the motor cortex of the cerebrum, controls complex motor activities. The brainstem controls involuntary visceral activities, such as breathing and heart rate. Those parts of the brain that control wakefulness, emotions, and homeostatic functions (such as temperature control) are also associated with the brainstem.

The spinal cord receives sensory information via dorsal roots of its spinal nerves and sends out motor information through the ventral roots of these nerves. Motor commands may be sent via the somatic division, which controls voluntary muscle activity, or the autonomic division, which controls involuntary activity. The autonomic division is divided into a parasympathetic branch responsible primarily for "housekeeping" activities, such as digestion and excretion, and a sympathetic branch primarily responsible for readying the body for meeting stressful situations. Many organs of the body receive stimulation from both parasympathetic and sympathetic neurons, which have opposing influences on the organ.

Key Terms

nerve growth factor (NGF) (p. 467)
neuron (p. 468)
excitatory neuron (p. 468)
inhibitory neuron (p. 468)
cell body (p. 468)
dendrite (p. 468)
axon (p. 468)
synaptic knob (p. 468)
sensory neuron (p. 470)
interneuron (p. 470)
motor neuron (p. 470)
neuroglial cell (p. 470)
Schwann cell (p. 470)
myelin sheath (p. 470)
nodes of Ranvier (p. 470)
potential difference (p. 471)
voltage (p. 471)

resting potential (p. 471)
threshold (p. 473)
action potential (p. 473)
saltatory conduction (p. 473)
synapse (p. 474)
synaptic cleft (p. 474)
neurotransmitter (p. 474)
synaptic vesicle (p. 474)
central nervous system (CNS) (p. 479)
brain (p. 479)
spinal cord (p. 479)
peripheral nervous system (p. 479)
nerve (p. 479)
reflex (p. 479)
reflex arc (p. 479)
sensory receptor (p. 479)
gray matter (p. 481)

white matter (p. 482)
ventricle (p. 482)
cerebrospinal fluid (p. 482)
cerebral hemisphere (p. 482)
cerebral cortex (p. 482)
learning (p. 483)
motor cortex (p. 485)
hypothalamus (p. 486)
limbic system (p. 486)
cranial nerve (p. 489)
spinal nerve (p. 489)
somatic division (p. 490)
autonomic division (p. 490)
sympathetic system (p. 490)
parasympathetic system (p. 490)
cephalization (p. 492)

Review Questions

1. Describe the various parts of a myelinated neuron and the role of each part in neural function.

2. Describe the effects of depolarization and hyperpolarization of a target cell on the inhibition or excitation of that cell.

3. What is the source of the energy needed to fire an impulse along a neuron? Why doesn't the strength of the impulse diminish as it travels along the neuron?

4. Describe the components required to mediate a simple reflex response, such as the knee-jerk. How can these same components become part of more complex circuits that involve conscious decisions, such as kicking a football?

5. Compare and contrast the following: somatic and autonomic divisions; sympathetic and parasympathetic systems; resting potential and action potential; the role of leak channels versus gated channels in generating an action potential; the roles of sodium and potassium gated channels; cranial and spinal nerves; myelinated and nonmyelinated axons.

Critical Thinking Questions

1. Suppose that antibodies against NGF injected into a chick embryo had no visible effect on the development of the animal's nervous system. What might you conclude about the role of NGF in development? How could you explain the results of experiments in which NGF injected into an embryo causes enlargement of the sympathetic nervous system?

2. Neurons typically have many short, highly branched dendrites and one axon, which is longer and thicker than the dendrites. How do these differences in structure reflect the differences in function between dendrites and axons?

3. Tetrodotoxin is a toxin present in the spines of puffer fish; it has the capability of blocking the function of gated sodium channels. What effect do you suppose this substance would have on the formation of nerve impulses? On the state of muscle contraction in the body? Why?

4. Transmission of nerve impulses is often referred to as an "electrochemical" phenomenon. Explain what is meant by this term.

5. Most actions controlled by simple reflex arcs involve protective responses or actions that must be completed frequently. Why is it adaptive that these responses do not depend upon the central nervous system?

6. What effects would you predict would result from severe injury to the following parts of the central nervous system: medulla, cerebellum, ventral root of a spinal nerve to the arm, reticular formation, motor cortex, motor cortex on the right side of brain, lower two-thirds of the spinal cord?

7. Explain how the existence of two autonomic systems permits fine-tuning of behavior.

8. Much of what is known about the functioning of the human nervous system has been learned by studying simpler animals, such as the snail, squid, and grasshopper. Why is it possible to learn about the human nervous system from studying these organisms that appear very different from us?

Additional Readings

Alkon, D. A. 1989. Memory storage and neural systems. *Sci. Amer.* July:42–50. (Intermediate)

Aoki, C., and P. Siekevitz. 1988. Plasticity in brain development. *Sci. Amer.* Dec:56–64. (Intermediate)

Kalil, R. E. 1989. Synapse formation in the developing brain. *Sci. Amer.* Dec:76–85. (Intermediate)

Kandel, E., and J. Schwartz. 1985. *Principles of neural science*, 2d ed. New York: Elsevier. (Intermediate–Advanced)

Levi-Montalcini, R. 1988. *In praise of imperfection.* New York: Basic Books. (Introductory)

Marx, J. 1991. Mutation identified as a possible cause of Alzheimer's disease. *Science* 251:876–877. (Intermediate)

Robertson M. 1992. Alzheimer's disease and amyloid. *Nature* 356:103. (Intermediate)

Selkoe, D. J. 1991. Amyloid protein and Alzheimer's disease. *Sci. Amer.* Nov:68–78. (Intermediate)

Tuomanen, E. 1993. Breaching the blood-brain barrier. *Sci. Amer.* Feb:80–84. (Intermediate)

Wallace, B. C. 1991. *Crack cocaine.* New York: Brunner/Mazel. (Introductory)

The brain. 1990. Time-Life Books. (Introductory)

CHAPTER
◂ 24 ▸

Sensory Perception: Gathering Information about the Environment

---
**STEPS
TO
DISCOVERY**
Echolocation: Seeing in the Dark

Echolocation:
Seeing in the Dark

*F*or hundreds of years, naturalists have wondered how bats are able to fly through caves in total darkness or through wooded fields at night, capturing fast-flying insects they can't possibly see. Near the end of the eighteenth century, Lazarro Spallanzani, an Italian biologist, caught a number of bats in a cathedral belfry, impaired their vision, and let them go. Returning to the belfry at a later time, Spallanzani found the bats back in their roost at the top of the cathedral. Not only did the bats return home, but their stomachs were full of insects, indicating that they had been able to capture food without their sense of sight. In contrast, bats whose ears had been plugged with wax were reluctant even to take flight and frequently collided into objects when they tried to fly. Hanging objects in front of the bats' mouths

also impeded the animals' flying abilities. Spallanzani was baffled: How could an animal's ears and mouth be of more importance than its eyes in guiding flight?

The question remained unanswered until 1938, at which time, Donald Griffin, an undergraduate at Harvard University, brought a cage of bats to one of his physics professors, George Pierce. Pierce had developed an apparatus for converting high-frequency sound waves into a form that could be heard by the human ear. After placing the cage of bats in front of the sound detector, Griffin and Pierce became the first people to "hear" the bursts of sounds emitted by bats. When the bats were set free in the physics laboratory, these same high-frequency sounds could be heard as a bat flew in a straight direction toward

By placing invisible barriers in the path of porpoises, biologists discovered that these animals use echolocation to navigate

the microphone. For Griffin, this was the initial step in a long, illustrious career centered around studying the role of sound in animal navigation.

The mechanism bats use to fly through a pitch-black obstacle course relies on the same principles first employed by the Allied navy to hunt for German submarines during World War II. That mechanism is sonar. The basic principle of sonar is simple: Sound waves are emitted from a transmitter; they bounce off objects in the environment; and the "echoes" are detected by an appropriate receiver. The elapsed time between the emission of the sound pulse and its return provides precise information about the location and shape of the object. Most bats employ a remarkably compact and efficient sonar system to detect insect prey and to avoid colliding with objects during flight. High-frequency sounds—those beyond the range of the human ear—are produced by the bat's larynx (voicebox), and the echoes are received by the animal's large ears. Impulses from the ears are then sent to the brain, where they are perceived as a detailed mental picture of the objects in the environment. Griffin coined the term "echolocation" to describe this phenomenon.

Echolocation by bats is an adaptation appropriate for their nocturnal (nighttime) peak of activity. Using echolocation, a bat flying at night can detect and avoid a wire as thin as 0.3 millimeters in diameter or can home in on a tiny, moving insect with unerring accuracy. These feats become even more impressive when you realize that the echo returning to the bat from such objects is about 2,000 times less intense than is the sound emitted. Moreover, a bat can pick out this faint echo in a crowd of other bats, each sending its own pulses of sound waves.

During the early 1950s, studies by Winthrop Kellogg, a marine biologist at Florida State University, and others, demonstrated that bats were not the only mammals capable of navigating by echolocation. One day, Kellogg and Robert Kohler, an electronics engineer, were out in a boat in the Gulf of Mexico when they saw a school of porpoises swimming toward them. The scientists lowered a microphone connected to a speaker and tape recorder into the water. As the porpoises swam toward them, all that could be heard above water was the sound of the animals exhaling through their blowholes. " . . . but the underwater listening gear told a very different story." According to Kellogg, the animals were emitting sequences of underwater clicks and clacks "such as might be produced by a rusty hinge if it were

opened slowly. . . . By the time the group was about to make its final dive, the crescendo from the speaker in our boat had become a clattering din which almost drowned out the human voice."

To study the echolocating capabilities of these animals, Kellogg persuaded Florida State University to build a special "porpoise laboratory" at its nearby marine station. There, investigators discovered that porpoises possessed a sophisticated sonar system they could use to distinguish between similarly shaped objects, to avoid invisible barriers (such as sheets of glass), to swim through an elaborate obstacle course, to locate food, and to pick up objects off the bottom of their enclosure.

It may surprise you to learn that you probably have some capability to echolocate as well. Over the past hundred years or so, many reports have been written about the remarkable ability of many blind people to detect the presence of obstacles in their path without the use of a cane. The first carefully controlled experiment on this subject was conducted in 1940 by Michael Supa and Milton Kotzin, a pair of graduate students at Cornell University, one of whom was blind himself. Subjects that were either blind or sighted but blindfolded were asked to walk down a long wooden hallway. They were told that a large, fiberboard screen *might* be placed at some random site along their way and were asked to report when they detected the presence of the screen. The subjects were able to detect the screen within 1 to 5 meters of its presence. In those cases when the screen was absent, none of the subjects reported its presence. In contrast, when the subjects' ears were tightly plugged, they collided with the screen in every trial. The authors of the experiment concluded that the echo of the sound emitted from the subjects' footsteps on the floor provided information about the location of the screen. A later report documented the case of a young blind boy who was able to avoid obstacles while riding a bicycle by making clicking sounds and listening to the echoes.

around objects.

*C*onsider for a moment what it would be like if you were suddenly deprived of information from your sense organs. Our concept of the world is largely based on information gathered by our eyes, ears, nose, mouth, and the surface of our skin. Without our sense organs, we could not find food, avoid danger, or communicate with others. We would know absolutely nothing about the outside world. Our sense organs also provide us with many of the pleasures in life — the taste of food, the sounds of music, and the touch of another person. Collecting sensory data satisfies a basic human need. In fact, in experiments where people are placed in an environment that is devoid of sensory stimulation — a dark, silent, constant-temperature chamber — the subjects quickly become restless and irrational and often begin to hallucinate.

We also depend on information gathered by the sense organs in our internal environment that are required to maintain homeostasis. Without information from sensors located in muscles, joints, tendons, and internal organs, we could not walk, stand, or digest food; we would have no awareness of our body's state of well-being. Without information from sensors that detect chemicals and pressure in the walls of our arteries, we would not be able to maintain proper cardiovascular or respiratory function. In fact, very little in our body would work properly.

The properties of sense organs vary widely among different animals. Our "picture" of the outside world is shaped largely by our sense of sight, but many other animals rely on very different types of sensory organs (Figure 24-1). Consider, for example, the shark, which can find prey buried in the sand by detecting tiny electric potentials generated by the muscles of the buried animal; the bloodhound, which can follow the trail of a specific person hours after he or she has passed; or the rattlesnake, which can strike in total darkness, guided only by the heat emitted from its warm-blooded prey.

▼ ▼ ▼

THE RESPONSE OF A SENSORY RECEPTOR TO A STIMULUS

The first step in the chain of events leading to sensory awareness is the interaction between a *stimulus* and a *sensory receptor*. A stimulus is any change in the internal or external environment, such as a cold breeze or an empty stomach, and a sensory receptor is a cell that can be activated by that change. Some receptors, such as those of the eyes and ears, are specialized cells that respond to a stimulus and transmit the information to a sensory neuron. Other receptors, such as the stretch receptors of the muscles (page 479), are part of the sensory neurons themselves. Each type of stimulus activates only one type of sensory receptor (Table 24-1), which include:

- *Mechanoreceptors,* which respond to mechanical pressure and detect motion, touch, pressure, and sound.
- *Thermoreceptors,* which detect changes in temperature and react to heat and cold.
- *Chemoreceptors,* which are activated by specific chemicals, such as those that induce a particular taste or smell. Chemoreceptors also monitor concentrations of critical nutrients (glucose, amino acids) or respiratory gases (oxygen and carbon dioxide) in the blood.
- *Photoreceptors,* which respond to light.
- *Pain receptors,* which respond to excess heat, pressure, and irritating chemicals or chemicals that are released from damaged or inflamed tissue.

The structure of each sensory receptor enables the receptor to respond specifically to a particular type of stimulus, such as light of certain wavelengths or particular chemicals. The interaction between a stimulus and a receptor elicits some change in the receptor, which, in turn, evokes an alteration in the ionic permeability of the plasma membrane. This change alters membrane potential at the site of stimulation, which may trigger an action potential in a sensory neuron. For example, photoreceptors located in the retina of your eye contain membrane pigments that absorb light energy. Absorption of light energy causes a change in the molecular conformation of the pigment, leading to a change in membrane voltage. This alteration can trigger an impulse in a neuron of the optic nerve, generating electrical activity in those parts of the brain concerned with vision. The result is the perception of light.

We learn about conditions in the external environment and within our own bodies through receptors, which are activated by specific types of stimuli: light, pressure, chemicals, and so forth. The perception of a stimulus occurs in the brain. (See CTQ #2.)

SOMATIC SENSATION

Somatic sensation (*soma* = body) is a sense of the physiological state of the body. The information is gathered by somatic sensory receptors located in the skin, skeletal

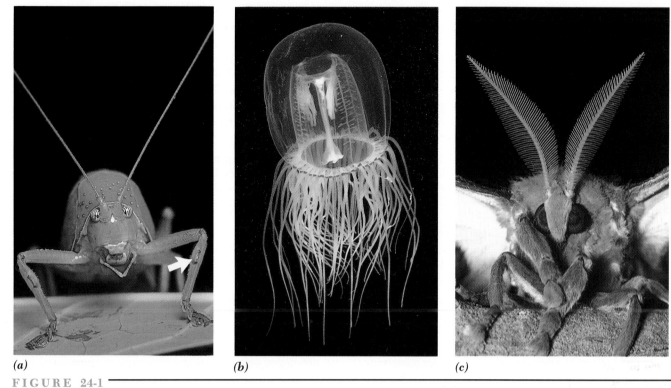

(a)   *(b)*   *(c)*

FIGURE 24-1

A gallery of invertebrate sense organs. *(a)* The hearing organ of this katydid is located on its legs. Sound produces vibrations of the exoskeleton. *(b)* Motion detectors *(statocysts)* in the rim of the bell of this jellyfish allow the animal to maintain its orientation as it swims. *(c)* Chemoreceptors in the antennae of this male atlas moth allow it to detect the smell of a female moth of the same species from a distance of 11 kilometers.

TABLE 24-1

TYPES OF SENSORY RECEPTORS

Example of Receptor	*Example of Function*
1. *Mechanoreceptors*	
Pacinian corpuscle in deep layers of skin	Sense of deep touch
Stretch receptors in skeletal muscles	Maintenance of posture
Hair cells of vestibular apparatus and cochlea of inner ear	Sense of balance and hearing
2. *Thermoreceptors*	
Ends of sensory neurons in skin	Monitor external temperature
Neurons in hypothalamus	Monitor internal temperature
3. *Chemoreceptors*	
Receptor cells in arteries	Monitor blood oxygen level
Receptor cells in taste buds	Sense of taste
Receptor cells in olfactory epithelium	Sense of smell
4. *Photoreceptors*	
Rod cells in retina	Low-light vision
Cone cells in retina	bright-light, color vision
5. *Pain receptors*	
Ends of sensory neurons	Awareness of tissue damage

muscles, tendons, and joints. Somatic sensory receptors inform the CNS about pressure, stretch, temperature, and whether or not a stimulus is intense or threatening—judgments that often lead to the perception of pain. Information from these receptors is required for bodily movements and for maintaining homeostasis.

Although the skin is not considered a specialized sensory organ like the eye or the ear, it is generously endowed with several types of sensory receptors, all of which are the tips of sensory neurons (Figure 24-2). Different nerve endings are specialized to respond to warmth, cold, painful stimuli, or pressure (touch). Some of the pressure receptors of the skin, such as the Pacinian corpuscle shown in Figure 24-2, contain a multilayered capsule that resembles an onion. The capsule insulates the sensitive nerve ending so that weak stimuli, such as pressure exerted by the clothes you are wearing, fail to trigger neural impulses.

Information gathered by somatic sensory receptors in the skin travels through the central nervous system until it reaches two narrow strips of cortex, one on each side of the brain (Figure 24-3). These strips are called the **somatic sensory cortex.** The map of the **somatic sensory cortex** has been obtained by studying individuals undergoing neurosurgery. For example, touching a person on the fingers, stomach, and toes produces electrical activity in three distinct and predictable areas of the brain's sensory cortex. Conversely, if a neurosurgeon directly stimulates a few cells in this part of the brain using an electrified probe, a person reports feeling warmth or tingling in the corresponding part of the body. Those parts that are most sensitive to stimulation and have the greatest number of sensory receptors, such as the fingers and lips, are represented by a disproportionately large number of neurons in the somatic sensory cortex (Figure 24-3). The pathways by which tactile (touch) information is processed in the brain is discussed in The Human Perspective: Two Brains in One.

Somatic sensory receptors gather information on the state of the body, which is used primarily to direct bodily movements and to maintain homeostasis. (See CTQ #3.)

VISION

▮▶ Vision is a sense that is based on light rays reflected into our eyes from objects in the external environment. To most people, vision is the richest form of sensory input. The complexity of our eyes and the large amount of brain tissue devoted to sight attest to the evolutionary importance of this sense. We inherited a keen sense of sight from our primate ancestors who lived in trees and got around by jumping from branch to branch. One miscalculation could result in a long fall to the ground. In this environment, natural selection favored those animals with better eyesight.

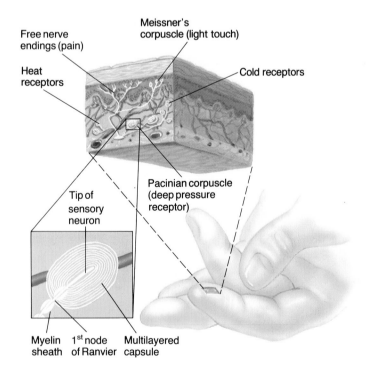

Free nerve endings (pain)

Heat receptors

Meissner's corpuscle (light touch)

Cold receptors

Pacinian corpuscle (deep pressure receptor)

Tip of sensory neuron

Myelin sheath   1st node of Ranvier   Multilayered capsule

FIGURE 24-2

Human skin is the residence of a variety of somatic sensory receptors that respond to mechanical, chemical, and thermal stimuli. These receptors are particularly numerous in the skin of the fingertips, lips, and genitals, where they provide heightened sensitivity. Several mechanoreceptors, such as the Pacinian corpuscle, consist of a free sensory nerve ending surrounded by a capsule whose layers of connective tissue resemble the layers of an onion. The contents of the capsule suppress slight mechanical stimuli, preventing initiation of a neural impulse. In contrast, Meissner's corpuscles, which are located closer to the surface and lack the multilayered capsule, are activated by light touch.

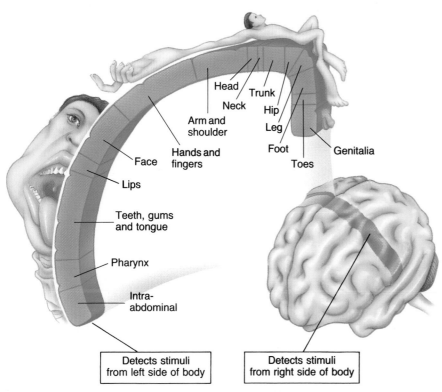

RIGHT SOMATIC SENSORY CORTEX

Head
Neck
Trunk
Hip
Leg
Foot
Toes
Arm and shoulder
Hands and fingers
Genitalia
Face
Lips
Teeth, gums and tongue
Pharynx
Intra-abdominal

Detects stimuli from left side of body

Detects stimuli from right side of body

FIGURE 24-3

The somatic sensory cortex.   This strip of gray matter is located just behind the motor cortex (see Figure 23-13) on both sides of the brain. Information from sensory receptors in the skin, muscles, tendons, and joints is transmitted to the somatic sensory cortex, where it is processed and used in directing bodily movements and maintaining homeostasis. The area of cortex devoted to each part of the body is indicated.

The main component of the human eye—the eyeball (Figure 24-4)—is roughly spherical in shape and is situated within a protective socket of the skull. The wall of the eyeball can be divided into three complex layers, each with distinct functions.

1.   The outer layer consists of the sclera in the rear and the cornea in the front. The **sclera** is a tough, connective-tissue capsule that protects the eye, whereas the **cornea** is a delicate, transparent window, whose remarkable structure lets light rays pass into the eye unobstructed.

2.   The middle layer consists of the choroid in the rear and the iris in the front. The **choroid** is a dark tissue that contains large numbers of blood vessels which supply nutrients and oxygen to all parts of the eye. The **iris** is a disklike, muscular structure with a circular opening, the **pupil,** through which light passes into the eyeball's interior. Pigments in the iris block the passage of stray light and give the eye its distinctive coloring.

3.   The inner layer is evident only in the rear of the eye, where it forms the **retina,** a blanket of photoreceptors that provides the screen on which visual images are projected.

THE LENS: FOCUSING IMAGES ON THE RETINA

The interior of the eyeball contains the **lens** (Figure 24-4), which focuses incoming light rays onto the retina. The focusing abilities of the lens are derived from the glass-like proteins of which it is composed. Contraction of the *ciliary muscles* changes the shape of the lens so that objects at different distances from the eye can be focused on the retina. If this change in lens shape (called *accommodation*) did not occur, the image of nearby objects would be focused behind the retina and would appear blurry. As we age, changes in the structure of the lens proteins makes the lens less elastic. Consequently, our eyes become less able to focus on close objects, which explains why so many people need reading glasses after the age of about 45. In some older individuals, the lens becomes opaque, causing a *cataract*, which can be treated surgically.

THE RETINA: A LIVING PROJECTION SCREEN

The retina contains two different types of photoreceptor cells, the *rods* and *cones* (Figure 24-4, inset), names that

FIGURE 1

Apparatus for testing the functions of the corpus callosum. In this experiment, the subject can pick up the objects on the table with one hand or the other or can touch an object with the feet but cannot see the objects being manipulated.

As you will recall from the previous chapter, the largest part of the human brain is divided into right and left cerebral hemispheres, which are connected to one another by the corpus callosum, a "cable" approximately 200 million nerve cells thick (see Figure 23-11). During the first half of the twentieth century, scientists observed that accidental damage to the corpus callosum in people who suffered from epileptic seizures led to dramatic improvement in the person's epileptic condition. In the 1940s, William Van Wagenen, a neurosurgeon at the University of Rochester, attempted to treat a group of epileptic patients by surgically severing their corpus callosum. The patients improved and showed virtually no ill effects from the operation. In fact, neurological and psychological tests indicated no significant loss of brain function.

Roger Sperry of the California Institute of Technology had studied the neural basis of consciousness and was convinced that earlier investigators had failed to discover the behavioral effects of severing the corpus callosum because they hadn't examined the function of each hemisphere independently of the other. In the 1960s, Sperry devised a procedure for testing these split-brain patients so that the func-

tions of each half of the cerebrum could be analyzed separately. He controlled experimental conditions so that the sense organs of "split-brain" patients were stimulated on only one side of their body. For example, Sperry would ask the subjects to pick up an object with only one hand while sitting at an apparatus that prevented them from seeing the object (Figure 1). If the subjects picked up a cube with their right hand and were asked to name the object, they were able to say it was a cube. They could also write the word "cube" on a piece of paper. If they picked up the same object with the left hand, however, the subjects were unable to identify the object either verbally or in writing. This was clearly different from a "normal person" with an intact corpus callosum, who had no difficulty identifying the object or writing the object's identity, regardless of which hand held it.

Sperry had an explanation for this seemingly strange behavior in split-brain subjects: The speech and writing centers of the brain are located in the left cerebral hemisphere, which receives information from the sense organs of the right side of the body. Consequently, when the subjects picked up an object with their right hand, sensory information about the shape of the

object was sent to the left cerebral hemisphere which, because it contains the speech and writing center, allowed the subjects to describe the object. In contrast, when the same object was picked up by the left hand, the information went to the right cerebral hemisphere, which had no way of "informing" the speech center in the left hemisphere of the experience. Consequently, the subjects could not verbalize the shape, often guessing wildly at its identity. Yet, they could accurately point to a picture of a cube among a gallery of different-shaped objects, indicating an awareness of the object's identity. Remarkably, they might even draw a picture of a cube and, at the same time, verbally guess that they were holding a sphere or a rod. Such responses were accompanied by signs of frustration from the subjects, indicating that the right side of the brain recognized that this was an erroneous verbal response. Sperry concluded that, without an intact corpus callosum, each hemisphere is unaware of the sensory and mental experience of its partner. In recognition of his work on the brain, Sperry was awarded the 1981 Nobel Prize in Medicine and Physiology.

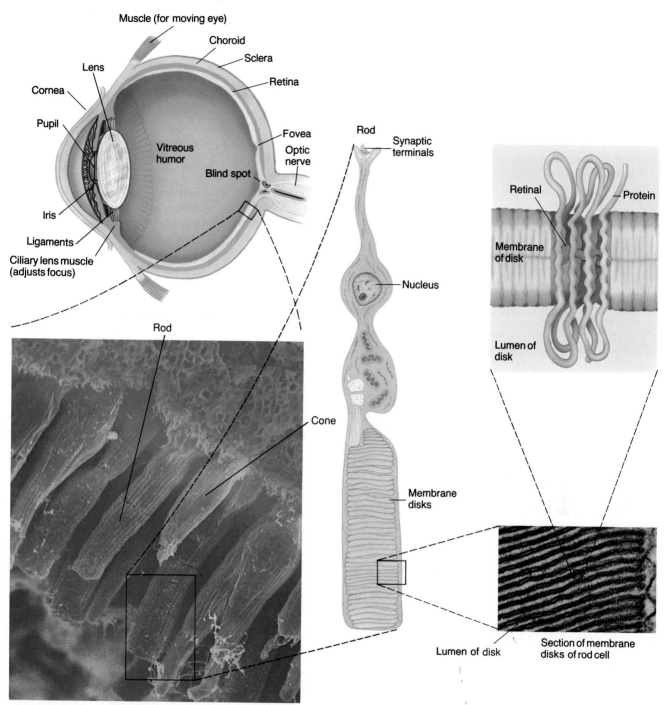

FIGURE 24-4

The human eye. Light enters through the cornea and then travels through the pupil, the hole in the middle of the colored iris. Pupil size changes to regulate the amount of light that enters the eye. Light rays are focused on the retina by the lens, whose shape is changed by *ciliary lens muscles* acting on ligaments. The retina has two kinds of photoreceptors: rods and cones. Activation of the photoreceptors initiates impulses, which are conducted along neurons that make up the *optic nerve*. The light-absorbing pigments of the photoreceptors are embedded in membranous disks of the photoreceptor cells. The acute light sensitivity of rods allows us to see in very dim light. Cones allow us to see color. The *fovea* is the site in the retina of the greatest concentration of cones. The hole through which the optic nerve exits the inner surface of the eye lacks photoreceptors and creates a small "blind spot" in the visual field. The rear chamber of the eye is filled with a viscous fluid, the *vitreous humor.* Upper right inset courtesy of J. David Rawn, *Biochemistry,* ©1989, p. 1066. Reprinted by permission of Prentice Hall, Englewood Cliffs, New Jersey.

reflect their microscopic shape. Both types of cells contain pigment molecules that absorb incoming light. Each pigment molecule consists of a protein linked to a molecule of *retinal,* which is synthesized from vitamin A (thus making carrots, which are rich in vitamin A, good for your eyesight). Like the light-absorbing pigments of plant chloroplasts (Chapter 8), photopigments of the rods and cones are embedded in membrane disks. *Rhodopsin,* the best-studied visual pigment, worms its way back and forth across the membranes of rods (most magnified inset of Figure 24-4). The absorption of a single photon of light (the smallest unit of light energy) can cause an alteration in the shape of the retinal portion of a single pigment molecule, triggering a sequence of changes that culminate in the closure of several hundred ion channels and a change in membrane potential of about 1 millivolt. The additive changes in potential that result from the absorption of a number of photons lead to the initiation of an action potential in a sensory neuron of the *optic nerve* leading to the brain.

Rods are much more sensitive to light than are cones and function primarily under low-light conditions. The human retina contains over 100 million rods, which are concentrated toward the peripheral regions of the retina. Rods provide us with night vision, which is characterized by a lack of color and sharpness but a heightened sensitivity. The high sensitivity of the rods can be appreciated when you walk out of bright sunshine into a darkened theater. At first you are virtually blind because only your cones—which respond to bright light—are functioning; the pigment molecules of the rods have been temporarily inactivated by the bright light. As you spend more time in the dark theater, the rod pigments are reactivated, and your eyes become *dark-adapted,* enabling you to see the outlines of people in the theater. The dilation of the pupil, which allows more light to reach the retina, is another response of the dark-adapted eye. Even with these dark adaptations, humans have relatively poor night vision compared to nocturnal animals, such as cats and owls. Many nocturnal animals have a mirrorlike layer (*tapetum*) behind their retina that reflects light back out of the eye, giving the rods a "second chance" to absorb light rays. This mirrorlike layer is the reason the eyes of many animals appear to glow at night (see Figure 24-10).

Unlike rods, cones are relatively insensitive to light and are essentially useless under conditions of low light intensity. Cones are concentrated in the center of the human retina, where they provide a highly detailed image of the visual field. Unlike rods, cones also provide information on the wavelength of the light, which we perceive as color. Three different types of cones are distinguished by the structure of their pigments and the wavelengths of light they absorb. It is thought that the color we perceive in a particular part of the visual field depends on the ratio of impulses generated by the blue-, green-, and red-absorbing cones in the corresponding area of the retina. The various types of colorblindness can be explained as a loss of one of the three types of cones.

Impulses triggered by the rods and cones of the retina travel to the **visual cortex** of the occipital lobe of the brain (Figure 23-12). The visual cortex was mapped with considerable precision around the turn of the century by a Japanese physician, Tatsuji Inouye. Inouye interviewed soldiers from the Russo-Japanese War who had recovered from bullet wounds to the back of the head. He was able to correlate blind spots in the soldier's visual field with the location of brain damage determined by the entrance and exit wound of the high-velocity Russian bullets.

The eye collects light rays from the external environment and focuses them as an image on the blanket of photoreceptors that comprise the retina. Impulses from the receptors are interpreted by the brain as a detailed mental picture. (See CTQ# 4.)

HEARING AND BALANCE

The human ear contains two separate sensory systems, one that governs hearing, and the other that allows us to maintain our balance. The sensory receptors for both of these senses are **hair cells**—mechanoreceptors that lie in the interior portion of the ear (Figure 24-5a) and contain very fine "hairs." When these hairs are bent or displaced by movements or pressure, a change in the membrane potential of the hair cell is generated, which may launch an impulse along a sensory neuron carrying the information to the brain. Depending on the part of the ear in which the impulses are generated, we may perceive a particular sound or a feeling of motion.

FIGURE 24-5

The human ear. (*a*) The *outer ear* consists of the ear flap (or *pinna*) and the channel leading to the tympanic membrane, which collects sound waves. The *middle ear* is a small chamber composed of three interconnected ear bones (often called the hammer, anvil, and stirrup) which transmit and amplify sound waves. The *inner ear* contains the cochlea (and its receptors for hearing) and the vestibular apparatus (and its receptors for balance). Motion of the fluid in the cochlea is initiated when the stirrup pushes in at the *oval window.* Waves are transmitted through the cochlear fluid until they reach the *round window,* where they are dissipated. The first electron micrograph shows several rows of hair cells along a portion of the cochlea, and the second micrograph shows a single hair cell. The hair cells translate vibrations into neural impulses that travel to the hearing centers of the brain. (*b*) Pete Townsend of the "Who" has experienced severe hearing loss as a result of exposure to amplified music.

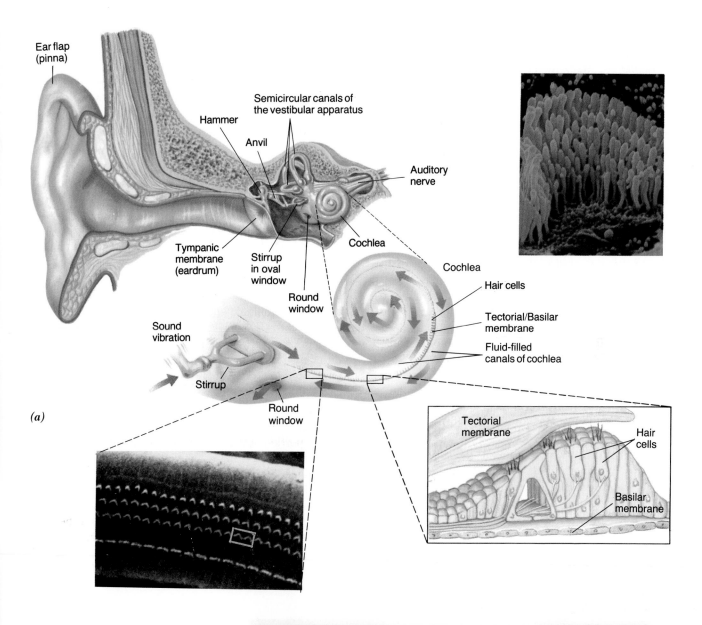

(a)

(b)

TRANSLATING VIBRATIONS INTO SOUND

Sound vibrations travel outward in waves from a source, much like ripples in a pond. When a sound wave strikes a solid object, it sets up vibrations in the object. This is evident in the vibrations of a wall in response to the bass sounds from a nearby speaker and is the basis for our sense of hearing. Sound waves are collected by the outer ear flaps, where they pass through the canal of the outer ear and strike the **tympanic membrane** (the eardrum), which oscillates at the same frequency as do the incoming sound waves. These oscillations are transmitted through the middle ear by the movements of three tiny, interconnected bones (often called the hammer, anvil, and stirrup) which amplify the vibrations by a factor of about 20. The middle ear also contains a number of muscles that contract by reflex action when the ears are exposed to unusually loud sounds. Contraction of these muscles reduces the intensity of the vibrations of the middle-ear bones, helping to protect them from damage.

Vibrations enter the inner ear from the stirrup and pass along the **cochlea,** an elongated spiral structure that resembles the shell of a garden snail. The cochlea contains two fluid-filled canals that are separated by a thin **basilar membrane.** Nearly 20,000 hair cells are anchored to the basilar membrane. A cluster of sensory "hairs" project from the outer surface of each hair cell; the tips of these hairs are embedded in a delicate structure, called the **tectorial membrane.** Because of this arrangement (Figure 24-5a), the hair cells are, in essence, suspended between the basilar and tectorial membranes. Sound waves entering the cochlea induce vibrations in the basilar membrane, causing movement of the bases of the hair cells relative to the tectorial membrane, which does not vibrate. The bending of the hairs activates the hair cells, leading to a change in the voltage across their plasma membrane and to the initiation of impulses in a sensory neuron of the *auditory nerve* leading to the auditory centers of the brain. Loud noise or music can produce damage to the hair cells and subsequent deafness (Figure 24-5b).

TRANSLATING FLUID MOVEMENTS INTO A SENSE OF BALANCE

In addition to the cochlea, the inner ear contains the **vestibular apparatus,** which gathers information about the position and movement of the head. This information enables us to maintain balance (even if we are standing on our head), to detect up from down (even with our eyes closed), and to feel changes in the body's position (Figure 24-6). Unlike our other major sensory systems, we are usually not aware of the activity of our vestibular apparatus. This lack of perception can abruptly change if we have the misfortune of developing an inner-ear malady, in which case barrages of impulses from our vestibular apparatus to our brain may cause us to feel like the room is "swimming around" simply by lifting our head off the pillow.

The vestibular apparatus contains three fluid-filled **semicircular canals,** whose walls are lined with hair cells that function as motion receptors. When you turn your head, the walls of the semicircular canals move faster than does the fluid they contain. This movement bends the sensory hairs, which triggers an impulse, informing the brain of the motion. Because the three canals are perpendicular to one another (Figure 24-5), motion in any direction can be

FIGURE 24-6
Balance. The sense of balance being put to the extreme test.

detected. Excessive motion, such as that which occurs to a small boat in choppy seas, overstimulates the hair cells of the vestibular apparatus, which can cause nausea and vomiting.

> **Our senses of hearing and balance derive from mechanoreceptors that are located in separate parts of the inner ear. These receptors respond to movements of fluid which are initiated either by sound vibrations in the air or by changes in orientation of the head. (See CTQ# 5.)**

CHEMORECEPTION: TASTE AND SMELL

Chemoreception in most animals, including humans, constitutes two distinct senses—**taste** and **smell,** or **olfaction.** Taste is stimulated by *direct* contact with compounds present in food. In contrast, chemicals that excite the sense of smell arrive from a *distant* source, carried in the air (or in the water, in the case of aquatic animals). In humans, the senses of taste and smell are largely utilized in the selection of palatable food. In addition to heightening our enjoyment of food, these senses protect us from various dangers. For example, toxic plants often evoke a bitter taste that makes them distasteful; food in an advanced state of microbial decomposition is usually detected as spoiled by its aroma before we even taste it; and the acrid smell of smoke alerts us to the presence of fire.

The mechanism of chemoreception is generally poorly understood. Chemoreceptor cells contain receptor proteins in their plasma membranes that specifically engage the chemicals being sensed. There are probably a wide variety of different receptor proteins, each having an affinity for distinct chemical groups. Since not everyone is able to detect the same tastes or scents, there are probably genetic differences in the structure of these proteins within the population. The recent identification of the genes for chemoreceptor proteins has opened the door to a better understanding of the mechanism of taste and smell.

TASTE

The receptors for taste reside in the 10,000 or so *taste buds* (Figure 24-7a) located on the surface of the tongue and other sites within the mouth. Dissolved chemicals enter the

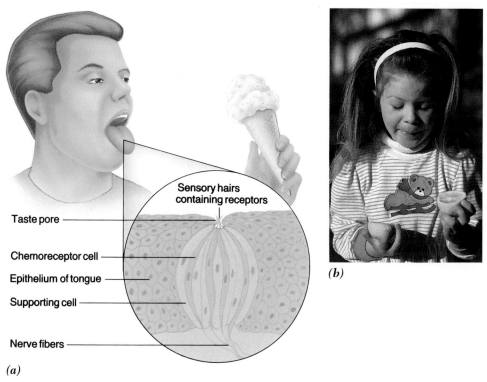

Sensory hairs containing receptors

Taste pore

Chemoreceptor cell

Epithelium of tongue

Supporting cell

Nerve fibers

(a)

(b)

FIGURE 24-7

Taste buds. *(a)* Taste buds are present within the epithelium of the tongue. Each taste bud has a pore through which dissolved substances enter and subsequently contact receptor proteins on the "hairs" of chemoreceptor cells. Nonsensory *supporting cells* help maintain the structure of the taste bud. *(b)* The citric acid of a lemon evokes one of the primary tastes—sour.

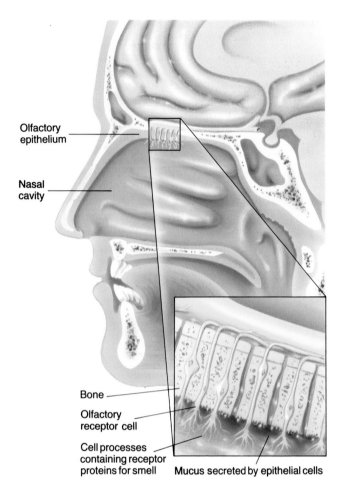

FIGURE 24-8

Olfaction. The olfactory receptor cells are located in patches within the inner lining of the nasal cavity. The receptor proteins themselves are located on processes that extend into a layer of mucus that lines the nasal chamber. Activation of the receptor cells sends impulses along fibers of the olfactory nerve into a nearby portion of the brain, called the olfactory bulb. Information is then transferred from the olfactory bulb to the olfactory centers of the cerebral cortex, where odor is perceived.

Labels: Olfactory epithelium; Nasal cavity; Bone; Olfactory receptor cell; Cell processes containing receptor proteins for smell; Mucus secreted by epithelial cells

taste bud through an open pore and interact with sensory hairs projecting from each receptor cell. There are only four *primary tastes:* salty, sweet, sour, and bitter (Figure 24-7b). These primary tastes become combined in various ways to give us our rich discrimination of flavors. The perception of each primary taste can be elicited by chemicals with very different structures. For example, the sweetest substance yet discovered is thaumatin, a protein isolated from the berry of an African shrub. Thaumatin is several hundred times sweeter than the dipeptide aspartame (Nutrasweet), which is much sweeter than is the disaccharide sucrose (table sugar). Even though taste and smell constitute distinct senses, the taste of our food is greatly affected by its smell. This becomes evident when you suffer from a bad cold; you lose the ability to discriminate among subtle taste differences.

SMELL

The sensory receptors for smell are located within the olfactory epithelium that lines the interior wall of the nasal cavity (Figure 24-8). Even humans, whose sense of smell is relatively poor compared to other mammals (such as mice or dogs), can discern thousands of different types of sub-

stances and can detect them at extremely low concentrations. Methyl mercaptan, for example, the substance added to natural gas (which is otherwise odorless) to give it the characteristic aroma that alerts us to gas leaks, can be detected at a concentration of one part methyl mercaptan per billion parts of air. For many substances, a single molecule impinging on an olfactory receptor cell can trigger an impulse to the CNS.

Our senses of taste and smell derive from chemoreceptors present in the mouth and the lining of the nasal cavities, respectively. Our perception of a taste or smell results from the particular combination of chemoreceptors activated. (See CTQ# 6.)

THE BRAIN: INTERPRETING IMPULSES INTO SENSATIONS

In the previous chapter, we mentioned that all impulses traveling along a neuron exhibit the same strength. Consequently, neurons cannot conduct stronger impulses in re-

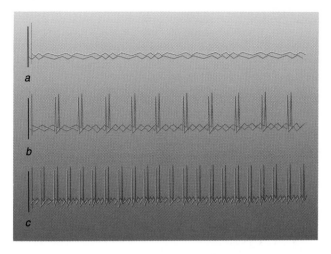

F I G U R E 24-9
The strength of a stimulus is generally encoded in the frequency of the impulses carried by the neuron. These recordings, taken from a living neuron, show a neuron at rest *(a)*, after exposure to a weak stimulus *(b)*, and after exposure to a stronger stimulus *(c)*. Each spike represents an action potential moving across a point on the neuron's membrane.

sponse to increased intensity of stimulation. Yet, we are clearly capable of detecting differences in the strength of a stimulus. If you put your fingers into a bathtub of water, for example, you will learn more than whether or not the water is hot. You are able to perceive a *range* of thermal differences, from unpleasantly cold to scalding hot.

The ability to make sensory discriminations depends on several factors. For example, a stronger stimulus, such as scalding water, activates more nerve cells than does a weaker stimulus, such as warm water. It also activates "high-threshold" neurons that would remain at rest if the stimulus were weaker. Stimulus strength may also be *encoded* in the frequency by which action potentials are launched down a particular neuron; the stronger the stimulus, the greater the frequency (Figure 24-9).

DETERMINING THE NATURE AND LOCATION OF A STIMULUS

Why does an impulse create pain when delivered from a pain receptor, whereas the exact same type of impulse creates sight when delivered by a neuron from a photoreceptor? Furthermore, how do we know where the pain is felt, even without looking at the part of the body experiencing it? Just think about the last time a doctor drew a blood sample from your thumb. Even without looking at the damage, you know what has happened and the exact spot where the needle entered. Your brain has been notified of the disturbance and translates the incoming impulses as "sharp pain in the tip of the left thumb." The neurons triggered by this stimulus carry no such message, however; all they can transmit is an action potential sweeping along the neuron. The nature and exact location of the disturbance are determined by the specific portions of the brain that are stimulated. Sensory input from the eyes, ears, nose, and somatic sensory receptors of the skin travel to separate locations within the cerebral cortex (see Figure 23-12). These are the sites in the brain where our perceptions of sight, hearing, smell, and touch are derived. Any time the cells in a particular portion of the brain are activated, the same sensation is perceived in the same location in the body; the brain perceives the sensation, not the finger. This phenomenon is readily demonstrated by electrode stimulation during open-skull surgery. If the electrode stimulates a particular site in the brain, the patient will experience a very real sensation of a sharp object piercing his or her unassaulted thumb, for example, just as though the finger were really being pierced.

The type of sensation perceived by the brain depends on three factors: (1) the specific region of the cerebral cortex that is stimulated; (2) the number of brain cells that are stimulated; and (3) the frequency with which impulses arrive along various neurons. (See CTQ# 7.)

EVOLUTION AND ADAPTATION OF SENSE ORGANS

Humans don't use their eyes to help them hear or their ears to help them see. Unlike a nervous system or a digestive system, whose parts are functionally linked, sense organs function and evolve relatively independent of one another. As a result, sense organs exhibit a much greater diversity in form and function among various animals than do most other types of physiological structures. For example, sense organs that detect light have appeared over and over again throughout the course of evolution, producing animals with a great variety of light-sensitive organs (Figure 24-10). Yet, despite the diversity among sense organs, obvious similari-

◁ B I O L I N E ▷
Sensory Adaptations to an Aquatic Environment

■▶ In this chapter, we have focused on human sense organs, which are adapted for life in a terrestrial environment. A brief examination of some of the sense organs found in fishes will illustrate how aquatic environments select for sense organs having very different properties.

VISION

If you have ever tried to find an object on the bottom of a pool, you are aware that humans don't see very well underwater. Our poor underwater eyesight can be blamed in part on our eyes and in part on the limitations that water places on vision. For a vertebrate living in a terrestrial habitat, light rays are bent (*refracted*) as they pass from air across the cornea into the watery tissue of the eye; this bending of light is part of the focusing process. In contrast, light rays are not bent as they pass from the aquatic environment into the eye of a fish. Consequently, virtually all of the focusing by the fish's eye must be accomplished by the lens, which is much harder and more spherical than is the lens of a terrestrial vertebrate. Because of its hardness, the shape of a fish's lens cannot be changed by muscles; instead, light rays are focused by a back and forth motion of the lens within the eye. The reason we see so poorly underwater is that our cornea is essentially eliminated as a focusing element. Furthermore, light rays cannot travel nearly as far through water as they can through air; thus, fish cannot use their eyes as distance receptors as do terrestrial vertebrates.

HEARING

A fish's ears consist exclusively of an inner-ear compartment; there is no outer or middle ear. In most fishes, sounds are probably transmitted to the receptors in the inner ear by vibrations of the bones of the skull, which is a relatively insensitive mechanism. In fact, the structure of a fish's inner ear suggests that the ear primarily provides information on body position, not hearing. It is thought that the original function of the vertebrate ear was to maintain body equilibrium; it became a hearing organ secondarily.

CHEMORECEPTION

Due to the limitations imposed on their visual sense, most fishes rely primarily on their sense of smell. Substances dissolved in the water inform a fish of the presence of food, a potential mate, a toxic substance, or even a predator. For example, some fishes that travel in large schools emit an *alarm substance* from their skin in the presence of danger, which alerts other fish to disperse.

MECHANORECEPTION

Water is an excellent conductor of vibrations or pressure disturbances. Fishes have evolved a unique type of sensory system— the *lateral line*—to monitor such stimuli. The system consists of canals that run down the flank of the fish and across its head. The canals contain sensory cells with hairlike projections that are activated by the flow of water. The lateral line can be thought of as a system for sensing "distant touch." Using this system, fishes can detect waves hitting a distant shore or waves generated by a distant boat. Closer to home, the system can inform the fish of objects in its immediate surrounding which affect the flow of water around the animal. For example, under normal circumstances, a fish will avoid a finger placed in its midst. If the lateral line system is covered so that the fish cannot receive sensory information, the finger is ignored, and the fish can be gently pushed with it.

ELECTRORECEPTION

Certain eels, catfish, and rays are able to deliver a strong electric shock—up to 600 volts—to anyone who happens to come too near. Other fishes have much weaker electric organs that are used not as weapons but as sensory systems and a means of communication. These fish generate an electric current that flows from the tail to the head. Objects that come within the fish's electric field will interfere with the flow of electric current through the water; this change is detected by sense organs in the fish's head. Impulses from these sensory receptors provide information on the nature and location of objects in the water. These electrolocation systems are particularly important to species that live in murky waters, where vision is highly limited.

FIGURE 24-10

Some striking adaptations for photoreception. *(a)* Stalked eyes of a mantis shrimp. *(b)* The compound eyes of a robber fly are composed of thousands of individual lenses that give the fly an extremely broad visual field. *(c)* Only their eyes betray the silent nocturnal approach of this flotilla of alligators closing in on an observer's spotlight. A special layer of tapetum cells in the alligator's retina reflects light, creating the telltale orange-red glow. (Inset: An alligator's eye in daylight.)

ties exist among many of the components, particularly the receptor cells. For example, all photoreceptors contain pigments capable of absorbing light and opening ion channels in the membrane. On the surface, the sound-receptor organs of a grasshopper (Figure 24-1a) and a human could not be more different: The organ is constructed out of a portion of the exoskeleton on the grasshopper's leg, while the human organ is embedded deep within the skull. Yet, both sound detectors contain a mechanoreceptor cell that depends on vibrations of a part of the body surface. Grasshoppers sense vibrations of their thin, outer exoskeleton, while humans detect vibrations of their thin eardrum.

▐▶ Consider for a moment the different types of challenges facing animals that live in different environments. An earthworm or mole that lives underground is exposed to an entirely different set of environmental stimuli than is a fish or shrimp that lives in the sea. Similarly, an aquatic animal faces entirely different conditions than does a lizard or a bird living in a terrestrial habitat. Since sense organs monitor changes in an animal's environment, they must be able to respond to the types of stimuli to which the animal is exposed. In fact, sense organs play a key role in adapting an animal to a particular environment. For example, earthworms have chemoreceptors that measure the acidity of the soil; fish have mechanoreceptors that detect vibrations in the water; and the eyes of a lizard are adapted to focusing light rays from the air (rather than from water). The role of the environment in shaping the properties of sense organs is discussed in the accompanying Bioline: Sensory Adaptations to an Aquatic Environment.

Similarly, each animal has sense organs that are adapted to the animal's particular mode of existence in that environment. For example, an owl, which feeds in the dark, has such a keen sense of *hearing* that it can attack and kill a mouse while blindfolded. In contrast, a hawk, which feeds during the day, has such a keen sense of vision that it can spot a mouse while flying hundreds of feet in the air. Echolocation in bats is one of the best examples of sensory adaptation. Bats live in a world of darkness, roosting in pitch-black caves and picking minute insects out of the air while flying at night. Eyes are of minimal use to bats and play a minimal role in their lives. In contrast, a bat's ears are unaffected by darkness, making echolocation an ideal primary sense.

The type of sense organs an animal possesses can be correlated with the animal's habitat and mode of existence. Despite the diverse structure of sense organs, the receptor cells often share common features that may reflect common ancestry and/or common selective pressures. (See CTQ# 8.)

REEXAMINING THE THEMES

Relationship between Form and Function

↻ Sensory organs are devices whose structure allows them to detect a particular type of stimulus. The human ear, for example, contains a large flap for collecting sound waves, an external channel for focusing sound waves on the eardrum, and an elaborately constructed mechanical apparatus for amplifying remarkably small vibrations. A closer look at the actual receptor cells within a sense organ reveals a similar relationship between structure and function: Receptors in the ear contain clusters of "hairs" that are sensitive to bending; receptors in the eye contain membrane pigments whose structure allows them to absorb particular wavelengths of light; and receptors in the nose and mouth contain proteins whose structure allows them to bind specific molecules in the air or food.

Biological Order, Regulation, and Homeostasis

↻ Maintaining the order and constancy of the internal environment requires sense organs that monitor conditions both inside and outside an animal's body. If the environment becomes too warm, too salty, or too acidic, sensors scattered around the body inform the central nervous system of these changing conditions, providing the information necessary for the body to initiate an appropriate homeostatic response.

Unity within Diversity

◉ Regardless of the type of animal, photoreceptors contain light-sensitive pigments; chemoreceptors contain proteins that combine with specific chemicals; mechanoreceptors contain elements that become deformed when pressure is applied; and so forth. Yet, despite this unity, the organs that house the actual receptor cells in different animals are often very different in structure. For example, the hearing organ of a grasshopper, the eye of a fly, or the olfactory organs of a moth, all have very different structures than do their counterparts in the human body.

Evolution and Adaptation

▐▶ Each sense organ has evolved in response to the selective pressures at work in the particular environment in which an animal lives. For example, humans and most other mammals possess sense organs that are adapted to a terrestrial environment. Our eyes and ears are adapted to collecting sound waves from the air, which is why we have such a terrible time seeing and hearing underwater. Similarly, humans evolved as a species that was active during the day and slept at night. Natural selection favored individuals whose eyes were adapted for conditions of relatively bright light, which is why we see sharp, color images during daylight hours and only dim shadows on a dark, moonless night. In contrast, the eyes and ears of mammals that live underwater, or that have their peak activity at night, function optimally under aquatic or nocturnal conditions, respectively.

SYNOPSIS

Sense organs collect information about conditions in the external and internal environment and pass the information on to the central nervous system. Each type of sensory receptor is specialized to respond to a particular type of stimulus. The interaction between a stimulus and a receptor induces a change in the receptor, which leads to a voltage change across the receptor cell plasma membrane, which may initiate an action potential in a sensory neuron leading to the CNS.

Humans rely on several different types of sensory stimuli. Somatic sensation monitors conditions on the surface and within the body, providing information re-

quired for movement and homeostasis. Vision is a sense that is based on light rays reflected into our eyes from objects in the external environment. Light rays pass through an outer, transparent cornea and are focused by a glasslike lens onto a living "screen" that contains photoreceptors. The brain interprets impulses from these receptors as a visual image of the environment. The retina contains two types of receptor cells: rods, which help provide an image under conditions of low light levels, and cones, which provide a highly detailed, multicolored image when environmental light levels are high. The inner ear contains a pair of sense organs—the cochlea and the vestibular apparatus—which provide us with the sense of hearing and balance, respectively. Both senses derive from hair cells whose membrane voltage changes when sensory hairs projecting from the receptor cell are displaced. Displacement of hairs in the cochlea occurs when sound vibrations are transmitted from the environment into the basilar membrane of the cochlea, while displacement of hairs in the vestibular apparatus results from fluid movements that occur when the head is moved. Taste and smell are senses that depend on the stimulation of chemoreceptors located in taste buds on the tongue and in the inner lining of the nasal cavity, respectively. In both cases, the chemoreceptor cells contain membrane protein receptors that combine with specific chemicals—either as dissolved food matter (taste) or as airborne particles (smell).

The strength, location, and nature of the stimulus are interpreted by the brain. Stimulus intensity may be coded in the number of neurons stimulated and the frequency of impulses propagated. The cerebral cortex contains different regions devoted to somatic sensation, sight, hearing, and smell. The location and nature of the stimulus are determined by the particular sites within these areas that ultimately receive the impulses. In humans, the two cerebral hemispheres function relatively independently of each other and communicate across the corpus callosum.

Key Terms

mechanoreceptors (p. 500)
thermoreceptors (p. 500)
chemoreceptors (p. 500)
photoreceptors (p. 500)
pain receptors (p. 500)
somatic sensation (p. 500)
somatic sensory cortex (p. 502)
sclera (p. 502)

cornea (p. 503)
choroid (p. 503)
iris (p. 503)
pupil (p. 503)
retina (p. 503)
lens (p. 503)
hair cell (p. 506)
tympanic membrane (p. 508)

cochlea (p. 508)
basilar membrane (p. 508)
tectorial membrane (p. 508)
vestibular apparatus (p. 508)
semicircular canal (p. 508)
taste (p. 509)
smell (p. 509)
olfaction (p. 509)

Review Questions

1. Trace the events that occur in the eye and the central nervous system that allow you to see the outlines of people seated in a dark theater.

2. How do you know where a stimulus is coming from in the body? Or the strength of the stimulus? Or its nature?

3. Why wouldn't a person with a severed corpus callosum notice any loss of abilities in his or her daily activities?

4. Name the sensory structure(s) in humans that contain(s): hair cells, photoreceptors, mechanoreceptors, thermoreceptors, chemoreceptors, olfactory receptors, semicircular canals, tympanic membrane, and somatic receptors.

Critical Thinking Questions

1. What control might the graduate students working on blind or blindfolded human subjects have performed to be certain that sound reception was the basis for avoidance of the screen?

2. The Chinese describe the senses as "the gateway to the mind." Is this an accurate description? Explain your answer.

3. Diagnose the area of injury in each of the following cases: (1) a stroke patient is paralyzed on the right side and cannot speak or write; (2) following an accident, a truck driver can no longer perceive distance; (3) after surgery, a patient can pick out specific shapes but cannot name them correctly.

4. How, and why, would each of the following affect a person's vision? (1) cataracts in both eyes; (2) injury to the ciliary muscles; (3) vitamin A deficiency; (4) loss of elasticity of the lens; (5) damage to the visual cortex of the brain; (6) blindness in one eye.

5. Otosclerosis is a disease that limits the movement of the bones in the middle ear. Why would this condition affect hearing? Can you devise some surgical procedure that might correct the problem?

6. What is the justification for lumping the senses of taste and smell together as "chemoreception?"

7. How can you explain the facts that: (1) images are projected onto the retina upside-down, but we "see" objects right side up; (2) people who become able to see after being blind from birth do not immediately have a three-dimensional view of the world; (3) we can be fooled by optical illusions?

8. Which sense(s) would you expect would be highly developed in each of the following, and why?: a hawk, a monkey, an owl, a bat, a mole, a salmon.

Additional Readings

Barinaga, M. 1991. How the nose knows: Olfactory receptor cloned. *Science* 252:209–210. (Intermediate–Advanced)

Borg, E., and S. A. Counter. 1989. The middle-ear muscles. *Sci. Amer.* Aug:74–81 (Intermediate)

Freeman, W. J. 1991. The physiology of perception. *Sci. Amer.* Feb:116–125. (Intermediate)

Glickstein, M. 1988. The discovery of the visual cortex. *Sci. Amer.* Sept:118–127. (Introductory)

Griffin, D. R. 1974. *Listening in the dark.* New York: Dover. (Introductory)

Hudspeth, A. 1983. The hair cells of the inner ear. *Sci. Amer.* Jan:54–66. (Intermediate)

Kandel, E., and J. Schwartz. 1985. *Principles of neural science.* New York: Elsevier. (Intermediate–Advanced)

Kellogg, W. N. 1961. *Porpoises and sonar.* Chicago: University of Chicago Press. (Introductory)

Konishi, M. 1993. Listening with two ears. *Sci. Amer.* April:66–73. (Intermediate)

Parker, D. 1980. The vestibular apparatus. *Sci. Amer.* Nov:118–130. (Intermediate)

Shreeve, J. et al. 1993. The mystery of sense. *Discover* June:35–85. (Intermediate)

Stryer, L. 1987. The molecules of visual excitation. *Sci. Amer.* July:42–50. (Intermediate)

CHAPTER

25

Coordinating the Organism: The Role of the Endocrine System

STEPS TO DISCOVERY
The Discovery of Insulin

THE ROLE OF THE ENDOCRINE SYSTEM

THE NATURE OF THE ENDOCRINE SYSTEM

The Discovery of Hormones

Types of Hormones

A CLOSER LOOK AT THE ENDOCRINE SYSTEM

The Hypothalmic-Pituitary Control Pathway

The Adrenal Glands

The Thyroid Gland

The Parathyroid Glands

The Pancreas

Gonads: The Testes and Ovaries

Prostaglandins

ACTION AT THE TARGET

Cell Surface Receptors

Cytoplasmic Receptors

EVOLUTION OF THE ENDOCRINE SYSTEM

BIOLINE
Chemically Enhanced Athletes

THE HUMAN PERSPECTIVE
The Mysterious Pineal Gland

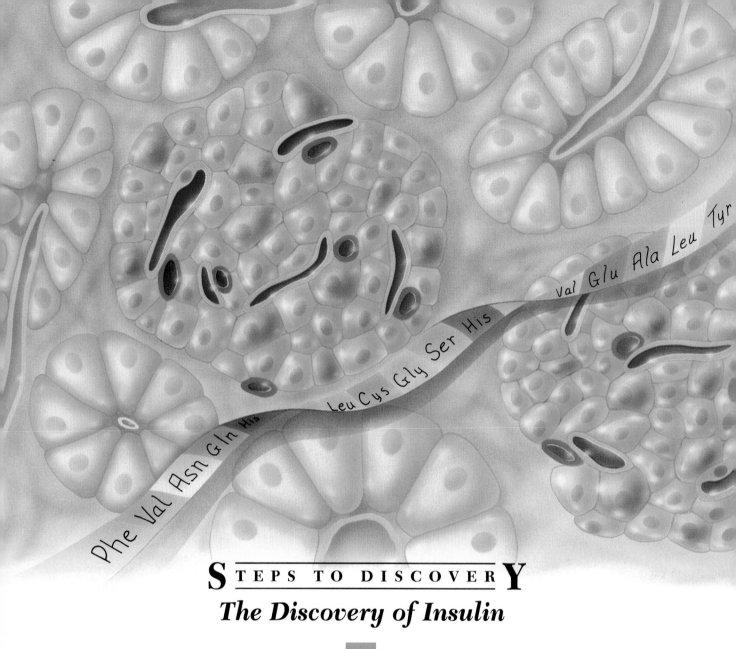

Val Glu Ala Leu Tyr

Leu Cys Gly Ser His

Phe Val Asn Gln His

STEPS TO DISCOVERY

The Discovery of Insulin

*D*iabetes remains a common and potentially fatal disease, one that can destroy the kidneys, damage the heart, disrupt circulation (leading to the amputation of toes and feet), and cause blindness. Yet, diabetes is treatable; today, most diabetics lead normal lives. Until the 1920s, however, diabetes claimed its victims rapidly and horribly. Diabetics excreted enormous quantities of urine; possessed an unquenchable thirst; and generally "wasted away" due to their inability to utilize the nutrients they consumed. Today, daily injections of a simple protein, insulin, help control diabetes.

The story begins in 1869, when Paul Langerhans, a German doctoral student, discovered that the pancreas had two very different types of secretory cells. One group, the *acinar cells*, produced digestive enzymes that were shipped through ducts from the pancreas to the small intestine. The other group of cells was clustered into islands (later named

the *islets of Langerhans*), which possessed neither ducts nor any detectable avenues for the export of the substance these glands secreted; their function was unknown.

Nearly 20 years later, two German physiologists, Oscar Minkowski and Joseph von Mering, attempted to determine the functions of the pancreatic digestive enzymes by removing the pancreas from a dog. Following removal of the glands, the dog, who had previously been housetrained, began urinating frequently on the floor. Minkowski and von Mering were aware that frequent and large-volume urination was one of the key symptoms of diabetes, as was a high level of sugar in the urine. They analyzed the dog's urine and found that it had an elevated sugar content. Removal of the pancreas had made the dog diabetic. The scientists traced the induced effect to the absence of a product from the islets of Langerhans. Yet, for the next 30

A stained section of the pancreas reveals islands of endocrine cells (reddish), which include cells that produce insulin. The

years, all attempts to isolate the product of these cells proved unsuccessful. This would soon change when, in 1920, a young doctor lay in bed struggling with insomnia. Unable to sleep, he devised a plan.

The doctor was Frederick Banting, a 29-year-old Canadian surgeon. As part of his preparation for a lecture he was to deliver to a medical class at the University of Western Ontario, Banting had just read a paper describing an autopsy on a person who had died from an obstruction of the major duct leading from the pancreas. The acinar cells of the pancreas had degenerated, leaving the islets of Langerhans unharmed. As Banting tossed around in bed, the contents of the article circulated through his mind. On a pad of paper lying next to his bed he wrote: "Diabetes. Ligate [tie off] pancreatic ducts of dogs. Keep dogs alive till acini degenerate leaving islets. Try to isolate the internal secretion of these to relieve glycosuria [sugar in the urine]." Banting had concluded (as had others) that the inability to isolate the product of the islets of Langerhans was due to the destruction of the substance by the digestive enzymes of the acinar cells. Banting intended to tie off the pancreatic duct, allowing the acinar cells to degenerate. He hoped that this would allow the antidiabetes factor of the Langerhans cells to be extracted in an undigested state.

Many research projects begin with a scientist reading a paper that arouses his or her curiosity or sparks an idea for a new experimental approach to a nagging question. But Banting was not a research scientist, and he had no lab. Instead, he traveled to the nearby University of Toronto to discuss his plan with John Macleod, an authority on carbohydrate metabolism and diabetes. Banting asked to work in Macleod's laboratory, but Macleod resisted initially; he didn't think the experiment would work, nor did he have confidence in Banting's abilities. After all, even experienced researchers had failed to isolate the secretion of the islets. Banting was persistent, however, and he finally convinced Macleod to give him a bit of lab space and a few dogs to work with. Macleod also asked his physiology class for a volunteer to help out in the summer research project. A 21-year-old medical student named Charles Best came forward. Macleod soon left for an extended holiday in his homeland of Scotland, leaving Banting and Best to carry out what he thought would be a fruitless project.

Together, the two inexperienced researchers worked out a technique to ligate the duct of the pancreas; they then waited several weeks to allow the dog's gland to degenerate. Meanwhile, the researchers prepared several test subjects —dogs whose pancreas had been removed, causing them to become diabetic. Toward the end of July, Banting and Best prepared an extract from the ligated, shrivelled pancreas and injected the extract into a diabetic dog. The urine and blood of the dog showed a dramatic decrease in sugar content. After repeating the procedure several times over the next few weeks, the researchers noted a clear pattern: Injection of the pancreatic extract temporarily relieved the symptoms of diabetes.

By the time Macleod returned in September, Banting and Best were convinced that they were on the verge of a treatment for diabetes. Macleod remained skeptical, however, insisting that the researchers conduct additional experiments. Slowly, the findings were confirmed, and Macleod turned his entire lab over to the study of the pancreatic secretion. He invited James Collip, a visiting biochemist, onto the project to help develop improved procedures for extracting the antidiabetic substance, which the group had named *insulin*. By the end of 1921, the group's work had not yet been published, but the newspapers got wind of the story, and insulin became front-page news around the world.

In January 1922, a 12-year-old boy lay dying of diabetes in a Toronto hospital bed. The Toronto group extracted insulin from beef pancreas, and the boy's physician injected the insulin into the patient, producing astounding results. After receiving daily injections of the extract, the boy left the hospital and resumed a normal life. In the words of M. Bliss, Banting's biographer, "Those who watched the first starved, sometimes comatose, diabetics receive insulin and return to life saw one of the genuine miracles of modern medicine. They were present at the closest resurrection of the body that our secular society can achieve. . . ." Within the year, insulin was being prepared commercially by Eli Lilly, a U.S. pharmaceutical company, and thousands of diabetics were soon receiving the life-saving medication.

In 1923, the Nobel Prize in Medicine and Physiology was awarded to Banting and Macleod, a choice that probably caused more controversy than any other presented. Banting immediately announced that he would share his prize money with Best, whom he felt should have been the corecipient, rather than Macleod. Macleod, not to be outdone, announced that he would share his prize money with Collip. Macleod and Banting remained bitter enemies for the rest of their lives.

ribbon illustrates part of the amino acid sequence of an insulin molecule.

*T*hings could have gone terribly wrong. For days, the child received no calcium in her diet. The absence of this essential mineral in the blood could have triggered a number of life-threatening breakdowns: Neurons could have failed to release their neurotransmitters; muscles might have locked up in painful spasms; blood could have failed to clot, resulting in fatal hemorrhages. But these disasters never materialized; they were prevented by four small glands buried in the child's neck. The pea-sized glands sensed the drop in blood calcium and sounded a molecular alarm, releasing tiny amounts of a chemical—a *hormone*—that stirred the body into corrective action. Some calcium was withdrawn from the bones, the body's calcium bank, and was poured into the bloodstream. At the same time, the hormone slowed excretion of the mineral, while enhancing its absorption from the intestine. The deficit was managed without misfortune.

Yet, the child was not out of danger. There was still the risk that a backlash from these calcium-boosting measures might elevate the concentration of calcium to dangerously high levels in the blood. Fortunately, before that could happen, another gland began releasing its hormone, one that prevents calcium excesses by *reversing* the effects of the first hormone. The constant increase and decrease in the output of these two glands maintains a stable concentration of calcium in the blood, even during temporary deficiencies. This is the essence of homeostasis; without such a system, wild vacillations in calcium concentrations would cause a breakdown in biological order and quickly eradicate any chance of life.

THE ROLE OF THE ENDOCRINE SYSTEM

Like the nervous system (Chapter 23), the **endocrine system** is a communications network, one that uses chemical messengers, called hormones, rather than electrical impulses, to carry information. The hormone molecules are secreted by "ductless" glands into the tissue fluids surrounding the gland. From there, the hormones diffuse into the bloodstream and are carried to distant parts of the body, where they interact with specific *target cells*—cells that have receptors for that particular hormone. When hormone molecules bind to the receptors of a target cell, the hormone evokes a dramatic change in the target cell's activity. For example, your gonads secrete steroid hormones—testosterone, if you are a male, and estrogen, if you are a female. These hormones circulate in the blood, but only target cells, such as those that make up your reproductive tract, bind the hormone and are affected by it. If these hormone molecules should stop being produced, as occurs in women during menopause, the reproductive tract loses its ability to carry out reproductive activities.

↻ Like the nervous system, the endocrine system is also a command-and-control center. Animals face a constant threat of disruptions. Fluctuations in hundreds of critical chemicals occur every minute and must be corrected quickly before they disrupt homeostasis and interfere with

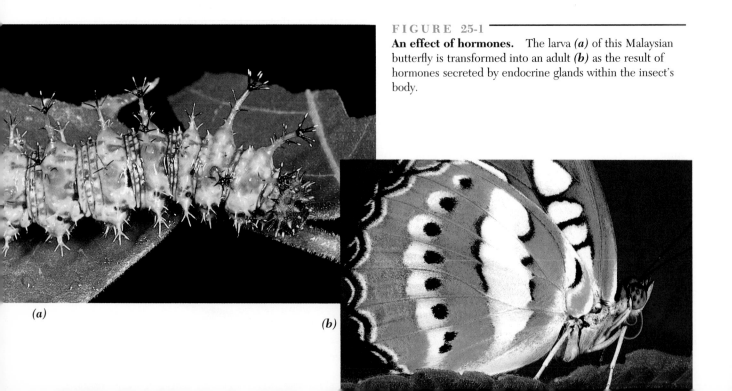

FIGURE 25-1

An effect of hormones. The larva *(a)* of this Malaysian butterfly is transformed into an adult *(b)* as the result of hormones secreted by endocrine glands within the insect's body.

(a)

(b)

vital processes. The endocrine system is responsible for detecting many such variations and releasing hormones whose actions help maintain a stable internal environment. Unlike the nervous system, which evokes rapid responses that typically require muscle contraction, the endocrine organs stimulate slower responses that require changes in metabolic activities of target cells.

Not all endocrine functions maintain homeostasis. Some hormones evoke changes that move the body in a new direction, away from a stable, homeostatic condition. For example, hormones "tell" an animal when to become reproductively mature, when to produce sperm, when to release eggs, and, in mammals, when to expel a fully developed fetus and produce milk. In some cases, hormones have the power to transform an animal into a different form that bears little resemblance to its former appearance (Figure 25-1).

The endocrine system regulates bodily functions through the action of hormones—chemical messengers that circulate in the blood and act on specific target cells, changing their metabolic activity. (See CTQ #2.)

THE NATURE OF THE ENDOCRINE SYSTEM

The vertebrate endocrine system consists of a collection of disconnected glandular tissues positioned around the body (Figure 25-2). Some of these tissues exist as physically separate glands, including the thyroid and adrenal glands. Others are components of a larger organ that has multiple functions, such as the pancreas or gonad. Regardless of their location, all endocrine cells secrete hormones that travel to their targets via the bloodstream.

THE DISCOVERY OF HORMONES

The existence of hormones was first glimpsed in 1902 by two English physiologists, W. M. Bayliss and E. H. Starling, in their study on the control of pancreatic secretions. In addition to secreting insulin and other hormones, the pancreas produces digestive juices that pass through ducts into the intestine. These juices only flow when food materials enter the intestine, which suggested to biologists at the time that nerves informed the pancreas of the arrival of food in

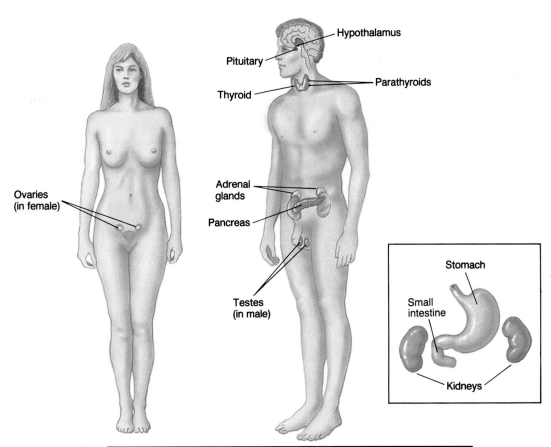

FIGURE 25-2

The major endocrine glands, plus some endocrine cells that are dispersed in nonendocrine organs (inset). Endocrine glands are defined as ductless glands; their secretions diffuse into the bloodstream, where they are carried to a distant target cell. (The endocrine activities of the digestive tract, kidneys, and gonads are discussed in Chapters 27, 28, and 31, respectively.)

the intestine. Bayliss and Starling found otherwise. In one of their numerous experiments, they joined together the bloodstreams of two dogs and observed that the pancreas of both animals would secrete their juices when only one animal was being fed. The researchers concluded that a *chemical* message was being sent from the intestine to the pancreas via the circulatory system. They called this blood-borne chemical messenger *secretin.* Bayliss and Starling further suggested that secretin was probably just one of a number of hormones that acted as chemical messengers within the body.

TYPES OF HORMONES

In 1970, fewer than 30 hormones had been identified. Today, some 200 possible hormones produced by various cells scattered throughout the body are being investigated. Some of the best-studied human hormones are listed in Table 25-1. Most of the known hormones fall into two broad categories: (1) amino acids, derivatives of amino acids, peptides, or proteins; and (2) steroids, all of which are constructed around the same type of multiringed, molecular skeleton (see Figure 4-11).

⚠ Similar types of molecules serve as hormones in most animals, although they may evoke very different types of responses. For example, estrogen causes a thickening of the wall of the human uterus, whereas ecdysone, a molecule of similar structure, causes a lobster to shed its outer skeletal covering. Not only do hormones have a similar structure in diverse animals, their mechanism of action at the cellular and molecular level is virtually identical in all animals, as we will discuss later in the chapter.

TABLE 25-1

THE MAJOR ENDOCRINE GLANDS AND THEIR HORMONES

Gland	Hormone	Regulates
Anterior pituitary	Growth hormone (GH)	Growth; metabolism
	Thyroid-stimulating hormone (TSH)	Thyroid gland secretions
	Adrenocorticotropic hormone (ACTH)	Adrenal cortex secretions
	Prolactin	Milk production
	Gonadotropic hormones: Follicle-stimulating hormone (FSH); Luteinizing hormone (LH)	Production of gametes and sex hormones by gonads
Posterior pituitary	Oxytocin	Milk secretion; uterine motility
	Antidiuretic hormone (ADH) or vasopressin	Water excretion; blood pressure
Adrenal cortex	Cortisol	Metabolism
	Aldosterone	Sodium and potassium excretion
Adrenal medulla	Epinephrine and norepinephrine	Organic metabolism; cardiovascular function; stress response
Thyroid gland	Thyroxine (T-4) Triiodothyronine (T-3)	Energy metabolism; growth
	Calcitonin	Calcium in blood
Parathyroids	Parathyroid hormone (PTH)	Calcium and phosphate in blood
Ovaries	Estrogen and progesterone	Reproductive system; growth and development; female secondary sex characteristics
Testes	Testosterone	Reproductive system; growth and development; male secondary sex characteristics; sex drive
Pancreas	Insulin and glucagon	Metabolism; blood-glucose concentration
Kidneys (see Chapter 28)	Renin	Adrenal cortex; blood pressure
	Erythropoietin	Erythrocyte production
Gastrointestinal tract (see Chapter 27)	Gastrin	Secretory activity of stomach-small intestine; liver; pancreas; gall bladder
	Secretin	
	Cholecystokinin	Release of digestive enzymes
Pineal gland	Melatonin	Biological cycles; sexual maturation

"Reins and Spurs": Regulating Hormone Concentration

⊘ Hormones exert their effects in extraordinarily low concentrations, and their targets are acutely sensitive to even slight endocrine variations. In many cases, the precise concentrations of a hormone are regulated by:

- *Negative feedback of the hormone itself.* In many cases, the concentration of a hormone directly influences its own production. Some endocrine cells produce hormones only when the levels of those hormones in the blood drop below the homeostatic level (Figure 25-3). The drop in concentration of a particular hormone directly or indirectly spurs the gland to secrete more; consequently, the hormone concentration in the blood rises. Elevated concentrations have the opposite effect: They pull the reins in on hormone secretion before an excess amount can accumulate.

- *Negative feedback based on the hormone's effect.* Some glands are sensitive to the changes their hormones elicit. For example, the calcium-sensitive glands we discussed earlier in the chapter stabilize concentrations of calcium in the blood. In this form of negative feedback, the gland responds to the eventual *effect* (in this case, the concentration of blood calcium) rather than to concentrations of the hormone itself.

- *Inactivation or removal.* Hormones do not persist indefinitely. Most are removed by the kidneys and excreted; others are enzymatically inactivated or transformed into other compounds. If it is not replaced by continued secretion, the hormone slowly diminishes in concentration.

In many cases, endocrine cell activity is regulated by concentrations of blood-borne substances that stimulate or inhibit the production and secretion of hormones. (See CTQ #3.)

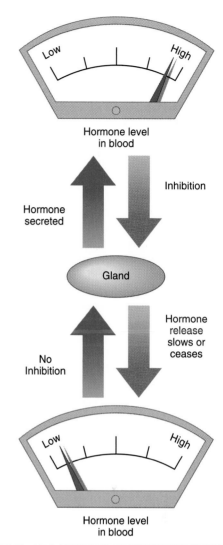

FIGURE 25-3

Negative feedback provides one mechanism for regulating hormone concentration in the blood. Higher concentrations of a hormone decrease the production and release of that hormone, reducing levels to the homeostatic value. As the concentration falls, production and release of the hormone increases, raising levels to the homeostatic value.

A CLOSER LOOK AT THE ENDOCRINE SYSTEM

Although endocrine cells have recently been found almost everywhere in the human body, a few organs stand out as major sources of hormones. In this section, we will examine each of these major endocrine glands (Table 25-1). We will begin with those parts of the endocrine system that have a close relationship with the nervous system. The nervous and endocrine systems work hand in hand as an integrated team, supervising bodily functions. In their role as regulatory systems, the nervous and endocrine systems collaborate on some activities and work independently on others.

 Throughout the animal kingdom, from flatworms to mammals, biologists have discovered the existence of cells that look very much like neurons but act like endocrine cells. The form and function of these so-called **neurosecretory cells** make them ideally suited to serve as a "bridge" between the two major regulatory systems of the body. Neurosecretory cells look and behave like neurons in many ways (see Figure 25-5). Like neurons, neurosecretory cells receive information from other nerve cells and have elongated axons that conduct neural impulses. When the impulses reach the tip of the axon, they stimulate the release of stored materials from vesicles. Unlike neurons, however, the substance released from the tip of a neurosecretory cell is a hormone, which diffuses into the bloodstream, rather than a transmitter, which diffuses across a synaptic cleft. For example, oxytocin, the hormone that induces labor during childbirth, is a product of neurosecretory cells in the brain.

THE HYPOTHALMIC-PITUITARY CONTROL PATHWAY

⟳ The endocrine system has no true central coordinating system comparable to the brain of the nervous system. The endocrine system's pituitary gland has been called the "master gland," however, because it controls the activities of so many other endocrine elements. The master gland has a master of its own—the hypothalamus—which, you will recall from Chapter 23, is part of the brain (see Figure 25-2). The hypothalamus and the pituitary gland are functionally related, as illustrated in Figure 25-4. Information about the state of the body arrives at the hypothalamus from two sources: (1) blood flowing through the hypothalamus provides information on the concentration of various chemicals in the bloodstream; and (2) sensory information arriving by neurons provides information on the conditions of various parts of the body. This information is used to control pituitary secretions.

The pituitary is actually two glands in one: the **posterior pituitary** and the **anterior pituitary.** These two parts of the pituitary gland are controlled by the hypothalamus in very different ways, but both mechanisms utilize neurosecretory cells (Figure 25-5).

1. The posterior pituitary manufactures no hormones of its own but stores and secretes two hormones that are produced by neurosecretory cells that originate in the hypothalamus (Figure 25-5a).

2. The anterior pituitary is a true endocrine gland, manufacturing at least six separate hormones. These pituitary hormones are secreted in response to *releasing factors* that are produced by neurosecretory cells of the hypothalamus (Figure 25-5b). These releasing factors are produced in extremely small amounts; their purification and characterization were monumental tasks, requiring the extraction of hundreds of tons of sheep brains. The work was conducted by Roger Guillemin of Baylor University and Andrew Schally of Tulane University and earned the scientists the 1977 Nobel Prize.

   When the hypothalamus senses a need for one of the anterior pituitary hormones, an impulse passes down the axon of the neurosecretory cell, and the appropriate releasing factor is discharged into tiny blood vessels. The blood then carries the releasing factor to the nearby anterior pituitary, where it triggers the secretion of pituitary hormones. The hypothalamus also produces *inhibiting factors* that prevent excessive secretion of specific hormones by the anterior pituitary.

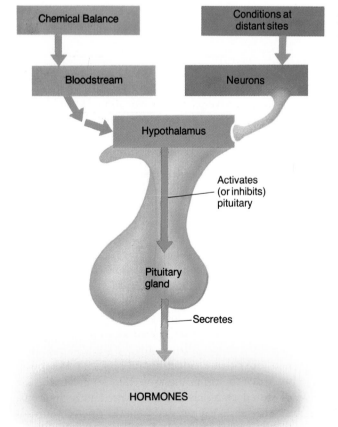

FIGURE 25-4

The major neuroendocrine connection. Input from both the bloodstream and the nervous system influences the hypothalamus in its regulation of the secretory activity of the pituitary gland. The anatomic nature of the hypothalmic-pituitary relationship is illustrated in Figure 25-5.

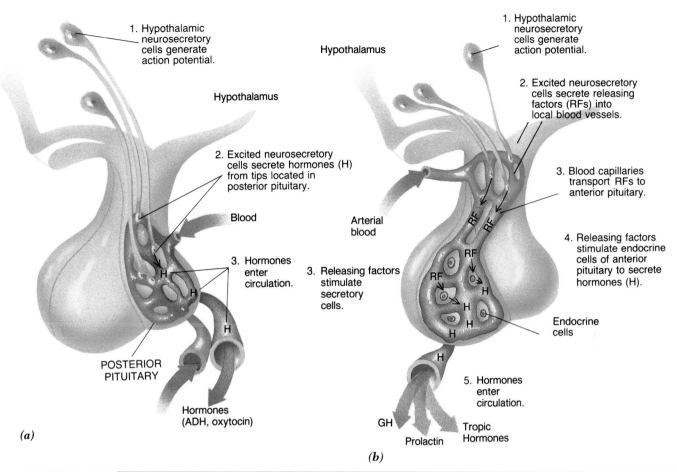

FIGURE 25-5
The pituitary-hypothalmic linkage. *(a)* The posterior pituitary is a storage depot for two hormones —ADH and oxytocin—which are produced in the cell bodies of hypothalmic neurons. Hormones are enclosed within vesicles in the cell body and are transported down the length of the axon to the posterior pituitary; from there, the hormones are released into the bloodstream. Once in circulation, these hormones are carried to distant sites, where they act on target cells. *(b)* Secretion of the various hormones of the anterior pituitary is controlled by releasing and inhibiting factors, which are secreted from the hypothalamus by neurosecretory cells into local blood capillaries that carry the factors to the anterior pituitary. Hormones produced and secreted by the anterior pituitary circulate through the bloodstream and act on target cells around the body.

Hormones of the Posterior Pituitary

The hormones released by the posterior pituitary (Figure 25-5*a*) are both peptides. The first, **antidiuretic hormone (ADH)**, acts on the kidneys to reduce the loss of water during urine formation (Chapter 28). In females, the second hormone, **oxytocin,** triggers uterine contractions during childbirth. Oxytocin also stimulates the release of milk when the breast is stimulated by nursing.

Hormones of the Anterior Pituitary

Of the six known hormones produced by the anterior pituitary (Figure 25-6), four are **tropic hormones;** that is, they regulate the activity of other endocrine glands. The remaining two are direct-acting hormones. One of these, **prolactin,** promotes the production of milk by mammary glands in the breast. The other, **growth hormone (GH),** encourages body growth by stimulating bone elongation and accelerat-

ing fat breakdown and protein synthesis in a wide variety of target cells.

Growth hormone (also called *somatotropin*) is produced predominantly during childhood and adolescence and plays a key role in promoting normal body growth. When the anterior pituitary ceases its GH output, overall growth stops. This typically occurs sooner in girls than in boys, accounting for the shorter average height of women, compared to men. Interestingly, most growth takes place while sleeping, when bursts of GH secretions usually occur. Extremes in height (giantism or certain kinds of dwarfism) are due to overproduction or underproduction of GH during childhood (Figure 25-7). Children who fail to grow at normal rates due to a deficiency of GH secretion can now be treated with growth hormone that is produced by recombinant DNA technology (Chapter 16). Excessive secretion of GH *in an adult*, which is typically due to a pituitary tumor,

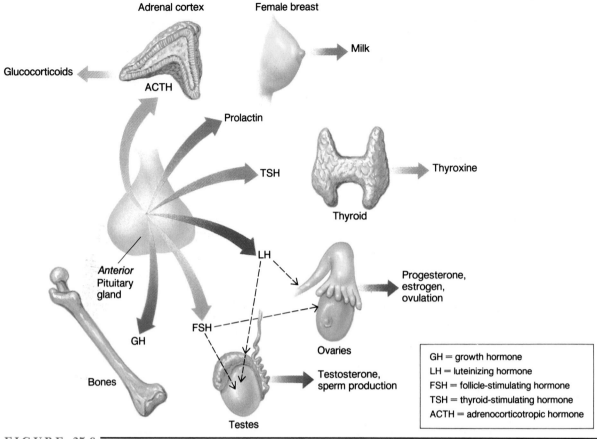

FIGURE 25-6

Hormones of the anterior pituitary. Four of the six hormones (ACTH, TSH, LH, and FSH) are tropic hormones, which stimulate other endocrine glands to produce hormones. Of the two direct acting hormones, prolactin stimulates milk production, while GH stimulates protein synthesis and the growth of bones.

leads to a condition known as *acromegaly*, in which certain parts of the body, including the hands, feet, jaw, and nose, become enlarged (Figure 25-8). Because it also produces increased muscle mass, GH has been used by some body-builders and athletes (see Bioline: Chemically Enhanced Athletes).

The four tropic hormones of the anterior pituitary stimulate endocrine glands to produce other hormones. **Adrenocorticotropic hormone (ACTH)** stimulates the cortex (outer layer) of the adrenal glands to secrete its hormones (glucocorticoids and mineralocorticoids). **Thyroid-stimulating hormone (TSH)** activates secretion of thyroid hormones by the thyroid gland. The other two tropic hormones act on the gonads and are called **gonadotropins**. Both gonadotropins are identical in males and females. They are **follicle-stimulating hormone (FSH)** and **luteinizing hormone (LH)**. FSH and LH promote gamete development and stimulate gonads to produce sex hormones, as discussed in Chapter 31. In the following discussion, we will see how two of the tropic hormones—ACTH and TSH—influence their target glands.

THE ADRENAL GLANDS

Perched atop each kidney is an adrenal gland (*ad* = upon, *renal* = kidney). Like the pituitary gland, each adrenal gland is essentially composed of two independent glands that produce very different hormones (Figure 25-9).

The Adrenal Cortex

The gland's outer layer—the **adrenal cortex**—secretes a family of steroid hormones. These include:

* *Glucocorticoids*, which promote the conversion of amino acids to glucose and its uptake by the brain. The most important glucocorticoid is *cortisol* (also known

FIGURE 25-7
Circus side shows often included a pituitary dwarf and giant.

(a)   *(b)*   *(c)*   *(d)*

FIGURE 25-8
Acromegaly. Photos were taken at *(a)* age 9 years, *(b)* age 16 years, *(c)* age 33 years, and *(d)* age 52 years. The effects of excess growth hormone production in the adult are seen by age 33 as an enlarged lower jaw and nose and thickened lips, features which become exaggerated with age.

◁ B I O L I N E ▷
Chemically Enhanced Athletes

In 1896, Ellery Clark leaped nearly 21 feet to win the long-jump event in the Olympic Games; in the 1936 Olympic Games, Jesse Owens won the event with a jump of more than 26 feet; today, the world record is more than 29 feet. While modern athletes are superior to any in history, their achievements have been tainted over the past decade by the discovery that too many athletes rely on drugs to maximize their physical potential. The disqualification of gold-medal winner Ben Johnson from the 100-meter dash in the 1988 Olympics, and his subsequent explusion from competition for life in 1993, exemplifies the severity of the problem. The reason for Johnson's explusion: Highly sensitive tests detected the presence of illegal drugs in the athlete's urine.

Topping the list of banned substances taken by some athletes are steroids, commonly referred to as "anabolic" steroids because they tend to increase biosynthesis (anabolism), especially protein synthesis.

The most common of these steroids is the male sex hormone testosterone and its synthetic derivatives. These drugs help build muscle, restore energy, and enhance aggressiveness. Their use has grown so common, even among high-school athletes, that suppliers of black-market drugs distribute printed advertising brochures and order forms at local gyms.

The dangers of steroid abuse are considerable. A steroid surplus in the blood, produced by taking external sources of testosterone, causes the body's own testosterone production to drop due to negative feedback and causes the gonadal cells that produce the male hormone to waste away. Other side effects include lowered sperm counts, penis atrophy, reduced resistance against infectious disease, and potentially serious heart irregularities.

When a man's body is flooded with testosterone, some of the hormone is converted to the female hormone estrogen,

often stimulating female-type breast enlargement. Women taking testosteronelike steroids experience the opposite effects: The male hormones stimulate the development of facial hair and other masculine features. Prolonged exposure to steroids suppresses the immune system in both sexes, compromising the body's ability to defend itself against many infectious diseases. Steroid supplements also increase the likelihood of cancer.

In recent years, growth hormone has become commercially available through recombinant DNA technology and is being taken by some athletes to increase muscle mass. Excess growth hormone is broken down within the body within an hour or so, so that it is virtually undetectable by testing procedures. Like steroids, excess growth hormone carries with it serious health risks, including irreversible overgrowth of certain bones (a characteristic of acromegaly, illustrated in Figure 25-8), arthritis, and enlargement of the heart.

---

as hydrocortisone), whose level is highest during periods of stress, such as that which occurs following a severe physical injury or a period of emotional trauma. Secretion of cortisol helps maintain normal blood glucose levels and fuels brain activity. Cortisol secretion is stimulated by ACTH released from the pituitary. At very high concentrations, cortisol suppresses the body's normal response to injury, including inflammation. Consequently, cortisol and related synthetic compounds are highly effective in treating persistent inflammation, such as that caused by arthritis or bursitis, and severe allergies or asthma. Glucocorticoids may have serious side effects, however, including suppression of the body's ability to fight infections and inducement of hypertension and ulcers, and, thus, must be used with great caution. Extended periods of stress may take their toll on the body as a result of the increased secretion of cortisol.

- *Mineralocorticoids*, which regulate the concentrations of sodium and potassium in the blood. The most important mineralocorticoid is *aldosterone*, which stimulates the kidney to reabsorb sodium into the blood, maintaining homeostatic levels of this important ion. If the level of sodium in the blood rises, it reduces aldosterone secretion, increasing excretion of the ion in the urine.

↻ The adrenal cortex is essential for survival. Without aldosterone, for example, the body quickly loses sodium ions needed for neuron activity, muscular contraction, and blood-pressure stability. Without glucocorticoids, glucose concentrations in the blood plummet, crippling cellular energy metabolism. A person whose adrenal cortex fails to produce sufficient levels of these hormones suffers from *Addison's disease*, which is characterized by extreme weakness, weight loss, and impaired heart and kidney function and is treated by administering synthetic adrenal cortical steroids. John F. Kennedy suffered from Addison's disease prior to the campaign for the presidency in 1960; diagonosis and treatment of the condition transformed a sick, weak-looking candidate into a healthy, vigorous campaigner.

The Adrenal Medulla

The "fight-or-flight" response discussed on page 491 includes activation of the **adrenal medulla,** the core of the adrenal gland (Figure 25-9). When stimulated by sympathetic nerve fibers, the adrenal medulla secretes two hormones: **epinephrine** (also called adrenalin) and **norepinephrine** (or noradrenalin). These hormones jolt the body into readiness to escape or confront an emergency. When confronted with an emergency, a person's metabolic rate increases and heart rate accelerates; additional glucose and oxygen are shunted to voluntary muscles, increasing muscular performance; blood vessels to the skin constrict, helping reduce the loss of blood through injured tissues; and red blood cells pour into the bloodstream from their reserves in the spleen, enhancing oxygen transport and replacing blood cells that are lost during bleeding. Epinephrine and norepinephrine also inhibit activities that are of little help in an emergency, such as contraction of muscles in the digestive tract.

THE THYROID GLAND

In the early twentieth century, many people found themselves afflicted with enormous swellings in the front of their necks. This condition, known as *simple goiter,* is the result of an enlargement of the **thyroid gland,** a butterfly-shaped gland that lies just in front of the windpipe (trachea) in humans (Figure 25-10). The swelling is often accompanied by lethargy, hair loss, slowed heart beat, and mental sluggishness. Goiter initially occurred frequently in geographic regions where natural sources of dietary iodine were scarce. The link between iodine deficiencies and the disease was not established until 1916, at which time it was shown that adding iodine to the diet, such as in iodized table salt, prevented the formation of goiters.

Why should iodine be related to thyroid function? The reason becomes evident when we examine the structure of two hormones produced by the thyroid gland. These hormones—**thyroxin** and **triiodothyronine**—are amino acids that contain iodine. Together, they are called *thyroid hormone.* When dietary iodine is inadequate, the thyroid cannot produce enough thyroid hormone. Since thyroid gland activity is regulated by negative feedback through the pituitary hormone TSH, the low concentration of thyroid hormone in the blood cannot feed back to inhibit TSH

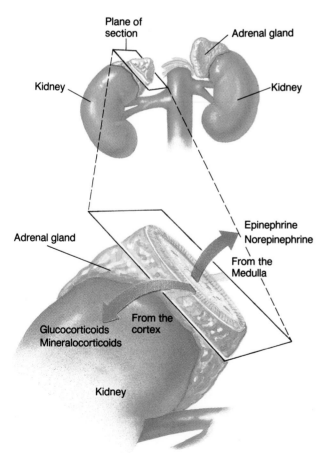

Plane of section

Adrenal gland

Kidney

Kidney

Adrenal gland

Epinephrine
Norepinephrine

From the Medulla

Glucocorticoids
Mineralocorticoids

From the cortex

Kidney

FIGURE 25-9

The adrenal glands are located in the fatty tissue adjacent to the anterior ends of the kidneys. The adrenal cortex produces an array of steroid hormones, including glucocorticoids and mineralocorticoids, and a small amount of sex hormones, while the adrenal medulla produces two related amino acid derivatives: epinephrine and norepinephrine.

FIGURE 25-10
The thyroid and parathyroid glands produce hormones that regulate metabolic rate and blood calcium levels respectively.

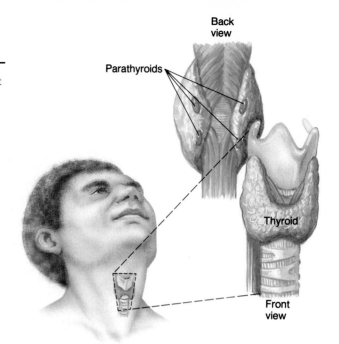

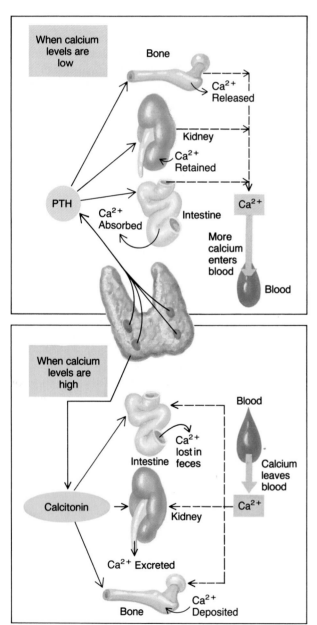

FIGURE 25-11
Balancing calcium. A low calcium (Ca^{2+}) concentration in the blood stimulates PTH secretion by the parathyroid glands (shown in green) and inhibits calcitonin secretion by the thyroid gland (in red). PTH leads to the demineralization of bone; the retention of calcium by the kidney; and the absorption of calcium by the intestine. All of these responses trigger an increase in blood calcium levels. A high concentration of calcium has the opposite effects, decreasing calcium levels in the blood.

production. As a result, the high levels of TSH cause the thyroid to enlarge into a large lump, or goiter, in its perpetual "attempt" to produce enough thyroid hormone.

☀ Normally, thyroid hormone enhances oxidation in the mitochondria of various target cells, which increases energy availability and metabolic rate. Consequently, a person with a lowered output of thyroid hormone—whether due to a decreased availability of iodine or to a sluggish thyroid gland—experiences fatigue and low energy levels. Such thyroid hormone deficits are called *hypothyroidism.* Extreme hypothyroidism during childhood produces *cretinism,* a type of dwarfism that is characterized by severe mental deficiency. Cretinism can be prevented by early treatment with thyroid hormone supplements. The same treatment is effective in correcting the sluggish metabolic rate of adult hypothyroid victims. An excess of thyroid hormone *(hyperthyroidism)* may lead to *Grave's disease,* which is characterized by hyperactivity, weight loss, nervousness, insomnia, irritability and, in extreme cases, *exophthalmia,* in which the eyeballs bulge from the sockets.

After decades of studying thyroid function, a third thyroid hormone, **calcitonin,** was discovered in the early 1960s. Calcitonin regulates blood calcium levels in cooperation with another set of endocrine structures, the parathyroid glands.

THE PARATHYROID GLANDS

You probably think of your bones as stable, inert structures. In reality, the inorganic salts that give bones their hardness are continually being dissolved and redeposited by cells embedded in the bone tissue. This is the major reason why calcium is an essential dietary substance; it is needed to replace the calcium that is removed from bone and lost in the urine. Maintaining the calcium concentrations in the blood requires the cooperation of the thyroid gland and four tiny **parathyroid glands.** The parathyroids were discovered only at the end of the last century, when it was shown that the muscular spasms and convulsions that sometimes followed thyroid removal were due to the loss of the adjoining parathyroid glands, not the thyroid.

🔄 When calcium levels in the blood are low, the parathyroid glands secrete **parathyroid hormone (PTH),** which acts on the bones, kidneys, and intestines to restore normal calcium concentrations (Figure 25-11, *top*). Under PTH influence, calcium is withdrawn from bones, and the released mineral is absorbed into the bloodstream; kidneys retain calcium; vitamin D is activated; and calcium absorption from the intestines is enhanced. As blood calcium levels normalize, negative feedback decreases the level of PTH secretion. In contrast, if blood calcium levels should rise to abnormally high levels, the thyroid gland is stimulated to secrete calcitonin, which exerts the opposite effects of PTH (Figure 25-11, *bottom*).

THE PANCREAS

The pancreas is predominantly an *exocrine* gland; that is, a gland that secretes its products into a duct. The exocrine products of the pancreas are digestive enzymes. The pancreas also contains tiny endocrine centers, called **islets of Langerhans,** which secrete several protein hormones into the blood. One of these hormones—insulin—is secreted when the concentration of glucose in the blood begins to exceed normal levels, usually as sugar floods the bloodstream following a meal (Figure 25-12, *left*). Insulin acts on numerous organs of the body to stimulate the cellular uptake of glucose, which is necessary in initiating the utilization of the sugar. Insulin also directs the conversion of surplus glucose into glycogen for storage in the liver and muscles. This conversion prevents the loss of surplus sugar since excess sugar that remains in the blood is excreted in urine.

Insulin can do too good a job, however, and deplete the blood of glucose. When the concentration of blood sugar begins to drop below normal, the islets of Langerhans alter their secretory priorities and increase the secretion of **glucagon,** another pancreatic hormone (Figure 25-12, *right*). Glucagon promotes glycogen breakdown in cells where it is stored and elevates glucose concentration in the blood, especially during times of stress, when increased cellular and physical activity is likely to expend greater amounts of energy.

As we discussed in the chapter opener, a deficiency in insulin production can lead to *diabetes mellitus;* complications can include cardiovascular damage, kidney failure, blindness, and susceptibility to life-threatening infections. Because they are unable to take up glucose from the blood, the cells of untreated diabetics must turn to other sources of energy, such as protein and fat reserves. Consequently, some diabetics may become emaciated. Diabetics may also be very thirsty or dehydrated because increased blood glucose levels promote frequent urination.

In approximately 15 percent of the cases, diabetes appears during childhood as the result of the destruction of the insulin-producing cells of the pancreas, which apparently results from either a viral infection or an attack by the person's own antibodies (Chapter 30). In these cases, which are classified as *juvenile-onset,* or *Type 1,* diabetes, the person produces little or no insulin and must be treated by daily injection of the hormone. Advanced delivery systems that continually release metered supplies of the hormone are beginning to replace daily injection routines, thereby avoiding undesirable fluctuations in the hormone's concentration associated with periodic insulin shots.

The majority of diabetics are classified as having the less severe, *adult-onset,* or *Type 2,* form of the disease, in which insulin levels may be normal, but the target cells fail to respond to the hormone because of insulin receptor abnormalities. Even though Type 2 diabetics may not ex-

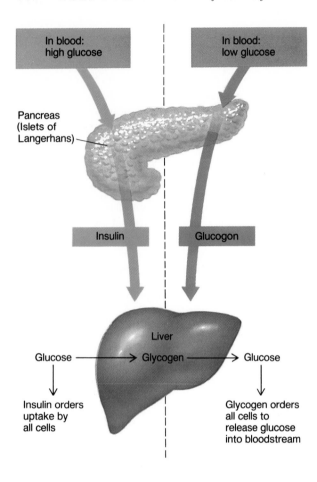

Pancreas (Islets of Langerhans)

Insulin

Glucogon

Liver

Glucose → Glycogen → Glucose

Insulin orders uptake by all cells

Glycogen orders all cells to release glucose into bloodstream

In blood: high glucose

In blood: low glucose

FIGURE 25-12

Control of blood sugar concentration by the islets of Langerhans in the pancreas. When blood glucose levels are high, insulin is secreted, promoting glucose uptake and storage by the liver and muscle cells as glycogen. When blood glucose levels are low, glucagon is secreted, promoting the breakdown of glycogen into glucose and its subsequent release into the bloodstream.

hibit extreme urinary and metabolic derangements, it is still essential that they take steps to maintain normal blood-sugar levels, or life-threatening complications may develop. Type 2 diabetes can often be controlled if the patient adheres to strict dietary recommendations.

GONADS: THE TESTES AND OVARIES

When stimulated by pituitary gonadotropins, the gonads—the male **testes** and the female **ovaries**—secrete the powerful steroid sex hormones **testosterone** and **estrogen,** respectively. These substances promote the development of the reproductive tracts and secondary sex characteristics that distinguish genders, such as deeper voices and facial hair, which is stimulated by testosterone, and breast enlargement, which is triggered by estrogen. The activities and regulation of the sex hormones are discussed in more detail in Chapter 31. The role of another endocrine structure that affects maturation of the reproductive organs is discussed in The Human Perspective: The Mysterious Pineal Gland.

PROSTAGLANDINS

Unlike the hormones described thus far, **prostaglandins** are secreted by endocrine cells scattered throughout the body; their activities seem as diverse as are their sources.

Prostaglandins are modified fatty acids and are quite different in structure from other hormones. Prostaglandin activities affect reproduction, digestion, respiration, neurological function, pain perception, inflammation, and blood clotting. Especially high concentrations of prostaglandins are found in *semen,* the fluid discharged during male orgasm. The name "prostaglandin" is derived from the presumption that these substances were produced in the prostate gland, which is situated at the base of the male urethra. The prostaglandins in semen cause uterine muscles to contract, which helps propel sperm toward the egg. In women, the contraction of uterine muscles responsible for cramping during menstruation has also been traced to prostaglandins. Other prostaglandins promote inflammation and stimulate pain receptors. The pain-relieving and anti-inflammatory effects of aspirin and ibuprofen are probably due to the ability of these drugs to inhibit prostaglandins.

Many of the body's endocrine glands are part of a hierarchical control system. The hypothalamus sits at the top of the hierarchy, collecting information from the nervous system and the bloodstream and sending out commands via neurosecretory cells to the pituitary gland. In turn the pituitary gland releases tropic hormones that control various peripheral endocrine tissues. Other endocrine glands are regulated by blood-borne substances whose concentration is directly affected by the hormone's action. (See CTQ #5.)

◁ THE HUMAN PERSPECTIVE ▷
The Mysterious Pineal Gland

Embedded deep within the brain is a tiny, pinecone-shaped organ, the **pineal gland,** whose function has been a source of speculation for over 2,000 years. As late as the seventeenth century, the philosopher Descartes designated the pineal gland as the "seat of the soul." One of the first important findings regarding the biological function of the pineal gland was made in 1898 by Otto Heubner, a German physician, who reported on a 4-year-old boy who had undergone premature puberty and then died. Autopsy results indicated that the boy had died from a tumor of the pineal gland. This finding, and similar reports from others, led Heubner to propose that the pineal gland normally produces a hormone that suppresses sexual development during childhood. If production of that chemical should cease—if the gland is destroyed by a tumor, for example—the inhibition is removed, and sexual maturation occurs prematurely.

Heubner's hormone was not isolated and identified until 1958. At that time, Aaron Lerner of Yale University purified the substance from approximately 200,000 cattle pineal glands. He named the hormone *melatonin.* Although melatonin is still thought to be a factor involved in inhibiting the onset of sexual maturation in mammals, another role of the hormone has captured the recent attention of researchers. For years, it was known that the pineal gland in amphibians and reptiles was a light-sensitive organ, whose hormonal secretions controlled the darkening of the skin. The pineal gland of many amphibians and reptiles is directly exposed to light, whereas the pineal gland of mammals is covered by a thick layer of cranial bone. Yet, a number of studies have indicated a relationship between light exposure and the activity of a mammal's pineal gland. For example, when rats are exposed to constant illumination, the level of melatonin synthesis rapidly drops as much as 80 percent, compared to animals kept in continuous darkness.

In humans, melatonin concentrations in the blood are highest at night and lowest in the day. Daily variations in the synthesis of melatonin may, in turn, regulate a variety of human daily rhythms, including sleep, motor activity, and brain waves. Studies of people who fly across several time zones indicate that it takes several days for the normal rhythm of melatonin to reestablish itself, suggesting that the familiar problem of "jet lag" may be due to the time needed to reset the pineal "clock." The pineal gland has also been implicated as a cause of seasonal depression, or seasonal affective disorder syndrome (SADS). As winter approaches, and the days become shorter, some people become tired, sad, and depressed, only to rebound when the days become longer in the spring. The onset of the "winter blues" correlates with a seasonal rise in melatonin production and, in some cases, can be treated by exposing depressed subjects to periods of bright light, which decreases the pineal's output of melatonin.

ACTION AT THE TARGET

In order for a cell to respond as a target to a particular hormone, the cell must have a protein receptor whose structure allows it to bind that hormone specifically. Depending on the particular hormone, the receptor may be located at the cell surface or within the cytoplasm. The position of the receptor within the cell is a key factor in determining the mechanism of action of the corresponding hormone.

CELL SURFACE RECEPTORS

Most of the hormones that act via cell surface receptors are water-soluble substances (amino acids, peptides, and proteins) that cannot passively diffuse through the plasma membrane and enter the cytoplasm. (The prostaglandins are an exception to this rule; despite their lipid-soluble structure, prostaglandins also act at the cell surface.) All of these hormones act without entering the target cell.

The hormone itself can be considered a "first messenger." It binds to a receptor on the outer surface of the target cell's plasma membrane (Figure 25-13), promoting a change at the membrane and the release of a **"second messenger,"** which enters the cytoplasm and actually triggers the response. The best-studied and most widespread second messenger is a small molecule called *cyclic AMP.* We will illustrate this type of mechanism using the pancreatic hormone glucagon.

When glucagon binds to a glucagon receptor located at the surface of the plasma membrane, the hormone changes the shape of the receptor, which activates an enzyme located on the inner surface of the membrane (Figure 25-

14a). This enzyme, called *adenylate cyclase*, converts ATP to cyclic AMP (cAMP), one of the most universally important molecules associated with cellular regulation. Cyclic AMP then diffuses into the cytoplasm, where it activates protein kinases—enzymes that attach phosphates to other enzymes (page 129). When glucagon binds to a liver cell, causing the synthesis of cAMP, the activated protein kinase activates an enzyme that splits glycogen into its glucose monomers. As a result, glucagon secretion increases the concentration of glucose in the blood. The involvement of two messengers—a hormone and cAMP—amplifies the original signal. The binding of one glucagon molecule at the cell surface promotes the synthesis of thousands of cAMP molecules inside the cell. As a result of this amplification, a very small concentration of hormone in the blood can produce a rapid, massive response within a target cell.

But how does such a mechanism explain the specificity of hormones? If glucagon, calcitonin, parathyroid hormone, epinephrine, the pituitary-releasing factors, and several other hormones do no more than increase the cytoplas-

mic cAMP concentration in target cells, why don't they all cause the body to change in the same way? There are two parts to the answer to this question. First, different target cells have different hormone receptors on their surface. If the only receptors on a cell happen to be epinephrine receptors, for example, only epinephrine will cause the cAMP-mediated response. Second, cAMP has different effects in different target cells, depending on which enzymes are present in that cell. For example, when liver cell enzymes are activated by cAMP, glucose is released into the bloodstream, but when the proteins found in the cells of the adrenal cortex are activated by cAMP, glucocorticoids are produced and released.

CYTOPLASMIC RECEPTORS

Steroid hormones have a very different mechanism of action (Figure 25-14b). These hydrophobic hormones diffuse through the plasma membrane, where they bind to a specific cytoplasmic receptor molecule. The receptor–hormone complex then enters the nucleus, where it becomes a gene regulatory protein that binds to a particular DNA sequence. Binding of the receptor–hormone complex activates the genes responsible for the hormone-induced changes (page 305). Again, the specificity of the receptor molecule determines which cells will respond as targets to particular hormones. Only those cells that have the appropriate hormone receptors can be triggered by a particular hormone. Thyroid hormones also act via cytoplasmic receptors, but their precise mechanism of action remains a subject of research.

A target cell responds to a particular hormone because it has a receptor that is capable of binding that hormone as well as the machinery to respond to the hormone-receptor complex. The type of response triggered by the hormone depends on the unique array of components present in the target cell. (See CTQ #6.)

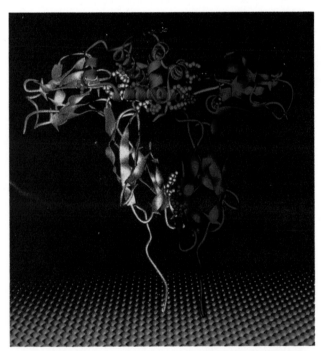

FIGURE 25-13

A hormone-receptor complex. Computer model of human growth hormone (red) bound to two receptor molecules (blue and green), which are projecting outward from the plasma membrane (the pebbled reddish surface). This binding event generates a signal that is transmitted through the membrane and activates the cell as illustrated in Figure 25-14.

EVOLUTION OF THE ENDOCRINE SYSTEM

Among the simpler invertebrates, such as flatworms, roundworms, and echinoderms (sea stars and sea cucumbers), hormones are produced primarily by neurosecretory cells, suggesting that the first endocrine cells to evolve were modified neurons. It is a relatively small evolutionary step from a typical neuron secreting a chemical transmitter substance into a synaptic cleft to a neurosecretory cell secreting a hormone into the bloodstream. Yet, the evolution of the neurosecretory cell provided new opportunities to coordinate events occurring in the outside world with the animal's own internal activities. Information relayed from surface

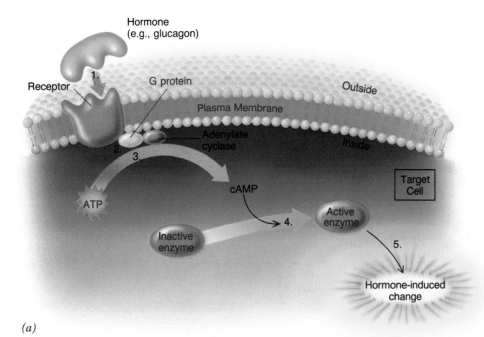

(a)

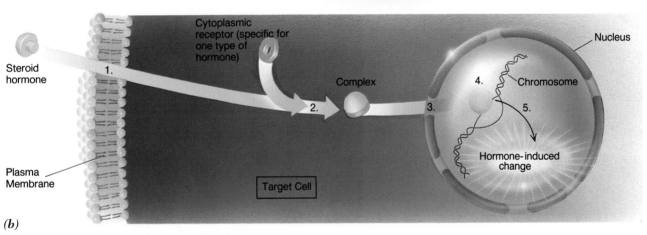

(b)

FIGURE 25-14 ──────────────────────

Mechanisms of hormone action.   **(a) Second messenger model:** (1) Hormone binds to surface receptor. (2) Binding activates a G protein, which activates adenylate cyclase. (3) Activated adenylate cyclase converts ATP to cAMP (a "second messenger"). (4) cAMP activates (or inhibits) specific enzymes. (5) Activated enzymes produce specific changes in the cell. *(b) Steroid hormones:* (1) Hormone readily passes through the plasma membrane and (2) reacts with a protein receptor molecule. (3) The hormone-receptor complex enters the nucleus and (4) attaches to a specific site on the chromosome. (5) Attachment activates those genes responsible for the hormone-induced change, such as the genes required for construction of a hair shaft by a previously dormant hair follicle on an adolescent boy's chin.

receptors could be transmitted through the nervous system to neurosecretory cells; chemical messengers could then regulate various internal physiological activities. As invertebrates became more complex, they developed more elaborate endocrine systems that included separate endocrine glands as well as neurosecretory cells. Arthropods (such as lobsters, spiders, and insects) and mollusks (especially squids and octopuses) have complex endocrine systems analogous to those of vertebrates.

⊙ The endocrine glands and their hormones are quite similar among all vertebrates, from fish to mammals. Insulin and glucagon regulate blood-sugar levels in a catfish and in a cat; thyroid hormones influence the metabolic rate of dogs and dolphins; epinephrine induces a life-saving fight-or-flight response in people and pigeons. There are exceptions to this unity, however. For example, prolactin maintains the balance of salt and water concentrations in lower vertebrates but has shifted in function to the control of milk production in mammals.

The chemical structure of many different polypeptide hormones is remarkably similar among different species. For example, even though evolution has created different functions for FSH, LH, and TSH, these hormones share a similar amino acid sequence in vertebrates and are undoubtedly derived by evolution from a single protein that was present in some prevertebrate ancestor.

While the mechanisms by which hormones act have remained remarkably uniform among diverse animals throughout their evolution, the types of metabolic processes that are regulated by hormones vary greatly from one type of animal to another. (See CTQ #7.)

REEXAMINING THE THEMES

Relationship between Form and Function

꘡ Neurosecretory cells essentially are neurons that secrete hormones rather than neurotransmitters. These cells combine the structure and function of both neurons and endocrine cells; in doing so, they form a bridge between the two great regulatory systems of the body. As part of the nervous system, neurosecretory cells receive impulses from other neurons and conduct their own impulses down an axon. As part of the endocrine system, neurosecretory cells respond to stimuli by secreting hormones into the bloodstream.

Biological Order, Regulation, and Homeostasis

○ Along with the nervous system, the endocrine system is the primary regulatory agency within an animal's body. Whereas the nervous system is geared for rapid responses, such as those requiring muscle contraction and movement, the endocrine organs regulate slower responses that require changes in the metabolic activities of target cells. Endocrine organs regulate such diverse activities as protein and polysaccharide synthesis, body growth, reproductive maturation, blood-cell formation, and the ionic content of the blood.

Acquiring and Using Energy

☀ One of the simplest measures of energy utilization is metabolic rate: the amount of oxygen consumed per unit of body weight per minute. Metabolic rate can be directly related to energy utilization. In vertebrates, metabolic rate is affected by a number of hormones including those that affect glucose metabolism (e.g., glucagon, epinephrine, insulin, and glucocorticoids) and those that affect mitochondrial function (thyroid hormone). For example, daily doses of thyroid hormone can transform a person with a hypothyroid condition from a lethargic, sluggish individual to one with a normal metabolic rate.

Unity within Diversity

⊙ Even though different animals possess different hormones that evoke different responses, the basic structure and mechanism of action is very similar in all species. For example, both insects and mammals produce steroid hormones that bind to the DNA and activate nearby genes. Yet, a steroid hormone in an insect may cause metamorphosis from a pupa to an adult, while a steroid hormone in a mammal may cause the reabsorption of sodium from the blood or the development of breasts. Similarly, neurosecretory cells are found throughout the animal kingdom, but they regulate totally different types of functions in different animals.

Evolution and Adaptation

▮▶ The endocrine system provides some of the best evidence for the common ancestry of all vertebrates. Consider that a hormone, such as insulin, has virtually the same structure and function from one end of the vertebrate scale to the other. Common ancestry explains how you and a catfish have acquired so many similar hormones.

S Y N O P S I S

Hormones are chemical messengers that are secreted into the bloodstream via ductless glands of the endocrine system. Hormones act on specific target cells that possess receptors for that hormone, eliciting a specific metabolic response in the target cell. Some hormones maintain homeostatic conditions; others cause irreversible bodily changes, including metamorphosis.

Hormone concentrations are often regulated by negative feedback. As the concentration of the hormone (or its effect) increases, the output of the hormone decreases. The effects of hormones are short-lived, because hormones are regularly excreted or enzymatically destroyed.

The double-lobed pituitary is the "master gland" of the endocrine system. The hormones of the posterior pituitary—antidiuretic hormone (ADH) and oxytocin—are products of neurosecretory cells which originate in the hypothalamus. These neurosecretory cells release their hormone secretions from their tips following the arrival of neural impulses. ADH regulates water reabsorption in the kidney, and oxytocin regulates uterine contractions and the release of milk from the mammary glands. The anterior lobe of the pituitary consists of endocrine cells that produce at least six hormones: four tropic hormones (ACTH, FSH, LH, FSH), prolactin, and growth hormone (GH). The secretion of each of these hormones occurs in response to factors released by neurosecretory cells from the hypothalamus. Prolactin induces milk production, and growth hormone stimulates body growth.

The adrenal cortex secretes steroid hormones that regulate sugar metabolism and stabilize sodium and potassium concentrations in the blood. **The adrenal medulla** secretes epinephrine, which boosts metabolic activity and prepares the body to cope with perceived danger, emergencies, or stressful situations.

The thyroid gland secretes thyroid hormone, which regulates the body's overall rate of metabolism, and calcitonin, which lowers the concentration of calcium in the blood. This latter effect is countered by the parathyroid hormone, which helps balance blood calcium concentrations.

The pancreas secretes insulin and glucagon. Insulin promotes glucose absorption by cells and conversion to glycogen, decreasing blood glucose. Glucagon promotes glycogen breakdown, increasing blood glucose.

Prostaglandins are produced by secretory cells that are scattered throughout the body; they regulate activities associated with reproduction, blood clotting, uterine contraction, inflammation, and pain perception.

Hormone action depends on the position of the receptor within the target cell. Most hormones bind to specific receptor molecules on the target cell's surface, causing the release of a second messenger (most often cyclic AMP) into the cytoplasm. Cyclic AMP activates a protein kinase, which, in turn, modifies the activities of particular enzymes, depending on the cell. Steroid hormones bind to cytoplasmic receptor molecules. The resulting complex attaches to specific DNA sequences in the chromosomes and turns on specific genes, whose products trigger the response.

Key Terms

endocrine system (p. 520)
neurosecretory cell (p. 523)
pituitary gland (p. 524)
posterior pituitary (p. 524)
anterior pituitary (p. 524)
antidiuretic hormone (ADH) (p. 525)
oxytocin (p. 525)
tropic hormone (p. 525)
prolactin (p. 525)
growth hormone (GH) (p. 525)
adrenocorticotropic hormone (ACTH) (p. 526)

thyroid-stimulating hormone (TSH) (p. 526)
gonadotropin (p. 526)
follicle-stimulating hormone (FSH) (p. 526)
luteinizing hormone (LH) (p. 526)
adrenal cortex (p. 526)
glucocorticoid (p. 526)
mineralocorticoid (p. 528)
adrenal medulla (p. 529)
epinephrine (p. 529)
norepinephrine (p. 529)
thyroid gland (p. 529)

thyroxin (p. 529)
triiodothyronine (p. 529)
parathyroid gland (p. 531)
parathyroid hormone (PTH) (p. 531)
islets of Langerhans (p. 531)
glucagon (p. 531)
testes (p. 532)
ovaries (p. 532)
testosterone (p. 532)
estrogen (p. 532)
prostaglandin (p. 532)
second messenger (p. 533)

Review Questions

1. Many hormones have antagonistic hormones that reverse their effects. Name an antagonist for each of the following hormones, and describe how each pair works together: parathyroid hormone, insulin, and prostaglandin (as a promoter of inflammation).

2. Why is cAMP called a "second messenger"? What is the first messenger? How is it possible that a variety of different hormones can all utilize the same second messenger and evoke markedly different responses?

3. How does the anterior pituitary differ from the posterior pituitary? (Consider the different hormones secreted by these two parts of the pituitary gland and the relationship between the secretion of these hormones and the hypothalamus.)

4. Name two organs that produce steroid hormones, and describe the effects of the hormones from these two sources.

Critical Thinking Questions

1. If Banting and Best had injected far too strong an extract from a ligated pancreas during their first trials on dogs, what effect do you think the injection of excess insulin would have had on these diabetic animals?

2. Both the nervous and endocrine systems perform the same general function—transfer of information—but by different mechanisms. Complete the chart below, comparing these two systems.

Characteristic	Nervous System	Endocrine System
type of message		
message carried by		
area affected by message		
speed of responses		
duration of responses		

3. Why is it important to regulate the concentrations of hormones precisely?

4. In one of Bayliss and Starling's early experiments on the control of pancreatic secretions, the researchers attempted to cut the nerves between the intestine and the pancreas. Why do you suppose they would have carried out this experiment? What results do you think they would have found when their test animals were fed?

5. Describe two major links between the endocrine and nervous systems.

6. Explain how the endocrine system achieves feedback, amplification, and specificity.

7. How are hormones evidence of the unity of life? What evidence is there for a common origin for nervous and endocrine systems? What reasons can you suggest to explain why the two diverged to form separate systems?

Additional Readings

Atkinson, M. A., and N. K. Maclaren. 1990. What causes diabetes. *Sci. Amer.* July:62–71. (Intermediate)

Bliss, M. 1982. *The discovery of insulin.* Chicago: Univ. of Chicago Press. (Introductory)

Guillemin, R., and R. Burges. 1972. The hormones of the hypothalamus. *Sci. Amer.* Nov:24–33. (Intermediate)

Hadley, M. 1988. *Endocrinology*, 2d ed. Englewood Cliffs, NJ: Prentice Hall. (Intermediate)

Lehrer, S. 1979. *Explorers of the body (Banting & Best).* New York: Doubleday. (Introductory)

Linder, M. E., and A. G. Gilman. 1992. G proteins. *Sci. Amer.* Dec:108–115. (Advanced)

Reiter, R. J., ed. 1984. *The pineal gland.* New York: Raven. (Advanced)

Turner, C. D., and J. T. Bagnara. 1976. *General endocrinology*, 6th ed. Philadelphia: Saunders. (Intermediate—Advanced)

Weissmann, G. 1991. Aspirin. *Sci. Amer.* Jan:84–91. (Intermediate)

Witzmann, R. 1981. *Steroids: Keys to life.* New York: Van Nostrand-Reinhold Co. (Intermediate)

CHAPTER
◂ 26 ▸

Protection, Support, and Movement: The Integumentary, Skeletal, and Muscular Systems

STEPS
TO
DISCOVERY
Vitamin C's Role in Holding the Body Together

THE HUMAN PERSPECTIVE
Building Better Bones

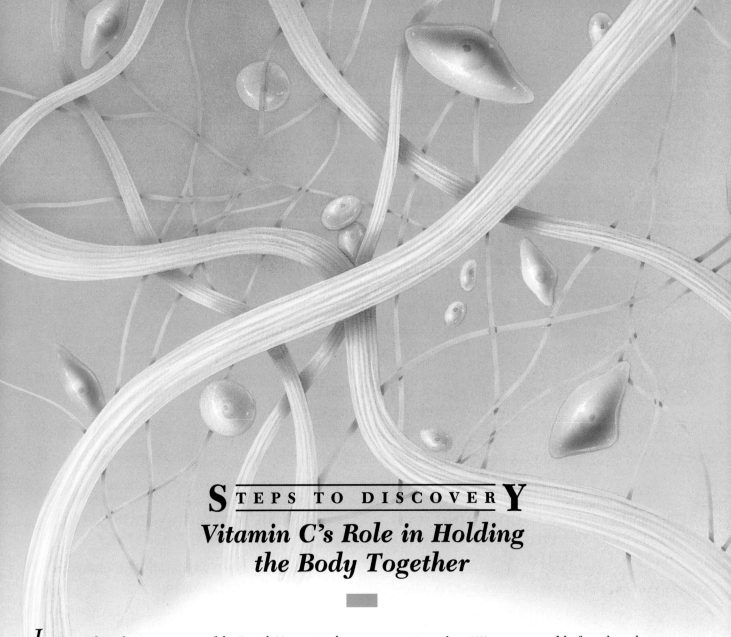

STEPS TO DISCOVERY

Vitamin C's Role in Holding the Body Together

In 1744, Admiral George Anson of the British Navy set sail with approximately 1,000 men aboard ship. When he returned to England less than a year later, only 144 men remained alive. More than 85 percent of the crew had died of a sailor's disease called *scurvy*. Three years later, James Lind, a surgeon in the British Navy, published a treatise on scurvy, concluding that the dreaded disease was caused by an imbalanced diet. Sailors on these expeditions subsisted almost exclusively on preserved meats and fish. Lind discovered that the addition of citrus fruits, such as lemons and limes, to the sailors' diets would totally prevent the disease (it also earned British sailors the nickname "limeys").

More than 150 years passed before the substance responsible for preventing scurvy was isolated, first from lemon juice, then from cabbage, and, finally, from the adrenal gland of laboratory animals. The structure and properties of the substance, which had been given the name *vitamin C,* were described in the late 1920s and early 1930s, primarily by the Hungarian-born biochemist Albert Szent-Gyorgyi and the British carbohydrate chemist Walter Haworth. Together, Szent-Gyorgyi and Haworth named vitamin C *ascorbic acid,* for its antiscurvy activity. They both received Nobel prizes for their work in 1937.

Although vitamin C and scurvy are obvious topics in a

Loose connective tissue containing scattered cells (fibroblasts), thinner elastic fibers, and thick collagen bundles that are held

discussion of nutrition, it is less evident how they relate to the subjects of this chapter, namely skin, bones, and muscles. All of these parts of the body contain a large amount of connective tissue. Victims of scurvy typically suffer from inflamed gums and tooth loss, poor wound-healing, brittle bones, and internal bleeding. All of these are consequences of serious defects in the formation and maintenance of connective tissues throughout the body. The properties of connective tissues are determined largely by the properties of the *extracellular matrix*, which contains polysaccharides and proteins secreted by cells into their surroundings. The strength and resiliency of connective tissues often depend on a single protein, collagen. Examination of the connective tissues from laboratory animals suffering from scurvy revealed a marked reduction in the number of collagen fibers that normally fill the spaces between the cells. It appeared that there was some relationship between ascorbic acid and the formation of collagen fibers.

Over the next few decades, a number of laboratories turned their attention to the structure and synthesis of collagen. The great strength of the collagen fibers in the extracellular matrix depends in large part on the formation of cross-linking chemical bonds that bind the collagen polypeptides together into strong fibers. The formation of these cross-links requires that some of the amino acids (specifically lysine and proline) in the collagen polypeptide chains are first modified by the addition of a hydroxyl (—OH) group. During the 1960s, scientists found evidence that ascorbic acid deficiency, which is the cause of scurvy, decreases the amount of cross-linking that occurs between collagen polypeptides, thereby weakening the entire fabric of the body's connective tissues. Subsequent research revealed that ascorbic acid is, in fact, a coenzyme that is

required by the enzymes which add hydroxyl groups to the amino acids. A person suffering from scurvy could not modify his or her amino acids to produce the cross links that strengthen collagen fibers.

Ehlers-Danlos syndrome (EDS) is another condition that has been traced to an inability to add hydroxyl groups to collagen. This inherited disorder is characterized by poor wound-healing, tissues that bruise and bleed, extremely flexible joints, and highly extensible skin (which led some EDS sufferers to work as "rubber men" in circus side shows). In 1972, Sheldon Pinnell and his co-workers at Harvard University discovered that a common form of EDS was due to a mutation in an autosomal recessive gene (page 343) that ordinarily directed fibroblasts to produce the enzyme that added a hydroxyl group to the amino acid lysine. A person who is homozygous recessive for this gene fails to form collagen molecules with normal lysine cross-links, a condition that leads to the various symptoms of EDS. Ehlers-Danlos syndrome is less severe than is scurvy, however, because other cross-links (those involving the amino acid proline) are still able to form. In scurvy victims, neither lysine nor proline crosslinks are formed, creating a much weaker collagen fiber.

together by covalent cross linkages.

Skin, bones, and muscles constitute more than 65 percent of your body mass and largely determine your physical appearance, from your body stature to your facial features. Your skin, bones, and muscles protect you from the environment, support you against the effects of gravity, and enable you to move toward food and away from danger. We will begin our discussion of these systems with a look at an animal's first line of defense: its outer body surface.

▼ ▼ ▼

THE INTEGUMENT: COVERING AND PROTECTING THE OUTER BODY SURFACE

The **integument** is the outer body covering of an animal; in vertebrates, it is the **skin.** Depending on the functions it performs, the integument may be soft, flexible, and permeable (as in an earthworm or a frog) or coarse, stiff, and impermeable (as in a lizard or a fish). Whatever its nature, the integument is strategically located at the boundary between the living animal and its environment. Consequently, the integument must act as a protective barrier; it helps shield the individual's delicate, moist, internal tissues from a changing and often harsh environment that might otherwise infect the body with bacteria, freeze the body's fluids, evaporate the body's water, or mutate the body's genes.

THE HUMAN INTEGUMENT: FORM AND FUNCTION

Your skin is a biological cooperative of the four tissue types: epithelial, connective, muscle, and nerve (Figure 26-1). Thus, skin is an organ. In fact, it is the largest organ of your body. Human skin consists of two distinct layers: the outer **epidermis** and the inner **dermis.** These layers have very different structures that reflect their different functions.

The Epidermis: An Outer, Protective Layer

The epidermis is a protective epithelium, approximately 0.1 millimeter thick. It consists of many layers of cells that are formed by mitosis in the deepest layer of the epidermis and then move toward the body surface. As they approach the surface, the cells become flattened, and their cytoplasm becomes filled with filaments of the tough, resistant protein *keratin* (page 801). By the time they reach the surface, the cells have been transformed into a dead, outer layer of keratin, making your skin airtight, watertight, and resistant to bacteria and most chemicals. Pigmentation of the skin is due to the presence of dark granules that eventually reside within the dead epidermal cells.

Fingernails and toenails are remnants of the *cuticle*— the narrow, sensitive flap of living tissue that is located at the back rim of the nail. As the cells of the outer layer of the cuticle die and form nails, they acquire a special hardness due to additional covalent linkages that form between the keratin molecules.

The Dermis: A Complex, Inner Layer

Beneath the epidermis lies the dermis (Figure 26-1), which consists of dense connective tissue and a rich supply of blood vessels, nerve fibers, and smooth muscle cells. The border between the dermis and the epidermis is characterized by hills and valleys, which increase the hand's gripping ability and form the basis of a person's fingerprints. Bundles of dermal collagen fibers give the skin its strength and cohesion as a continuous thin layer. Elastic fibers provide the skin with the elasticity that allows it to snap back when stretched. The dermal blood vessels provide nutrients and oxygen to the overlying epidermis, which lacks its own blood supply. These vessels also play a key role in maintaining the body's constant temperature by carrying warm blood to the body's surface, where heat can be lost to the environment. Blood flow into the dermis can range from a bare trickle, when heat must be conserved during cold external conditions, to 50 percent of the body's blood supply when heat loss is needed to cool the body. The pinkness of the skin in light-skinned individuals is due to blood flow through the dermis. In dark-skinned individuals, the presence of the dermal vessels is obscured by overlying epidermal pigmentation.

Hair: Protecting and Insulating the Skin

Most mammals have considerably more hair (fur) than do humans. A thick layer of hair provides an excellent cover for protecting the body against abrasion and insulating it against heat loss. Human evolution, which probably occurred in a warm, tropical environment, was accompanied by the loss of body hair, which is thought to have been a result of natural selection favoring those individuals who were better able to lose excess body heat.

Hairs consist of dead, keratinized cells similar to those of the outer layer of the skin. Each hair is formed within a living **follicle** (Figure 26-1). Attached to each follicle is a small, smooth muscle that, when contracted, causes the hair to "stand on end." In furrier mammals, such as bears, the contraction of these smooth muscles increases the insulation value of the coat. In humans, contraction of these muscles simply causes an indentation of the skin, producing "goose bumps."

The Skin's Various Secretions

Your skin contains large numbers of glands, whose secretions find their way to the surface of your body. All skin

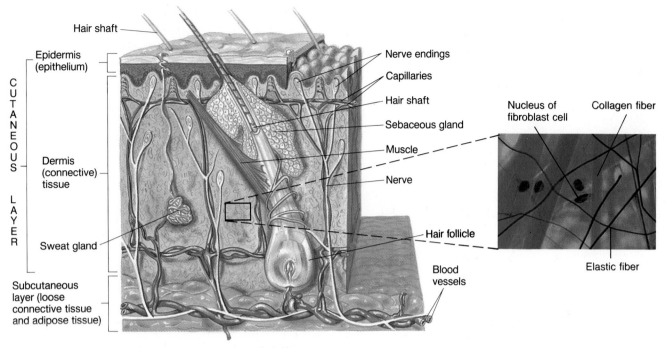

FIGURE 26-1

Human skin. The outer epidermis consists of a superficial layer of dead cells that cover the underlying living epithelial cells. Below the epidermis lies the dermis, where connective tissue predominates. Embedded in the dermis are blood vessels, muscles, nerves, and the basal portions of glands and hair follicles, which form in the epidermis and then sink into the deeper layers of the skin. Hairs are formed from epidermal cells that are generated deep within the follicle. The cells become filled with keratin; they then die and become part of the *hair shaft*. As new layers of dead, keratinized cells are added at its base, the hair is forced upward, increasing in length. Skin (the *cutaneous layer*) is firmly secured to the underlying layer (the *subcutaneous layer*) by connective tissue. Inset shows the connective tissue of the dermis contains scattered cells (fibroblasts) and extracellular collagen and elastic fibers.

glands are **exocrine glands;** that is, they release their secretory products through a duct (in contrast to the "ductless" endocrine glands discussed in Chapter 25). The glands of the skin include two broad types: sebaceous glands and sweat glands. **Sebaceous glands** produce a mixture of lipids *(sebum)* that oil the hair and skin, keeping them pliable. **Sweat glands** are distributed over most of the skin's surface, where they secrete a dilute salt solution, whose evaporation cools the body. A second type of sweat gland is restricted to just a few sites, including the anal and genital regions, nipples, and armpits. These glands appear after puberty and secrete a fluid that contains protein and other organic molecules. Although this fluid is odorless as it is initially secreted, the molecules are eventually broken down by skin bacteria into products that give the human body some of its characteristic odors.

As you can see, the skin is a dynamic, living organ. Its importance is dramatically evident when you consider that burns to as little as 20 percent of the body can be fatal if not treated rapidly. The cause of death in such burn cases is dehydration, which results from the loss of water through the damaged, previously waterproof, body cover.

EVOLUTION OF THE VERTEBRATE INTEGUMENT

All vertebrate integuments are built according to the same two-layered, epidermal–dermal plan, but the two layers can have a very different structure and function, depending on the habitat and lifestyle of the particular animal (Figure 26-2).

The earliest vertebrates were jawless, bottom-dwelling fishes that were clothed in heavy, bony armor that protected them from predators. The plates of bone that formed the fish's armor shields were located in the dermis; hence, they are called *dermal bone*. During subsequent evolution, fishes moved away from the ocean bottom, becoming more buoyant and mobile. The thick plates of dermal bone along the sides of the body were no longer adaptive and became reduced to the thin, familiar bony *scales* that are scraped away when a fish is "cleaned."

As vertebrates moved out of the water and onto the land, the integument became adapted to terrestrial habitats. The bony scales of the ancestral fishes were lost, and the dermis became a more fibrous, flexible layer. In

(a)

FIGURE 26-2

Contrasting integuments. The yellow spotted salamander (an amphibian) has thin, moist, permeable skin *(a)*, while the iguana (a reptile) has thick, dry, impermeable skin *(b).*

(b)

amphibians—animals that live both in water and on land—the skin is usually moist and permeable, which facilitates oxygen absorption across the body surface (Figure 26-2a). Among reptiles and other land vertebrates, the epidermis has become a tough, impervious layer that prevents water loss in harsh, dry, terrestrial environments (Figure 26-2b). The role of oxygen uptake in land vertebrates with impermeable integuments was taken over by the lungs. In birds, the epidermis gave rise to a new structure—feathers—which are composed largely of keratin.

The integument, which resides at the outer body surface, must protect the animal from the environment while facilitating certain types of exchanges with the environment, such as the loss of heat. Integuments are highly variable in structure because the properties of the body surface must be adapted to the animal's habitat and mode of existence. (See CTQ #2)

THE SKELETAL SYSTEM: PROVIDING SUPPORT AND MOVEMENT

Most animals possess a rigid form of support which either completely surrounds the body as a protective encasement or forms a system of living girders inside the animal. Even those animals that lack rigid supporting structures still have a way of doing battle with gravity. These animals employ "hydrostatic skeletons." As you read about the various types of skeletons, notice how the properties of each are derived from its particular structure.

HYDROSTATIC SKELETONS: USING FLUID TO MAINTAIN RIGIDITY

Although an earthworm doesn't have a single bone in its body, it can push the tip of its body through compact soil, a feat that requires the front end of the earthworm's body to remain highly rigid. As it burrows, the worm maintains rigidity by generating pressure inside a closed, fluid-filled chamber. It does so by contracting muscles that encircle the chamber, forming a **hydrostatic skeleton.** You can envision how a hydrostatic skeleton works by filling a balloon with water and exerting pressure by squeezing one end. The balloon becomes rigid and capable of supporting itself against the force of gravity. Sea anemones and corals possess a similar fluid-filled internal chamber (Figure 26-3) that acts as a hydrostatic skeleton to maintain the upright stature of the animal in the face of gravity.

EXOSKELETONS: PROVIDING SUPPORT FROM OUTSIDE THE BODY

Anyone who has cracked open a crab or lobster leg at the dinner table, or found the discarded husk of an insect, or

Gastrovascular cavity

FIGURE 26-3
A hydrostatic skeleton allows these coral polyps to stretch tall against the pull of gravity. By contracting ring-shaped muscles that encircle an internal, fluid-filled chamber (the *gastrovascular cavity*), these animals generate hydrostatic (water) pressure that elongates the animal and prevents it from collapsing.

examined a clam shell is familiar with **exoskeletons**—the hard, nonliving external coverings that are secreted by the outer epidermal layer of the animal's body. Exoskeletons provide both protection and support. Some animals, notably snails, clams, oysters, and scallops, are enclosed in a protective, rocklike **shell** that is composed largely of calcium carbonate. Another type of exoskeleton is a **cuticle**—a nonliving, hardened, outer layer that conforms to the surface of the animal's body (Figure 26-4a). Cuticles, which consist largely of protein and the polysaccharide chitin, cover the bodies of arthropods, including insects, spiders, and crabs. These hardened exoskeletons serve a skeletal function in arthropods by supporting the weight of the soft tissues of the body. The outer casings of the limbs consist of

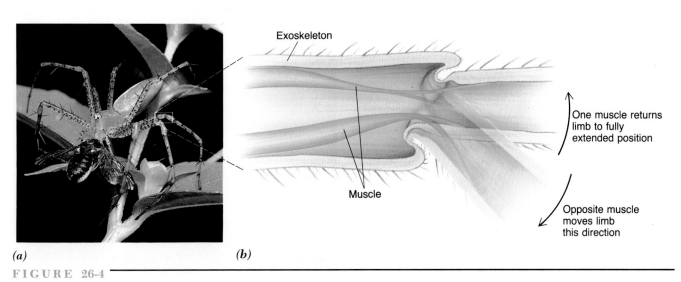

Exoskeleton

One muscle returns limb to fully extended position

Opposite muscle moves limb this direction

Muscle

(a) *(b)*

FIGURE 26-4

The arthropod exoskeleton. *(a)* The body of an arthropod is covered by a cuticle that conforms to its contour. In insects and spiders, the cuticle consists largely of protein and the polysaccharide chitin. In crustaceans, such as crabs and lobsters, the cuticle is hardened by salts. The limbs of arthropods are covered by tubular segments of cuticle that are joined by flexible joints. The mobility of the spider and the fly in this photo is provided by dozens of joints in their exoskeletons. *(b)* The muscles that move the appendages of an arthropod are attached to the inner surface of the exoskeleton.

tubular sections of cuticle which are connected to one another by thin, flexible joints. The limbs are operated by bundles of muscle tissue attached to the inner surface of the cuticle (Figure 26-4*b*). Without a rigid exoskeleton, muscular contraction would merely distort the soft tissue, and the animal wouldn't be able to move. The disadvantages of exoskeletons are discussed in Chapter 39.

ENDOSKELETONS: PROVIDING SUPPORT FROM WITHIN THE BODY

Endoskeletons (Figure 26-5) are support structures that reside inside the animal's body. Although found in a few invertebrates, such as sponges and sea stars, the endoskeletons of vertebrates are the most complex and versatile. Vertebrates are supported by internal skeletons that are composed of two distinct types of connective tissues: bone and cartilage.

Bone

Bone combines two properties: strength and light weight. These properties are rarely found together in a single material. One cubic inch of bone can withstand a pressure of 19,000 pounds, the weight of ten compact automobiles. Bone has been called "living concrete," but it is four times stronger than concrete and much lighter. For example, a person's skeleton adds only 14 percent to the total weight of the body.

A bone's strength is a product of its architecture. Like other connective tissues, the bulk of a piece of bone consists of extracellular matrix: materials secreted by cells into the space that surrounds the cells. The extracellular matrix of bone is woven from flexible cables consisting of strands of collagen fibers and is hardened by crystals of calcium phosphate. Together, the protein and mineral salts make bone both hard and resilient. Collagen fibers act as reinforcing rods, similar to the steel rods found in reinforced concrete (and to the cellulose microfibrils of the plant cell wall, page 108). Without collagen, your bones would be so brittle that they would shatter under your body's weight. Without calcium, your bones would bend as though they were made of rubber.

The most striking difference between bone and reinforced concrete is that bone is alive, even in its most rocklike parts. Within the solid mass of mineral and collagen, living bone cells thrive. These cells, called **osteocytes** (*osteo* = bone, *cyte* = cell), are engulfed in the calcified matrix that they manufacture. The structure of bone is described in Figure 26-6*a*.

Bones are generally hollow, a property that actually increases their strength. A solid bone would not only be extremely heavy but would fracture more easily. Each bone has a solid, rocklike *portion*, called **compact bone,** which usually surrounds a honeycombed mass of **spongy bone** (Figure 26-6*a*). The hollows within spongy bone are filled with **red marrow,** the soft tissue that produces red blood cells. **Yellow marrow,** which stores fat, is located in the hollow core of the long bones, such as those of the arms and legs.

The structure and function of a bone can be illustrated by examining the femur, the long bone of the human thigh.

(a)

(b)

FIGURE 26-5

Endoskeletons are found in echinoderms *(a)* and vertebrates *(b).* The skeleton of the sea urchin in *(a)* lies just beneath the outer epidermis and consists of a large number of fused plates made of calcium carbonate and protein.

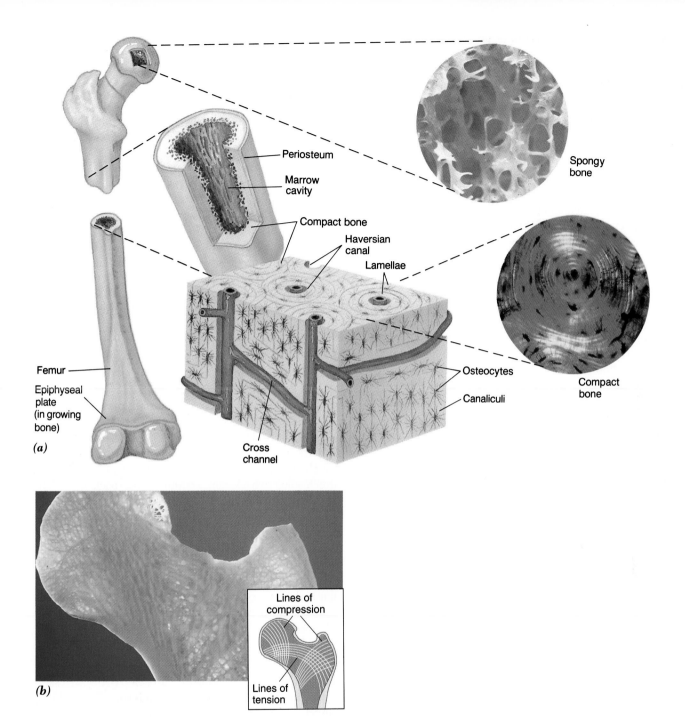

Periosteum

Marrow
cavity

Compact bone

Haversian
canal

Lamellae

Spongy
bone

Osteocytes

Canaliculi

Compact
bone

Femur

Epiphyseal
plate
(in growing
bone)

(a)

Cross
channel

Lines of
compression

Lines of
tension

(b)

FIGURE 26-6

The architecture of bone.   *(a)* Solid-looking, compact bone consists of living cells, called *osteo-cytes,* engulfed in an extracellular matrix. This matrix is deposited as concentric cylinders called *lamellae.* The result is a greatly strengthened structure. Osteocytes obtain their nutrients from blood vessels that are threaded through channels in the bone tissue, called *Haversian canals,* which are the lifeline of compact bone. The tiny chambers that house the osteocytes are connected by cross channels and *canaliculi*—microscopic channels that extend outward like a spider web from the central canal. Canaliculi channel nutrients and growth-regulating hormones from the blood to every osteocyte in the Haversian system and evacuate their metabolic wastes. Spongy bone consists of thin, bony elements that surround marrow-filled chambers. The long bone of the leg, shown at the left, grows during childhood and adolescence at *epiphyseal plates* located near each end. The epiphyseal plate is the last part of the bone to become mineralized when elongation of the bone ceases in adulthood. The entire bone is covered by a connective tissue sheath, the *periosteum.* *(b)* The head of the femur contains spongy bone, whose elements are aligned to resist the stress placed on the femur by the body's weight. The photograph of a section of the femur shows these thin, bony plates, while the accompanying diagram shows the lines of stress. Some of the stress lines result from crushing (compression) forces, while others result from pulling (tension) forces.

The head of the femur consists of spongy bone, whose microscopic structure is related precisely to its weight-bearing function. This is illustrated in Figure 26-6*b*. Notice how the thin, bony elements form precisely along the lines of stress that develop in this bone as it bears the weight of the upper body. Not only does the microscopic structure of a bone conform to the pressures it bears, but the shape of the bone itself can be correlated with its function. This is why the bones of related animals differ in noticeable ways; each is shaped to facilitate the particular mechanical activities of that animal (as illustrated by the bones of the forearms of vertebrates shown in Figure 34-3).

Although the bones of adults no longer increase in size, bone is constantly being disassembled and rebuilt throughout an individual's life (see The Human Perspective: Building Better Bones). The constant interplay between bone assembly and disassembly enables our bodies to strengthen those bones that receive the most use and to diminish the size of those whose services are in less demand. In other words, physical activity produces larger, stronger bones, just as it enlarges and strengthens muscles. Disuse of bones does just the opposite: It causes *atrophy* (wasting). Even enclosing a leg in a cast for a few weeks can diminish the size of a bone.

Cartilage

If you were to examine the bones as they initially form in a human embryo, you would find that most of them aren't bones at all but are composed of another structural material: cartilage. Just as bone consists of osteocytes, **cartilage** consists of cells called **chondrocytes,** that secrete an extracellular matrix that envelopes the cells. The matrix of cartilage also contains large numbers of collagen fibers, which are embedded in an amorphous material composed of protein-polysaccharide. Like bone, cartilage provides strength and resilience, but its lack of mineral deposits keeps cartilage flexible. To gain a sense of the tough, yet flexible, character of this "building material," wiggle your outer ear or the end of your nose, both of which are composed largely of cartilage.

THE HUMAN SKELETON

By the time you are born, much of the cartilage present in the early embryo has been transformed into about 350 partially hardened bones. As you grow, many of these bones fuse with one another, producing an adult skeleton that consists of 206 individual bones, linked together by various types of joints. Each bone in the body participates in a specific body movement, supports a particular part of the body's weight, or protects an internal organ from damage. The bones of the mammalian skeleton can be divided into two functional groups: the axial skeleton and the appendicular skeleton (Figure 26-7).

The Axial Skeleton

Bones aligned along the long axis of the body—the skull, vertebral column, and rib cage—comprise the **axial skeleton.** This structure assumes the skeleton's protective functions.

The precious 3-pound mass of nervous tissue in your head is enclosed in the *cranium,* an unyielding vault of 8 bones, some of which evolved from the dermal bone of our early aquatic ancestors. The **skull,** which is composed of all the bones of your head, includes the cranium and 11 additional bones. At birth, the individual bones of the cranium are held together by flexible membranes that allow the head to compress a bit as it passes through the birth canal. By the end of a child's second year, these vulnerable "soft spots" have been replaced by strong, interlocking lines of fusion between adjacent bony plates.

The spinal cord is protected by the **vertebral column,** or backbone. The flexible backbone consists of 33 bones, called *vertebrae,* which are arranged in a gracefully curved line and are cushioned from one another by disks of cartilage. These disks take on particular importance in humans since walking upright on two legs causes the backbone to bear the entire weight of the upper body. If not for these disks, the vertebrae would grind one another to dust. Sitting in a chair can place even greater stress on the vertebrae, often leading to back pain, a common drawback of a sedentary human life.

The **rib cage** embraces the chest cavity and protects its vital organs. Ribs extend from the vertebrae and form a "cage" by attaching to the *sternum* (the breastbone) in the front. The lower two pairs of ribs, called "floating ribs," remain free at their outer ends. Contrary to popular belief, men and women have the same number of ribs: 24. The marrow found in ribs is one of the most prolific producers of red blood cells.

The Appendicular Skeleton

The movable limbs (*appendages*) attached to the axial skeleton comprise the **appendicular skeleton** (Figure 26-7), which forms a system of levers, providing mobility and dexterity. The **pectoral girdle** holds the arms to the axial skeleton. It consists of four bones: two **scapulae** (shoulder blades), and a pair of **clavicles** (collar bones). The ability of the scapulae to "skate over" the surface of the rib cage contributes to the impressive dexterity of the human arm. The two bones (the *ulna* and *radius*) in the forearm allow you to rotate your hand. One of the most extraordinary collection of bones anywhere in the animal kingdom lies at the end of the wrist of humans, apes, and monkeys. The hand combines strength with dexterity enabling a person to crush objects with the same hand that is delicate enough to insert a contact lens into the eye.

The **pelvic girdle** (or **pelvis**) receives the weight of the upper body from the vertebral column and transmits it either to the bones of the legs or to the surface on which you are now sitting. The central opening of the pelvis is larger in

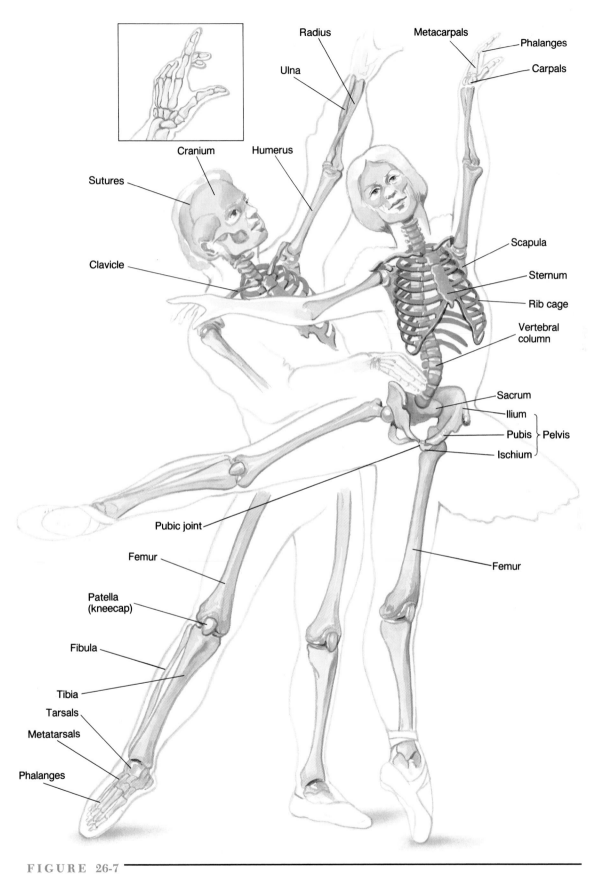

FIGURE 26-7

The human skeleton. Bones of the axial portion are shown in red; those of the appendicular portion are in blue.

◁ THE HUMAN PERSPECTIVE ▷
Building Better Bones

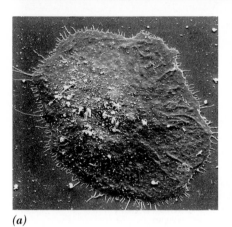

(a)

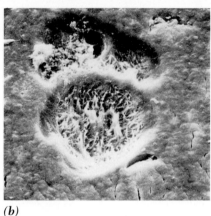

(b)

FIGURE 1

Repair and remodeling of bone begins with the dissolution of a portion of the existing matrix by large, multinucleated osteoclasts *(a)*. A moonlike crater is etched into bone by the action of an osteoclast *(b)*.

Chances are, when you think of a bone, you picture a dead remnant of a once-living organism. But bones are not dead; they are dynamic, living tissues that require continual maintenance and repair. Maintenance of bone occurs in two stages. First, a portion of existing bone matrix is broken down; second, new bone matrix is deposited, replacing that which was removed. Neither of these complex steps is understood well, and both are regulated by a variety of factors, including hormones- (calcitonin, parathyroid hormone, and mineralocorticoids), growth factors, vitamin D, and substances produced by various cells. An intensive research effort is under way to understand the nature and roles of many of these bone-related factors.

Bone matrix is disassembled by *osteoclasts,* specialized cells that emerge from the bone marrow under mysterious circumstances and migrate to sites where bone is being remodeled (Figure 1*a*). Upon its arrival, the osteoclast secretes acid, which dissolves the calcium salts of the bone matrix, and collagenase, an enzyme that digests the collagen fibers. In about 10 days, the osteoclast has finished its work, creating a microscopic crater (Figure 1*b*) that is ready to be filled in by a newly arriving *osteoblast*. The osteoblast secretes collagen and a number of factors that promote bone mineralization, the precipitation of calcium-phosphate salts.

One of the motivations for researchers to study bone maintenance is to develop better treatments for *osteoporosis,* a bone-weakening condition that occurs predominantly in postmenopausal women. A person with osteoporosis may have a "hunched-over" appearance and is very susceptible to bone fractures, particularly of the hip or vertebrae. Osteoporosis occurs when the breakdown of bone material by osteoclasts exceeds its reformation by osteoblasts, resulting in a net bone loss. Osteoporosis is particularly common in women who undergo early menopause or who have had their ovaries removed as part of a hysterectomy in early life. The cessation of estrogen production by the ovaries appears to be a major cause of the condition. Osteoporosis is best treated by the administration of the sex hormone estrogen, which appears to bind to receptors in the osteoblasts, stimulating them to increase the amount of bone deposition. The mechanism by which this phenomenon occurs remains unknown.

women than in men to accommodate childbirth. Although the foot is not as dextrous as the hand, the 26 bones that make up each foot are arched to withstand tremendous forces. An arch is the most efficient structure for supporting weight.

Joints

If the skeleton were a single, solid piece of bone, it would be stronger, but movement would be impossible. Strength must be compromised somewhat to provide mobility, producing weaker points that are capable of movement. The site where two bones come together is called a **joint.** In general, the more mobile the joint, the weaker it is. Joints *(sutures)* that join the plates of the skull, for example, are very strong but are virtually immobile after age 2. In contrast, the shoulder joint is the most mobile joint of the body. Any athlete with a shoulder dislocation can painfully testify to both its flexibility and its vulnerability. Mobile joints provide various forms of movement. A *hinge joint*, such as that of the elbow and knee, allows back-and-forth movement, much like the hinge on a door. A *ball-and-socket* joint is the most mobile of all. Two such joints provide the 360° shoulder motion that allows you to swing your arms in vertical circles (Figure 26-8).

FIGURE 26-8
The mobility of the shoulder joint is dramatically revealed in this photo of a golfer's swing illuminated by a strobe light.

The adjoining bones of a joint are held together by strong straps of connective tissue, called **ligaments.** For example, the *anterior* and *posterior cruciate ligaments* hold the bones of the upper and lower legs together at the knee joint. These ligaments are often torn during contact sports; their repair is usually accomplished by *arthroscopic surgery,* whereby an orthopedic surgeon inserts instruments into the knee joint through an incision only a few centimeters long. The progress of the operation is followed on a television screen using a picture taken with the aid of an optical fiber inserted into the knee joint. Recovery usually occurs in days rather than the weeks required after conventional knee surgery.

Highly movable joints, such as the knee, also need lubrication and cushioning to prevent the bones from crunching against each other. This is the function of the **synovial cavities** (Figure 26-9) that contain fluid that separates the surfaces of the adjoining bones and lubricates their movements against one another. The knee and a few other joints are equipped with additional bags of lubricant called *bursae.* These sacs may become inflamed, producing the pain of *bursitis.*

If an animal consisted solely of cells, it would be unable to maintain its shape or to support itself against the force of gravity. These supportive functions are performed by skeletal elements, whose properties are derived from extracellular materials. Animal skeletons range from simple, fluid-filled cavities to hardened, jointed structures that form an external covering or an internal framework. (See CTQ #3)

THE MUSCLES: POWERING THE MOTION OF ANIMALS

Muscle is a highly specialized tissue with one basic function: to generate a pulling force. Without muscles, a skeleton can do nothing more than support an animal. But muscles do much more than move parts of the skeleton. Muscles move eyelids and tongues; pump internal fluids through circulatory pipelines; propel nutrients through the digestive thoroughfare; discharge wastes; squeeze secretory products out of glands; suck oxygen into the lungs of air-breathing animals; and generate powerful jets of water that can propel an

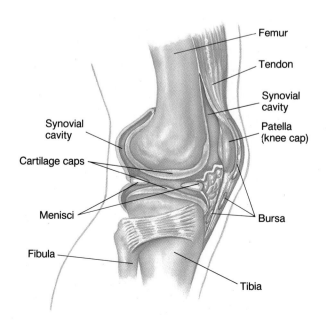

FIGURE 26-9
Lubricated for life. The knee is an example of a highly movable joint in which the ends of the bones are covered with a layer of protective cartilage and enclosed in a fluid-filled synovial cavity. The knee contains two C-shaped cartilaginous *menisci,* which cover the end of the tibia and protect against forces exerted from the side.

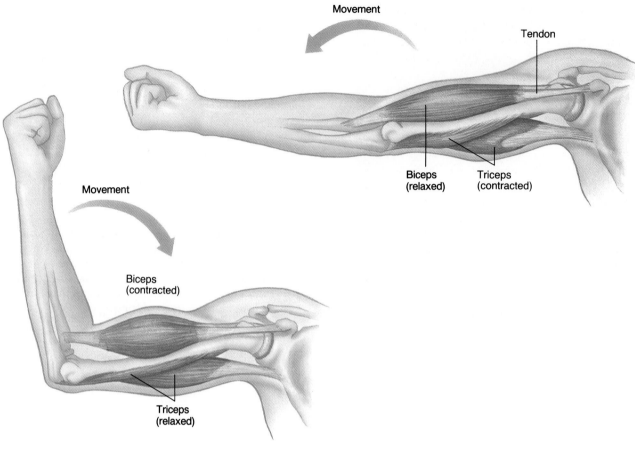

FIGURE 26-10

Biceps and triceps: muscles in opposition.   Antagonistic pairing of muscles allows the motion of
one muscle, such as the biceps, to be reversed by the action of the other muscle, the triceps.

animal out of danger. Although their shape and role in body
mechanics may differ dramatically, all muscles have several
common attributes. They contain many of the same *contractile* proteins and generate a pulling force by a similar
molecular mechanism.

Muscles require tremendous amounts of ATP to fuel
their activities; a person running at full speed burns about
1,000 Calories per hour, which is equivalent to the amount
of energy contained in a 6-ounce chocolate bar. Muscles are
superbly efficient in the use of this ATP. Muscle cells convert 35 to 50 percent of the energy released into mechanical
energy, making them much more efficient than most auto-
mobile engines. The remainder of the energy is released as
heat, which can warm a body to uncomfortable levels.

Vertebrates are equipped with large amounts of muscle. In fact, about half your body weight consists of three
types of muscle: *skeletal, smooth,* and *cardiac muscle*. These
three types of muscle tissue differ in physical appearance,
the types of jobs they perform, the tissue to which they are
attached, their speed of contraction, and the manner in
which they are excited into action. The most familiar type

are the skeletal muscles that bulge beneath the skin and give
bodybuilders their characteristic contours.

SKELETAL MUSCLE: RESPONDING TO VOLUNTARY COMMANDS

Skeletal muscles are under voluntary control; they can be
consciously commanded to contract. Skeletal muscle makes
up nearly 40 percent of a man's body and nearly 23 percent
of a woman's. These muscles are known for their speed of
contraction: A single muscle "twitch" lasts less than a tenth
of a second. Skeletal muscles derive their name from the
fact that most of them are anchored to the bones they move.
The muscle tapers at its end, forming a dense connective-
tissue cord, or **tendon** (Figures 26-10), that attaches the
muscle to the bone.

Skeletal muscles are often categorized as either *flexor
muscles*, whose contraction causes a joint to bend, or *extensor muscles*, whose contraction causes a joint to straighten.
For example, your *biceps*—the large skeletal muscle of the
front of the upper arm—is a flexor muscle whose contrac-

tion forces a bone in your forearm to bend at the elbow (Figure 26-10, *lower left*). The biceps cannot move the forearm *away* from the upper arm, however. Muscles only shorten and *pull;* they cannot push. An opposing muscle must be used to provide the opposite movement so that the limb is not stuck in one position. The biceps is paired with the *triceps,* the extensor muscle along the back of the upper arm that has just the opposite effect of the biceps, causing the arm to straighten at the elbow (Figure 26-10, *upper right*). Most skeletal muscles are arranged in such *antagonistic pairs,* allowing one muscle to reverse the effects of the other. Neural orders that command a muscle to contract are accompanied by simultaneous, inhibitory instructions to its antagonist; thus, when one muscle contracts, the other automatically relaxes.

In spite of their name, not all skeletal muscles move bones. Some nonskeletal activities, such as closing the eyelids and restricting the flow of urine out of the bladder, also require the voluntary control afforded by skeletal muscle tissue. Most of the body's 600 skeletal muscles mobilize parts of the bony substructure, however (Figure 26-11).

The Structure of Skeletal Muscle Cells

If we define a cell as the contents enclosed within a continuous plasma membrane, then the vertebrate skeletal muscle cell is highly unorthodox (Figure 26-12). A single, cylindrically shaped muscle cell may be 100 micrometers thick and run the entire length of a bulky muscle, such as the biceps in your arm. Furthermore, each cell may contain thousands of nuclei; therefore, a skeletal muscle cell is more appropriately called a **muscle fiber.** Muscle fibers have multiple nuclei because each fiber is a product of the fusion of large numbers of mononucleated "premuscle" cells in the embryo. A skeletal muscle, such as the biceps, consists of bundles of muscle fibers; each bundle is enclosed in a connective-tissue sheath.

Skeletal muscle cells may have the most highly ordered structure of any cell in the body. A cross section of a muscle fiber (Figure 26-12) reveals it to be a cable made up of hundreds of thinner, cylindrical strands, called **myofibrils,** which are separated from one another by cytoplasm. Each of the fiber's myofibrils consists of a linear array of contractile units called **sarcomeres,** each of which is endowed

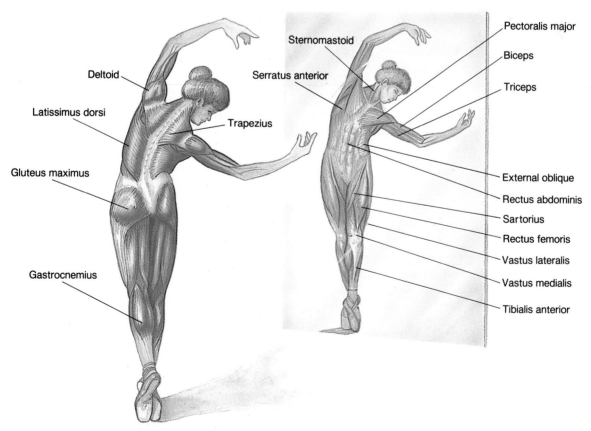

FIGURE 26-11
600 "engines." A few of the 600 skeletal muscles that power voluntary movements in the human body.

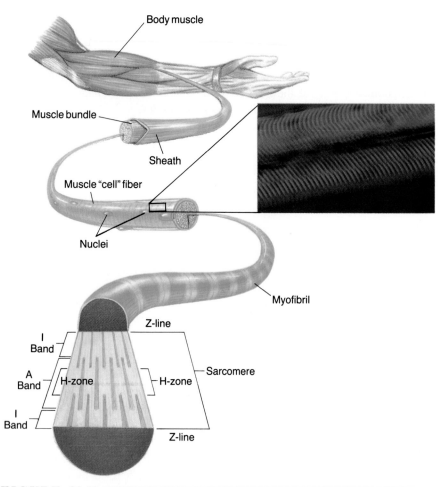

FIGURE 26-12

Skeletal muscle is composed of bundles of parallel, multinucleated cells (muscle fibers). Each muscle fiber is covered by a continuous plasma membrane (called the *sarcolemma*) and is packed with contractile myofibrils. Each myofibril consists of a linear array of sarcomeres—the contractile units of the muscle fiber. Adjacent sarcomeres are separated from one another by dark Z *lines*. Between the Z lines are several dark bands and light zones which make up the sarcomere. The banding pattern results from the overlapping array of contractile protein filaments. Each sarcomere has a pair of lightly staining *I bands* located at its outer edges; a more densely staining *A band* located between the outer I bands; and a lightly staining *H zone* located in the center of the A band. The I band contains only thin filaments, the H zone only thick filaments, and that part of the A band on either side of the H zone represents the region of overlap and contains both types of filaments.

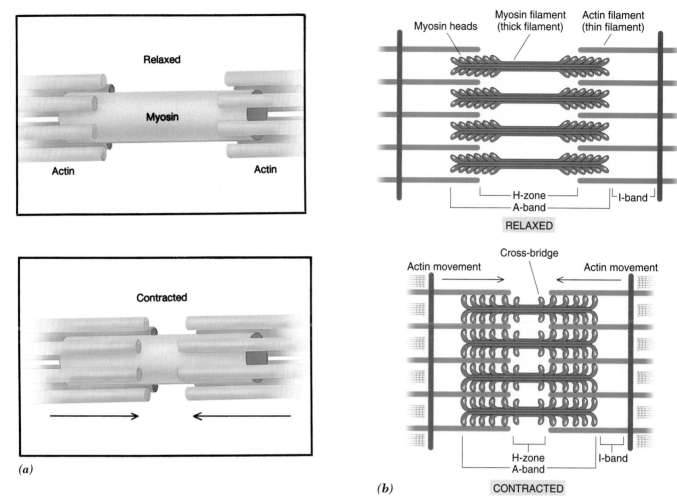

FIGURE 26-13

The mechanism of muscle contraction. *(a)* This three-dimensional model shows the sliding of thin filaments over a central, thick filament. *(b)* The change in banding pattern within a sarcomere during contraction results from the sliding of thin, actin-containing filaments over central, myosin-containing thick filaments. As a result, the sarcomere shortens in length, as evidenced by the decrease in width of the I band and H zone.

with a characteristic pattern of bands and lines (Figure 26-12). Examination of muscle sarcomeres with the electron microscope shows the banding pattern to be the result of the partial overlap of two distinct types of filaments, referred to as **thin filaments** and **thick filaments** (Figure 26-12, 26-13).

During the 1930s and 1940s, two proteins—**actin** and **myosin**—were found to constitute as much as 90 percent of the dried content of a muscle mass. When myosin was extracted from skeletal muscle fibers, the thick filaments disappeared. Subsequent extraction of actin from the fibers caused the disappearance of the thin filaments. Investigators concluded that the thick filaments were composed of myosin, and the thin filaments of actin. It was soon discovered that, in addition to forming the thick fibers, myosin was

an enzyme that could hydrolyze ATP, releasing the stored energy required for muscle contraction. These findings set the stage for the formulation of a hypothesis of the mechanism of muscle contraction.

Sliding Filaments and Molecular Ratchets

An important clue to the mechanism underlying muscle contraction came from observations of the banding pattern at different stages in the contractile process (Figure 26-13). In the 1950s, Hugh Huxley and Jean Hanson of University College in London proposed that muscle contraction resulted from the sliding of the thin actin filaments toward the center of the sarcomere.

The nature of the force that drove the thin filaments across the sarcomere remained a mystery until the discov-

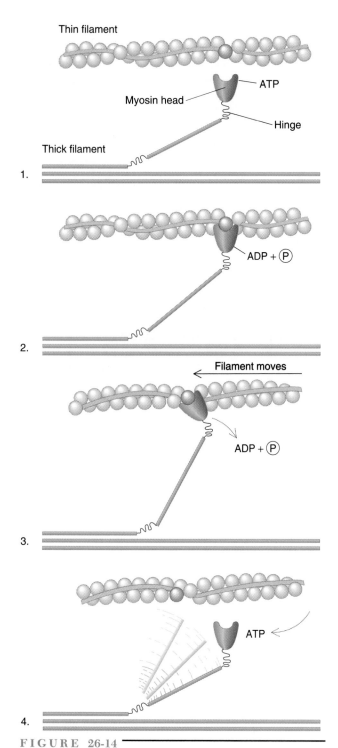

Thin filament

Myosin head

ATP

Hinge

Thick filament

1.

ADP + Ⓟ

2.

Filament moves

ADP + Ⓟ

3.

ATP

4.

FIGURE 26-14

The molecular basis for filament sliding. (1) In its relaxed state, the myosin molecules of the thick filament remain unattached to the surrounding thin filaments. (2) Contraction occurs when bulbous "heads" projecting from the thick filament attach themselves to the actin molecules of the thin filament, forming cross-bridges that bend (3) and cause the thin filaments to slide toward the center of the sarcomere. (The attachment of the myosin heads to the thin filament is controlled by the concentration of calcium ions, as discussed later.) Following its bending, the myosin head is released (4) and is free to reattach to the thin filament. This action of the myosin head is repeated until the sarcomere is fully contracted.

ery of bulbous "heads" projecting from the ends of the thick myosin filaments (Figure 26-13, 26-14). In the relaxed muscle fiber, the heads fail to make contact with the nearby, thin filaments. Each myosin head is loaded with a molecule of ATP, however, which is energetically "cocked" like a spring, ready to be triggered. When the command arrives, the myosin heads attach to the actin molecules, forming a "cross-bridge" between the thick and thin filaments. These cross-bridges, which form simultaneously along the entire muscle fiber, provide the driving force responsible for muscle contraction (Figure 26-13, 26-14).

Once the cross-bridges form, ATP molecules are hydrolyzed, and the energy released causes the myosin heads to bend toward the center of the sarcomere. The movements of the myosin heads serve as a power stroke that slides the thin actin filaments a perceptible distance over the myosin. The heads immediately release the actin and snap back to their original positions, recocked by the energy of another molecule of ATP. The heads then reattach themselves to the actin filament at a new site further along the thin filament's length and generate another power stroke, moving the filament a bit closer toward the center of the sarcomere (Figure 26-14). A fully contracted sarcomere has shortened by about 35 percent of its original length, requiring each cross-bridge to repeat its bending movement 50 to 100 times within a fraction of a second. This ratchet mechanism of muscle contraction is analogous to a team of rowers propelling a boat forward with each power stroke, then lifting the oars out of the water so that they can return to their starting position after each cycle. Next time you are engaged in strenuous exercise, remember that virtually all of the energy you are expending is being used to bend billions of tiny ratchets within the sarcomeres of your muscles.

In the absence of ATP, the myosin heads cannot be recocked. Instead, they remain attached to the actin filament, locking the muscle fiber in a state of prolonged rigidity. Shortly after death, when cells can no longer generate ATP, all the muscles in the body stiffen, producing the state of *rigor mortis*.

Unleashing the Fiber's Potential

Now that we have described what happens when a muscle contracts, let's take a look back at the events that occur when an impulse is transmitted from a motor neuron to a skeletal muscle fiber. The acetylcholine molecules released by the synaptic knobs of a motor neuron bind to receptors on the plasma membrane of the muscle cell, causing an increase in the permeability of sodium ions. The influx of sodium in turn causes a depolarization of the muscle membrane and initiates an action potential in the muscle cell. Unlike in a neuron, where an action potential remains at the cell surface, the impulse generated in a skeletal muscle cell is propagated deep into the interior of the cell along membranous folds, called **transverse (T) tubules** (Figure 26-15). The T tubules terminate in very close proximity to a system of cytoplasmic membranes that make up the **sarco-**

plasmic reticulum (SR), which functions as a storage site of calcium ions.

The importance of calcium ions in muscle contraction was first shown by Lewis Heilbrunn in 1947, when he injected a solution of calcium into a muscle fiber and watched the fiber suddenly contract. In the relaxed state, the level of calcium ion within the cytoplasm surrounding the myofibrils is very low. The arrival of an action potential along the T tubules causes an increase in the permeability of the SR to calcium ions, which diffuse out of their membranous containers and over the short distance to the myofibrils. These calcium ions bind to the thin filaments, exposing sites on the actin molecules to attachment of the myosin bulbs.

That is all it takes. As soon as cross-bridges form, filament sliding is initiated, and the muscle fiber shortens. After contraction has occurred, and the impulses have stopped, the free calcium ions are pumped back into the chambers of the SR, the sites on the actin molecules are hidden, and the muscle fiber again relaxes.

Regulating the Strength of Contraction

The same muscles that generate enough power to lift a 400-pound barbell can also gently pick up a newborn baby without hurling it against the ceiling. The strength by which a muscle contracts depends primarily on the number of muscle fibers that are stimulated which, in turn, depends on the number of neurons that carry impulses into the muscle tissue. In general, the more neurons that are activated, the more muscle fibers that shorten, and the stronger the contraction.

You are born with essentially all the skeletal muscle cells you will ever have. The myofibril content of each fiber reflects the muscle's usage, however. If you work out lifting weights or engage in an activity in which you use your strength, the muscles that are worked will develop more myofibrils. As a result, the diameter of each muscle fiber will increase, as will the thickness of the whole muscle and the force it can exert. Conversely, if you become bedridden for a period of time and are unable to use your muscles, they shrink in size (atrophy). If muscle inactivity persists long enough, the diminished size and strength of a muscle can be more or less permanent. Muscle fibers die and are replaced with connective tissue or fat. Even though dead fibers cannot be replaced, exercise can stimulate the surviving fibers to produce more muscle proteins and to grow in diameter, gaining enough strength to replace the lost cells and restore the muscle to full capacity.

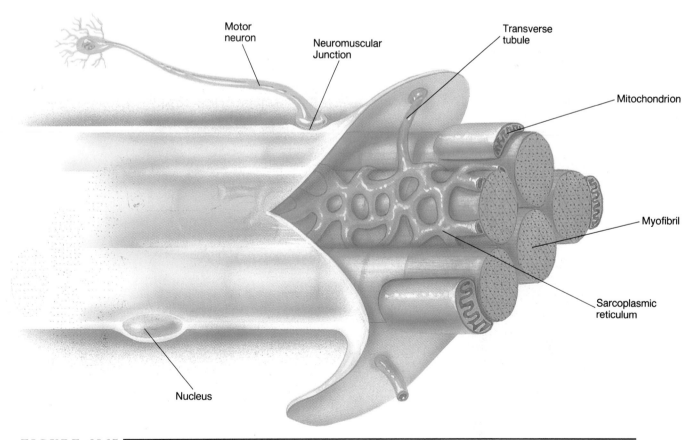

FIGURE 26-15

The anatomy of a muscle fiber. Calcium is housed in the elaborate network of internal membranes that make up the sarcoplasmic reticulum (SR). When an impulse is delivered by a motor neuron, the calcium escapes from the SR and binds to the thin filament, allowing cross-bridges to form between actin and myosin molecules and initiating filament sliding.

SMOOTH MUSCLE: RESPONDING TO INVOLUNTARY COMMANDS

Unlike skeletal muscle, **smooth muscle** is neither striated nor multinucleated, nor is it anchored to bone. It is not that smooth muscles lack actin and myosin filaments; rather, these contractile proteins are present in a much less ordered arrangement. Their interaction still generates the force of contraction, however. Smooth muscle consists of spindle-shaped cells (cells with tapering ends) that may be present in small clusters or as part of muscle sheets that surround the body's hollow organs (Figure 26-16).

Smooth muscle is *involuntary* in that its contraction is regulated by the autonomic nervous system (page 490) and is thus independent of conscious control. Smooth muscle in your urinary bladder, for example, automatically contracts in response to the internal pressure exerted on the walls of a full bladder. Fortunately, we have a backup skeletal muscle that is under voluntary control and closes off the exit to prevent voiding urine at inopportune moments. Other functions of smooth muscle include control over the diameter of blood vessels, the diameter of the pupil of the eye, and the state of erection of the hairs of the skin.

CARDIAC MUSCLE: FORMING THE BODY'S PUMP

Cardiac muscle tissue, which forms the muscular wall of the heart (Figure 26-17), consists of a type of muscle cell that has a unique combination of properties. Like smooth muscle, cardiac muscle is under involuntary control. Like skeletal muscle, cardiac muscle cells are striated, due to the alignment of the contractile filaments of the sarcomeres. Unlike most skeletal muscles, cardiac muscle does not function anaerobically (page 178). If the oxygen supply to the heart is blocked, as may occur during a heart attack, heart damage may ensue.

Unlike other muscle tissues, cardiac muscle cells are lined up end to end, joined to each other by a dense band, or **intercalated disk.** When viewed in the electron microscope, the intercalated disk is seen as a complex, communicating junction (page 110) that contains fine channels that run between the cytoplasm of adjacent cells. The role of these channels in heart function is discussed in detail in Chapter 28.

> **The forces required for bodily movements are generated by the contraction of specialized muscle cells brought about by interactions between two proteins: actin and myosin. Ordered movements of actin- and myosin-containing filaments are driven by the energy that is released from ATP hydrolysis. (See CTQ #5.)**

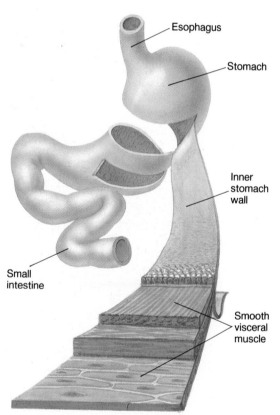

FIGURE 26-16

Smooth, involuntary muscle is often present as sheets of uninucleated, nonstriated muscle cells, such as those found in the digestive tract (pictured here) or along the ducts of the urinary tract. Waves of contraction (*peristalsis*) pass along these muscle sheets, generating a moving constriction that pushes the contents of the channel along the tract.

Labels on figure: Esophagus; Stomach; Inner stomach wall; Small intestine; Smooth visceral muscle

BODY MECHANICS AND LOCOMOTION

Now that we have seen how bones, joints, and muscles function individually, we can better understand how these components of the skeletomuscular system work together to perform mechanical work. Humans use levers, such as the wheelbarrow depicted in Figure 26-18*a*, to lessen the amount of force required to lift and move an object (in this case, a load of dirt). In this example, the wheel is the *fulcrum*, or pivot site; the dirt is the load to be lifted; and energy is applied as an upward force against the handle. The body uses bones as levers, and joints as fulcrums, to accomplish similar functions. When you stand on your tiptoes, for example, an upward force is generated by the muscle in

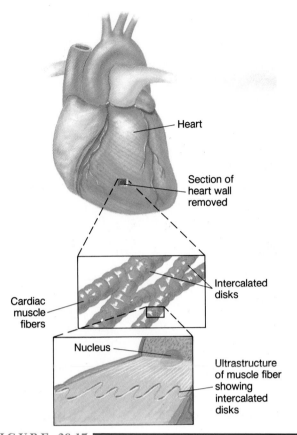

FIGURE 26-17
Cardiac muscle consists of striated, mononucleated cells that are interconnected by membranes that form intercalated discs.

your calf using the ball joint of the toes as the fulcrum (Figure 26-18*b*) and the bones of the foot as a lever. The load, in this case, is the entire weight of your body.

Muscles, bones, and joints work together to move animals from place to place. Animal locomotion is so diverse, it defies brief summation (Figure 26-19). Animals swim, fly, glide, crawl, climb, run, burrow, and hop over and through their environment. Of all the various styles of locomotion, bipedal (two-legged) movement by humans is one of the most unusual. The only contact between our body and the ground occurs at the bottom of two relatively small, bony feet. Unlike in other animals, the center of gravity in humans is elevated to a position far above the ground, which is unstable and precarious. The transition from four-legged to two-legged locomotion was accompanied by rapid evolutionary changes in the skeleton. Included among these were the expansion of the feet, a shift in the articulation between the femur and the pelvis (to allow for a vertical posture), and the development of an S-shaped curve in the lower (lumbar) region of the back.

The movement of a part of the body is accomplished by a force (generated by muscle contraction) acting on levers (represented by bones) that pivot at a fulcrum (one of the body's joints). (See CTQ #6.)

FIGURE 26-18
Bones serve as the body's levers. Pushing a wheelbarrow *(a)* and standing on your tiptoes *(b)* utilize similar types of levers. In both cases, the fulcrum (pivot point) is at one end of the system, the load to be lifted is in the middle, and the force needed to move the load is exerted on the opposite end. Both the wheelbarrow and the bones of the foot act as levers that allow this type of mechanical work to be performed.

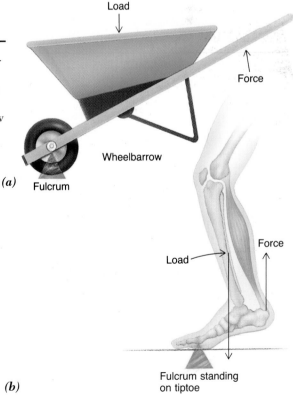

FIGURE 26-19

Examples of Animal locomotion. *(a)* Cuttlefish swimming, *(b)* a basilisk lizard running over the water's surface, and *(c)* lacewings flying.

(a)

(b)

(c)

EVOLUTION OF THE SKELETOMUSCULAR SYSTEM

▌▶ The skeleton within your body can be traced back through a series of fossils to the skeletons of primitive fishes that lived more than 400 million years ago. Earlier, we noted that these fishes were covered with plates of bone that formed within the dermis of the skin. While nearly all of this dermal bone has disappeared over the course of vertebrate evolution, a number of the bones in your jaw and skull

are derived from these dermal bony plates present in ancestral fish. Other parts of your skull and jaws, as well as the bones in your middle ear, are derived from bones that once supported the gills of these ancient fish (Chapter 39). In contrast, the bones in your limbs have more recent evolutionary roots; limbs first appear in the fossil record in primitive amphibians, soon after vertebrates made their way onto the land. The bones of these limbs were derived from bones that supported the fins of the fishes from which these amphibians evolved.

The skeletal system of vertebrates probably arose in primitive fishes, whereas the muscles of vertebrates reveal an even earlier origin. Actin and myosin, the major proteins of vertebrate muscle tissue, are found in virtually all eukaryotes, from yeast and fungi to trees and mammals. Muscle cells, which are specialized for contraction, can be found in the most primitive multicellular animals—the sponges —and thus must have appeared very early in the course of animal evolution.

Next time you dine on a trout and are carefully pulling the bony skeleton away from the meat, you might remember that this is the form from which the skeletal muscles in your body probably evolved—as a "wall" of muscle tissue on each side of a fish's body. As vertebrates moved out of the water, and their skeletons became reorganized by natural selection in many different ways, the solid wall of skeletal muscle became divided into discrete muscle masses capable of moving individual bones. The shape and position of the muscles of an animal are closely correlated with the peculiarities of its skeleton and the specialized movements it makes.

Muscle tissue is present throughout the animal kingdom, whereas bone is unique to vertebrates. The particular shapes and arrangements of the muscles and bones of an animal can be correlated with the types of movements the animal performs. (See CTQ #7.)

REEXAMINING THE THEMES

Relationship between Form and Function

Bones are shaped to carry out particular functions; their internal architecture reflects the specific stresses to which they are subjected. Bones are moved by skeletal muscles, whose cells contain an elaborate array of overlapping protein filaments that convert these cells into pulling machines. A key component in this machinery is myosin, a long, rod-shaped molecule with bulbous heads that project from one end. The receptor sites in the actin filaments fit perfectly with the myosin heads. The bending of the myosin heads creates the power stroke that causes the actin filaments to slide over the myosin rods, shortening the entire length of the cell.

Biological Order, Regulation, and Homeostasis

Contraction of a skeletal muscle cell is regulated by the concentration of calcium ions in the myofibrils. In the relaxed state, calcium ions are sequestered within the membranous compartments of the SR. When an impulse speeds along a motor neuron and reaches a muscle cell, the SR membranes become permeable to calcium ions; the calcium concentration around the contractile filaments rises; calcium ions bind to the thin filaments, exposing sites on the actin molecules to attachment of the myosin bulbs; and the fiber shortens. When nerve impulses stop arriving, calcium ions are once again sequestered, and contraction ceases.

Acquiring and Using Energy

The large amount of energy required to fuel muscle activity derives from one simple molecular activity: bending the heads on the myosin molecules that push the actin filaments toward the center of the sarcomeres of the muscles that are contracting. ATP molecules bind to the myosin heads before they attach to the actin filaments and are subsequently hydrolyzed as the actin filaments are slid over the myosin. This mechanism of contraction is so efficient that up to 50 percent of the energy released by ATP hydrolysis is converted to mechanical energy, the remainder is released as heat.

Unity within Diversity

All types of vertebrate integuments are built of the same two layers—an epidermis and a dermis—yet the properties of the integument vary greatly from one animal to another. The bones of vertebrates have a similar molecular composition and a similar microscopic construction, yet the overall shape of each bone varies greatly. Even within a given person, individual bones range from the tiny elements of the wrist to the massive, long bones of the leg. Similarly, the muscles of vertebrates (and all animals) are built of the same types of contractile proteins and operate by a very similar molecular mechanism, yet the particular movements they promote vary greatly both within an animal and among diverse species.

Evolution and Adaptation

A specific organ, derived from a particular part of an early embryo, can become modified dramatically over the course of evolution as it meets the needs of different types of animals living in different types of habitats. The vertebrate integument, for example, ranges from thick and scaly in most fishes, to moist and permeable in most amphibians, to dry and horny in most reptiles, to feather covered in birds and hair covered in mammals. Even among related animals, there is great variation. For example, gorillas are covered with hair, while humans are almost devoid of these epidermal derivatives.

SYNOPSIS

An animal's body surface is covered by a protective integument. In humans and other vertebrates, the skin is composed of an outer epidermis and an inner dermis. The outer, protective layer of the human epidermis consists of dead, keratinized cells that are continually sloughed and replaced. The dermis contains connective tissue, which provides support and skin cohesion, and blood vessels, which nourish the skin and play a role in heat conservation or heat loss. Sebaceous glands keep the skin pliable. Sweat glands secrete a dilute salt solution, whose evaporation cools the body. The properties of the vertebrate integument vary, depending on the type of animal and the habitat in which it lives.

Skeletal systems provide the rigidity that is required for support and movement. Some animals use only internal water pressure to support the body; others have a rigid exoskeleton that is external to the animal's living tissues. This may be a calcified shell or a hardened cuticle of protein and chitin. Still other animals have an internal endoskeleton: Sponges have needlelike spicules, sea stars and sea urchins have calcified plates, and vertebrates have bones.

The human skeleton consists of 206 bones, each with a unique shape that allows it to perform a particular function. Most bones arise in the embryo as cartilaginous structures. They are then converted to bone by the deposition of calcium phosphate to the protein–polysaccharide matrix which surrounds the living cells. Bones are typically hollow and contain both compact and spongy regions. The axial skeleton includes the skull, which protects the brain; the vertebral column, which protects the spinal cord; and the rib cage, which protects the organs of the chest cavity.

The appendicular skeleton consists of the paired pectoral and pelvic girdles, which attach the limbs to the axial skeleton, and the bones of the limbs themselves. The bones of the skeleton are connected by joints that possess varying degrees of flexibility.

Muscles are composed of specialized cells that contract (shorten) and generate a pulling force. Vertebrate muscle is divided into skeletal, smooth, and cardiac types. Skeletal muscle responds to voluntary commands and is primarily responsible for moving portions of the skeleton. Skeletal muscles are composed of large, multinucleate cells (fibers) containing myofibrils that have a markedly striated appearance. The structural unit of the myofibril is the sarcomere, which contains overlapping sets of thick (myosin-containing) and thin (actin-containing) filaments. When a muscle fiber is activated to contract, the thin filaments of the sarcomeres slide over the thicker filaments. The force required for this motion is fueled by ATP hydrolysis and is generated by the bending movements of the bulbous heads of the myosin molecules when they are attached to the adjacent, thin filaments. The attachment of the myosin heads to the actin molecules of the thin filaments is triggered by calcium ions, which are released from the sarcoplasmic reticulum of the muscle cell following the arrival of a nerve impulse. The strength by which a muscle contracts depends on the number of fibers that contract, which depends on the number of neurons that carry impulses into the muscle. Smooth muscles, which lack striations, mediate involuntary movements, such as the constriction of blood vessels and the closure of the pupil of the eye. Cardiac muscle, which is striated, makes up the wall of the heart.

Key Terms

integument (p. 542)
skin (p. 542)
epidermis (p. 542)
dermis (p. 542)
exocrine gland (p. 543)
sebaceous gland (p. 543)
sweat gland (p. 543)
hydrostatic skeleton (p. 544)
exoskeleton (p. 545)
cuticle (p. 545)
endoskeleton (p. 546)
bone (p. 546)
osteocyte (p. 546)
compact bone (p. 546)
spongy bone (p. 546)

red marrow (p. 546)
yellow marrow (p. 546)
cartilage (p. 548)
chondrocyte (p. 548)
axial skeleton (p. 548)
skull (p. 548)
vertebral column (p. 548)
rib cage (p. 548)
appendicular skeleton (p. 548)
pectoral girdle (p. 548)
scapulae (p. 548)
clavicle (p. 548)
pelvic girdle (p. 548)
pelvis (p. 548)
joint (p. 550)

ligament (p. 551)
synovial cavity (p. 551)
muscle fiber (p. 553)
myofibril (p. 553)
sarcomere (p. 553)
thin filament (p. 555)
thick filament (p. 555)
actin (p. 555)
myosin (p. 555)
transverse (T) tubule (p. 556)
sarcoplasmic reticulum (SR) (p. 557)
smooth muscle (p. 558)
cardiac muscle (p. 558)
intercalated disk (p. 558)

Review Questions

1. Compare and contrast skeletal muscle and cardiac muscle; skeletal muscle and smooth muscle; osteoclasts and osteoblasts; outer and inner layers of the human epidermis; sebaceous and sweat glands; endocrine and exocrine glands; compact and spongy bone; red and yellow marrow; hydrostatic skeleton and exoskeleton; cartilage and bone.

2. What properties do collagen and cellulose molecules have in common? How is the role of collagen in bone similar to that of cellulose in a plant cell wall?

3. Describe how the integument has changed over the period of vertebrate evolution. Contrast the appendicular skeleton of a fish with that of a human. Contrast the musculature of a fish and a human.

4. Describe the events that occur in a muscle fiber following the arrival of an excitatory nerve impulse. Carry the discussion through to the point where the fiber returns to its original, relaxed state.

Critical Thinking Questions

1. Explain, in terms of enzyme activity, why Ehlers-Danlos syndrome would produce less severe effects than would extreme ascorbic acid deficiency.

2. In addition to protecting the body against the environment, the skin has to receive information from and exchange materials with the environment. How is the skin organized and structured to perform these somewhat contradictory functions?

3. What would you expect to remain if a chicken leg bone were soaked in an acid, such as vinegar, for several days? How would it differ from an untreated bone? What if, instead of acid, you subjected the bone to very high heat in a dry oven?

4. Consider two terrestrial animals, one with an exoskeleton (such as an insect) and the other with an endoskeleton (such as a dog). As these two animals increase in size, which skeleton would you expect to represent a greater and greater percentage of overall body weight? What effect do you think this might have on the size each of these animals could attain?

5. If smooth, instead of striated, muscle were attached to the bones of the skeleton, what effects would this have on the movement of the body?

6. Imagine that your right elbow is resting on the table and your right hand is lifting a heavy weight. Considering the role of the biceps in bending your arm (see Figure 26-10), how do the relative positions of the fulcrum, load, and force differ in this case, compared to the example of standing on your tiptoes (Figure 26-18)?

7. Why are all large animals with exoskeletons aquatic? Why are there no insects as large as elephants (or even dogs)? Why are the largest animals with an exoskeleton (the whales) aquatic?

Additional Readings

Alexander, R. M. 1983. *Animal mechanics.* London: Blackwell. (Intermediate–Advanced)

Diamond, J. 1993. Building to Code. *Discover* May:92–98. (Intermediate)

Gans, C.. 1980. *Biomechanics: An approach to vertebrate biology.* Ann Arbor: Univ. of Michigan. (Intermediate–Advanced)

Hildebrand, M. 1989. Vertebrate locomotion: An introduction. *BioScience* 39:764–765. (Intermediate)

Huxley, H. 1969. The mechanism of muscular contraction. *Science* 164:1356–1366. (Intermediate–Advanced)

Langley, L. L., I. R. Telford, and J. B. Christensen. 1980. *Dynamic anatomy and physiology,* 5th ed. New York: McGraw-Hill. (Intermediate)

Levin, R. M. 1991. The prevention of osteoporosis. *Hosp. Pract.* May:77–97. (Advanced)

Luciano, D. S., A. J. Vander, and J. H. Sherman. 1978. *Human structure and function.* New York: McGraw-Hill. (Intermediate)

Weiss, L. ed. 1988. *Cell and tissue biology: A textbook of histology.* Baltimore: Williams & Wilkins. (Advanced)

CHAPTER

◂ 27 ▸

Processing Food and Providing Nutrition: The Digestive System

STEPS TO DISCOVERY

The Battle Against Beri-Beri

THE DIGESTIVE SYSTEM: CONVERTING FOOD INTO A FORM AVAILABLE TO THE BODY

THE HUMAN DIGESTIVE SYSTEM: A MODEL FOOD-PROCESSING PLANT

The Mouth and Esophagus: Entry of Food into the Digestive Tract

The Stomach: A Site for Storage and Early Digestion

The Small Intestine: A Site for Final Digestion and Absorption

The Large Intestine: Preparing the Residue for Elimination

EVOLUTION OF DIGESTIVE SYSTEMS

Adaptations to Different Dietary Behaviors

BIOLINE

Teeth: A Matter of Life and Death

THE HUMAN PERSPECTIVE

Human Nutrition

STEPS TO DISCOVERY
The Battle against Beri-beri

Unlike so many other human diseases, beri-beri was probably not an ancient scourge, but primarily a product of the Industrial Revolution. The first clearly documented cases of this nervous disorder, which is characterized by fatigue, muscle deterioration, and possible paralysis, appeared in Asia in the nineteenth century. The disease became prevalent among prisoners and soldiers stationed in the Dutch East Indies in the 1880s. The Dutch government dispatched a team of scientists to look into the problem. Among the members of the team was a medical officer named Christiaan Eijkman.

During the 1880s, etiology (the study of disease) was dominated by the findings of Louis Pasteur and Robert Koch, who had been instrumental in proving that diseases are often caused by "germs" that grow in the body. Unfortunately, Pasteur's and Koch's contemporaries believed that this "germ theory" applied to all diseases; that is, they believed that all diseases were attributed either to bacterial infections or to the toxins produced by bacteria. Eijkman spent 4 fruitless years trying to isolate the bacterium responsible for beri-beri.

One day in 1896, a sudden development provided an unexpected breakthrough in Eijkman's research. For no apparent reason, the chickens that Eijkman was using as

Experiments on chickens revealed that rice kernels contain a vitamin required for normal metabolism. The red bar

experimental animals developed a nerve disease whose symptoms resembled those of human beri-beri. Many of the animals died, but after 4 months, the chickens that had survived the disorder had recovered completely. Upon investigating the matter, Eijkman discovered that the chickens began to recover after a new animal keeper had stopped feeding them leftovers from the military hospital, which consisted largely of polished rice—rice that had been processed by a steam mill until the outer hulls had been removed.

Eijkman used this information in a dietary experiment. He fed some of the chickens a diet of polished rice; these chickens soon developed symptoms of the disease. In contrast, control animals that were fed whole rice remained healthy. Furthermore, the afflicted group could be cured if they were fed either whole rice or polished rice to which the outer hulls had been added. Eijkman concluded that the disease was not due to a bacterial infection but to a dietary deficiency, providing the first evidence that a disease could be caused by the absence of some trace component of the diet. The idea that a disease could result from a dietary deficiency did not initially gain widespread acceptance, however. Eijkman and a colleague, Gerrit Grijns, tried to isolate the substance from the rice hulls which corrected the deficiency, but all they learned was that the factor could be extracted from the hulls in water.

In 1911, Casimir Funk, a chemist working at the Lister Institute in London, succeeded in purifying a substance from the hulls of rice. Funk believed the substance was the same ingredient that Eijkman had found to reverse the symptoms of beri-beri. The substance was an amine (one containing an amino [—NH$_2$] group), which led Funk to coin the word "vitamine," meaning an *amine* that was *vital* to life. Later work showed that the substance crystallized by Funk was not, in fact, the same one that was active against beri-beri. The name caught on, however, and remained the common term used to describe organic substances that are

required by the body in trace amounts. Most vitamins, in fact, contain no amine groups.

Finally, in 1926, two Dutch chemists, B. C. Jansen and W. Donath, working in Eijkman's old lab in the East Indies, developed a procedure for purifying the anti-beri-beri factor from rice bran. Crystals of the substance were sent back to Holland, where Eijkman confirmed that this single chemical compound was effective against the nervous disorder exhibited in birds. However, when Jansen and Donath determined the chemical formula for the substance, they overlooked an important feature: the presence of a sulfur atom. This oversight set back the effort to determine the structure of the compound, which had been named vitamin B$_1$. The presence of the sulfur atom wasn't discovered until 1932; the correct structure was published in 1936 by Robert Williams, an American chemist who had been working on the problem for over 20 years. Within a year, Williams had worked out a complex procedure for synthesizing the compound, which he named thiamin. Soon, thiamin was being manufactured and became available as a widespread vitamin supplement.

The last major step in the story of vitamin B$_1$ was the discovery of its biological action. In 1937, two German biochemists, K. Lohmann and P. Schuster, found that thiamin was a coenzyme involved in the reaction that converts pyruvate to acetyl CoA prior to its entry into the Krebs cycle (page 181). Later investigations revealed that virtually all vitamins act in conjunction with enzymes in carrying out one or more crucial metabolic reactions. It was the *failure* to catalyze these reactions that led to the symptoms of the deficiency diseases, such as beri-beri. For his work in establishing the existence of dietary deficiency diseases, Eijkman was awarded the 1929 Nobel Prize in Medicine and Physiology.

and bent arrow pinpoint the reaction that is blocked when the substance is absent from the diet.

*F*or some animals, obtaining food is a relatively simple task. Tapeworms, for example, have no mouth or digestive tract; they simply attach to the wall of an animal's intestine and absorb digested nutrients across their outer body surface. For these animals, eating is not necessary. Unfortunately, as humans, we can't enjoy the same advantage. We can't recline in a bathtub of oatmeal for breakfast and chicken soup for lunch and simply soak up the nutrients. Any such endeavor would only end in starvation. Even if the nutrients could penetrate the epithelial layer of the skin and enter the bloodstream, most of the molecules would be too large to cross the plasma membranes and enter the cells. The molecules would only be wasted since the value of food is unleashed only inside living cells. To meet this requirement for life, most animals possess a team of specialized organs that constitute the **digestive system.**

▼ ▼ ▼

THE DIGESTIVE SYSTEM: CONVERTING FOOD INTO A FORM AVAILABLE TO THE BODY

At what point does food enter the body? Most people would answer, "as soon as you put it in your mouth," or "when you swallow it." Yet, eating does not actually introduce food into your body. When substances are in the stomach or intestines, they are still outside of you, just as your finger poking through the hole of a doughnut remains outside the pastry. The **digestive tract,** or *gut,* is actually a tubelike continuation of the animal's external surface into or completely through its body. The walls of the tract might be likened to an absorbant version of the skin, one that forms a barrier between the external environment and the internal tissues of the body. Because of this barrier property, the digestive tract can safely provide residence for a large number of bacteria that would be dangerous to the interior, living tissues of the body.

To enter the body itself, nutrients must be absorbed across the epithelium that lines the digestive tract. **Digestion** prepares food to do just that. Digestion is the process of disassembling large food particles into molecules that are small enough to be absorbed by the cells that line the digestive tract. Ultimately, these molecules enter the cytoplasm of every cell in the body, where the nutritive value of food is harvested. Eating only initiates the digestive process, launching food on a journey through tunnels and chambers, where digestion and absorption occur.

An animal's diet consists largely of macromolecules— proteins, polysaccharides, lipids, and nucleic acids— that have been synthesized to the specifications of the organism being eaten. To be of use to an animal, these macromolecules must first be disassembled (digested) and then absorbed into the animal's body, where the nutrients can be used for synthesis of new macromolecules. Digestion and absorption are functions of the digestive system.

THE HUMAN DIGESTIVE SYSTEM: A MODEL FOOD-PROCESSING PLANT

The human digestive tract is approximately 9 meters (30 feet) long. It consists of the *mouth, esophagus, stomach, small intestine, large intestine,* and *anus,* plus a variety of accessory organs (Figure 27-1). Each part of the digestive tract is structurally adapted to carry out a particular phase of the digestive process. In fact, the human digestive tract is a model food-processing plant for the stepwise disassembly and absorption of food material. During its journey through the digestive tract, ingested food matter is mixed with various fluids; churned and propelled by the musculature of the wall; broken down by enzymes that are secreted by various glands; and absorbed by cells that line the digestive channel. The indigestible residues are then eliminated from the tract through the anus. All of these complex processes are regulated by the coordinated action of both the nervous and endocrine systems (Chapters 23 and 25).

The wall of the digestive tract (Figure 27-1) is composed of several layers, including an inner glandular epithelium (*mucosa*), layers of circular and longitudinal smooth muscle, and a connective tissue sheath (*serosa*). Glandular cells in the epithelium secrete mucus, enzymes, ions, and other substances. Contraction of the muscle layers help break up congealed food matter, which is mixed with secreted fluids and moved through the tract. We will begin our journey through the digestive system as food enters the first part of the digestive tract: the mouth.

THE MOUTH AND ESOPHAGUS: ENTRY OF FOOD INTO THE DIGESTIVE TRACT

Digestion of food begins in the **mouth,** where food is cut and ground by the teeth. This action makes the food matter easier to swallow and increases its access to digestive enzymes (see Bioline: Teeth: A Matter of Life or Death). While in the mouth, the macerated food is mixed with **saliva:** the secretion that initiates chemical digestion. Saliva is produced by three pairs of **salivary glands,** whose ducts open into the mouth. (It is the salivary glands that become swollen during a mumps virus infection.) Saliva contains the

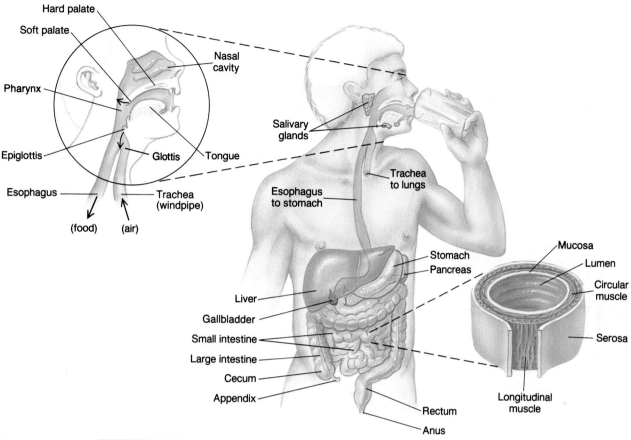

FIGURE 27-1

The human digestive system. In addition to the tubular digestive tract, the digestive system includes accessory organs (the pancreas, liver, and gallbladder) that aid digestion. The mucosa and musculature of the digestive tract are shown in the intestinal cross section. The inset depicts structures involved in swallowing. During swallowing, the *soft palate* elevates and closes off the nasal cavity, while the *glottis* (the opening to the windpipe and lungs) is sealed by the *epiglottis,* leaving the esophagus the only open passageway for the mass of chewed food (bolus).

enzyme amylase that initiates digestion of starch and helps protect teeth from decay by breaking down and dislodging starchy food particles that get trapped between teeth. Saliva also contains *mucin,* the major protein of mucus, which acts as a lubricant. This feature is best appreciated when we try to swallow something that has not been adequately covered by the slippery fluid. Mucin also binds the macerated food together into a cohesive mass, called a **bolus.**

Swallowing is essentially a two-step process. The first step is voluntary: The tongue pushes the bolus to the back wall of the oral cavity (the *pharynx*), where it stimulates sensory receptors that initiate the second step. This step involves a series of reflex, muscular contractions that push the bolus into the **esophagus,** the tubular channel that leads to the stomach. During the swallowing process, the openings to the respiratory and nasal passages are automatically closed to ensure that food is kept out of these nondigestive pathways (see inset, Figure 27-1). This explains why

you can't breathe while swallowing. The swallowing reflexes occasionally fail to maintain the proper sequence, and food or liquids accidentally enter the airways. The explosive coughing reflex that is triggered by such misdirection protects the respiratory tract, which gracefully accepts only gases (Chapter 29).

The walls of the esophagus contain muscle layers that contract in a rhythmic manner, sending successive waves of contraction, or **peristalsis,** down its length (Figure 27-2). Peristalsis constricts the channel of the esophagus, pushing the bolus ahead of the traveling wave and through a thickened muscular band, called the **cardiac sphincter** (see Figure 27-5), and into the stomach. The cardiac sphincter opens automatically as the wave reaches the bottom of the esophagus and then quickly closes so that the acidic contents of the stomach do not back up and injure the esophagus. Occasionally, this backup occurs anyway, causing "heartburn" or "acid stomach," which is felt neither in the

◁ B I O L I N E ▷
Teeth: A Matter of Life and Death

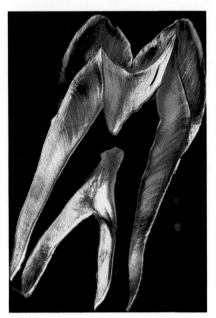

FIGURE 1

The elephant's heart was in excellent condition. So were the animal's other vital organ systems. The giant was free of infection and disease. Yet, she was dying, slip-ping toward the same end that has claimed her kind for thousands of generations. Although surrounded by food, the elephant was starving to death.

The coarse diet of plants on which an elephant dines steadily grinds away the animal's teeth. During the elephant's lifetime, its worn-out *molars* (which grind food) are periodically replaced so that eating can continue. Only six sets of teeth can be produced, however; by the age of 60, the last set has usually been ground flat. The otherwise healthy elephant can no longer chew its food and literally starves to death.

Teeth provide an excellent example of the relationship between form and function. Animals whose diets include large amounts of plant material (such as that of the elephant) require teeth that can crush the tough cell walls, releasing the intracellular nutrients. The teeth must have broad-ridged surfaces that grind together, as do our molars (Figure 1). Dogs, cats, and other carnivores that feed almost exclusively on meat have little need for teeth with grinding surfaces since there are no cell walls in their diets to crush. In these animals, grinding surfaces have been replaced by sharp, cutting edges that rip and slice the food into smaller pieces. The knifelike, cutting edges of the front teeth (*incisors*) are specialized for cutting off bites of food. Four pointed teeth, called *canines* (or "fangs"), are instrumental in securing and killing prey. The reduction in the length of canine teeth during human evolution reflects our different strategy for capturing and handling prey. Primitive humans used weapons and their hands rather than their teeth for this purpose.

Animals that swallow their food whole do not use their teeth to assist digestion. For example, snakes have backward-curving teeth that are used for capturing and holding prey. The sharp teeth of sharks provide a fascinating, if not chilling, example of teeth that are adapted strictly for tearing off chunks of meat to be swallowed without being chewed. Birds, which lack teeth altogether, have a muscular grinding organ, called the *gizzard* (Figure 27-8c). With the help of swallowed stones, the gizzard macerates food as it passes through the digestive tract.

---

heart nor in the stomach, but *near* the heart, in the esophagus.

THE STOMACH: A SITE FOR STORAGE AND EARLY DIGESTION

The **stomach** continues the mechanical breakdown of food solids and limited enzymatic digestion, which were begun in the mouth and esophagus. The average human stomach can hold and store about a liter (nearly a quart) of material. The stomach contents are churned to a pastelike consistency and are mixed with *gastric juice* (*gastro* = stomach), forming a solution called **chyme.** Gastric juice is produced by secretory cells that are located in pits in the wall of the stomach (Figure 27-3). *Hydrochloric acid* (HCl), one of the compounds that makes up gastric juice, lowers the pH of the stomach contents to around 2.0. This extremely acidic environment kills most microbes found in food, including many of those that could cause illness. Stomach acid also denatures (opens up) highly folded protein chains so that they can be more easily attacked by protein-digesting enzymes.

Although most enzymatic digestion occurs in the small intestine, protein digestion begins in the stomach with the action of the enzyme *pepsin*. To prevent the stomach from becoming its own next meal, the protein-digesting enzyme is secreted as an inactive precursor, *pepsinogen,* which is converted to an active pepsin molecule by the hydrochloric acid of the stomach. The living tissue that lines the stomach is protected from both acid and pepsin by a thick layer of alkaline mucus, which is secreted by specialized glandular cells found in the stomach wall. When protective mechanisms fail, the stomach may begin digesting portions of itself, causing painful and dangerous *peptic ulcers.* For decades, peptic ulcers have been treated by administration of antacids. Persons suffering from these lesions, however,

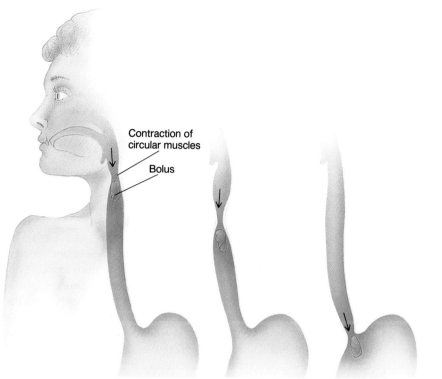

FIGURE 27-2

Movement by peristalsis. Contraction of a layer of circular muscle in the wall of the tubular esophagus sweeps down the length of the organ, pushing the bolus into the stomach. Similar types of peristaltic waves force ingested material through the entire length of the digestive tract.

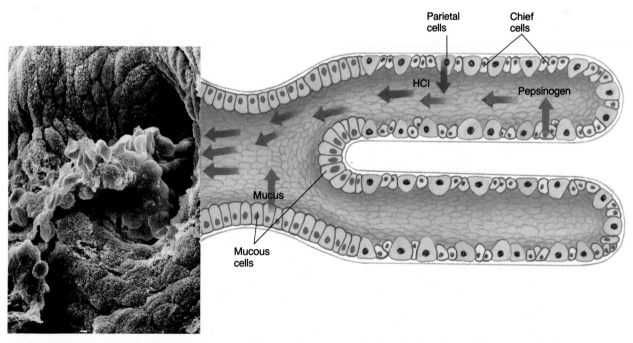

FIGURE 27-3

Gastric juice combines the secretory products from several different types of gland cells, which are located in pits (inset) that open into the stomach chamber. Within the wall of the pit are *mucous cells* that secrete mucus, *chief cells* that secrete pepsinogen (the inactive precursor of the enzyme pepsin), and *parietal cells* that secrete hydrochloric acid (HCl).

have found this therapy ineffective; while antacids help to alleviate symptoms, over 95 percent of patients experience a relapse within two years. Within the past few years, a number of studies have shown that, while stomach acidity may be a factor in the development of peptic ulcers, the primary cause is infection by the acid-resistant bacterium *Helicobacter pylori*. Ulcer patients treated with antibiotics that kill the bacteria are much less likely to suffer relapses than those treated solely with antacids. Don't be surprised if, one day, a vaccine becomes available to prevent ulcers. According to one estimate, the eradication of *H. pylori* could cut as much as 80 percent of the cost of gastrointestinal medicine.

Hunger and the Control of Gastric Secretions

⟳ Food-seeking behaviors, such as those that overcome many of us on a late-night outing, usually follow sensations of *hunger*: the body's message to your conscious mind that nutrients need to be replenished. During the hours following a meal, the concentration of glucose in the blood is high enough to stimulate neurons, called *glucoreceptors*, in the hypothalamus. Activated glucoreceptors inhibit the brain's *hunger center*, which is also located in the hypothalamus. The sensation of hunger reoccurs when glucose levels fall slightly, signaling glucoreceptors to release their inhibitory

grip on the hunger center. Yet, glucoreceptors cannot be the sole factor responsible for regulating hunger since the sensation subsides as soon as our stomachs become full, long before glucose from the meal is absorbed into the bloodstream. This rapid inhibition of hunger is probably the direct result of stomach distension, or swelling, which stimulates pressure-sensitive sensory neurons in the stomach wall.

The control of the secretion of gastric juices illustrates the complex communication that occurs during physiological processes (Figure 27-4). The first phase of gastric secretion is stimulated by nerve impulses that reach the stomach from the brain as a result of the smell, taste, or even thought of food. When food actually enters the stomach, two new types of signals are generated, which lead to a marked increase in the secretion of gastric juice. One of the signals is carried by sensory neurons from the stomach to the brainstem, which responds by sending impulses down autonomic motor fibers. This stimulates the digestive-gland cells of the stomach wall to release their products. The other signal is a chemical message that is sent by the hormone gastrin, which is released by endocrine cells located in the stomach lining. The message is carried locally in the blood vessels of the stomach wall to the stomach's glandular cells, triggering these cells to release gastric juices. By sending these nervous and endocrine signals, the stomach ensures that gastric juices will be secreted when they are necessary.

Only a few small molecules, such as aspirin and alcohol, enter the bloodstream through the stomach wall. This explains the rapid onset of their effects. Most nutrients are not absorbed until they enter the small intestine.

THE SMALL INTESTINE: A SITE FOR FINAL DIGESTION AND ABSORPTION

Peristaltic waves moving along the wall of the stomach repeatedly push small quantities of chyme through a muscular valve (the *pyloric sphincter*) and into the **small intestine** (Figure 27-5). The small intestine consists of 6 to 7 meters (about 21 feet) of highly coiled muscular tubing, about 2.5 centimeters (1 inch) in diameter. During their stay in the small intestine, macromolecular food substances are digested into small, organic molecules, such as simple sugars, amino acids, and nucleotides, all of which are absorbed into the bloodstream.

Digestion of materials within the small intestine requires the cooperative activities of several major organs and their secretions, including intestinal juice, pancreatic secretions, and emulsifying materials from the liver and gallbladder.

Intestinal Secretions

When the inflow of chyme from the stomach stretches the intestinal wall, the action triggers a neural reflex response, whereby the cells of the intestinal lining secrete *intestinal juice* and mucus. Under normal conditions, the small intestine secretes about 2 to 3 liters of intestinal juice per day.

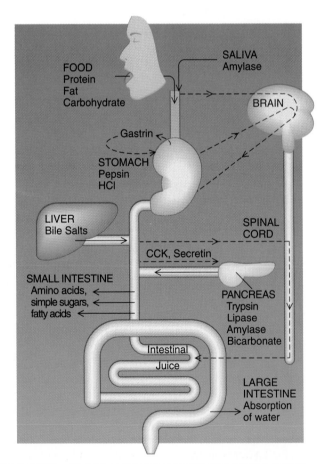

FIGURE 27-4

Control of the processes of digestion by the nervous and endocrine system.

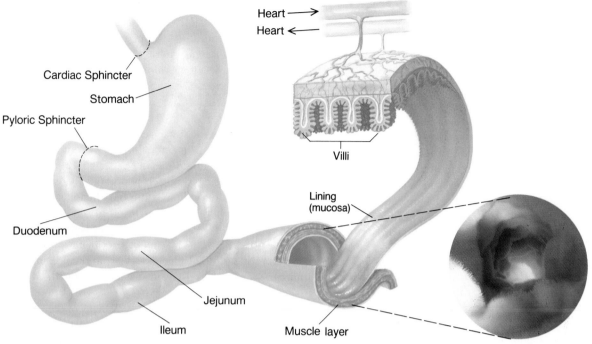

Heart →
Heart ←

Cardiac Sphincter

Stomach

Pyloric Sphincter

Villi

Lining
(mucosa)

Duodenum

Jejunum

Ileum

Muscle layer

FIGURE 27-5

Anatomy of the small intestine. Approximately the first 30 centimeters of the small intestine forms a C-shaped curve, called the *duodenum*. The *jejunem* comprises the next 1.5 meters of the small intestine, and the *ileum* constitutes the remaining 2.5 meters of the tube. The surface area of the intestinal lining is increased by the folding of the intestine and by the presence of villi, which are shown in the inset photo. The average person sends about 725 kilograms (330 pounds) of solid food and 700 liters (650 quarts) of ingested liquids through this tunnel every year.

This fluid is needed to dissolve the molecules for digestion and to facilitate absorption across the intestinal epithelium. Cholera, one of the most dreaded human diseases, results from a bacterial toxin that greatly increases the fluid released by these intestinal cells. When stimulated by the cholera toxin, the intestine can pour out over 1 liter of fluid per hour, most of which is simply lost through the digestive tract as diarrhea. Unless this loss of fluid is made up by drinking or intravenous entry, patients are at high risk of dying from dehydration. The bacteria that cause cholera are usually acquired by drinking water that is contaminated by the feces of other people with the disease.

Pancreatic Secretions

As we discussed in Chapter 25, the pancreas (review Figure 27-1) is a gland that manufactures digestive enzymes and discharges them directly into the *duodenum*, the first region of the small intestine. The pancreas also releases sodium bicarbonate, an alkaline substance that helps neutralize the severe acidity of the chyme entering the small intestine from the stomach. Within the pancreatic secretion are enzymes that digest all four major types of macromolecules:

*   *Trypsin, chymotrypsin,* and other such proteolytic enzymes degrade proteins to amino acids.

*   *Nucleases* break down nucleic acids to nucleotides and nitrogenous bases.

*   *Carbohydrases* hydrolyze complex carbohydrates to disaccharides or simple sugars.

*   *Lipases* split triglycerides into fatty acids and glycerol.

The secretion of pancreatic enzymes and bicarbonate is stimulated by two hormones: *cholecystokinin (CCK)* and *secretin* (Figure 27-4). These hormones are secreted into the blood by endocrine cells in the wall of the small intestine in response to the inflow of chyme from the stomach.

↻ The enzymes produced by the pancreas are potentially dangerous molecules since they have the ability to destroy most of the body's own materials. The pancreas is protected from some of these enzymes by two mechanisms: (1) synthesis of the enzymes in an inactive state, and (2) synthesis of the enzymes along with the inhibitors that keep them inactive until they reach the intestine. The reality of the threat is revealed by a rare condition called *pancreatitis,* in which the duct from the pancreas to the intestine becomes blocked, and the secretions accumulate to high concentrations. This accumulation can cause the pancreas to digest itself rapidly with its own secretions, a condition that can prove fatal.

Liver and Gallbladder Secretions

The common approach to removing baby oil or motor grease from your hands or animal fat from a pan is to wash your hands or the pan with soap and water. Soap contains detergent molecules that are both hydrophobic and hydrophilic—one end is soluble in fat, and the other end is soluble in water. Because of their structure, detergents can surround fat molecules and suspend them in water. As a result, the grease or fat comes off your hands or the pan and becomes *emulsified* (suspended) in the surrounding water. A similar approach is taken by your digestive system in dealing with fats in your diet. Suppose you eat a pizza topped with double cheese. In order for the fat molecules in the cheese to be efficiently hydrolyzed by the pancreatic lipases in the intestine, large globules of fat must be broken apart into much smaller clusters. This process is accomplished by **bile salts,** which are produced in the liver and stored in the **gallbladder** (see Figure 27-1), a small sac that empties its contents into the intestine through a duct. Bile salts are similar in structure to the detergents present in soap. In the presence of bile salts, fat globules are reduced to stable, microscopic droplets that can be more efficiently attacked by lipid-digesting enzymes.

In addition to bile salts, the fluid of the gallbladder also contains the breakdown products of hemoglobin molecules of aging red blood cells that are pulled from the circulation and destroyed by the liver. Diseases that affect liver function, such as hepatitis, often interfere with the normal processing and discharge of these products, which remain in the blood and cause the skin to take on a yellowish, or *jaundiced,* appearance.

Absorption of Products Across the Small-Intestinal Wall

The first step in the absorption of the small, digested food molecules is their movement from the lumen (the space inside a tube) of the small intestine into the epithelial cells that form its lining. The inner surface of the intestinal wall has a specialized structure that makes it ideal for efficient absorption. The "velvety" texture of the intestinal lining is provided by fingerlike projections, called **villi** (Figures 27-5, 27-6), which protrude from the entire surface. Villi increase the absorptive surface of the intestine in the same way that the texture of terry-cloth towels enables them to soak up much more water than can smooth cloth towels.

Each villus (singular of villi) is covered with smaller projections, called **microvilli** (inset, Figure 27-6), which further increase the surface area of the small intestine enormously. Together, intestinal villi and microvilli create an interior surface area that is 150 times greater than the surface of your entire skin. Packed inside your abdomen is an intestinal surface equivalent to the surface area of a tennis court!

Each villus is laced with a rich network of capillaries surrounding a single, centrally located lymphatic vessel

known as a **lacteal.** The lacteal absorbs the products of lipid digestion, such as the fatty acids that are produced by lipase digestion of the cheese on the pizza you consumed hours earlier. From the lacteal, microscopic fat droplets are transported through a series of lymphatic vessels that eventually drain into a large vein in the neck (Figure 27-6 and Chapter 28). Most of these fat molecules are removed from the bloodstream by adipose (fat) cells scattered throughout the body, adding to the body's fat content.

What about the remainder of the ingredients in that pizza, such as the glucose in the starchy dough, and the amino acids in the protein-rich pepperoni? Most nonfatty nutrients diffuse directly into the blood capillaries of the intestinal villi, where they are carried to the liver and are removed from the bloodstream. The liver is the body's primary metabolic regulatory center, producing waste products, controlling blood glucose levels, and releasing substances into the bloodstream, as needed by the body's tissues.

Some nutrients undergo the last stages of digestion as they pass across the plasma membrane of the epithelial cells. Lactase, for example, is an enzyme in the plasma membrane which severs the disaccharide lactose (milk sugar) into two simple sugars. Many people cannot drink milk or eat dairy products without developing severe diarrhea because they lack the enzyme lactase. Consequently, milk sugar remains undigested in the large intestine, upsetting this organ's osmotic balance by creating a gradient that draws an excessive amount of water into the intestinal lumen.

THE LARGE INTESTINE: PREPARING THE RESIDUE FOR ELIMINATION

By the time digested food material has made its long journey to the end of the small intestine, virtually all of its nutrients have been removed, along with most of the water. The nutrient-depleted chyme is now propelled by peristalsis into the next part of the digestive tract, the **large intestine** (Figure 27-7).

The most important functions of the large intestine are the reabsorption of water from the digestive tract and the conversion of the remaining contents into a mass, called *feces.* Pressure-sensitive neurons detect when solids accumulate in the terminal portion of the large intestine (the **rectum**) and respond by provoking a *defecation reflex:* Impulses from the large intestine travel to the spinal cord and back to the muscles of the rectum, causing the muscles to contract and force the feces out through the anus. Because one of the two anal sphincters is under voluntary control, we can consciously delay expulsion of feces until an appropriate time.

Projecting from the large intestine is a short, blind (dead-ended) pouch, the *cecum,* from which a small blind tube, the *appendix,* extends (Figure 27-7). The cecum and appendix have no known function in humans. They are

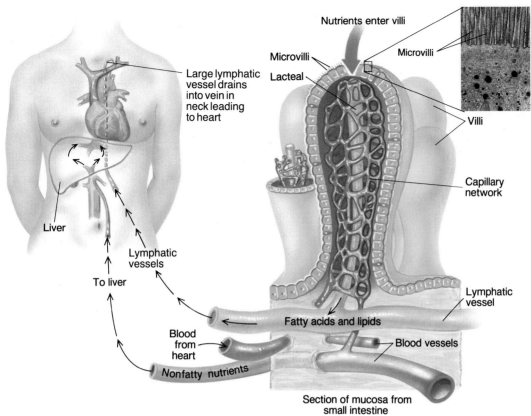

FIGURE 27-6

Structure and function of the small intestine's villi. Nutrients pass from the lumen of the small intestine into capillaries and a central lacteal located within each villus, a projection of the intestinal lining. Tiny projections, called microvilli, extend from the end of each epithelial cell in the villi, further increasing the absorbing surface. Substances absorbed from the small intestine into the bloodstream are carried to the liver, which detoxifies many harmful molecules and stores various nutrients, such as glucose. From the liver, nutrients enter the general circulation.

vestigial reminders of our evolutionary descent from mammals whose enlarged ceca (plural of cecum) aided digestion of plant material. Inflammation of the appendix leads to *appendicitis,* which, if untreated, can cause the appendix to rupture, spilling bacteria into the abdominal cavity and creating a potentially fatal infection.

The Large Intestine as a Home for Bacteria

Huge numbers of bacteria reside in the healthy human digestive tract. In fact, the number of resident microbes in the large intestine is so large that bacteria constitute almost half the dry weight of human feces. These intestinal bacteria metabolically attack organic substances in chyme and use them as nutrients, often producing unpleasant-smelling byproducts. These organisms are not freeloaders, however; they manufacture vitamin K, biotin, folic acid, and other nutrients we absorb and utilize. Although these services are

not required when the diet is well balanced, intestinal bacteria contribute enormously to our well-being by competing with potentially dangerous microbes for the body's limited space and nutrients. This becomes apparent when we destroy our normal bacterial flora by the extended use of antibiotics, which can cause digestive dysfunction (such as extended bouts of diarrhea) and yeast infections.

In providing nutrients for the body, the digestive system must grind ingested materials, disassemble macromolecules, absorb fluid and digested products, and eliminate undigestible residues. These processes occur in a stepwise manner, as food material passes through a series of chambers whose walls are specialized for secreting, churning, absorbing, and sending neural and hormonal signals to associated glands to release digestion-related substances. (See CTQ #3.)

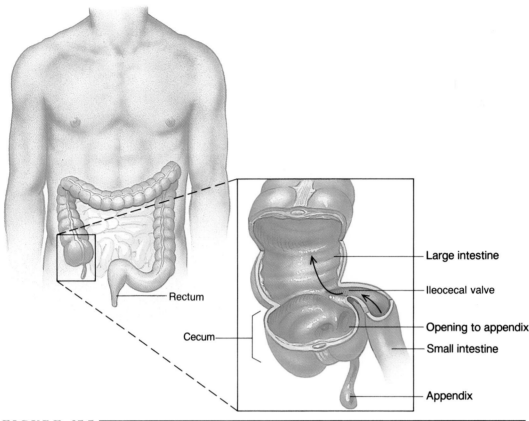

Large intestine

Ileocecal valve

Opening to appendix

Small intestine

Appendix

Rectum

Cecum

FIGURE 27-7

The large intestine receives unabsorbed wastes from the small intestine through the *ileocecal valve*. The large intestine in humans is about 1.5 meters long; the final 18 centimeters is the *rectum*. The large intestine, except for the rectum, is called the *colon*. The short cecum and appendix in humans is all that remains of larger chambers that housed cellulose-digesting micro-organisms in our ancestors.

EVOLUTION OF DIGESTIVE SYSTEMS

▐▶ As with most other organ systems, the evolution of increasingly complex animals was accompanied by the appearance of more elaborate digestive systems. The most primitive form of digestion in animals is best revealed by observing the process in many of the larger protists. Figure 27-8*a* shows one ciliated protozoan engulfing another ciliated protozoan by phagocytosis (page 145), a process known as **intracellular digestion.** Such prey are incorporated into food vacuoles and digested by lysosomal enzymes; the molecular products are then absorbed into the cytoplasm. Many of the simpler animals, including sponges, sea anemones, and flatworms also utilize intracellular digestion. Food is taken into the body of the animal, where it is engulfed by cells and disassembled within food vacuoles.

Most animals—including arthropods, annelids, and vertebrates—utilize **extracellular digestion,** whereby food material is hydrolyzed within a digestive tract by enzymes that are secreted into the tract by various digestive glands. The small-molecular digestive products are then absorbed, and the undigested residue, such as arthropod exoskeletons or plant cell walls, passes out of the body.

Some of the simpler animals—including jellyfish, sea anemones and flatworms—have an **incomplete digestive system** (Figure 27-8*b*), which consists of a blind digestive chamber with only one opening to the outside environment. This opening serves two functions: It is both an entrance for food and an exit for undigested residues. In contrast, most animals require the specialization afforded by a **complete digestive system,** which includes a tube with openings at both ends: a mouth for entry and an anus for exit (Figure 27-8*c*). In humans, food moves sequentially through different compartments, each region performing its specialized tasks of maceration, digestion, absorption, dehydration, and elimination. Complete digestive tracts can be thought of as "disassembly lines"—assembly lines that operate in reverse.

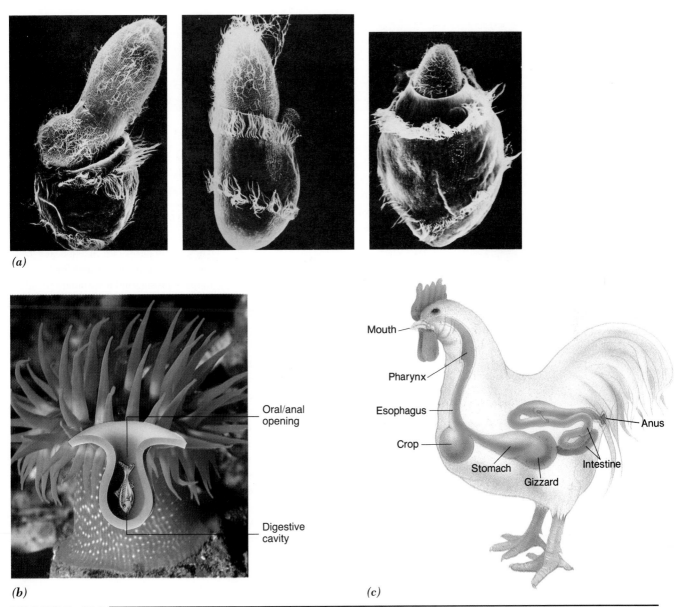

(a)

(b)

(c)

FIGURE 27-8

Digestive diversity. *(a)* Intracellular digestion. The protozoan *Didinium* engulfs a larger *Paramecium* for dinner. *(b)* Incomplete digestive tract of a sea anemone; food enters and residues exit through the same opening. *(c)* Complete digestive tract of a bird. Food is pushed to the back of the mouth (pharynx) and swallowed into the esophagus; from there, it passes into the crop, where it is stored. Food is ground, with the help of swallowed stones, in the bird's muscular gizzard.

ADAPTATIONS TO DIFFERENT DIETARY BEHAVIORS

▐▶ An animal's feeding apparatus and digestive system are adapted to the type of food the animal regularly consumes. The human digestive system is specialized for digesting a variety of foods, ranging from chunks of meat to fibrous vegetables and fruits. Many animals have more specialized diets. For example, a number of mammals, including ant-eaters and pangolins, feed exclusively on ants (Figure 27-9*a*). Each of these ant-eating species has powerful claws that can dig up anthills; elongated snouts that can extend into an ant nest; and long, sticky tongues that trap ants. Many aquatic animals, including barnacles, clams, and blue whales, feed on tiny organisms suspended in the water (Figure 27-9*b*). These *filter feeders* have some type of straining device or sticky, mucous-covered surfaces that screen or trap tiny food particles suspended in the water.

◁ THE HUMAN PERSPECTIVE ▷
Human Nutrition

FIGURE 1

Controversy continues to rage over such topics as the impact of dietary sugar, cholesterol, and saturated fats on our health. Nutritionists generally agree, however, that the healthiest diets are those that balance carbohydrates, triglyceride lipids (fats and oils), and proteins (Figure 1). Foods that provide these three groups of molecules should also contain enough energy, organic building blocks, vitamins, and minerals to satisfy the needs of the average person.

CALORIES AND ENERGY

◐ You may recall from the discussion of oxidative metabolism that subunits from all of the various macromolecules (amino acids, sugars, fatty acids, and nucleotides) are fed into the same metabolic pathways that provide ATP for the cell (page 188). Consequently, the body can fulfill its energy needs from any of these macromolecules. It is estimated that an average per-son engaged in a relatively sedentary lifestyle requires about 2,500 calories per day to maintain his or her body at a stable level. A person who engages in frequent strenuous activity, such as a professional athlete, may need over 6,000 calories per day. Since people differ in their metabolic set-points (page 76), the number of calories that will maintain one person's weight may cause another person either to gain or lose pounds.

CARBOHYDRATES

Carbohydrates provide the most readily available form of glucose and, therefore, the most rapid, readily available form of usable energy. This is why runners, for example, eat large amounts of pasta (which is high in carbohydrates) the night before a marathon. Glucose is the "all-purpose" energy source; it is also the only one that can be used by all brain and nerve cells. When these cells are deprived of energy, the resulting temporary *hypoglycemia* (low blood sugar) may cause unclear thinking, clumsiness, depression, or muscle tremors due to diminished neurological function.

Not all carbohydrates are easily digestible. Cellulose, for example, the polysaccharide of plant cell walls, resists disassembly in the human digestive tract. Yet, cellulose (such as that found in celery stalks) promotes health by providing bulk-fiber, which assists in the formation and elimination of feces. Low-cellulose diets may cause constipation and have been linked to colon cancer. For this reason, foods that are rich in both polysaccharides and undigestible bran, such as fruits, grain products, and legumes, are recommended as part of a balanced diet. In contrast, the simple carbohydrates that are found in sugary products, such as candy and soft drinks, provide calories but lack the fiber necessary for healthy digestion.

LIPIDS

Because of their highly reduced state, fats and oils are rich sources of energy (page 128). Two fatty acids, linolenic and linoleic acid, are *essential fatty acids,* since they are needed but cannot be manufactured in the body and are thus required in the diet. These fatty acids are needed for cell-membrane construction and for the synthesis of prostaglandin hormones. In addition to being fattening, diets rich in saturated fats and cholesterol (found in butter, animal fats, and eggs) may predispose susceptible individuals to circulatory disease by increasing the likelihood of deposition of lipids in the walls of arteries. In contrast, unsaturated fats, which are found in most vegetable oils, may increase the risk of cancer because breakdown of the fatty acids in these molecules produces free radicals (page 56). Nutritionists generally agree that an ideal diet should be low in fat; that is, fat should provide less than 30 percent of food calories.

PROTEINS

Dietary protein is needed to supply the amino acids from which we assemble our enzymes, antibodies, hormones, and various other types of proteins. Protein can be obtained from virtually any food, including meat, cheese, eggs, and vegetables. We can manufacture all but eight amino acids metabolically. The absence of even one of these eight *essential amino acids* prevents the synthesis of all proteins. All of these required amino acids can be synthesized by plants and microorganisms. In fact, it is thought that during the early evolution of animal life, the enzymes needed for the synthesis of these particular amino acids were lost because these molecules were present in adequate supply within the diet. Plant proteins are *incomplete* because they provide essential amino acids but not in the proportions necessary for proper nutrition. For this reason, vegetarians are advised to combine vegetables and grains in particular ways to ensure proper amino acid intake.

The prolonged absence of any of the essential amino acids in the diet can lead to severe protein deficiency and a condition known as *kwashiorkor*, which is one of the world's most serious health problems. We are all too familiar with pictures of listless, malnourished children with swollen bellies (a result of water retention) and arms and legs composed of just skin and bones. These symptoms are a result of the world's shortage of dietary protein.

VITAMINS

An organic compound is designated a *vitamin* when it is needed in trace amounts for normal health but cannot be synthesized by the body's own metabolic machinery and thus must be obtained in the diet. As humans, we must acquire 13 vitamins from our diet, or we run the risk of suffering vitamin-deficiency disorders, some of which can be fatal. A well-balanced diet normally provides all the vitamins needed.

Vitamins are usually divided into two groups, based on their solubility properties. Vitamins in the first group are water soluble; these include eight different vitamins of the B complex and vitamin C. Vitamins in the second group are insoluble in water but soluble in oil; these include vitamins A, D, E, and K. Most of these vitamins function as coenzymes that assist essential enzymatic reactions (Table 27-1). The required daily allowance of each vitamin is relatively low because these molecules are not consumed in the reactions they assist so each molecule is used again and again. Furthermore, the small amount that is required replaces that which is normally excreted.

Given this information, is there any value to a person with a balanced diet taking vitamin (and mineral) supplements?

TABLE 27-1

DIETARY ESSENTIALS IN HUMAN NUTRITION: VITAMINS

Designation	Major Mode of Action	Major Sources[a]	Symptoms of Deficiency[b]
Retinol (A)	Part of visual pigment, maintenance of epithelial tissues	Egg yolk, butter, fish oils; conversion of carotenes[c]	Nightblindness, corneal and skin lesions, reproductive failure
Calciferol (D)	Ca and P absorption, bone and teeth formation	Fish oils, livers; irradiation of sterols[c]	Rickets, osteomalacia
Tocopherols (E)	Antioxidant	Vegetable oils, green, leafy vegetables	In animals: muscular degeneration, infertility, brain lesions, edema[d]
Vitamin K	Synthesis of blood coagulation factors	Green, leafy vegetables; bacterial synthesis	Slowed blood coagulation
Thiamine (B_1)	Energy metabolism–decarboxylation	Whole grains, organ meats	Beri-beri, polyneuritis
Riboflavin (B_2)	Hydrogen and electron transfer (FAD)	Whole grains, milk, eggs, liver	Cheilosis, glossitis, photophobia
Nicotinic acid (niacin)	Hydrogen and electron transfer (NAD, NADP)	Yeast, meat, liver[e]	Pellagra
Pyridoxine (B_6)	Amino acid metabolism	Whole grains, yeast, liver	Convulsions, hyperirritability
Pantothenic acid	Acetyl group transfer (CoA)	Widely distributed	Neuromotor and gastrointestinal disorders
Biotin	CO_2 transfer	Eggs, liver; bacterial synthesis	Seborrheic dermatitis
Folic acid	One-carbon transfer	Leafy, green vegetables, meat	Anemia
Cobalamine (B_{12})	One-carbon synthesis; molecular rearrangement	Animal products, esp. liver; bacterial synthesis	Pernicious anemia
Ascorbic acid (C)	Hydroxylations, collagen synthesis	Citrus, potatoes, peppers	Scurvy

[a] Most vitamins, especially of the B group, occur in a multitude of foodstuffs and in all body cells.

[b] A variety of symptoms occur with certain vitamin deficiencies; vitamin deficiencies are frequently of a multiple nature, and symptoms similar to those described may have their origin in conditions not related to nutrition.

[c] Certain carotenes, found in green and yellow vegetables, are precursors of vitamin A. Certain sterols, including 7-dehydrocholesterol, which is synthesized in the body, are precursors of vitamin D.

[d] No well-defined syndrome is described for humans.

[e] Niacin is one of the end products of normal tryptophan metabolism.

Source: From P. D. Sturkie, *Basic Physiology,* New York, Springer-Verlag, 1981, p. 345.

This question continues to be debated, and no sweeping, satisfactory answer is currently available. Scientific studies in human nutrition are notoriously difficult to carry out in a controlled manner (as exemplified by the debate over cancer and vitamin C, page 39), and the effects of various eating habits may not become apparent for decades. On the one hand, there is no current scientific evidence to suggest that taking *large* amounts of vitamins is helpful, and there is clear evidence that taking *large* amounts of vitamins A and D can be injurious and even fatal. (There are documented cases, for example, of hungry individuals dying from consuming a bear's liver, which contains high concentrations of fat-soluble vitamins.) On the other hand, some biochemists and nutritionists argue that the capability of vitamins C and E to destroy free radicals (page 56) makes them valuable supplements *in moderate doses.* Clearly, the subject of vitamin supplements is an area in which more controlled research studies are required.

MINERALS

A dietary supply of *minerals* (Table 27-2) is just as important for proper nutrition as is that of vitamins. Without calcium and magnesium, for example, large numbers of enzyme-mediated reactions would simply not occur. Calcium is also needed for bone growth and muscle function. Iron forms the functional core of cytochromes, which are needed by all cells for aerobic respiration, and of hemoglobin, the oxygen-transporting protein in blood. Phosphorus is needed for ATP and nucleic acid synthesis. Sodium, potassium, and chloride are required for maintaining osmotic balance and for propagating nerve impulses. A number of other minerals, including iodine, copper, zinc, cobalt, molybdenum, manganese, and chromium, are required in such tiny amounts that they are referred to as *trace elements.*

TABLE 27-2
DIETARY ESSENTIALS IN HUMAN NUTRITION: MINERALS[a]

Designation	Major Functions	Major Sources	Symptoms of Deficiency[b]
Calcium (Ca)	Bone and teeth, nervous reactions, enzyme cofactor	Dairy products, leafy, green vegetables	Calcium tetany, demineralized bones
Phosphorus (P)	Bone and teeth, intermediary metabolism	Dairy products, grains, meat	Demineralized bones
Magnesium (Mg)	Bone, nervous reactions, enzyme cofactor	Whole grains, meat, milk	Anorexia, nausea, neurological symptoms
Sodium (Na)	Maintenance of osmotic equilibrium and fluid volume	Table salt[c]	Weakness, mental apathy, muscle twitching
Potassium (K)	Cellular enzyme function	Vegetables, meats, dried fruits, nuts	Weakness, lethargy, hyporeflexia
Chlorine (Cl)	Maintenance of fluid and electrolyte balance	Table salt[c]	d
Iron (Fe)	Hemoglobin, myoglobin; respiratory enzymes	Meat, liver, beans, nuts, dried fruit	Anemia
Copper (Cu)	Enzyme cofactor (cytochrome-c-oxidase)[e]	Nuts, liver, kidney, dried legumes, raisins	Anemia, neutropenia, skeletal defects
Manganese (Mn)	Enzyme cofactor, bone structure, reproduction	Nuts, whole grains	d
Zinc (Zn)	Enzyme cofactor (carbonic anhydrase)[e]	Shellfish, meat, beans, egg yolks	Growth failure, delayed sexual maturation
Iodine (I)	Thyroid hormone synthesis	Iodized table salt, marine foods	Goiter
Molybdenum (Mo)	Enzyme cofactor (xanthine oxidase)[e]	Beef kidney, some cereals and legumes	d
Chromium (Cr)	Regulation of carbohydrate metabolism (glucose tolerance factor)	Limited information available	d

[a] A human need for the following trace elements is possible but has not been unequivocally established: selenium (Se), fluorine (F), silicon (Si), nickel (Ni), vanadium (V), and tin (Sn). The need for sulfur (S) is satisfied by ingestion of methionine and cystine, and for cobalt (Co) by vitamin B_{12}.
[b] Except for Ca, Fe, and I, dietary deficiency in humans is either unlikely or rare.
[c] Many processed foods contain considerable amounts of sodium chloride.
[d] No specific deficiency syndrome described in humans.
[e] Examples of activity as enzyme cofactors.
Source: From P. D. Sturkie, *Basic Physiology*, New York, Springer-Verlag, 1981, p. 344.

(a)

(b)

FIGURE 27-9 ──────────

Feeding specialists. *(a)* Anteaters trap their tiny prey on the sticky surface of their tongue. *(b)* Barnacles have modified appendages that strain the surrounding water, collecting microscopic, organisms. *(c)* Some sea cucumbers—relatives of sea stars—collect microbes on their feathery tentacles, which are periodically placed in the mouth.

(c)

Some filter feeders, such as sea cucumbers, extend feathery or highly branched tentacles that harvest microscopic bits of food (Figure 27-9c). Filter feeders usually feed *continuously*, while most animals feed *discontinuously*, taking in food matter only when it is convenient and available.

Some animals paradoxically obtain nutrients from fibrous plants that they alone cannot digest. Cattle, antelopes, buffalo, giraffes, and other such *ruminant* animals possess additional stomach chambers, one of which is heavily fortified with cellulose-digesting microorganisms (Figure 27-10). In this chamber, called the *rumen*, bacteria and protozoa break down the otherwise indigestible cellulose fibers and use them for growth of more microorganisms. This growing crop of microorganisms is then digested by the animal as the contents of the rumen travel through the rest of the digestive tract. Horses and other grazing animals that lack rumens are able to utilize plant cellulose because they possess an elongated *cecum*, a sac that extends from the large intestine. The cecum serves as a fermenting vat in which microbes disassemble cellulose into products that are digestible by the animal.

Animals procure food in various ways: They strain it from their aqueous environment, extract it from mud, scrape it from rocks, swallow it whole or in large pieces, and so forth. The structure and function of an animal's digestive system are adapted to the particular feeding habits of the animal. (See CTQ #5.)

FIGURE 27-10
The cape buffalo; an example of a ruminant.

REEXAMINING THE THEMES

Relationship between Form and Function

The human digestive tract is a model food-processing plant for the stepwise disassembly of food and absorption of nutrients. Each part of the tract is ideally suited for particular activities. For example, the mouth contains a set of differentiated teeth that are specialized for cutting, tearing, and grinding food material to facilitate its digestion. The esophagus, with its thick, circular bands of muscle tissue, is specialized for pushing mouthfuls of food matter into the stomach. The pouchlike stomach is specialized as a food storage site, and its muscular walls churn the food matter into a liquified slurry. The small intestine, with its tremendous number of villi and microvilli, is ideally suited for absorbing small-molecular-weight nutrients that have been generated by the digestive enzymes that pour in from the pancreas. The water-reabsorbing properties of the large intestine (and its large population of bacteria) make this organ suited for preparing fecal residues for elimination.

Biological Order, Regulation, and Homeostasis

The passage of food through the human digestive tract is not a continuous activity; rather, it occurs periodically. The activities that take place in various parts of the digestive tract must be timed to correspond to the presence of food matter in each specific part of the tract at a particular time. Timing and coordination of digestive activities is controlled by the cooperative activities of both the nervous and endocrine systems. When food enters the mouth, stomach, and intestine, it triggers reflex responses and hormonal secretions, which initiate specific secretory and muscular responses. Food seeking behavior is initiated by sensations of hunger, which occur in response to decreased blood glucose and lack of pressure in the stomach.

Acquiring and Using Energy

The energy that is expended in fueling our activities is derived primarily from polysaccharides and lipids that are present in the food we eat. These macromolecules must first be broken down into their subunits, which are then absorbed across the wall of the digestive tract into the bloodstream. These breakdown products eventually reach all the individual cells of the body, where the energy is harvested by oxidation. While carbohydrates and lipids provide the most readily available sources of chemical energy, proteins and nucleic acids in our diets can also be broken down and oxidized for their energy content.

Evolution and Adaptation

Digestive tracts show a graded increase from simple to more complex systems. Simpler, multicellular animals, such as sea anemones and flatworms, have an incomplete digestive tract. In addition, they utilize intracellular digestion, whereby microscopic food particles are phagocytized and enzymatically broken down within food vacuoles. Most animals have a complete digestive tract which is open at both ends and allows food matter to pass in a single direction along a continuous "disassembly line," where it is subjected to stepwise processing. Even though animals may have similar digestive tracts, each digestive system is specifically adapted to the particular feeding habits of the animal. For example, elephants have huge, replaceable molars for grinding tough vegetation. Birds, which lack teeth, have a large crop for storing food, and a muscular gizzard for grinding it. Mammals that graze on grass often possess a chamber, such as a rumen or cecum, that harbors cellulose-digesting microorganisms.

SYNOPSIS

An overview of digestion. Large, complex food substances are dismantled into molecules small enough to pass through plasma membranes and enter cells, where their nutritional value is unleashed. A complete digestive system, such as that of humans, consists of a continuous digestive tract and various accessory organs. Ingested material is forced through the entire tract by peristalsis—waves of contraction of the layers of muscle tissue in the wall of the tract. The muscular and secretory activities that occur at different sites along the digestive tract are coordinated by the actions of both the nervous and endocrine system.

Each part of the digestive tract is specialized for particular activities that occur in a stepwise fashion as food matter is pushed along the tract. The mouth macerates food and covers it with lubricating mucus. Swallowing combines voluntary and involuntary responses that push the food bolus into the esophagus, where it is moved by peristalsis into the stomach and churned into chyme. Glands in the stomach wall secrete a fluid that contains acid and protein-digesting enzymes. Chyme passes into the small intestine, where its acidity is neutralized by bicarbonate ions from the pancreas; its macromolecules are digested by enzymes also produced in the pancreas. Most of the fluid in which the food matter is suspended is derived from secretions of the wall of the small intestine itself. The fat that is present in the food is emulsified in the form of microscopic droplets by bile salts that are produced in the liver and stored in the gallbladder. As the various macromolecules are enzymatically digested, their component subunits are absorbed across the wall of the intestinal villi into either blood capillaries or lymphatic lacteals. Most of these nutrients are carried to the liver, where they are removed from the bloodstream. Those materials that cannot be digested and absorbed pass into the large intestine, where the remaining water is reabsorbed and the insoluble residues are compacted (together with bacteria) into feces, which are eliminated.

Most digestive systems consist of a continuous tube that is open at both ends. Because it is continuous with the external environment, the space within the digestive tract is actually a part of the outside world. Although most digestive systems have a similar overall plan, each has special adaptations that are appropriate to the feeding habits of the particular animal.

A healthy human diet must contain a variety of components. Chemical energy is most readily supplied by carbohydrates and fats, but it can also be provided by proteins and nucleic acids, which also can be degraded through a cell's oxidative pathways. Of the 20 amino acids incorporated into proteins, eight cannot be synthesized from other compounds and must be obtained from the diet. Protein deficiency is the world's major cause of malnutrition. Two fatty acids are also required as dietary elements. Vitamins are organic compounds that function primarily as coenzymes but cannot be synthesized by the body; thus, they are required in the diet. The diet must also supply a number of minerals. Some of these, such as calcium, magnesium, iron, sodium, and potassium, play major roles in the body and must be present at high levels in the diet. In contrast, the roles of trace elements, such as iodine, copper, and zinc, are more limited.

Key Terms

digestive system (p. 568)
digestive tract (p. 568)
digestion (p. 568)
mouth (p. 568)
saliva (p. 568)
salivary glands (p. 568)
bolus (p. 569)
esophagus (p. 569)

peristalsis (p. 569)
cardiac sphincter (p. 569)
stomach (p. 570)
chyme (p. 570)
small intestine (p. 572)
bile salts (p. 574)
gallbladder (p. 574)
villi (p. 574)

microvilli (p. 574)
lacteal (p. 574)
large intestine (p. 574)
rectum (p. 574)
intracellular digestion (p. 576)
extracellular digestion (p. 576)
incomplete digestive tract (p. 576)
complete digestive tract (p. 576)

Review Questions

1. Why isn't food considered to be inside the body immediately after it is swallowed? With this consideration in mind, is human digestion intracellular or extracellular? At what point do nutrients actually enter the body?

2. Trace the fate of a mouthful of food through the entire digestive tract. Describe the changes that occur in each portion of the digestive tract, and discuss the activities of saliva, gastric secretions, intestinal secretions, pancreatic secretions, bile, and intestinal bacteria.

3. How are the various activities described in your answer to the previous question regulated by the endocrine and/or nervous system?

4. Compare and contrast the sites of the preliminary digestion of starch, protein, and fats; the role of CCK and secretin; the effects of high versus low blood glucose concentrations on hunger; the role of most vitamins, compared to minerals, such as calcium and iron; the role of molars in an elephant and a dog.

Critical Thinking Questions

1. Suppose you were studying nutrition and found that laboratory rats that were fed on a diet of raw carrots remained healthy, but those that were fed on a diet consisting only of cooked carrots developed a condition that caused them to lose their hair. What conclusion might you draw about this substance? What kind of experiment would you run to confirm your conclusion? What controls would you use? If you attempted to extract the substance from carrots, how could you determine which of your extracts contained the necessary ingredient?

2. Both mechanical and chemical processes are involved in the breakdown of most foods into small molecules that can be absorbed by cells lining the small intestine. What is the role of each type of activity, and why are both necessary?

3. What effects might each of the following have on an individual: mumps infection; eating slowly; removal of a cancerous stomach; secretion of excess stomach acid; reversal of peristalsis; gallstones; removing a portion of the small intestine; appendectomy; colostomy (removal of the colon); long treatment with antibiotics?

4. Why do you think elevating your head when you sleep may help prevent heartburn in the middle of the night? How do you suppose people are able to swallow food even when they are standing on their heads? Do you think these two observations contradict each other? If so, how can you resolve the apparent contradiction?

5. How do teeth reflect the varied diet of humans? How do human teeth differ from the teeth of carnivores, such as lions? Herbivores, such as deer?

Additional Readings

Alper, J. 1993. Ulcers as an infectious disease. *Science* 260:159–160. (Intermediate)

Campbell-Platt, G. 1988. The food we eat. *New Sci.* May:1–4. (Introductory)

Davenport, H. W. 1978. *Physiology of the digestive tract*, 4th ed. Chicago: Year Book Medical Publishers. (Intermediate–Advanced)

Hamilton, E. 1982. *Nutrition: Concepts and controversies.* Menlo Park: West. (Introductory)

Makhlouf, G. M. 1990. Neural and hormonal regulation of function in the gut. *Hospital Practice* Feb:79–98. (Advanced)

Moog, F. 1981. The lining of the small intestine. *Sci. Amer.* May:154–176. (Intermediate)

Scrimshaw, N. S. 1991. Iron deficiency. *Sci. Amer.* Oct:46–53. (Intermediate)

Schutz, Y., et al. 1991. Unsolved and controversial issues in human nutrition. *Experentia* 47:166–193. (Intermediate–Advanced)

Readings from Scientific American. 1978. *Human nutrition.* New York: W. H. Freeman. (Intermediate–Advanced)

CHAPTER
◄ 28 ►

Maintaining the Constancy of the Internal Environment: The Circulatory and Excretory Systems

STEPS TO DISCOVERY
Tracing the Flow of Blood

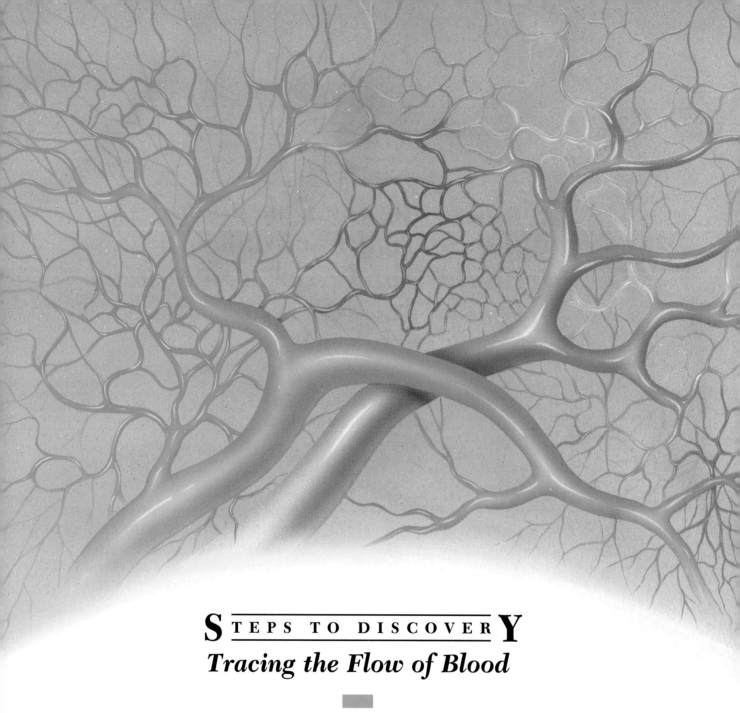

Tracing the Flow of Blood

*I*n 1628, a book entitled *An Anatomical Treatise on the Movement of the Heart and Blood in Animals* was written in Latin by an English physician and published in Germany. It was a small book, published on cheap, crumbling paper and filled with typographical errors. Yet, it has been hailed as the most important work ever published in the field of physiology. In this book, William Harvey, the son of a prosperous English merchant and a graduate of the University of Padua in Italy, described his experiments that led to a new concept of the organization of the human circulatory system.

Prior to Harvey's publication, the established views of blood circulation in humans had been formulated by a Greek named Galen, who had lived 1,400 years earlier and had served as the personal physician for the Roman Emperor Marcus Aurelius. Galen died in 201 A.D., but his views were still accepted in the sixteenth century, and his authority remained unchallenged. A brilliant, although dogmatic, scientist, Galen had made a number of important discoveries. But, he concluded that blood was formed in the liver then flowed to the heart and through the arteries en route to the tissues, where it was entirely absorbed. The

Contrary to the early view that veins and arteries were independent, blood flows through a circuit leading from the heart, into

blood utilized by the tissues was replaced by new blood from the liver. Galen envisioned the veins as a system of vessels independent of the arteries; blood in the veins simply ebbed back and forth within the same vessels, much the way the tide moves in and out along the shore. The fact that this view survived for 1,400 years, is an indication of the stagnation of science throughout the Middle Ages, which were also known as the Dark Ages.

In order to test Galen's hypothesis of blood flow, Harvey attempted to measure just how much blood is actually pumped out of the heart in a given amount of time. He measured the internal volume of the heart of a cadaver and multiplied this figure by the number of times the heart beats per minute. Using this method, Harvey concluded that it took the heart nearly 30 minutes (which is actually a considerable overestimate) to pump out an amount of blood equivalent to the body's entire blood supply. It was inconceivable that the tissues could actually absorb this amount of blood so rapidly, or that the liver could resupply the blood so quickly. Harvey proposed that, rather than being absorbed by the tissues, blood circulated through the body along some type of "circular" pathway. According to his hypothesis, blood left the heart through the arteries, passed into the various tissues, and then returned to the heart through the veins.

To support his hypothesis, Harvey demonstrated that blood could flow in only one direction through a given vessel. In one simple, but convincing, demonstration, Harvey pressed his finger against one of the major veins of the forearm then moved his depressed finger along the vein toward the individual's hand (away from the heart), pushing the blood out of the vein. If blood flowed in both directions, as was the prevailing view of the time, then the vein should rapidly refill with blood. If the vein carried blood in only one direction—back to the heart—the vein should remain empty. The results were clear: The vein remained empty of

blood until Harvey removed his finger. Galen's 1,400-year-old hypothesis of blood flow had been disproved with this simple, but elegant, demonstration.

One of Harvey's great frustrations was his inability to demonstrate just *how* blood flowed from the arteries into the veins. Harvey hypothesized that the tissues contain tiny vessels that complete the circuit from the arteries to the veins, but he had no way of demonstrating the existence of such vessels. The vessels linking the arterial and venous circulation were finally discovered in 1661, 4 years after Harvey's death, by the Italian anatomist Marcello Malpighi, whom we met in Chapter 19 in reference to his experiments on circulation and transport in plants. Using a newly developed instrument, the microscope, Malpighi prepared a thin piece of tissue from the lungs of a frog. He allowed the tissue to hang in the air to dry and then observed the blood vessels with a microscope; the red color of the blood vessels contrasted strikingly against the light background. Malpighi saw the larger arterial vessels branching into smaller vessels and finally giving rise to short, minuscule vessels that merged at the other end into the venous circulation. In this image, Malpighi had discovered the link between the arteries and veins; he named these tiny linking vessels *capillaries,* the Latin word for "hairlike." Malpighi confirmed his observations by examining living tissues, such as the wall of the urinary bladder. He traced the blood as it flowed through the arteries and into capillaries; the blood never spilled out into the spaces of the tissue. Malpighi had demonstrated that blood flowed in a unidirectional, continuous, and uninterrupted cycle.

arteries (red), through microscopic capillaries, into veins (blue), and back to the heart.

*J*ust as the towns and cities of a country are connected by roads and rails, the various parts of your body are connected by an extensive system of "living tubes," or vessels, that provide a continuous route for the movement of blood. Like a road or rail system carrying trucks or trains, the blood picks up and delivers materials as it courses through the body: Nutrients are picked up as the blood passes along the wall of the small intestine and are removed by the liver and, ultimately, delivered to all the tissues of the body; oxygen is picked up as the blood passes through the lungs and is removed by the body's cells; hormones are picked up as the blood passes through the various endocrine glands and are carried to their respective target organs; waste products are swept up as the blood passes through the tissues and are removed by the kidneys, lungs, and liver for disposal.

Blood vessels, the blood that flows through them, and the heart whose contractions propel the blood, make up an animal's **circulatory system.** The circulatory system is more than just a conduit for the movement of materials from one organ to another; it is the means by which complex, multicellular animals maintain homeostasis. If ions or other solutes become too concentrated in the body, for example, the blood carries them to the kidney for elimination. If the tissues become too acidic, the blood provides the buffering agents that lower the hydrogen ion concentration. If the tissues become depleted of oxygen or steeped in carbon dioxide, the blood carries the message to the nervous system to stimulate deeper or more rapid breathing. In birds and mammals, the blood also plays a crucial role in maintaining a constant body temperature; the blood carries heat either to the body surface, where it can be dissipated into the environment, or deep into the body, where it can be conserved. The bloodstream also plays a vital role in an animal's defense against disease-causing organisms (Chapter 30).

The bulk of a complex, multicellular animal is made up of fluid, which is distributed among three different fluid compartments: intracellular, interstitial, and vascular. The **intracellular compartment** is the space present within all of the animal's cells—the fluid of the cytoplasm and nuclei. The **interstitial compartment** is the space between and surrounding all of the cells: the extracellular fluid. The **vascular compartment** is the space present within the vessels and chambers of the circulatory system: the fluid of the bloodstream. The spatial relationship among the three compartments is depicted in Figure 28-1. Because of its position, the interstitial fluid acts as a "middleman" between the other two compartments. Substances such as oxygen, glucose, and hormones are able to reach the cells by diffusing from the blood through the interstitial fluid. Conversely, substances such as carbon dioxide and nitrogenous waste products are able to move in the opposite direction: from the cells, through the interstitial fluid, to the bloodstream. These exchanges are essential for maintaining the homeostatic condition of the organism.

▼ ▼ ▼

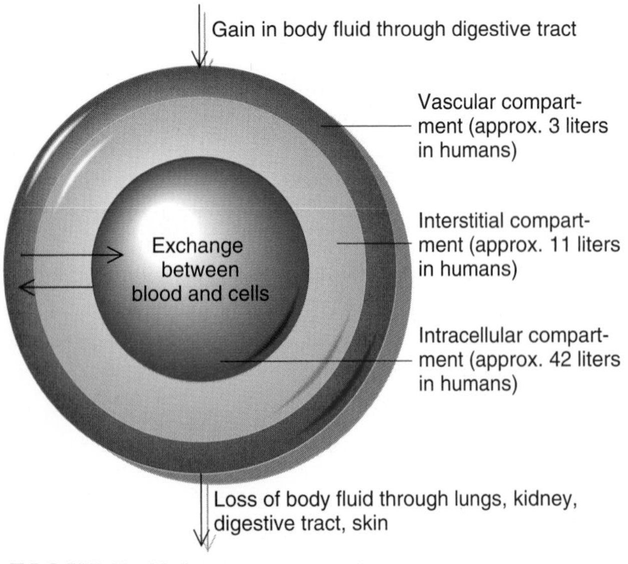

THE HUMAN CIRCULATORY SYSTEM: FORM AND FUNCTION

The human circulatory system, or **cardiovascular system** (*cardio* = heart, *vascular* = vessels), consists of blood vessels, the heart, and blood.

BLOOD VESSELS

The human cardiovascular system contains tens of thousands of kilometers of tubing, which assures that every cell of the body is within diffusion distance of a capillary. Blood vessels are divided into five basic types: arteries, arterioles, capillaries, venules, and veins (Figure 28-2), differing in form and function.

Gain in body fluid through digestive tract

Vascular compartment (approx. 3 liters in humans)

Interstitial compartment (approx. 11 liters in humans)

Exchange between blood and cells

Intracellular compartment (approx. 42 liters in humans)

Loss of body fluid through lungs, kidney, digestive tract, skin

FIGURE 28-1

The spatial interrelationship of the intracellular, interstitial, and vascular fluid compartments. All substances exchanged between the bloodstream and the body's cells must pass through the interstitial fluid that surrounds the cells.

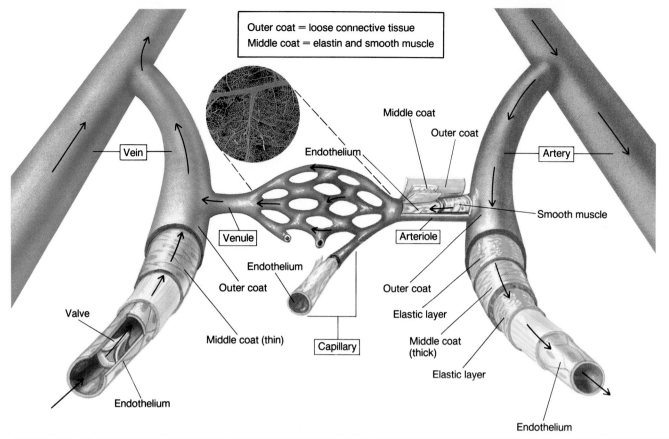

FIGURE 28-2

The form and function of the body's blood vessels are closely related. Each of the five categories of vessel has a distinct internal anatomy. The inner lining of all the vessels consists of a layer of flattened, "interlocking" endothelial cells which make up the *endothelium.* The walls of the capillaries—the thinnest of the vessels—consist only of a single, thin endothelial cell layer. The inset shows the delicate latticework of a capillary bed. (Note: this drawing is not to scale.)

Arteries: Delivering Blood Rapidly Throughout the Body

Blood is pumped out of the heart into **arteries**—large vessels that function as conduits, rapidly carrying blood to all parts of the body. Arteries typically are large in diameter and have complex walls that contain concentric rings of elastic fibers. When blood is pumped out of the heart, the walls of the arteries are pushed outward, increasing their fluid capacity. As the walls stretch, the rings of elastic fibers respond much as a rubber band does; that is, they recoil and exert pressure against the blood in the arteries. This is what we measure as **blood pressure.** To visualize this effect, imagine filling a rubber balloon with water. If you squeeze one end of the balloon with your hand, you are exerting pressure against the water. Since water can't be compressed into a smaller volume, the fluid pushes outward against the wall of the balloon, forcing it to elongate. This is the same principle underlying blood flow: The heart contracts, forcing blood into the arteries; the stretched arteries push against the blood, forcing it through the smaller vessels farther along the route. As the fluid is pushed out of the arteries, the diameters of the arteries decrease, and the

arterial pressure drops, until the next contraction abruptly returns the pressure to its maximum value.

Blood Pressure When you have your blood pressure taken, an inflatable cuff is strapped around your upper arm, providing a measure of the fluid pressure in the major arteries in that limb (Figure 28-3). Blood-pressure readings are expressed as two numbers, one over the other, such as 120/80 (the normal values for a young adult). The first number is the **systolic pressure,** as measured in millimeters of mercury. This is the highest pressure attained in the arteries as blood is propelled out of the heart. The pressure drops rapidly as blood is pushed out of the arteries and the diameter of the arteries decreases. Since the arteries don't have time to return to their resting (unstretched) diameter before the next contraction, the blood in the arteries remains under pressure from the stretched arterial walls. The second number is the **diastolic pressure,** or the lowest pressure in the arteries of the arm recorded just prior to the next heart contraction. If the observed systolic and diastolic values are higher than about 140/90, the person is said to have high blood pressure.

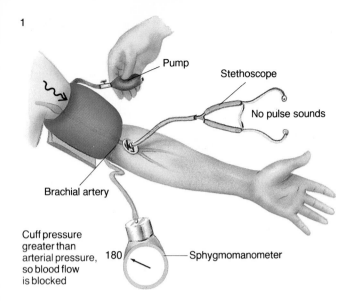

Pump

Stethoscope

No pulse sounds

Brachial artery

Cuff pressure greater than arterial pressure, so blood flow is blocked

180 — Sphygmomanometer

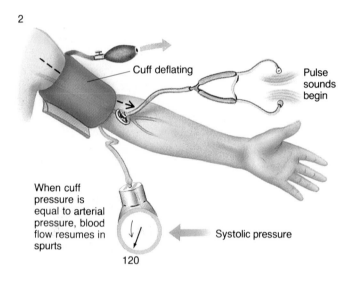

Cuff deflating

Pulse sounds begin

When cuff pressure is equal to arterial pressure, blood flow resumes in spurts

120

Systolic pressure

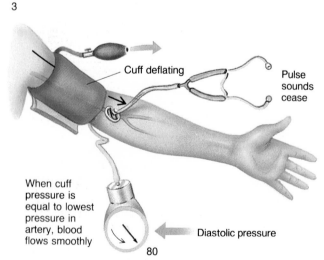

Cuff deflating

Pulse sounds cease

When cuff pressure is equal to lowest pressure in artery, blood flows smoothly

80

Diastolic pressure

FIGURE 28-3

Determining blood pressure using a *sphygmomanometer* and a stethoscope. The normal blood pressure in the brachial artery, the major artery of the arm, is 120/80.

While the causes of high blood pressure, or **hypertension,** are still poorly understood (see The Human Perspective: The Causes of High Blood Pressure), the effects of hypertension can become very evident if the condition continues untreated. The excessive pressure leads directly to a weakening of the walls of the arteries, increasing their chance of rupture. Increased arterial pressure also accelerates the buildup of fatty plaques on the walls of arteries (*atherosclerosis*) and a greatly increased likelihood that blood clots will develop in the vessels. Together, these deleterious effects on the body's arteries promote the rupture or blockage of cerebral vessels, causing a stroke and possible subsequent brain damage. Other risks include the rupture of vessels in the kidney, which can cause kidney failure, and blockage of the coronary arteries, which can cause a heart attack.

Arterioles: Regulating Blood Flow to the Tissues

The major arteries branch into smaller and smaller arteries, eventually giving rise to **arterioles:** small vessels whose walls lack elastic fibers but contain a preponderance of smooth muscle cells. The amount of blood that flows into a particular tissue depends largely on the diameter of the local arterioles. Arteriolar diameter is determined by the state of contraction of the smooth muscle cells in the walls of these vessels, which is regulated by the sympathetic nervous system (page 490). If an organ needs more oxygen, as does the heart of a person who is engaged in strenuous exercise, sympathetic stimulation to the muscle cells of the arterioles in the organ *decreases.* As a result, the muscle cells become more relaxed, and the arterioles increase in diameter—a response known as **vasodilation.** Conversely, those organs of the body that operate at a low activity level at a particular time, such as the digestive tract during a time of stress or exercise, receive less blood due to a temporary narrowing of the arterioles. The decrease in arteriole diameter is caused by an *increase* in sympathetic stimulation, causing the arteriole's muscle cells to contract—a response known as **vasoconstriction.**

Capillaries: Exchange between Tissue and Bloodstream

From the arterioles, blood passes through the **capillaries,** the smallest, shortest, and most porous channels of the vascular network. The lumen of these vessels is just large enough for red blood cells to move along in single file (Figure 28-4). The capillaries are the sites of exchange between the cells and the bloodstream. Your body contains about 40,000 kilometers of capillaries—enough to circumscribe the earth at its equator—creating an enormous surface area for the exchange of materials between the blood and the interstitial fluid of the tissues. The movement of water, nutrients, waste products, and dissolved gases in and out of the capillaries is illustrated in Figure 28-5. Red blood cells and most proteins are too large to penetrate the pores

◁ THE HUMAN PERSPECTIVE ▷
The Causes of High Blood Pressure

Approximately 20 percent of all Americans suffer from high blood pressure (hypertension) which, if untreated, can lead to kidney failure, heart attack, or stroke. Even though hypertension is one of the most thoroughly researched medical conditions, the underlying causes are still uncertain. We do know that high blood pressure usually arises as a result of a decreased diameter of arteries and arterioles. Most cardiologists believe that excessive vessel constriction results from a variety of factors that affect people differently, depending primarily on a person's genetic predisposition. Some of these are physiological risk factors discussed below; others are behavioral risk factors associated with stress, obesity, smoking, and consumption of dietary fats and alcohol.

In 1934, Harry Greenblatt of Case Western Reserve University published one of the most important experiments in modern medicine. Having constricted the major *(renal)* artery leading into one of the kidneys of a dog, Greenblatt found that the animal developed a markedly higher blood pressure. This experiment, among others, led to the discovery that the kidneys produced an enzyme, called *renin*, which is secreted into the blood, where it leads to the formation of another protein, called *angiotensin II*. Angiotensin II is a vasoconstrictor—a substance that acts on the smooth muscles of arterioles, causing them

to contract, decreasing their diameter, and increasing blood pressure. It was soon shown that injection of renin (or angiotensin II) into a test animal or a human causes temporary hypertension. These studies led to the widespread belief that elevated levels of renin and angiotensin II were a major cause of human hypertension. But humans suffering from chronic hypertension typically lack elevated levels of these blood proteins. The question of the role of these proteins remains unanswered. Some vascular researchers argue that elevated renin levels cause hypertension *early in life* and that even though the renin levels decrease over time, the hypertension remains. Others believe that the renin-angiotensin system is not an important cause of hypertension.

Another factor implicated as a cause of hypertension is salt, specifically sodium chloride (NaCl). The argument is based primarily on studies of sodium intake in different human populations. It appears that people from nonindustrialized societies that have very little sodium in their diet are virtually free of hypertension. When members of these groups change their diet to include sodium, however, as occurred among native Kenyans who joined the Kenyan army and began eating food high in sodium, hypertension appears within the population. It is difficult to extrapolate these findings to industrialized societies,

however. Why do some people who ingest large quantities of salt have normal, or even low, blood pressure, while others with similar diets suffer from hypertension? The answer may lie in our genes. It appears that some members of the population are less able to tolerate high-sodium-containing diets than are others. Moreover, once a person has developed hypertension, decreasing the salt intake appears to lower blood pressure only modestly. The question that remains is whether a person who is genetically prone to hypertension could avoid developing the condition if his or her salt intake is limited at an early age.

The latest research on the cause of hypertension has focused on a small protein called *endothelin* which is secreted by the cells that form the inner lining of blood vessels. Endothelin is the most potent vasoconstrictor discovered to date; it is ten times more potent than is angiotensin II, the runner-up. Studies suggest that endothelin functions as a local hormone, causing contraction of the same vessel that is responsible for secreting the substance. Even though endothelin has not yet been shown to be a factor in human hypertension, a number of pharmaceutical companies are racing to develop products that interfere with its action, hoping these drugs will prove to be effective antihypertensive agents.

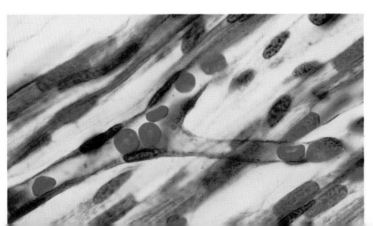

FIGURE 28-4

Red blood cells moving single file through a capillary. The walls of these microscopic vessels typically consist of a single layer of flattened cells, whose edges fit together in a manner resembling the interlocking pieces of a jigsaw puzzle (see Figure 28-2). Between the cells are openings, or pores, which, in most tissues, are large enough to allow the passage of small solutes but small enough to block the outward movement of proteins and blood cells. The thinness of the capillary walls is made evident by the fact that we can see the red blood cells so clearly.

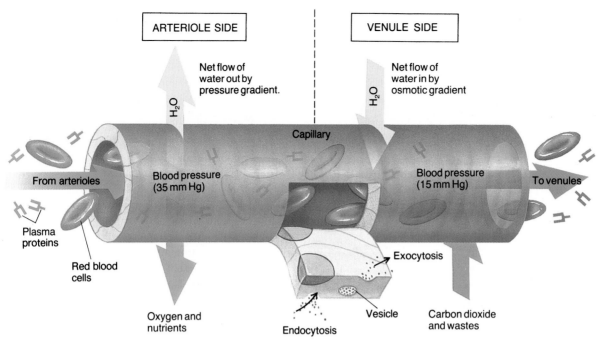

ARTERIOLE SIDE

VENULE SIDE

Net flow of
water out by
pressure gradient.

H_2O

Net flow of
water in by
osmotic gradient

H_2O

Capillary

Blood pressure
(35 mm Hg)

Blood pressure
(15 mm Hg)

From arterioles

To venules

Plasma
proteins

Red blood
cells

Exocytosis

Vesicle

Oxygen and
nutrients

Endocytosis

Carbon dioxide
and wastes

FIGURE 28-5

Action at the capillaries.   Some substances, including oxygen, carbon dioxide, and wastes, move between the interstitial fluid and the bloodstream by simple diffusion (red arrows) down their respective
concentration gradients. (Water movement is indicated by the yellow arrows.) Water is forced out of
the beginning portion of the capillaries in response to the steep pressure gradient between the blood,
at about 90 millimeters, and the surrounding fluid, which has virtually no pressure. As blood moves
through the capillary, its pressure drops rapidly, forcing less and less water out of the capillary. In
contrast, as the blood moves through the capillary, it gains in osmotic pressure due to the loss of
water. As a result, water is drawn back into the terminal portion of the capillary by osmosis that results
from the greater solute concentration within the blood compared to the surrounding fluid. In addition, some substances move into and out of the cells of the capillary wall by endocytosis and exocytosis.

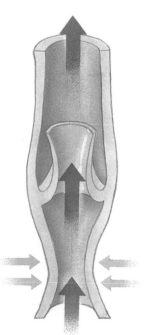

Blood
settles

(Closed
valve
prevents
backflow)

When squeezed
by skeletal
muscles

When skeletal
muscles relax

FIGURE 28-6

No turning back.   Blood flow is maintained through the veins by one-way
valves that open when pressure forces blood toward the heart (indicated by
the red arrows) and close shut when the flow is reversed.

between the cells that form the walls of the capillaries and thus remain in the blood vessel.

The movement of fluid back and forth across the porous walls of the capillaries is a key determinant of the balance of fluid between the interstitial and vascular compartments. Under some conditions, this balance can become greatly disturbed. For example, the rapid swelling that follows a physical injury results from an excess movement of fluid out of the damaged capillaries into the interstitial fluid, creating a condition known as *edema*.

Venules and Veins: The Return Trip to the Heart

Blood from capillaries collects in larger vessels, called **venules,** which empty their contents into large, thin-walled **veins;** from there, the contents return to the heart. Blood pressure in veins is very low, having been dissipated by travel through the narrow capillary passageways. Since most human veins direct blood flow upward to the heart, against the force of gravity, it may seem that blood would simply collect in these large, low-pressure vessels and go nowhere, but this is not the case. Unlike arteries, veins have virtually no elastic fibers or smooth muscle. Instead, they are thin walled and distensible, allowing some pooling of blood during periods of inactivity. However, skeletal muscle activity, such as that which occurs during walking, combined with the pressure that is exerted during breathing, squeezes veins and force their contents to travel toward the heart. This direction of blood flow is maintained by one-way flaps, or *valves* (Figure 28-6), which project from the walls of these vessels. Without the valves, blood would simply squirt back and forth in the veins as they are squeezed by skeletal muscles. In humans, the veins in the legs receive the greatest weight of blood. This weight can sometimes be so great that the walls of the veins become overdistended and the valves fail, causing the veins to *varicose* (enlarge), which accounts for the characteristic thick, blue appearance of varicose veins.

THE HUMAN HEART: A MARATHON PERFORMER

The major organ of the circulatory system—the **heart**—is a fist-sized, muscular pump that delivers an astonishing performance. The average person's heart works continually for more than 70 years. During this time, the human heart will beat 2.5 billion times and pump approximately 200 million liters (more than 50 million gallons) of blood. Every minute, the heart recirculates the entire volume of blood in the body—approximately 5 liters.

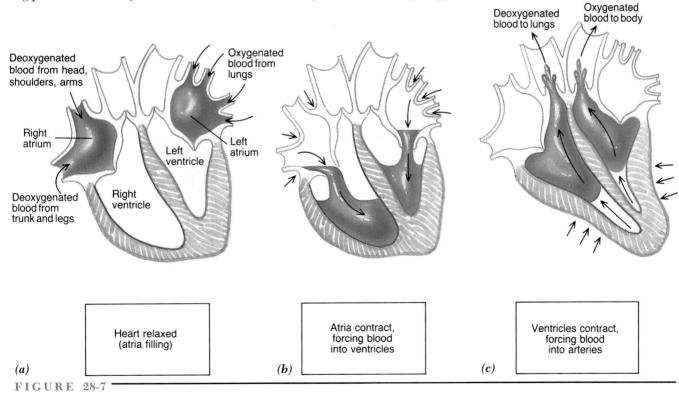

(a) Heart relaxed (atria filling)

(b) Atria contract, forcing blood into ventricles

(c) Ventricles contract, forcing blood into arteries

FIGURE 28-7

The movement of blood through the human heart. *(a)* When the heart is relaxed, blood flows from the veins into the two thin-walled atria. Deoxygenated blood from the body flows into the right atrium, and oxygenated blood from the lungs flows into the left atrium. *(b)* In the next stage of the cycle, the atria contract, forcing blood into the relaxed, thicker-walled ventricles. *(c)* In the last part of the cycle, the ventricles contract, forcing blood into the arteries. Oxygenated blood is forced out of the left ventricle into arteries that lead to the tissues of the body, while deoxygenated blood is forced out of the right ventricle into arteries that lead to the lungs. The average duration for the entire cycle in a person at rest is 0.83 seconds (72 beats per minute).

The human heart is a two sided, four-chambered model. Each side has a thin-walled **atrium** and a larger, thicker-walled **ventricle** (Figure 28-7). The left and right atria (plural of atrium) function as receiving stations for the blood that flows into the heart from the veins (Figure 28-7a). Blood from the left atrium flows into the left ventricle, while blood from the right atrium flows into the right ventricle (Figure 28-7b). Contraction of the walls of the ventricles forces the blood out of the heart and into the major arteries (Figure 28-7c).

Pumps and Vascular Circuits

The left and right sides of the heart are separated from each other by partitions that effectively divide the organ into two separate pumps, each of which powers blood through a different vascular *circuit* (Figure 28-8). In the **pulmonary circuit,** deoxygenated blood that has returned from its travels through the body's tissues is pumped from the right ventricle into the pulmonary arteries and on to the lungs, where it is oxygenated. Oxygenated blood returns to the left atrium of the heart via the pulmonary veins. (The

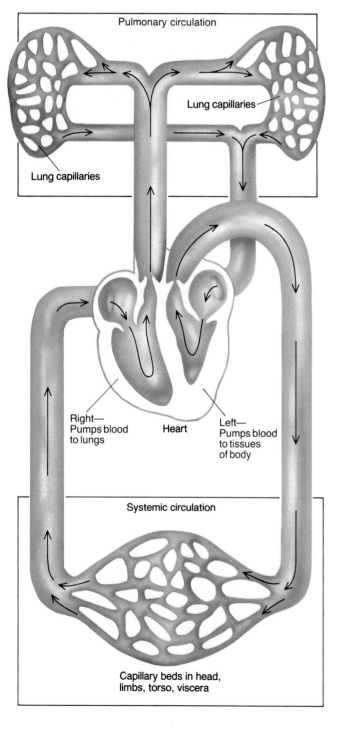

Pulmonary circulation

Lung capillaries

Lung capillaries

Right—
Pumps blood
to lungs

Heart

Left—
Pumps blood
to tissues
of body

Systemic circulation

Capillary beds in head,
limbs, torso, viscera

FIGURE 28-8

Our dual circulatory system. Deoxygenated blood is depicted in blue; oxygenated blood in red. In this schematic diagram, the right side of the heart drives the pulmonary circulation, pumping deoxygenated blood into the capillary beds of the lungs, where the blood picks up oxygen and returns to the left side of the heart. The oxygenated blood is then propelled from the left side of the heart through the systemic circulation. Oxygen is given up in the capillary beds of the tissues, and deoxygenated blood returns to the right side of the heart.

designation of a vessel as an artery or a vein is based on its position relative to the heart, not whether or not the blood is oxygenated or deoxygenated. Arteries carry blood from the heart; veins return it to the heart.) Blood returning to the left atrium from the lungs passes into the left ventricle, whose powerful contraction sends it out into the **systemic circuit,** where it nourishes the tissues of the body. In the systemic circuit (Figure 28-9), blood is initially pumped into the **aorta,** the largest artery of the body (approximately 2.5 centimeters, or 1 inch, in diameter), which feeds into many of the major arterial thoroughfares of the body.

Blood does not nourish the heart as it travels through its chambers; rather, large **coronary arteries** branch from the aorta immediately after the aorta emerges from the heart, providing cardiac muscle priority access to oxygen-rich blood. Oxygen is required to maintain the oxidation of glu-

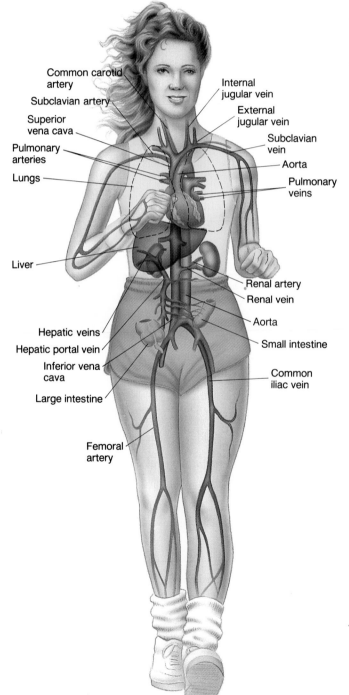

Common carotid artery
Subclavian artery
Superior vena cava
Pulmonary arteries
Lungs
Liver
Hepatic veins
Hepatic portal vein
Inferior vena cava
Large intestine
Femoral artery

Internal jugular vein
External jugular vein
Subclavian vein
Aorta
Pulmonary veins
Renal artery
Renal vein
Aorta
Small intestine
Common iliac vein

FIGURE 28-9

The major arteries and veins in the systemic circulation. As the *aorta* emerges from the heart, it travels in a headward *(anterior)* direction, then makes a sharp "U-turn," known as the *systemic* (or *aortic) arch,* and proceeds in a footward *(posterior)* direction through the thorax and into the abdominal cavity. As the aorta extends from the heart, it branches into a number of major arteries that provide blood to the body's organs. Blood from the body's tissues collects in a number of veins that feed into two large trunks, the *inferior* and *superior vena cavae,* which direct the flow of blood back to the heart.

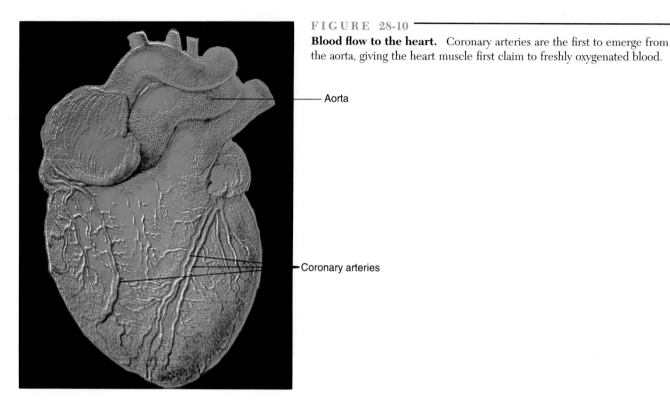

FIGURE 28-10

Blood flow to the heart. Coronary arteries are the first to emerge from the aorta, giving the heart muscle first claim to freshly oxygenated blood.

Aorta

Coronary arteries

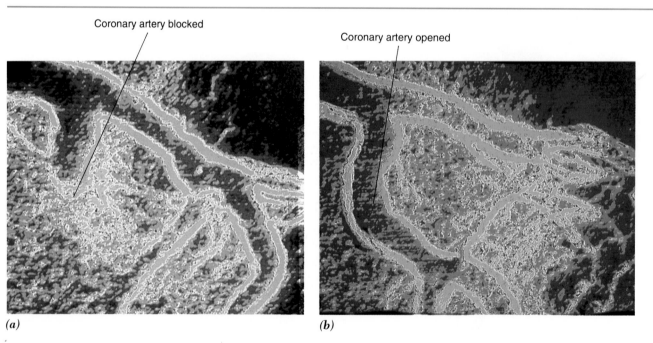

Coronary artery blocked

Coronary artery opened

(a)   *(b)*

FIGURE 28-11

Coronary artery disease. *(a)* An *angiogram* showing the inability of injected dye to flow normally through an obstruction in a coronary artery. *(b)* Angiogram showing the same coronary artery depicted in part *a* after being opened by angioplasty. *(c)* Instrument utilized in balloon angioplasty. After the tube is threaded into the blocked coronary artery, the tube is inflated near its tip, pushing the walls of the artery outward. *(d)* Diagrammatic results of coronary bypass surgery, whereby a portion of a leg vein has been used to shunt blood directly from the aorta to a coronary artery past the point of an obstruction.

cose, which provides the chemical energy that fuels the heart's contractions. The large coronary arteries are evident in the computer-enhanced view of the heart depicted in Figure 28-10. Obstructions in the coronary arteries can deprive the heart of the oxygen needed to keep heart tissue alive. Such obstructions are the primary cause of heart attacks. If a substantial portion of the heart is deprived of oxygen for a long enough period of time, the heart muscle ceases to function, and the person will die.

A person suffering from insufficient blood flow through the coronary arteries often receives warning signs of the condition, such as an inability to exert themselves or chest pains. The degree to which these arteries are "clogged" is revealed by *angiography,* a procedure in which a dye is injected into the bloodstream and its progress through the heart is monitored radiographically (Figure 28-11*a*). If one or more of the coronary arteries is shown to be largely obstructed, the patient is usually treated either by balloon angioplasty or coronary bypass surgery (Figure 28-11*b*). In *balloon angioplasty,* a tube is threaded into the blocked coronary artery. The tip of the tube is then inflated like a balloon (Figure 28-11*c*), pushing the walls of the artery outward. In *coronary bypass surgery,* a segment of

vein taken from the patient's leg is inserted between the aorta and the coronary artery at a point beyond the site of occlusion (Figure 28-11*d*). In this way, blood is able to flow from the aorta into the heart muscle, bypassing the blocked arterial vessel. Several blocked coronary arteries are usually bypassed in this way during a single operation.

The Heartbeat

The one-way movement of blood through the chambers of the heart and into the appropriate arteries is ensured by the opening and closing of two pairs of strategically located **heart valves** (Figure 28–12). The openings between the atria and the ventricles are guarded by a pair of *atrioventricular (AV) valves,* each of which consists of a flap of tissue that can open in only one direction. When the atria contract, the pressure of the atrial blood rises, causing the fluid to push against the valves. The valves are forced open, allowing blood to flow into the ventricles. When the atrial contractions cease, the pressure of the blood in the ventricles becomes greater than that in the atria. This pressure differential forces the atrioventricular valves to close, preventing the flow of blood back into the atria. The direction of flow out of the ventricles and into the major arteries is main-

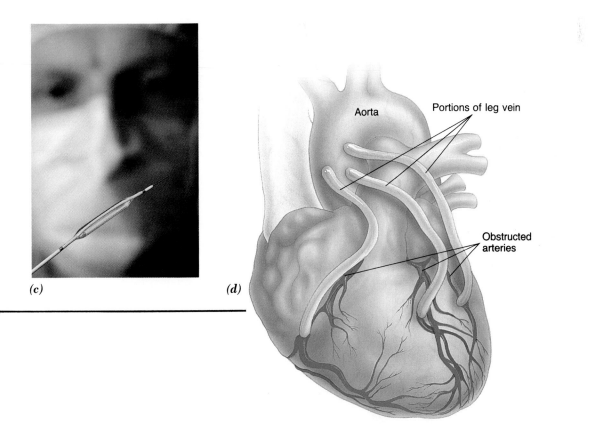

(c)　　　*(d)*

Aorta

Portions of leg vein

Obstructed arteries

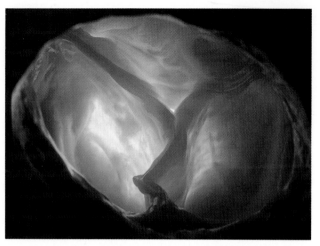

FIGURE 28-12

One of the heart valves that maintains the flow of blood through the chambers of the heart and into the major arteries.

tained by another pair of valves, the *semilunar valves*, which are located on each side of the heart.

If you listen to the heartbeat through a stethoscope, you will hear a "lub-dub" sound that repeats itself with each beat of the heart. The lower-pitched "lub" is produced by the closure of the AV valves at the onset of ventricular contraction. The higher-pitched "dub" is produced by the closure of the semilunar valves at the onset of ventricular relaxation. A number of conditions can cause a disruption in the smooth flow of blood through the heart, creating turbulence, which is heard as a *murmur*. A heart murmur may result if the opening through a valve is narrower than normal or if the valve is damaged, thereby allowing the backflow of blood. Heart murmurs may or may not be a reflection of a serious problem. Most serious heart-valve defects are either congenital or are the result of damage from rheumatic fever. If necessary, the valves can be replaced surgically with artificial ones.

Excitation and Contraction of the Heart

Embedded in the wall of the right atrium is a small piece of specialized cardiac muscle tissue, called the **sinoatrial (SA) node** (Figure 28-13), whose function is to excite contraction. If a healthy human heart is isolated, as is done during a heart transplant, the isolated heart can beat rhythmically on its own, without any outside stimulation. Each beat of an isolated heart is initiated by a spontaneous electrical discharge that originates every 0.6 seconds or so in the SA node. Because of its role in determining the rate the heart beats, the SA node is called the heart's *pacemaker*. A number of conditions can cause the heart to develop an abnormal *(arrhythmic)* heartbeat. If the condition is life-threatening, a surgeon will implant an electronic pacemaker in the chest. Artificial pacemakers deliver an electric shock at intervals that approximate the intrinsic cardiac rhythm.

Once an action potential is generated in the SA node, a wave of electrical activity spreads across the walls of the atria, passing from one cardiac muscle cell to the next

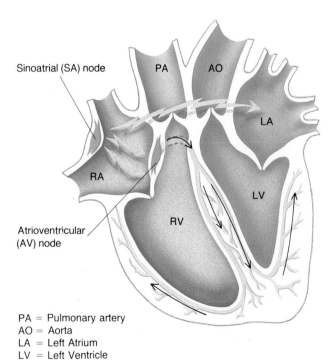

Sinoatrial (SA) node

Atrioventricular (AV) node

PA
AO
LA
RA
LV
RV

PA = Pulmonary artery
AO = Aorta
LA = Left Atrium
LV = Left Ventricle
RV = Right Ventricle
RA = Right Atrium

FIGURE 28-13

Synchronizing the beat. Impulses generated in the SA node (the pacemaker) sweep through the muscle cells of the two atria, causing them to contract in unison. When this impulse reaches the AV node, it launches a wave of electrical activity that is channeled through the ventricular walls, causing the two ventricles to contract just as the atria relax.

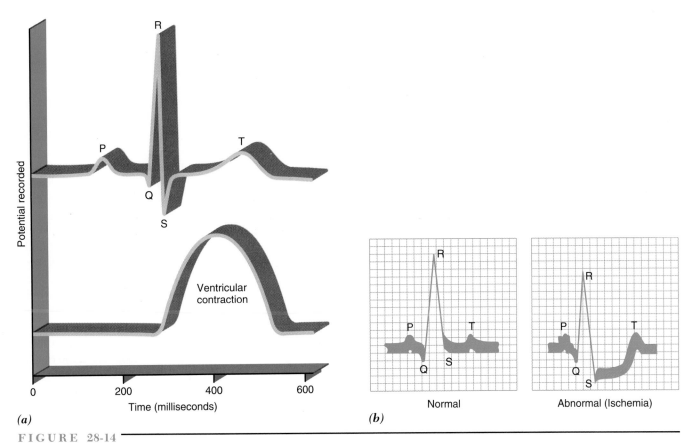

FIGURE 28-14

Diagnosing heart function. *(a)* An electrocardiogram from a normal heart and the corresponding period of ventricular contraction. Three distinct peaks are seen in the electrocardiogram: a P-wave, which results from the flow of current that sweeps across the atria at the start of the heart's contraction; a QRS complex, which results from the flow of current across the ventricles; and a T-wave, which results from the current associated with the return of the ventricles to the resting condition. *(b)* Abnormal cardiac cycles are often revealed by deviations from the normal tracing. Such deviations often show up during a stress EKG that is taken while the person is running on a treadmill. Ischemia results when the heart is not receiving sufficient oxygen due to blockage of the coronary arteries.

through communicating junctions (page 110) that link neighboring cells. The transmission of impulses through such junctions is extremely rapid, causing the entire atrial wall to contract in a synchronous, coordinated manner. As the wave of electrical excitation reaches the boundary of the atria, it cannot pass into the wall of the ventricles because the atria and ventricles are separated by a layer of connective tissue. Only one point of electrical connection exists between the two major parts of the heart—the **AV node** (Figure 28-13). The conduction of the electrical stimulus through the AV node is delayed by about 0.1 second, which allows time for the blood to flow from the atria to the ventricles on each side of the heart. Once the stimulus reaches the AV node, it spreads out over the ventricles, causing the ventricles to contract, forcing blood into the major arteries.

The electrical activity that occurs at different stages of a heartbeat is sufficiently intense to be detected by electrodes placed on the skin of the chest. The recordings of this activity—an **electrocardiogram (EKG)**—constitutes one of the most important diagnostic tools for the analysis of heart disease (Figure 28-14).

Regulating the Heartbeat

The rate at which the heart's pacemaker becomes excited is regulated by a number of factors, primarily by signals that arrive from the autonomic nervous system (Chapter 23). In general, impulses that reach the SA node over *parasympathetic nerves* decrease the rate of heart contraction, while impulses that arrive over *sympathetic nerves* increase it. Thus, at any given moment, the heart rate is determined by the balance between these antagonistic influences. The centers that control the cardiovascular system are housed in the hypothalamus and the medulla of the brainstem (page 486). If we become active, angry, or frightened, this information is passed down the brainstem to the medulla, increasing the sympathetic output to the SA node and decreasing the parasympathetic output, thereby increasing the heart rate.

COMPOSITION OF THE BLOOD

Human blood is a complex tissue (Table 28-1) that is composed of blood cells and cell-like components suspended in a clear, straw-colored liquid called **plasma.**

Plasma: The Soluble Phase of the Blood

The average person contains about 4.7 liters (10 pints) of blood, about 55 percent of which is plasma. Most (about nine-tenths) of plasma is water, the blood's solvent. The rest is composed of various dissolved substances, predominantly three plasma proteins:

- *Albumin*, which helps maintain osmotic conditions that favor recovery of water that has been forced out of capillaries and into the surrounding tissues (see Figure 28-5).
- *Globulins*, such as the *gamma-globulins* (including antibodies that contribute to immunologic defenses) and other globulins that assist in the transport of lipids and fat-soluble vitamins.
- *Fibrinogen*, which provides the protein network necessary for blood clotting.

Salts, vitamins, hormones, dissolved gases, sugars, and other nutrients make up just over 1 percent of plasma's volume.

Red Blood Cells

Erythrocytes, which are often called red blood cells due to their high concentration of red-colored hemoglobin molecules, transport about 99 percent of the oxygen carried by the blood. Oxygenated blood takes on a bright red color, whereas blood that is depleted of oxygen takes on a darker, purplish color that appears blue through blood-vessel walls. Erythrocytes are flattened, disk-shaped cells with a central depression that gives them the appearance of a doughnut with a depression instead of a hole (Figure 28-15). Not only are red blood cells the most abundant cells in the body (there are approximately 5 billion in each milliliter of blood), they are also the simplest. Human erythrocytes lack a nucleus, ribosomes, and mitochondria and constitute little more than nonreproducing sacks of oxygen-binding hemoglobin.

The human body produces red cells at the astonishing rate of 2.5 million every second, or about half a ton during an average person's lifetime. This massive construction project occurs in the red bone marrow, where undifferentiated precursors, called **stem cells,** proliferate and differentiate into erythrocytes. After circulating for about 4 months, aging erythrocytes are engulfed by scavenger cells of the liver and spleen.

TABLE 28-1
COMPONENTS OF BLOOD

Component	Percent	Function
Plasma	55	Suspends blood cells so they flow. Contains substances that stabilize pH and osmotic pressure, promote clotting, and resist foreign invasion. Transports nutrients, wastes, gases, and other substances.
White blood cells	<0.1	Allow phagocytosis of foreign cells and debris. Act as mediators of immune response.
Platelets	<0.01	Seal leaks in blood vessels.
Red blood cells	45	Transport oxygen and carbon dioxide.

Blood settles into three distinct layers when treated with substances that prevent clotting.

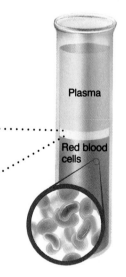

Plasma

Red blood cells

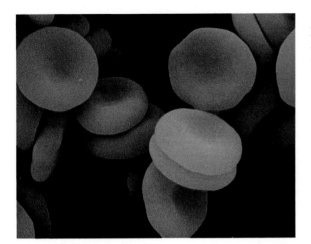

FIGURE 28-15

Red blood cells: biconcave sacks of hemoglobin. The flattened shape of erythrocytes keeps each of the 270 million hemoglobin molecules close to the cell's surface for rapid gas exchange.

The number of erythrocytes produced by the body fluctuates according to availability of oxygen. For example, a person becomes adjusted *(acclimated)* to higher elevations (Figure 28–16), where the air has a lower concentration of oxygen and other gas molecules, by producing more red blood cells, thereby compensating for the reduced amount of oxygen in each breath. The trigger for this response is *erythropoietin,* a hormone secreted by the kidney which initiates the transformation of stem cells in the bone marrow into erythrocytes. Low oxygen concentration stimulates erythropoietin formation, speeding up erythrocyte production until the oxygen supply in the tissues is restored to normal.

Lately, erythropoietin has been at the center of a major controversy in athletics. Now that it is available through recombinant DNA technology, some athletes are injecting themselves with erythropoietin to increase the number of erythrocytes in their blood, thereby supplying more oxygen to their tissues, which they believe improves their performance. Not only is this practice illegal, it may also be dangerous since the effects of excess levels of this highly active protein remain unknown.

White Blood Cells

If you examine a stained blood smear (Figure 28-17*a*), you will notice a small number of white blood cells, or **leukocytes** (*leuko* = white, *cyte* = cell), scattered among the erythrocytes. Leukocytes defend us against foreign microorganisms that could otherwise invade, overwhelm, and destroy our bodies. These blood cells also function as sanitary engineers, cleaning up dead cells and tissue debris that would otherwise accumulate to obstructive levels.

Five classes of leukocytes are clearly recognized: neutrophils, eosinophils, basophils, monocytes, and lymphocytes (Figure 28-17*b*). The first three types—*neutrophils, eosinophils,* and *basophils*—are characterized by their cytoplasmic granules. These cells protect the body by engulf-

FIGURE 28-16

At higher altitudes, an ice climber must become acclimated to the lower concentration of oxygen in the air.

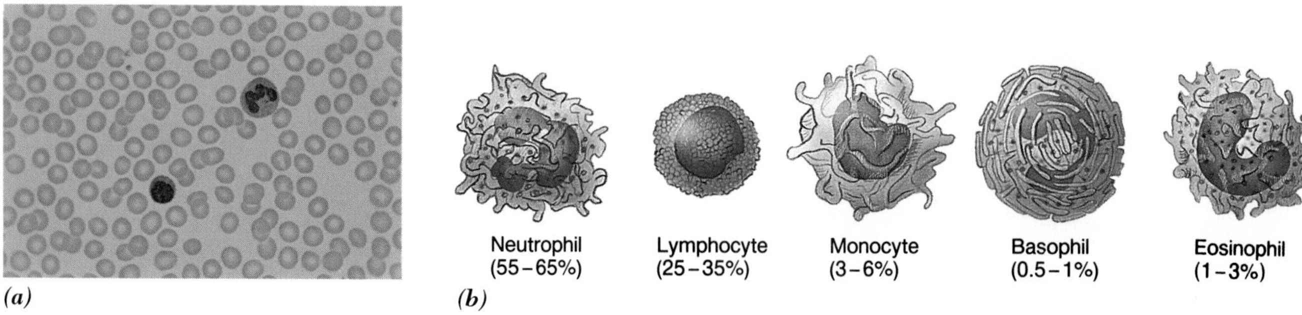

FIGURE 28-17

Human leukocytes and their relative abundance in the blood. *(a)* A blood smear showing two scattered leukocytes. *(b)* The five types of leukocytes. The red dots in the neutrophil, basophil, and eosinophil are granules that can be distinguished by different types of stains.

ing foreign intruders by phagocytosis (page 144). Basophils also function by releasing powerful chemicals that trigger an inflammatory response at the site of the wound or infection. These three white blood cells are distinguished by the types of stains that are used to make their granules visible. In contrast, *monocytes* are nongranular leukocytes that are attracted to sites of inflammation, where they too act as phagocytes, engulfing debris. Monocytes also give rise to phagocytic *macrophages* — huge cells that wander through the body's tissues, ingesting foreign agents. Macrophages also clean up "battlefields" that are littered with the cellular debris of combat between other phagocytes and invading microorganisms. Phagocytosis is not the only mechanism of protection afforded by leukocytes. The *lymphocytes,* a group of nonphagocytic white cells, are the "masterminds" of the immune system, the subject of Chapter 30.

The number and type of leukocytes in the blood can provide an indication of a person's health, which is why your physician may take a "blood cell count" when you are ill. Most infections stimulate the body to release into the bloodstream large numbers of protective leukocytes that are normally held in reserve, causing the "white cell number" to rise. Some disease agents affect only certain types of leukocytes. For example, the viral disease mononucleosis is characterized by an elevated number of monocytes, while infection with the hookworm parasite results in an elevated number of eosinophils. In contrast, certain viruses, including the virus responsible for AIDS, cause infections that *deplete* certain leukocytes to abnormally low levels (Chapter 30).

Factors Necessary for Blood Clotting

A hole in an injured vessel can lead to fatal blood loss if the leak is not quickly patched. In other words, a person could bleed to death. The blood is equipped with its own lines of defense that minimize blood loss and maintain ho-

meostasis. Immediately following injury to a vessel, circulating **platelets** — spiny fragments that are released from special blood cells — become trapped at the injury site. As platelets rapidly accumulate, they plug the leak, providing the first step in damage control. The aggregation of platelets triggers two additional responses at the site of injury: (1) the constriction of the damaged vessel by muscular contraction, and (2) the formation of a clot by coagulation, both of which further seal the vessel.

The resources for clotting circulate in the blood in an inactive form. **Fibrinogen** is a rod-shaped plasma protein which, when converted to the insoluble protein **fibrin,** generates a tangled net of fibers that binds the wound and stops blood loss until new cells replace the damaged vessel (Figure 28-18). The conversion of fibrinogen to fibrin involves the removal of several segments of the fibrinogen molecule, a reaction catalyzed by a proteolytic enzyme, called **thrombin.** Thrombin is also derived from an inactive precursor (*prothrombin*) by a reaction catalyzed by yet another proteolytic enzyme, **Factor X.** As a result of this "reaction cascade," a single molecule of Factor X can stimulate the rapid formation of a large number of fibrin molecules.

In some persons, the cascade fails to complete the formation of a clot. The most common clotting disorder is *hemophilia.* Most hemophiliacs have a defective gene that codes for an inactive version of **Factor VIII,** a protein that normally activates Factor X. Episodes of bleeding in hemophiliacs can now be treated with synthetic preparations of Factor VIII, which are produced by recombinant DNA technology. Such preparations eliminate the need for transfusions of blood products, which may be contaminated with viruses that cause AIDS or hepatitis.

While blood clots are one of the body's primary defense mechanisms, they are also responsible for the majority of deaths in industrialized countries. Most heart attacks are

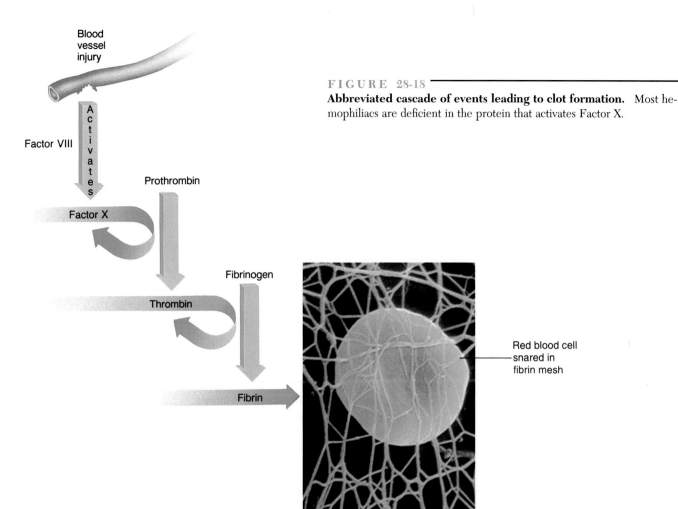

FIGURE 28-18

Abbreviated cascade of events leading to clot formation. Most hemophiliacs are deficient in the protein that activates Factor X.

Blood vessel injury

Activates

Factor VIII

Prothrombin

Factor X

Fibrinogen

Thrombin

Fibrin

Red blood cell snared in fibrin mesh

CLOT

due to the formation of a blood clot *(thrombus)* within a coronary artery at a site where the vessel has already been occluded by the buildup of plaque. While the plaque reduces the flow of blood to the heart, the blood clot totally obstructs the vessel, causing the rapid death of heart tissue which accompanies a heart attack. For this reason, the preferred treatment for heart-attack victims is the injection of a massive dose of a clot-dissolving substance, such as *tissue plasminogen activator (TPA)*. It is important that the "clot buster" be administered very soon after the onset of the attack, otherwise damage to the heart tissue due to oxygen deprivation becomes irreversible.

Maintaining human life requires the continual circulation of blood through a vast network of living vessels of varying structure and dimensions. Circulating blood carries substances and cells from one part of the body to another, maintaining the stability of the internal environment and facilitating the exchange of respiratory gases, wastes, and nutrients between the external environment and the living tissues. (See CTQ #1.)

EVOLUTION OF CIRCULATORY SYSTEMS

Before complex animals could become independent of the sea in which they evolved, they had to develop an internal fluid that mimicked the properties of sea water. In other words, they needed a "portable ocean." They also needed a system that circulated this fluid throughout the body, bathing all the cells in a life-sustaining solution, even while the animal lives on dry land. The evolution of this circulatory system did not occur in a single step.

Simpler, multicellular animals, including sponges, jellyfish, and flatworms, attain considerable size without a true circulatory system. This is possible because of the way these animals are constructed; that is, because of their *body plan*. The body plan of these simpler animals places virtually all cells close to the source of oxygen and nutrients. This is particularly well illustrated by the flatworm, whose flattened body shape allows oxygen to diffuse directly from the environment to all of the body's cells and allows waste products to diffuse in the opposite direction (Figure 28-

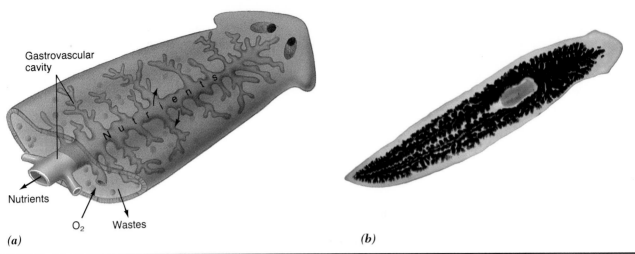

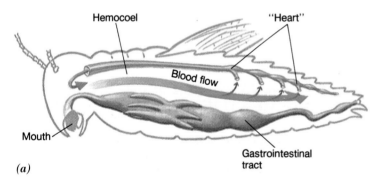

(a) Gastrovascular cavity — Nutrients — O₂ — Wastes

(b)

FIGURE 28-19

Flatworm form and function. (*a*) A flatworm's body is highly flattened, which brings every cell close enough to the outside medium to receive oxygen and expel wastes by diffusion. (*b*) The digestive chamber of a flatworm branches to form blind (dead-ended) passageways throughout the body. Nutrients are able to diffuse from the digestive channels directly to the body's cells. Because of its dual role as both digestive and circulatory system, the spacious, internal compartment of the flatworm is called a *gastrovascular cavity*.

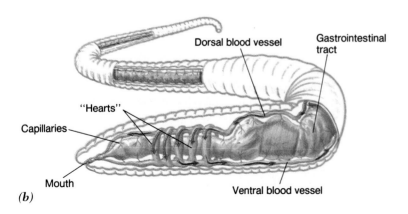

FIGURE 28-20

Circulatory strategies. (*a*) The open circulatory system of a grasshopper. Blood is pumped out of blood vessels into spaces (which form the *hemocoel*) that bathe the body's tissues. (*b*) The closed circulatory system of an earthworm; blood remains in vessels.

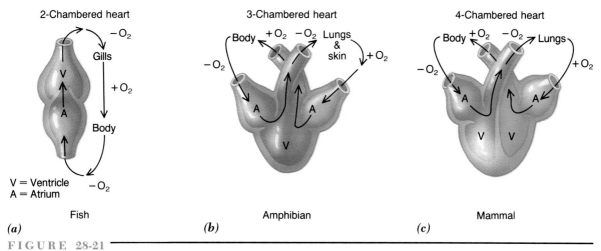

FIGURE 28-21

Evolution of the vertebrate heart. *(a)* Fishes have a two-chambered heart, consisting of a single atrium and ventricle, which pumps blood through a single vascular circuit. The blood flows from the heart to the gills, where it is oxygenated, and then on to the body, before returning to the heart. At first glance, this single, direct circuit in the fish seems to make more sense than does the more complex, double circuit in mammals, but it has a serious disadvantage: Blood that has passed through the capillary network of the gills will have lost nearly all of the pressure it had when it was pumped from the heart. Blood pressure is an important factor in determining flow rates, fluid exchange with the tissue, kidney function, and so forth, so that the drop in pressure that occurs in the gills seriously reduces the efficiency of the circulatory system. *(b)* Amphibians have a three-chambered heart—two atria and a single ventricle—and are the first vertebrates to have evolved separate pulmonary and systemic circuits. Oxygenated blood from the lungs enters the left atrium, and deoxygenated blood from the body enters the right atrium. Despite the presence of only a single ventricle, physiological studies suggest that the two streams do not become intermixed in the ventricular chamber and that the oxygenated blood from the lungs flows largely to the body, while the deoxygenated blood from the body flows largely to the lungs and skin, a major site of gas exchange in frogs. *(c)* The four-chambered heart of a mammal.

19*a*). Similarly, nutrients can diffuse to the body's cells directly from the highly branched digestive cavity (Figure 20–19*b*).

The evolution of large, more complex animals occurred together with the evolution of a transport system that could deliver oxygen and nutrients to all cells of the body and carry away waste products. Over the course of evolution, two basic types of circulatory systems have appeared: open and closed systems. Some animals, including arthropods (such as insects and spiders) and most molluscs (including snails and clams), have an **open circulatory system** (Figure 28-20*a*), whereby the blood is pumped through vessels that empty into a large, open space (*hemocoel*), in which most of the body's organs are immersed. The cells receive nutrients directly from the fluid in which they are bathed.

In contrast, the human circulatory system is a **closed circulatory system.** In a closed system, blood surges throughout the body in a *continuous* network of closed vessels. Many simpler animals, including the earthworm (Figure 28-20*b*), also have closed systems. In a closed sys-

tem, respiratory gases, nutrients, and waste products are exchanged across the porous walls of the finest vessels in the network—the capillaries.

Vertebrate circulatory systems provide an example of how an entire organ system can undergo gradual evolutionary change. The human heart, with its two separate pumping organs and two distinct circuits, evolved gradually, and some of the presumed intermediate stages have been preserved in the systems of other living vertebrates. The evolution of the vertebrate heart is described in Figure 28-21.

Diffusion is adequate for the movement of substances from one place to another as long as the distances traversed are small—less than a few millimeters. The evolution of larger organisms was accompanied by the gradual development of transport systems that could carry substances back and forth between the environment and the animal's deepest internal tissues. (See CTQ #2.)

THE LYMPHATIC SYSTEM: SUPPLEMENTING THE FUNCTIONS OF THE CIRCULATORY SYSTEM

In addition to the cardiovascular system, humans are equipped with a secondary network of fluid-carrying vessels and associated organs that make up the **lymphatic system** (Figure 28-22). Lymphatic vessels are a series of one-way channels that originate in the tissues as a bed of *lymphatic capillaries.* Lymphatic capillaries absorb excess interstitial fluid that fails to reenter the capillaries. The smaller lymphatic vessels fuse to form larger lymphatic vessels, which ultimately drain into the large veins of the neck. Fats that are absorbed from the intestine following digestion of a fat-containing meal are also collected and delivered to the bloodstream by lymphatic vessels (review Figure 27-6).

Before reentering the bloodstream, lymphatic fluid, or *lymph,* passes through a series of **lymph nodes** (Figure 28-22)—structures that help "purify" the fluid by removing foreign substances and any microorganisms that may be present. During infection, lymph nodes become engorged with white blood cells that are recruited to fight the microbial aggressor. "Swollen glands" (which is actually a misnomer since lymph nodes are not glandular) are therefore common signs of infection. The **spleen** is another lymphoid organ, which filters blood as well as lymph and serves as a reservoir of lymphocytes. Other lymphoid organs include the *tonsils,* which lie on either side of the throat, and the *thymus gland,* which is located at the base of the neck.

The lymphatic system is of great clinical importance because the lymphatic vessels are the common path taken by cancer cells that have been released from a tumor. These malignant cells travel to distant points in the body, where they become lodged and proliferate, forming secondary tumors. This is the reason that lymph nodes that lie in the vicinity of a malignant tumor are usually removed during surgery, along with the tumor itself. Examination of these lymph nodes provides a measure of whether or not the cancer has spread.

The vessels of the lymphatic network provide a specialized pathway for the recovery of excess tissue fluid and the movement of fatty food molecules and the body's protective white blood cells. (See CTQ #3.)

EXCRETION AND OSMOREGULATION: REMOVING WASTES AND MAINTAINING THE COMPOSITION OF THE BODY'S FLUIDS

One day you might eat a large bag of salty potato chips, while the next day you may choose relatively salt-free foods. Even though your diet may change drastically from day to

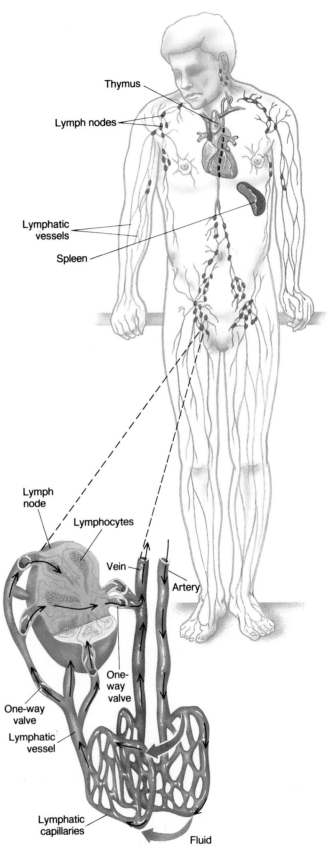

FIGURE 28-22

The lymphatic system. Lymphatic vessels capture and return excess fluid that is leaked from the capillaries of the bloodstream. Lymphoid organs include the lymph nodes, thymus, spleen, and tonsils.

day, the composition of your body fluids remains relatively constant. This, once again, is the essence of homeostasis. Maintaining the proper composition of the body's internal fluids requires two related activities: osmoregulation and excretion.

1. **Osmoregulation** is the maintenance of the body's normal salt and water balance. If you were not able to rid yourself of the excess salt you ingested after eating that bag of potato chips, the concentration of sodium, potassium, and chloride ions could increase to dangerous levels. Elevated potassium concentrations in the blood, for example, can lead to a fatal disruption of the rhythmic beating of the heart. In fact, numerous cases of physician and nurse "mercy killings" in hospitals have been attributed to administration of potassium chloride. Similarly, when you drink more fluid than your body needs to maintain its proper water content, excess water must be eliminated.

2. **Excretion** is the discharge of the body's metabolic waste products. The wastes that tend to accumulate at highest concentration and pose the greatest threat to the delicate physical and chemical balance required for life are products of the metabolic breakdown of nitrogen-containing compounds, notably proteins and nucleic acids. The nitrogen is released during metabolism as highly poisonous ammonia (NH_3). In humans, ammonia is quickly converted by the liver to **urea,** a relatively nontoxic nitrogenous molecule that is tolerated in the tissue fluids until it is eliminated by the kidney.

As we discuss later, many animals convert ammonia to *uric acid,* a nontoxic product that is excreted as an almost dry paste.

Humans have a single system that carries out both osmoregulation and excretion; this is not the case in all animals.

THE HUMAN EXCRETORY SYSTEM: FORM AND FUNCTION

The human excretory system is illustrated in Figure 28-23. The paired **kidneys** are biological cleaning stations that remove nitrogenous wastes as well as excess salts and water from the blood, forming **urine,** a solution destined for discharge from the body. From the kidneys, urine is moved by peristalsis down a pair of muscular tubes, the **ureters,** and into a holding tank, the **urinary bladder.** Discharge of urine from the bladder through the **urethra** is under control of neural commands.

The Structure of the Kidney

The functional unit of the kidney is the renal tubule, or **nephron** (Figure 28-24). Each nephron produces a small volume of fluid, or urine; together, the urine produced by the million or so nephrons of the kidneys makes up the fluid that you void several times a day. The nephron wall consists of a single layer of epithelial cells surrounding a central, hollow lumen. Closely entwined around each nephron is a system of blood vessels, which allows for the exchange of

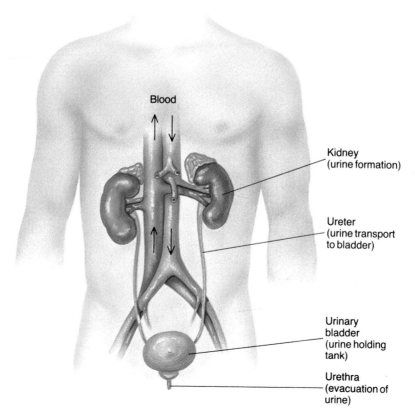

FIGURE 28-23

The human excretory system. The kidneys filter the blood, removing waste products and excess salts, forming urine. Ureters provide passageways to the urinary bladder, where urine is held until its release from the body through the urethra.

Blood

Kidney (urine formation)

Ureter (urine transport to bladder)

Urinary bladder (urine holding tank)

Urethra (evacuation of urine)

materials back and forth between the urinary fluid in the tubule and the bloodstream (Figure 28-24).

Even though the weight of both kidneys is only about 1 percent of the body's total weight, these organs receive approximately 25 percent of the body's total blood flow. Blood enters the kidney through the renal artery, which branches to form smaller vessels, each of which leads into a bundle of capillaries, called a **glomerulus.** Each glomeru-

lus is embedded in a blind, cup-shaped end of a nephron, called **Bowman's capsule** (Figure 28-24). From Bowman's capsule, the nephron begins a winding path, first as the **proximal convoluted tubule,** then into a long, U-shaped portion, called the **loop of Henle,** then into the **distal convoluted tubule,** and finally into a **collecting duct,** which drains the nephron's contents into the ureter. Notice in Figure 28-24 that part of each nephron (Bow-

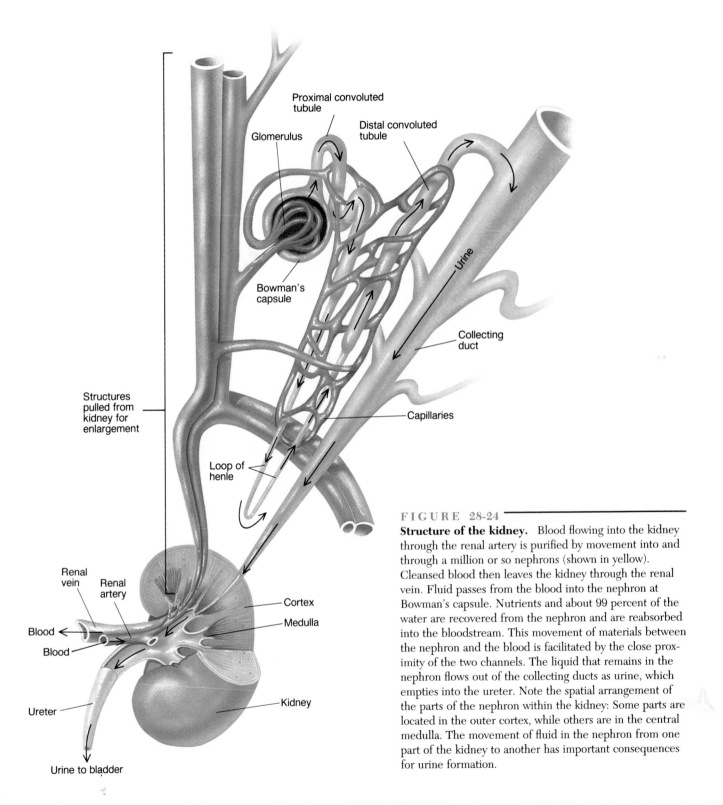

FIGURE 28-24

Structure of the kidney. Blood flowing into the kidney through the renal artery is purified by movement into and through a million or so nephrons (shown in yellow). Cleansed blood then leaves the kidney through the renal vein. Fluid passes from the blood into the nephron at Bowman's capsule. Nutrients and about 99 percent of the water are recovered from the nephron and are reabsorbed into the bloodstream. This movement of materials between the nephron and the blood is facilitated by the close proximity of the two channels. The liquid that remains in the nephron flows out of the collecting ducts as urine, which empties into the ureter. Note the spatial arrangement of the parts of the nephron within the kidney: Some parts are located in the outer cortex, while others are in the central medulla. The movement of fluid in the nephron from one part of the kidney to another has important consequences for urine formation.

man's capsule, and the proximal and distal convoluted tubules) is located in the outer *cortex* of the kidney, while the remainder (the loop of Henle and collecting duct) projects down into the inner *medulla* of the kidney. As you will see shortly, this structural feature of each nephron holds the key to the formation of urine.

Three Processes During Urine Formation

Urine formation occurs as the result of three distinct processes: glomerular filtration, tubular reabsorption, and tubular secretion.

Glomerular Filtration The capillaries of a glomerulus are particularly porous, and the blood in these vessels is present at unusually high pressure. During **glomerular filtration,** the pressure of the blood in the glomerulus provides the force that pushes fluid out of the capillaries (Figure 28-25, step 1), across the epithelial wall of Bowman's capsule, and into the lumen of the nephron. The first insight into glomerular filtration came in the early 1920s, when the physiologist Alfred Richards of the University of Pennsylvania developed a technique of withdrawing fluid from various portions of a single nephron of a frog kidney, using a

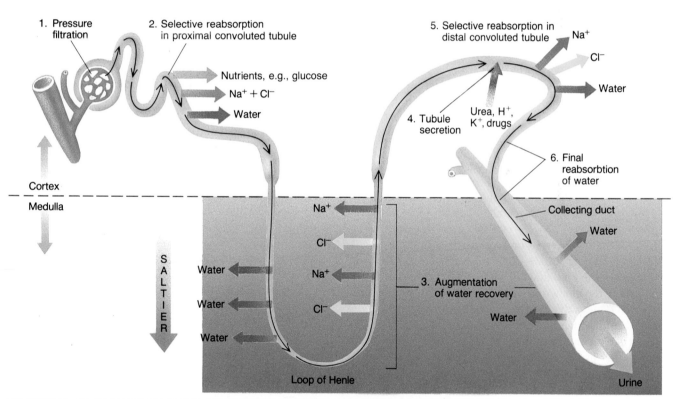

FIGURE 28-25

Urine formation in the nephron. (1) The fluid portion of the blood is forced into the proximal end of the nephron through Bowman's capsule by pressure filtration. (2) As the fluid passes along the proximal convoluted tubule, nutrients, salts, and water are recovered by reabsorption. The movement of water out of the proximal convoluted tubule by osmosis follows the active transport of salt. (3) The recovery of additional water is augmented by the loop of Henle, whose active transport of sodium establishes a steep osmotic gradient in the extracellular fluid of the medulla of the kidney (indicated by the increased shading). (4) As the fluid moves through the distal convoluted tubule, tubular secretion transfers hydrogen and potassium ions, along with waste products from the blood, into the nephron. (5) Additional salt is reabsorbed from the distal convoluted tubule, along with water, which flows outward by osmosis. (6) The fluid then enters the collecting duct, where it passes through the osmotic gradient established by the loop of Henle. As the fluid in the collecting duct moves into a saltier environment, more and more water is drawn out of the nephron, where it diffuses back into the bloodstream. The fluid that remains in the collecting ducts after it has passed through the medulla passes into the ureters as urine.

very fine pipette. More recently, the technique has been adapted to mammalian kidneys and has provided the best information we have about the processes occurring within the nephron.

Richards determined that the **glomerular filtrate**—the fluid that passes from the glomerulus into the lumen of the nephron—is similar to the blood of the glomerular capillaries in some respects and different in others. Unlike the blood, the glomerular filtrate lacks blood cells and proteins but otherwise has a similar concentration of solutes. These findings suggested that the passage of materials from blood to nephron is nonselective and determined solely by the size of the pores in the walls of the glomerular capillaries and Bowman's capsule. Proteins and cells are simply too large to pass through these pores, but virtually all other blood constituents, including valuable nutrients, water, and wastes, are forced out of the glomerular capillaries and enter the nephron.

The delicate structure of the glomerulus and the nephron is very sensitive to blood pressure. Too low a pressure decreases glomerular filtration, while too high a pressure can produce extensive, microscopic damage. This explains why a person with prolonged high blood pressure often experiences kidney failure. It is also one of the obvious reasons why hypertension should not go untreated.

Tubular Reabsorption The kidney is a remarkable purification plant. Approximately 900 liters of blood passes through the kidneys each day. Of this huge quantity, 20 percent (180 liters) is actually forced out of the million or so glomeruli into the adjacent nephrons. Of course, 180 liters of glomerular filtrate is considerably more fluid than you excrete as urine each day. In fact, the average output of urine is 1 to 2 liters per day, or approximately 1 percent of the daily glomerular filtrate. The remaining 99 percent is *reabsorbed* as the fluid flows down the nephrons. Most of the solutes of the glomerular filtrate are also reabsorbed; if they weren't, the body's store of essential substances would soon be discarded in urine. For example, the loss of glucose, which occurs in untreated diabetes (page 518), deprives a person of his or her readily available fuel supplies. The result is similar to having a hole in the bottom of your car's gasoline tank.

Glomerular filtration is a nonselective process, whereas **tubular reabsorption** is highly selective. The kidneys' "strategy" is to push everything into the nephron at its proximal end and then selectively remove those substances, such as salts, sugars, and water, that the body can't afford to lose, leaving behind waste products to be discharged in the urine. Let's look more closely at how reabsorption can be so selective.

As the water-rich, nutrient-laden glomerular filtrate moves along the proximal convoluted tubule, active transport systems in the membranes of the epithelial cells export glucose, salts, and other valuable nutrients out of the nephron (Figure 28-25, step 2). These substances move through the interstitial fluid and back into the bloodstream. As the solutes move into the interstitial fluid, they create an osmotic gradient that draws water out of the proximal convoluted tubule. In other words, the cells of the wall of the nephron expend energy to recover salts and nutrients, and water follows along passively by osmosis. Approximately 80 percent of the water in the glomerular filtrate is reabsorbed from the proximal portion of the nephron.

The efficiency of water recovery from the glomerular filtrate is boosted by the loop of Henle, whose role it is to establish a salt gradient within the kidney, which is needed to reclaim water from the distal portion of the nephron. To do this, ionized salt (Na^+ and Cl^-) is actively pumped from the ascending side of the loop (Figure 28-25, step 3), making the surrounding fluid very salty. The highest salt concentration develops in the inner medulla of the kidney, while the lowest salt concentration develops in the outer cortex. Establishing a steep salt gradient is an energy-expensive process, requiring ATP hydrolysis. This energy is expended so that water can be reabsorbed from the fluid of the collecting duct at a later stage (Figure 28-25, step 6) as it passes out of the kidney, toward the ureter.

Tubular Secretion The composition of urine is modified in the distal portion of the nephron by **tubular secretion**—the process whereby substances are transported from the blood into the fluid of the distal convoluted tubule (Figure 28-25, step 4). A variety of substances enter the nephron by secretion, including waste molecules, such as urea and urobilin (the yellow pigment derived from hemoglobin breakdown which gives urine its yellow color), certain ions (K^+ and H^+), and a number of medications, such as penicillin and phenobarbitol, and other drugs, such as the active substances found in marijuana and cocaine. Tubular secretion provides one of the final regulatory steps in maintaining homeostasis of the tissue fluids. For example, the kidney regulates blood pH by secreting more hydrogen ions when the blood becomes slightly acidic; this process is reversed when blood becomes too alkaline. The hydrogen ion concentration of the urine can vary more than 1,000-fold, depending on conditions.

Regulation of Kidney Function

🗘 Like most physiological activities, urine formation is regulated by both neural and endocrine mechanisms. Two hormones—*aldosterone* and *antidiuretic hormone (ADH)*—that act on distal parts of the nephron are particularly important. If you were to ingest a particularly salty meal, the increased level of sodium in your blood would trigger a decrease in the secretion of aldosterone by your adrenal cortex (page 528), leading to a decreased reabsorption of sodium by the distal convoluted tubule (Figure 28-25, step 5) and increased excretion of salt in the urine. Conversely, a drop in the blood's sodium concentration stimulates an increased secretion of aldosterone, leading to increased salt retention. A person suffering from untreated diseases of the

◁ B I O L I N E ▷
The Artificial Kidney

Kidney failure leads to many deleterious effects, including the retention of waste products, an abnormal concentration of salts and water, an altered pH, and the general deterioration of homeostasis. Until a few decades ago, kidney failure meant certain death. With the development of the *artificial kidney,* however, a person without kidney function can live a long life. The treatment is not without possible side effects, however, including fatigue and impotence (in men). Ideally, the use of the artificial kidney is a stopgap measure that keeps a person alive until he or she can receive a healthy kidney through transplantation.

The modern artificial kidney was developed largely by the efforts of Willem Kolff, a Dutch physician who emigrated to the United States after World War II and eventually settled at the University of Utah. During the use of an artificial kidney, blood is pumped out of an artery and into a long stretch of cellophane tubing. This tubing is immersed in a chamber filled with a salt solution of similar composition to normal blood plasma. The permeability properties of cellophane tubing, which is the same material that is used as sausage casing, are similar to that of the capillary wall; that is, small-molecular-weight solutes can pass through the pores, but proteins and blood cells are retained within the tubing.

The "purification" of the blood is accomplished by *dialysis,* a process that requires several hours and must be repeated two to three times a week. During dialysis, diffusible molecules pass back and forth between the tubing and the surrounding fluid. The net movements of a particular type of molecule depend on the relative concentration of the substance on the two sides of the cellophane membrane. For example, urea is present in the blood but is absent in the fluid of the dialysis chamber. Consequently, when the blood flows through the dialysis tubing, urea will diffuse out of the blood and into the chamber, cleansing the blood of this waste product. In contrast, the blood is likely to contain a reduced concentration of bicarbonate ions because these basic ions will have picked up much of the excess acid produced by the body since the last dialysis treatment. During dialysis, bicarbonate ions, which are present in the chamber's fluid, will enter the blood as it flows through the cellophane tubes, restoring the blood's buffering capacity. Blood is kept from clotting during dialysis by treatment with the anticoagulant heparin.

It is interesting to note that both the dialysis machine and the kidney bring about a similar end result but achieve it by very different means. The dialysis machine depends strictly on the nonspecific, passive diffusion of dissolved molecules, whereas the natural kidney utilizes highly specific transport processes and regulated changes in membrane permeability. That is why the natural kidney performs a much better job in a much more compact "apparatus" without having to carry around a large chamber of fluid that has to be changed on a daily basis.

adrenal cortex may produce little or no aldosterone and will excrete large quantities of both salt and water.

The volume of the blood is determined by its water content, which is regulated at the collecting ducts (Figure 28-25, step 6). Blood volume decreases when too little water is reabsorbed from the nephrons back into the bloodstream; this causes the blood to become too concentrated. Receptors in the hypothalamus detect the increase in the blood's osmotic strength and direct the posterior pituitary to release ADH, as we described in Chapter 25. ADH increases the permeability of the collecting ducts to water, allowing more water to be reabsorbed from the urine; this dilutes the blood while concentrating the urine. Alcohol interferes with the secretion of ADH, which is why beer induces more trips to the restroom than does drinking a comparable volume of water. In addition to directing the secretion of ADH, the hypothalamus alerts the cerebral cortex when the water content of blood is low. What you perceive consciously as thirst is your hypothalamus instructing you to consume more water.

EVOLUTION AND ADAPTATION OF OSMOREGULATORY AND EXCRETORY STRUCTURES

Some animals, including sponges, sea anemones, and sea stars, lack specialized organs for osmoregulation and excretion. It is presumed that the ionic regulation and excretion of wastes in these animals are handled by the cell membranes that line the body surface (the *integument*). Most animals—whether earthworms, crabs, or vertebrates—

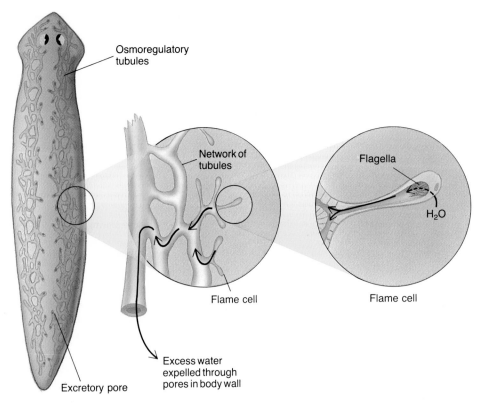

FIGURE 28-26

The osmoregulatory system of a flatworm. A network of osmoregulatory tubules run the length of the body on each side of the animal. Emerging from the tubules along their length are short canals that end blindly in a *flame cell.* These cells are named for the flickering motion exhibited by a tuft of flagella that projects from the cell into the lumen of the canal. The movement of the flagella is thought to suck fluid across the wall of the flame cell and into the lumen of the tubule. The contents of the fluid are altered by reabsorption and secretion, and the final product flows out of excretory pores located along the body.

possess some type of tubular organ whose mechanism of operation is similar to that found in humans. For example, flatworms contain an interconnected series of tubules that maintain the body's salt and water balance (Figure 28-26). As is the case in many animals, excretion of nitrogenous wastes in flatworms is handled by a separate mechanism.

▮▶ Osmoregulatory and excretory mechanisms are correlated with the type of habitat in which an animal lives. Invertebrates that live in marine habitats, such as a sea anemone or an octopus (Figure 28-27*a*), have very salty body fluids that are approximately equal in osmotic concentration to that of the surrounding sea. As a result, these animals neither lose nor gain water by osmosis. Therefore, they are not in need of an elaborate osmoregulatory system.

In contrast, marine vertebrates are generally much less salty than is their environment and thus tend to lose water by osmosis. Most marine fishes (Figure 28-27*b*) regain the

water they lose through osmosis by drinking the sea water they live in and excreting concentrated salt solutions from their gills. These animals produce virtually no urine, and many have very rudimentary kidneys. Marine sea birds and sea turtles (Figure 28-27*e,f*) must also obtain their water from the sea; these animals drink sea water and excrete concentrated salt solutions from salt glands located in their heads.

Animals that live in freshwater face just the opposite problem: Their environment is very high in water concentration and low in available salts. Consequently, freshwater animals tend to *gain* water by osmosis—water that must be expelled back into the environment. Freshwater fish and frogs (Figure 28-27*c,d*) have well-developed kidneys and produce a very dilute urine. The most serious problem faced by these animals is the loss of valuable salts, which are inevitably washed away in the large volume of urine. Freshwater animals possess highly effective active-trans-

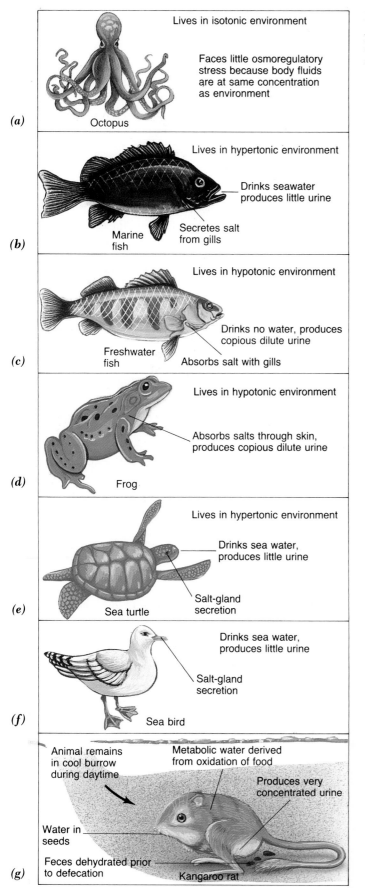

FIGURE 28-27
A diversity of osmoregulatory challenges. After: ANIMAL PHYSIOLOGY by R. Eckert. Copyright © 1988 by W. H. Freeman and Co. Reprinted by permission.

port mechanisms that pump salts back into their bodies, despite the low salt concentration of their environment: Freshwater fish have salt-absorbing gills, while frogs have salt-absorbing skin.

The most serious osmoregulatory problems are faced by terrestrial animals, particularly by those that live in dry, desert habitats. These animals possess physiological and behavioral adaptations that keep water loss to an absolute minimum. One of the best-studied water savers is the desert kangaroo rat (Figure 28-27g). This animal can live its entire life without ever drinking a drop of water: All the water the kangaroo rat needs is either present in the animal's food or formed as a byproduct of metabolic reactions (so-called *metabolic water*). Water conservation is accomplished by the animal's ability to produce an extremely concentrated urine (several times saltier than that which can be excreted by humans) and by its staying out of the desert heat. During the day, these animals remain in their relatively cool burrows; they emerge at night to carry out their feeding and social activities.

Nitrogenous Waste Products: Adaptations to the Environment

Aquatic and terrestrial animals also differ in the type of nitrogenous waste products they excrete. In a metabolic sense, the least expensive nitrogenous waste product is ammonia, which is formed directly when amino ($-NH_2$) groups are removed from amino acids. But ammonia is highly toxic and cannot be allowed to accumulate in body fluids. Aquatic animals are able to excrete ammonia since they can either allow it simply to diffuse into the environment through their body surface or to void it in a dilute urine. In contrast, terrestrial animals must conserve water and excrete urine with concentrated solutes. Ammonia is too toxic for this type of excretory practice. Instead, terrestrial animals convert the ammonia (at the expense of some ATP) to relatively nontoxic products that can be excreted in a concentrated state, such as urea (excreted by adult frogs and most mammals) or uric acid (the white, pasty material found in bird and reptile excrement).

Excretory/osmoregulatory systems maintain the constancy of the internal fluids by ridding the body of nitrogenous waste products, excess water and salts, and other undesirable substances that are produced by the animal or enter from the environment. The formation of an excretory fluid typically occurs within a tubule whose walls are specialized for the reabsorption of substances to be retained and the secretion of substances to be eliminated. (See CTQ #4.)

THERMOREGULATION: MAINTAINING A CONSTANT BODY TEMPERATURE

Humans normally maintain a constant body temperature of 37°C (98.6°F), regardless of whether they are walking on a desert path under a blazing sun or swimming in an ice-cold mountain stream. The homeostatic process by which mammals (and birds) maintain such a high body temperature is called **thermoregulation.** Thermoregulation requires that an animal follow one simple physiological rule: The amount of heat gained by the body must equal the amount of heat lost from the body.

GAINING HEAT TO KEEP THE BODY WARM

The heat present within an animal's body can derive from one of two sources: its own metabolism or the external environment. Animals that use metabolic heat to raise their body temperature are called **endotherms.** In contrast, animals whose body temperature derives from heat absorbed from the rays of the sun, the ground in their burrows, or the water in which they live, are called **ectotherms.** Birds and mammals are the only animals that are classified as true endotherms; all others are ectotherms. The reason for the rarity of endothermy is that it is very expensive metabolically. When you consider the dollar cost of maintaining the temperature of a house at about 70°F in the winter, you can appreciate the metabolic cost for an animal that lives outside during the same winter conditions and maintains its temperature at 98° to 100°F. This is the primary reason that birds and mammals (other than humans) are covered with feathers or fur; these materials provide a layer of insulation that retains body heat in the face of cold environmental temperatures.

Humans are not very well insulated. A slight drop in body temperature is immediately recognized by receptors in the hypothalamus of the brain which send out a message via sympathetic nerves to constrict the arterioles leading to the skin. This constriction causes the skin to become pale and cold, a response that conserves heat by reducing the flow of blood to the body's surface. If our body temperature continues to drop despite such vasoconstriction, more drastic measures are called for, and the brain sends out a signal to the body's muscles to contract involuntarily. This response, called **shivering,** represents the body's attempt to increase the output of metabolic heat in order to raise the body's temperature.

LOSING HEAT TO KEEP THE BODY COOL

Endotherms must also possess mechanisms to rid themselves of body heat as the temperature of the external environment rises or when the individual's level of physical activity increases. The first step in dispersing body heat is dilation of blood vessels in the outer portions of the body, which allows more blood to reach the body surface, where its heat radiates to the environment. If the body temperature continues to rise despite this conservative measure, additional heat must be lost by either sweating or panting. *Sweating*—the loss of body heat as water evaporates from the surface of the skin—is the primary mechanism of cooling in a variety of mammals, including antelopes, camels, and humans. Humans placed under conditions of extreme heat stress can produce approximately 4 liters of sweat in a single hour. It is only after the body becomes severely dehydrated that the flow of sweat decreases and the body temperature rises. Panting—which is the loss of body heat as water evaporates from the moist surfaces of the respiratory tract—is the primary mechanism of cooling in birds, dogs, and bears.

An animal's enzymes function optimally at a particular temperature, which varies from species to species. Endotherms tend to maintain their body at this optimal temperature by balancing the amount of heat gained—largely as a product of metabolism—with the amount of heat lost—largely by shunting warm blood to the body surface or by evaporative cooling. (See CTQ #5.)

REEXAMINING THE THEMES

Relationship between Form and Function

The microscopic structure of a blood vessel is closely correlated with its function. The larger arteries are wide in diameter, promoting rapid flow, and have concentric rings of elastic tissue, which allow the wall to maintain pressure on the blood when the heart is not contracting. Arterioles are endowed with a layer of smooth muscle cells, whose state of contraction determines which tissues will receive the greatest supply of blood. The capillaries are devoid of both elastic and muscle tissue. The extreme thinness and porosity of the capillary walls promote the exchange of substances between the blood and tissues. The veins have large diameters and thin, distensible walls, whereby large volumes of blood can return to the heart with very little external pressure, assisted by the action of one-way valves.

Biological Order, Regulation, and Homeostasis

The circulatory and excretory systems are both primary agents in maintaining homeostasis. The bloodstream serves as the conduit for carrying excess materials and body heat to sites specializing in their disposal. For example, excess heat is carried by the blood to the body surface; carbon dioxide is carried to the lungs; and nitrogenous wastes and excess salts are carried to the kidneys. The concentrations of various substances in the blood are maintained at appropriate levels by the processes of selective reabsorption and secretion that take place in the kidney. Kidneys balance blood pH, maintain the volume and osmotic strength of the blood, and eliminate nitrogenous wastes. Maintaining a constant body temperature, which keeps all of the body's enzyme systems working at maximum efficiency, is another key aspect of homeostasis.

Acquiring and Using Energy

Energy is required for virtually all of the activities described in this chapter. For example, pumping blood through the body requires large amounts of chemical energy. If supplies of chemical energy to the heart are reduced, as occurs when a person's coronary arteries become occluded, the heart cannot increase its pumping activities to meet the body's increased demands for oxygen, curtailing the person's ability to engage in strenuous activities. The reabsorptive and secretory activities of the kidney, which are required to produce urine, also depend on energy, which is used to drive the various ionic pumps located in the plasma membranes of the cells that line the nephrons. Thermoregulation can be the most energy-demanding activity of all. Up to 90 percent of a bird's or mammal's chemical energy may be expended in generating the heat needed to maintain an elevated body temperature.

Evolution and Adaptation

Organs that carry out osmoregulation are highly adapted to an animal's environment and may perform quite differently among closely related animals. For example, most marine fishes lose water to their environment, but they have no difficulty obtaining salts. These animals typically drink the surrounding sea water and excrete the excess salts through their gills; they have poorly developed kidneys. In contrast, freshwater fish tend to gain water and live in an environment where salts are scarce. These animals have well-developed kidneys and excrete a copious, dilute urine. They obtain the salts they need by active transport from their environment.

SYNOPSIS

Circulatory systems deliver a well-oxygenated nutrient solution to the body's tissues and carry away the wastes. The system also contributes to homeostasis by stabilizing pH and osmotic balance, delivering hormones, protecting against foreign intruders, and (in birds and mammals) resisting changes in body temperature.

The anatomy of blood vessels is well suited to their function. The design of the large, elastic arteries allows them to snap back after each surge from the heart, squeezing blood onward. To meet the needs of the tissues they service, muscular arterioles either increase or decrease their diameter. Capillaries are thin and porous and serve as sites of exchange with the surrounding interstitial fluid. Venules collect the blood leaving the capillaries and route it to veins. One-way valves assist the blood's return to the heart. Excess fluid returns to the bloodstream via the lymphatic system. Lymph is filtered through lymphatic organs, notably the spleen and lymph nodes.

The mammalian heart is equipped with one-way valves to keep blood flowing in a single direction. Acting as a double pump, the right side of the heart pumps blood to the lungs for oxygenation; the left side powers blood through the general (systemic) circulation. Although the heart will contract rhythmically without an external stimulus, neural impulses regulate the rate of contraction by stimulating the sinoatrial node, which initiates contraction of the atria and relays the stimulus to the rest of the heart.

Nearly half the volume of the blood consists of cells and cell fragments (platelets). Erythrocytes carry oxygen; granular leukocytes and monocytes phagocytize foreign intruders and tissue debris; platelets aid in clotting; and lymphocytes launch the immune response. The fluid phase (plasma) contains a number of proteins that are activated in a specific sequence, leading to the formation of a clot that consists of a tangled net of insoluble fibers of the protein fibrin.

Excretion and osmoregulation. The mammalian kidney has two roles: It maintains the body's salt and water balance, and it rids the blood of waste products generated by cellular metabolism. Wastes, excess solutes, and water are voided in the urine—the fluid produced in the nephrons of the kidney as a result of three processes: glomerular filtration, selective reabsorption, and secretion. Blood (minus its cells and proteins) is filtered out of the capillaries of the glomerulus into the proximal end of the nephron. From there, the fluid flows down the length of the tubule, while salts, sugars, and other nutrients are transported out of the nephron back into the bloodstream. Nitrogenous wastes, hydrogen ions, and excess potassium ions are secreted in the opposite direction. Water flows out of the collecting duct by osmosis in response to an osmotic gradient that is established in the interstitial fluid of the kidney by the salt-pumping cells of the loops of Henle. The hormones aldosterone and ADH act on the distal portion of the nephrons to regulate the amount of salt and water voided in the urine.

Birds and mammals maintain a constant, elevated body temperature by balancing the amount of heat gained and heat lost. Heat is generated primarily as a result of metabolism and is lost at the body surface. If the body temperature starts to drop, heat is retained by vasoconstriction of peripheral arterioles and by shivering. Heat is lost by vasodilation of peripheral arterioles and by evaporative cooling.

Key Terms

circulatory system (p. 588)
intracellular compartment (p. 588)
interstitial compartment (p. 588)
vascular compartment (p. 588)
cardiovascular system (p. 588)
artery (p. 589)
blood pressure (p. 589)
systolic pressure (p. 589)
diastolic pressure (p. 589)
hypertension (p. 590)
arteriole (p. 590)
vasodilation (p. 590)
vasoconstriction (p. 590)
capillary (p. 590)
venule (p. 593)
vein (p. 593)
heart (p. 593)
atrium (p. 594)
ventricle (p. 594)
pulmonary circuit (p. 594)
systemic circuit (p. 594)
aorta (p. 595)

coronary artery (p. 595)
heart valve (p. 597)
sinoatrial (SA) node (p. 598)
atrioventricular (AV) node (p. 599)
electrocardiogram (EKG) (p. 599)
plasma (p. 600)
erythrocyte (p. 600)
stem cell (p. 600)
leukocyte (p. 601)
platelet (p. 602)
fibrinogen (p. 602)
fibrin (p. 602)
thrombin (p. 602)
Factor X (p. 602)
Factor VIII (p. 602)
open circulatory system (p. 605)
closed circulatory system (p. 605)
lymphatic system (p. 606)
lymph node (p. 606)
spleen (p. 606)
osmoregulation (p. 607)
excretion (p. 607)

kidney (p. 607)
urine (p. 607)
ureter (p. 607)
urinary bladder (p. 607)
urethra (p. 607)
nephron (p. 607)
glomerulus (p. 608)
Bowman's capsule (p. 608)
proximal convoluted tubule (p. 608)
loop of Henle (p. 608)
distal convoluted tubule (p. 608)
collecting duct (p. 608)
glomerular filtration (p. 609)
glomerular filtrate (p. 610)
tubular reabsorption (p. 610)
tubular secretion (p. 610)
thermoregulation (p. 614)
endotherm (p. 614)
ectotherm (p. 614)
shivering (p. 614)

Review Questions

1. Name and discuss five essential functions of the circulatory system.

2. Name three locations where one-way valves are essential to fluid flow.

3. Explain why a two-chambered heart, such as that found in fishes, would be inadequate for humans and other mammals.

4. Discuss how, in closed circulatory systems, tissues get what they need from the blood without ever coming in direct contact with it. Include capillary anatomy and function in your answer.

5. Describe the various ways your kidney helps maintain homeostasis.

6. Draw and label a typical nephron, and relate each part to the three urine-forming processes.

7. Metabolism generates heat. How does blood flow prevent heat from building to lethal temperatures? What happens to heat after it is carried away by the blood?

Critical Thinking Questions

1. Suppose Galen had been right and blood really flowed back and forth in the veins. How would this have affected Harvey's experiment in which he used his finger to stop blood flow? How would the absence of valves in the veins of the arm have affected Harvey's experiment?

2. Redraw Figure 28–2 to represent circulation in an organism such as (a) the jellyfish, *without* a circulatory system, and (b) the crayfish, with an *open circulatory system.* Which body plan would you expect to find in an active, fast-moving animal: no circulatory system, an open system, or a closed system? Why?

3. The circulatory system has vessels that carry blood to the body tissues as well as from the tissues back to the heart. The lymph system, however, has only vessels leading back towards the heart. How can this be? Where does the lymph come from? Why must there be a means for emptying the lymph fluid into the blood system? Without this, what would happen?

4. Excretion in humans begins with nonselective filtration in the glomeruli. If you were designing an excretory system, would you make this part of the system selective or nonselective? What are the advantages of each approach? Do you think that the existing system is the most efficient design? Why, or why not?

5. Until recently, dinosaurs were considered to be ectotherms, like their relatives, the modern-day reptiles. But examination of the structure of fossil dinosaur bones reveals that they have characteristics usually associated with endothermy. What advantages would endothermy have given these reptiles? What disadvantages?

6. After a lengthy visit to a city at a high altitude we adjust to the lowered oxygen content of the air. Why is this capability referred to as *acclimation* rather than *adaptation?*

Additional Reading

Andrade, J. P., ed. 1986. *Artificial organs.* New York: VCH Publishers. (Advanced)

Golde, D. W., and J. C. Gasson. 1988. Hormones that stimulate the growth of blood cells. *Sci. Amer.* July:62–70. (Intermediate)

Golde, D. W. 1991. The stem cell. *Sci. Amer.* Dec:36–43. (Intermediate)

Grantham, J. J. 1992. Polycystic kidney disease I. Etiology and pathology. *Hospital Practice* March:51–59. (Advanced)

Lehrer, S. 1979. *Explorers of the body.* New York: Doubleday. (Introductory)

Robinson, T. F., S. M. Factor, and E. H. Sonneblick. 1986. The heart as a suction pump. *Sci. Amer.* June:84–91. (Intermediate)

Schmidt-Nielsen, K. 1964. *Desert animals: Physiological problems of heat and water.* New York: Oxford University Press. (Advanced)

Valtin, H. 1983. *Renal function: Mechanisms preserving fluid and solute balance in health, 2d ed.,* Boston: Little, Brown. (Advanced)

Weinberger, M. H. 1992. Hypertension in the elderly. *Hospital Practice* May:103–120. (Advanced)

Weisse, A. B. 1992. The alchemy of Willem Kolff, the first successful artificial kidney, and the artificial heart. *Hospital Practice* Feb:108–128. (Introductory)

CHAPTER

◂ 29 ▸

Gas Exchange:
The Respiratory System

STEPS TO DISCOVERY
Physiological Adaptations in Diving Mammals

THE HUMAN PERSPECTIVE
Dying for a Cigarette

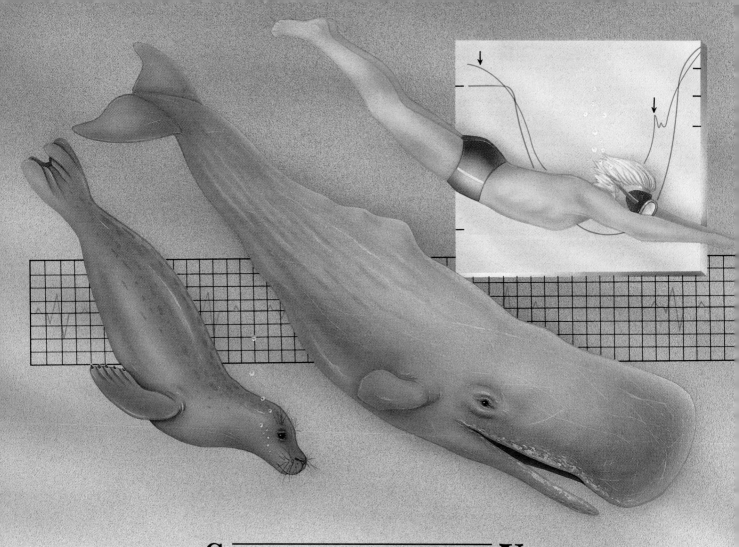

Physiological Adaptations in Diving Mammals

*I*n the South Pacific Islands, native pearl divers plunge into the ocean and search for oysters on the sea floor, often staying beneath the water for several minutes without the help of breathing devices. If you have ever experienced sensations of panic after being underwater for only a minute or so, you can appreciate the magnitude of this feat. But even the most accomplished pearl diver can scarcely approach the talents of a number of other mammalian divers. Consider the Weddell seal of the Antarctic, for example. This animal can remain below water for 70 minutes, searching for food or escaping predators. A sperm whale can dive to depths of 1,500 meters (nearly 1 mile) and remain submerged for as long as 2 hours.

How do these air-breathing mammals accomplish these feats? Perhaps these animals have huge lungs that fill with vast amounts of oxygen before diving. But that would be like our trying to swim to the bottom of a pool holding an inflated air mattress. In fact, just the reverse is true: Weddell seals and sperm whales have modest-sized lungs, and they *exhale* before a dive, thereby reducing their buoyancy. Exhaling also rids the animal of nitrogen-containing air, which can form nitrogen gas bubbles as the animal resurfaces at the end of the dive. Such bubbles can block the flow of blood through vessels, producing a potentially fatal condition known as the *bends*.

Some of the first insights into the physiological adapta-

Once submerged, scientists observed that whales, seals, and humans exhibit a diving reflex. This reflex is an adaptive mechanism

tions of diving mammals emerged in the 1930s from the studies of Per Scholander of the University of Oslo in Norway. Scholander and his colleagues brought seals into the laboratory, where the animals could be connected to instruments that monitored various physiological responses. The scientists had the seals simulate a dive by training the animals to hold their heads under water in a shallow pan. While this simulation is quite different from a true deep-sea dive, the seal's body had to adjust to a lack of fresh oxygen, just as it would have had to do in the sea environment. Scholander soon discovered that this exercise was accompanied by major physiological changes that permitted the diving seal to remain active in the absence of a fresh supply of oxygen.

Almost as soon as the seal's nose was submerged in the water, the rate of the animal's heartbeat fell dramatically, often to as low as one-tenth its normal rate. Because the reaction occurred so quickly, Scholander concluded that it was a reflex response triggered by submersion itself rather than by any metabolic changes that occurred as a result of a lack of oxygen. This cardiac response turned out to be only the initial phase of the so-called *diving reflex*. When blood flow was monitored, the investigators found that only the seal's brain and heart received their normal supply of blood —the same two organs that are most sensitive to oxygen deprivation in humans. The other parts of the seal's body —even the muscles that are required for swimming—were almost totally cut off from fresh oxygen-carrying blood.

How can muscles continue to work if they stop receiving oxygen? Recall from Chapter 9 that skeletal muscle tissue can continue to produce small amounts of ATP in the absence of oxygen as long as the pyruvic acid that is produced by glycolysis is converted to lactic acid. This is one of the paths of fermentation (page 178). Scholander found that the muscles of the submerged seal accumulated large quantities of lactic acid, a clear indication of energy production by fermentation. Normally, when a mammal exerts itself, as when you run long distances, the lactic acid appears in the bloodstream. In this case, however, because the blood flow to the seal's muscles was essentially shut off, the lactic acid remained in the muscle tissue. Once the seal resumed breathing air, blood flow to the muscles returned,

and lactic acid flooded into the general circulation and was oxidized.

After studying the diving reflex in seals, Scholander wondered whether humans might have a similar reflex response. To help answer this question, he recruited the native pearl divers of Australia and found that they showed the same physiological adaptations as did the seals, but to a lesser extent. Within about 20 seconds into their dive, the pearl divers' heart rate dropped, even though the divers were engaged in strenuous activity. As with the seals, lactic acid remained in the divers' muscles until the recovery period, at which time the lactic acid concentration in the blood rose dramatically. These results suggested that the distribution of blood in humans, as in seals, is altered following submersion.

In the 1960s and 1970s, Claude Lenfant of the University of Washington discovered that the blood of Weddell seals had a much greater capacity for oxygen transport and storage than did that of other mammals. In fact, 60 percent of the blood of diving seals is composed of oxygen-carrying red blood cells, as compared to 40 percent in the blood of humans, providing the seals with yet another physiological adaptation for diving.

Advances in electronic technology in recent years has allowed the study of diving mammals to move out of the laboratory and into the field. By attaching miniature computerized devices to seals living in their natural environment, the animals' physiological changes during diving can be monitored. The development of this type of instrumentation was pioneered by Roger Hill of the Massachusetts General Hospital. Although some of the earlier laboratory findings have needed revision, the basic physiological profile obtained from seals who held their noses in a pan of water also applies to those swimming dozens of meters below the water's surface.

that enables the animal to remain alive underwater for extended periods of time.

You can survive for weeks without food and for days without water but without oxygen you will last only a few minutes. Virtually all organisms depend on oxygen to keep the metabolic "fires" burning. Oxygen unleashes the chemical energy that is stored in food molecules, generating more than 90 percent of the ATP that drives energy-consuming activities. In Chapter 9, we referred to this process as *aerobic respiration.*

In this chapter, we will deal with two other aspects of animal respiration: the uptake of oxygen from the environment by a specialized set of organs that make up the **respiratory system,** and the transport of oxygen to the individual cells of the body. We also saw in Chapter 9 how the oxidation of biochemicals generates a waste product, carbon dioxide. The elimination of carbon dioxide from the body accompanies the acquisition of oxygen; therefore, the respiratory system is often called the **gas exchange system.**

▼ ▼ ▼

PROPERTIES OF GAS EXCHANGE SURFACES

◉ The exchange of respiratory gases—oxygen and carbon dioxide—occurs across a **respiratory surface**—a portion of the body surface that has a specialized structure suited for this particular physiological activity. Despite the fact that respiratory organs are found in all shapes and sizes, they all have respiratory surfaces with similar properties:

1. The respiratory surface must remain moist to avoid suffocation, since oxygen must dissolve in fluid before it can move across a cellular membrane. The moisture you see condensing as you exhale on a cold day is water that is lost from the surfaces that line your lungs (Figure 29-1).
2. The epithelial cells that make up the respiratory surface must be extremely thin and permeable; otherwise, oxygen could not cross the barrier from the environment into the body, and carbon dioxide would be unable to move in the opposite direction. The cells that constitute a respiratory surface are usually so flat that they are barely recognizable as cells under the microscope (Figure 29-2).
3. Getting oxygen across the respiratory surface is the first step in respiration, but some mechanism must be present for getting the absorbed oxygen to all the remote cells of the body where it is needed. If an animal is very small or very thin, such as a flatworm, the ab-

sorbed oxygen can simply diffuse through the body to all of the cells. In most larger animals, however, such as birds and mammals, gas exchange surfaces operate in conjunction with the circulatory system. The respiratory tissues of these animals abound with tiny blood vessels (capillaries) that receive oxygen as it diffuses across the respiratory surface from the environment (Figure 29-2). Oxygen is then transported from the site of acquisition, via the circulatory system, to the remainder of the body.

4. All movements of dissolved gases across biological membranes occur by simple diffusion; that is, there are no known transport mechanisms to aid in the passage. Consequently, the rate of oxygen uptake depends on the surface area across which the gas can diffuse. In general, the greater the surface, the more rapid the uptake.

SURFACE AREA AND RESPIRATION

Many animals, including jellyfish, earthworms, and frogs, respire *cutaneously;* that is, across virtually their entire body surface. Since the respiratory surface must be moist and thin, it follows that animals that respire cutaneously tend to be highly vulnerable to environmental conditions. This is evidenced by the familiar sight of a dehydrated earthworm that managed to make it half way across a sidewalk before succumbing to the loss of water across its thin, permeable outer surface. Because of the vulnerability associated with **cutaneous respiration,** evolution has favored animals

FIGURE 29-1

Breathing and water loss. Normally, you are not aware of the water that you lose with every breath, but the loss becomes evident when the exhaled water vapor condenses in cold air.

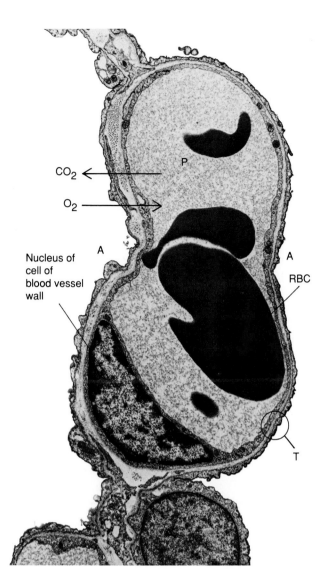

CO₂ ←

O₂ →

Nucleus of
cell of
blood vessel
wall

A   A

RBC

T

FIGURE 29-2

The microanatomy of gas exchange. Electron micrograph of a section through a portion of a human lung, showing the respiratory surface on both sides of a blood capillary. Note the extreme thinness of the barrier (T) between the air space in the lung (A, for alveolus) and the plasma of the blood (P). This barrier consists of a thin, flattened layer of epithelial cells of the lung and an equally thin, flattened layer of cells that line the capillary. Combined, these two cellular layers are approximately 0.2 micrometers thick. In comparison, the red blood cells (RBC) shown in the photograph are about 7 micrometers in diameter. The net movement of oxygen and carbon dioxide is indicated by the arrows.

whose delicate respiratory surfaces are restricted to particular sites on or within the body. This arrangement allows the rest of the body's surface to remain impermeable, as exemplified by the skin that covers your body or the exoskeleton of an insect or spider.

Since the amount of oxygen an animal can absorb from the environment depends on the area of body surface available for gas exchange, restricting the respiratory surface to a small part of the body creates a potential problem. The evolutionary solution has been an increase in the surface area available for absorption without a concurrent increase in the space required to house the enlarged surface. Your lungs, for example, fit conveniently within your chest cavity, yet they have a respiratory surface of 60 to 70 square meters, an expanse equivalent to that of a badminton court. We will see how this is accomplished shortly.

In order to absorb oxygen, an animal must expose a part of its delicate, living surface to the external environment. To facilitate gas exchange, respiratory surfaces are thin, moist, and permeable, and often have large surface areas (See CTQ # 2.)

THE HUMAN RESPIRATORY SYSTEM: FORM AND FUNCTION

The human respiratory system, which is typical of mammals, is depicted in Figure 29-3. The system consists of the following components:

1. a branched passageway, or **respiratory tract,** through which air is conducted, and
2. a pair of **lungs** through which oxygen enters the body and is absorbed into the bloodstream. Even though your lungs reside deep within your chest, like your digestive tract, they are actually a part of your body surface; that is, a surface that is exposed to the external environment.

THE PATH TO THE LUNGS

The entire passageway to the lungs is lined by a mucus-secreting epithelium. The blanket of mucus secreted by these cells keeps the surfaces of the airways moist so that even the

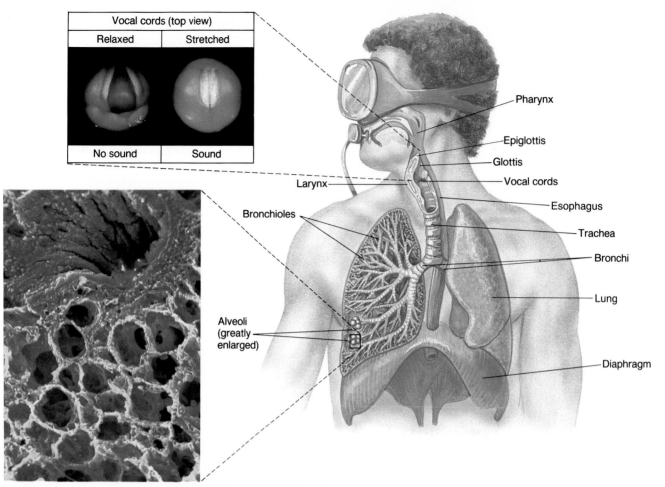

FIGURE 29-3

The human respiratory system. The human respiratory system consists of a system of airways that carry air into a pair of lungs. The airways are lined by a ciliated, mucus-secreting epithelium that moistens the air and traps microscopic debris. The top inset shows the vocal cords, which are located in the larynx. Sounds are produced when exhaled air passes through the vocal cords as they are stretched by muscles. The bottom inset shows a cluster of alveoli that fill the interior of the human lung with air pockets. Each lung is a sac that is filled with smaller sacs, thereby increasing the surface area by a factor of hundreds, with no increase in the space required to house it.

driest air is humidified by the time it reaches the gas exchange surface in the lungs. The sticky mucous layer also traps microbes and other dangerous airborne particles in the upper regions of the respiratory tract before they have a chance to enter the lungs and cause serious injury or infection, such as pneumonia. The mucus with its trapped particles is moved by hairlike cilia away from the lungs at the rate of about 2.5 centimeters per minute. Trapped microbes are moved into the throat, where they are swallowed and then killed by stomach acid.

Through the Nasal Cavity and the Airways

Air enters the respiratory system through either the mouth or *nostrils*—the openings into the **nasal cavity.** Air entering through the nostrils passes through a forest of nasal hairs, which traps inhaled particles, preventing them from entering the lungs. Air is warmed and moistened as it passes

through the nasal cavity to the **pharynx** (throat), a corridor that is shared by both the respiratory tract and the digestive tract. Food and liquids are routed to the esophagus and are kept out of the airways by a combination of anatomy and reflexes (Figure 29-3). The passage to the lungs carries the inhaled air through a small opening, called the **glottis,** and into the **larynx,** a short passageway that leads into the lower portion of the respiratory tract. During swallowing, the entire larynx moves up, forcing a flap of tissue, the **epiglottis,** to seal off the entrance to the lower airways. This reaction prevents food and liquids from mistakenly entering the airways on its route to the esophagus and stomach (page 569).

The human larynx is more than just a passageway for air. The sides of the larynx contain a pair of muscular folds —the **vocal cords**—from which the human voice emanates. The vocal cords are operated by the passage of air through the larynx. When we inhale, air silently rushes through the opening between the relaxed vocal cords (top inset, Figure 29-3). As air escapes from the lungs when we exhale, the stretched vocal cords vibrate, creating the sound of the voice. The loudness of the voice is determined primarily by the force of the exhaled air, while the pitch (highness or lowness of the note) is determined by the level of tension on the vocal cords, which is regulated by the contraction of muscles in the larynx. The sound is amplified by the hollow chamber that is formed by the laryngeal cartilage (the "Adam's apple" or "voicebox"). The larynx is also one of the targets of male sex hormones, which is why men have larger larynxes than do women, as evidenced by their deeper voices and more prominent Adam's apple.

Air passes through the larynx and into the tubular **trachea** (or windpipe), which descends into the chest. The trachea divides into a series of smaller and smaller tubes, or *bronchi*, which extend into the various regions of each lung. Both the trachea and the bronchi contain **C**-shaped bands of cartilage that provide structural reinforcement for these passageways, preventing the tubes from collapsing and im-

pairing the flow of air. The bronchi branch further to form a series of even smaller tubules, the **bronchioles.** Bronchioles lack cartilage, but they contain rings of smooth muscle fibers, whose state of contraction regulates the flow of air into the lungs. During an asthma attack or severe allergic reaction, the bronchioles can become clogged with mucus and may constrict due to prolonged muscular contraction.

Into the Lungs

As you have seen by now, the ultimate destination of inhaled air is the lungs. The lungs contain millions of microscopic pouches, called **alveoli** (Figure 29-3, bottom inset), which resemble the air pockets in a kitchen sponge. Due to their structure, alveoli are ideally suited for gas exchange. Each alveolus is a hollow, thin-walled sac that is richly surrounded by a network of capillaries (Figure 29-4). Alveoli are the sites of gas exchange in the lung. Gases readily pass through the thin walls of both the alveolus and the capillaries (see Figure 29-2) so that blood moving through the lung tissue can quickly pick up a fresh supply of oxygen

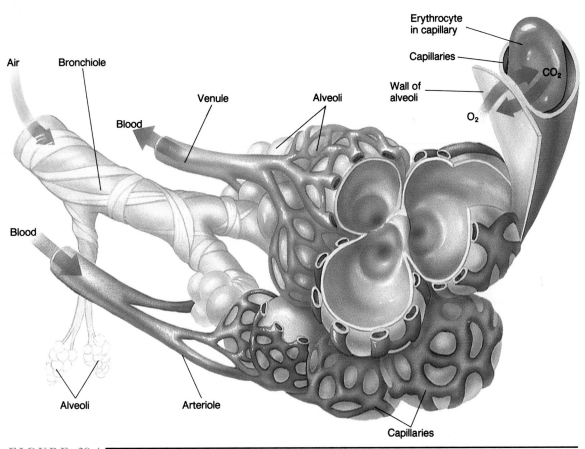

FIGURE 29-4

Alveoli, the lung's trade centers. Each alveolus is a thin-walled, bubblelike chamber (the size of a pinpoint) that is surrounded by capillaries. Oxygen and carbon dioxide gases quickly move in the directions indicated by the arrows. An electron micrograph of the alveolar–capillary interface is shown in Figure 29-2.

and unload its cargo of carbon dioxide. Freshly oxygenated blood (depicted in red in Figure 29-4) travels from the alveolar capillaries into larger vessels, until it enters the pulmonary vein, which carries the blood to the heart; from there, the oxygenated blood is pumped to the remainder of the body. The stale air in the lungs, which has been depleted of a portion of its oxygen, is forced out of the alveoli and back into the bronchioles, where it is expelled from the airways during exhalation.

If your lungs were mere hollow bags, the gas exchange surface would be less than half a square meter (about 5 square feet)—too small a surface to absorb enough oxygen to keep you alive under even the most restful situations. Instead, the spongy interior of your lungs houses more than 300 million alveoli, providing 60 to 70 square meters of an oxygen-collecting, carbon-dioxide-discharging surface. Because of their structure, the lungs are a very delicate tissue that can easily be damaged by inhaled pollutants (see The Human Perspective: Dying for a Cigarette?)

BREATHING: EXPANDING THE CHEST CAVITY

The lungs occupy a space—the **thoracic cavity**—situated within the chest. Cradled in the rib cage, the lungs are sealed in a waterproof, airtight, double-membraned sac, called the **pleura.** Occasionally the pleura becomes inflamed, a condition called *pleurisy,* causing fluid accumulation and painful breathing. A sheet of muscle tissue, the **diaphragm** (Figure 29-5), separates the thoracic cavity from the abdominal cavity. This enclosed configuration not only protects the lungs but also provides a mechanism for breathing.

In general, gases move down a pressure gradient, from a region of high pressure to one of low pressure. This simple principle explains the direction taken by winds, as well as the movement of air into and out of the lungs. The operation of the lungs occurs in a manner analogous to that of a bellows (Figure 29-5). While the pressure of air in the

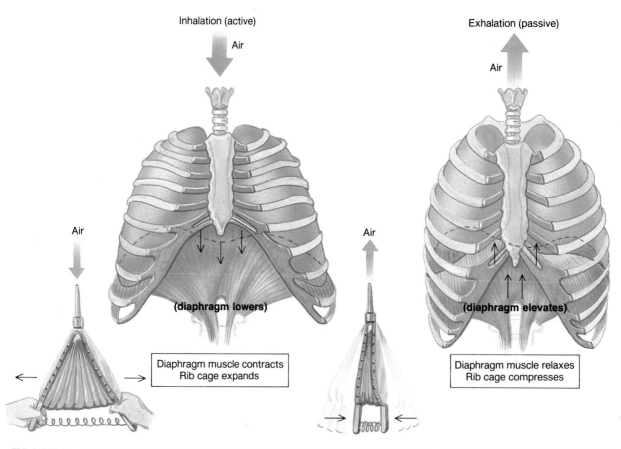

Inhalation (active)

Air

Air

(diaphragm lowers)

Diaphragm muscle contracts
Rib cage expands

Exhalation (passive)

Air

Air

(diaphragm elevates)

Diaphragm muscle relaxes
Rib cage compresses

FIGURE 29-5

Operation of the lungs. Contracting the diaphragm muscle (and intercostal muscles) expands the volume of the thoracic cavity, sucking air into the lungs. The inhaled air is expelled when these muscles relax. The operation of the lungs can be compared to that of a bellows, whereby air is forced in and out of the bellow's chamber.

atmosphere remains constant, the pressure within the lungs (or bellows) changes according to the lungs' (or bellow's) volume. When we inhale, the chest cavity expands in volume; the pressure inside each lung drops below that of the atmosphere; and air rushes into the lungs' chambers. The increase in lung volume during inhalation is not a result of the direct expansion of the lungs; rather, it occurs indirectly following the contraction of the muscular diaphragm and the *external intercostal muscles*, which lie between the ribs. The walls of the thoracic cavity contain elastic fibers that stretch during chest expansion. When the diaphragm and intercostal muscles relax, the lungs spring back to their smaller volume, causing the pressure inside each lung to rise above that of the atmosphere, forcing the stale air out of the lungs.

In humans, air is sucked down a pressure gradient through a series of increasingly narrow passageways and into the lungs—organs that contain millions of microscopic, thin-walled chambers, where respiratory gases can be exchanged with the bloodstream. The passageways to the lungs act to warm, moisten, and cleanse the inhaled air of debris. (See CTQ # 3.)

THE EXCHANGE OF RESPIRATORY GASES AND THE ROLE OF HEMOGLOBIN

Two types of gas exchanges are constantly occurring in your body: one between the alveoli and the capillaries in the lung, and the other between the capillaries and the various tissues of the body. These events are summarized in Figure 29-6. The principle underlying both types of exchange is a similar one: the passive diffusion of a dissolved respiratory gas from a region of higher concentration, across a permeable surface, to a region of lower concentration; that is, down a concentration gradient.

GAS EXCHANGE IN THE LUNGS: UPTAKE OF OXYGEN INTO THE BLOOD

Inhaled air is rich in oxygen (about 21 percent) and poor in carbon dioxide (about 0.4 percent). Blood returning from the tissues to the lungs carries the opposite complement of gases: a high concentration of carbon dioxide and a depleted oxygen cargo. As a result of these differences in concentration, oxygen in the lungs diffuses across the thin, cellular walls of both the alveolus and the capillary and into the bloodstream, while carbon dioxide diffuses in the opposite direction (Figure 29-6). This exchange is completed in about a quarter of a second, about the time it takes blood to flow through the site of exchange in a capillary.

If oxygen simply dissolved in the fluid of blood, the blood's oxygen-carrying capacity would be severely limited because oxygen is not very soluble in water. The capacity of the bloodstream to transport oxygen is greatly increased by the presence of the reddish, iron-containing protein **hemoglobin.** Each hemoglobin molecule contains four polypeptide subunits and binds four oxygen molecules, forming *oxyhemoglobin.* Hemoglobin allows human blood to carry 70 times more oxygen than it could otherwise. Put another way, the average person extracts 4,500 grams (about 10 pounds) of oxygen from 50,000 liters of air inhaled every day. Without hemoglobin, this number would plummet to about 60 grams of oxygen, only slightly more than we could obtain by breathing water, with equally fatal results.

The importance of hemoglobin in human blood is readily demonstrated by the fatal effects of carbon monoxide (CO), a gas that is present in high concentration in car exhaust fumes. Carbon monoxide molecules bind to the same sites on the hemoglobin molecule as do oxygen molecules, but they do so 200 times more tenaciously. As a result, the hemoglobin in the blood of a person exposed to high levels of carbon monoxide loses its ability to bind oxygen; if the person is not removed from the toxic environment, he or she will die as the result of oxygen deprivation.

While high concentrations of hemoglobin are needed to carry a large supply of oxygen, the protein cannot simply be dissolved in the blood plasma since it would thicken the fluid, impairing its flow through the vessels. This problem has been solved by encapsulating hemoglobin in **erythrocytes** (red blood cells). Each milliliter of your blood contains about 5 billion erythrocytes; your entire body contains a total of about 25 trillion.

GAS EXCHANGE IN THE TISSUES: RELEASE OF OXYGEN FROM THE BLOOD

Blood leaves the lungs carrying high concentrations of oxygen and low levels of carbon dioxide. When this oxygenated blood reaches a metabolically active tissue, it finds itself in an environment that is low in oxygen and high in carbon dioxide. As a result, oxygen diffuses into the tissues as carbon dioxide enters the blood; both gases move passively from areas of higher concentration to areas of lower concentration (Figure 29-6). The first oxygen molecules to move out of the blood are those that are simply dissolved in the plasma. As these dissolved oxygen molecules move out of the capillary, the concentration of oxygen dissolved in the plasma decreases, promoting the dissociation of oxygen molecules from their binding sites on hemoglobin molecules in the red blood cells. The oxygen molecules released from hemoglobin move out of the erythrocytes and dissolve in the plasma. From there, the dissolved oxygen molecules move out of the capillaries, promoting the release of additional oxygen molecules from the hemoglobin, and so forth. This is the process by which hemoglobin unloads its oxygen cargo in the tissues and readies itself to pick up more oxygen on its upcoming trip through the lungs.

◐ The exchange of gases in the tissues is self-regulating. Those tissues that are metabolizing more actively utilize

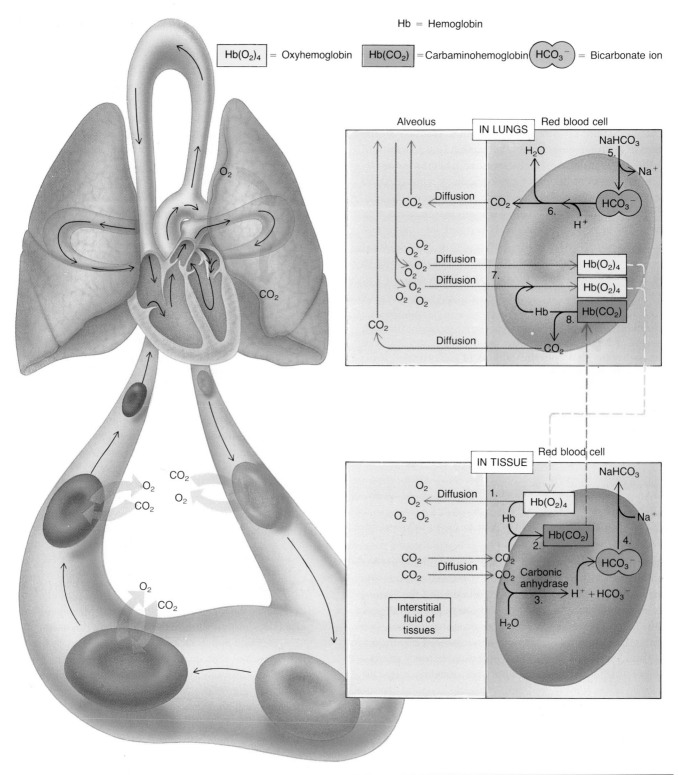

FIGURE 29-6

Transport and exchange of gases. In the tissues: (1) O_2 molecules are released from hemoglobin and diffuse into the cytoplasm of the red blood cell and then into the tissues. (2) As CO_2 diffuses from the tissues into red blood cells, some of it complexes with hemoglobin. (3) Most of the CO_2, however, reacts with water to form bicarbonate ions (HCO_3^-). The protons released by this reaction bind to hemoglobin, which promotes the release of additional oxygen. (4) Some bicarbonate diffuses from the cell and provides the blood with buffering capacity. **In the lungs:** (5) Bicarbonate ions diffuse into the red blood cell and (6) are converted back to CO_2, which diffuses into the alveolus. (7) As O_2 diffuses into the red blood cells, it complexes with hemoglobin, so a steep O_2 concentration gradient is maintained. (8) CO_2 molecules that were bound to hemoglobin molecules are released, freeing the hemoglobin molecules to pick up additional O_2.

more oxygen, thereby producing lower oxygen concentrations. The lower the oxygen concentration in the tissue, the steeper the gradient between tissue and blood, which favors the release of more oxygen from the bloodstream. According to this principle, those tissues most "in need" of oxygen receive the most oxygen from the passing blood.

We noted on page 62 that the blood contained buffers —particularly bicarbonate ions—that kept the blood from becoming too acidic or too alkaline. The bicarbonate ions in the blood are formed as the result of the following reaction between carbon dioxide and water that occurs when carbon dioxide is taken into the blood from the tissues:

$$CO_2 + H_2O \xrightarrow{\underset{\text{anhydrase}}{\text{carbonic}}} \underset{\underset{\text{acid}}{\text{carbonic}}}{H_2CO_3} \longrightarrow \underset{\text{proton}}{H^+} + \underset{\underset{\text{ion}}{\text{bicarbonate}}}{HCO_3^-}$$

This first reaction is catalyzed by the enzyme *carbonic anhydrase*, which is present in red blood cells. Carbonic acid, the product of the reaction, dissociates into two ions, a hydrogen ion (H^+) and a bicarbonate ion (HCO_3^-). Hydrogen ions lower the pH, which affects the shape of hemoglobin molecules, causing them to lose their grip on oxygen. This phenomenon is known as the *Bohr effect*. Those tissues that are metabolizing most actively not only have a lower oxygen concentration but also a higher carbon dioxide concentration, which leads to a lower pH. Since hemoglobin releases more oxygen at a lower pH, metabolically active tissues have a "built-in" mechanism for extracting additional oxygen from the passing bloodstream.

As the blood passes through the tissues, it gives up its oxygen and takes up carbon dioxide (Figure 29-6). Most of the carbon dioxide simply dissolves in the blood, forming bicarbonate, but a portion of the carbon dioxide molecules becomes bound to hemoglobin for the trip back to the lungs. When the blood reaches the lungs, the process of gas exchange is reversed from that which occurs in the tissues. The more oxygen that is removed during the previous passage of the blood through the body, the greater the number of hemoglobin molecules that will be lacking their full complement of oxygen molecules, and the more oxygen that will be picked up from the alveoli. The lungs have a remarkable capability for gas absorption. Even if you are running at top speed, and your blood is virtually depleted of its oxygen content, the blood will be fully resupplied with oxygen during its short, rapid passage through the lungs.

Oxygen and carbon dioxide move into and out of the bloodstream by simple diffusion in response to differences in their respective concentrations. These concentration gradients arise as the result of the consumption of oxygen and the formation of carbon dioxide within the tissues. The uptake of oxygen into the blood as it courses through the lungs is greatly facilitated by the presence of the oxygen-binding protein hemoglobin. (See CTQ # 4.)

REGULATING THE RATE AND DEPTH OF BREATHING TO MEET THE BODY'S NEEDS

During the course of a typical day, the respiratory system must make rapid and dramatic changes in its level of activity. One minute, you might be resting quietly on a park bench, and the next minute you might be running after a bus. This shift from a resting to a running state results in a great increase in the utilization of oxygen. How can the blood increase its rate of oxygen delivery twentyfold or more in less than a minute?

Three major mechanisms are available to supply the body with an increased supply of oxygen, each of which becomes quite evident during strenuous exercise (Figure 29-7): You breathe more rapidly; you breathe more deeply; and the rate of your heartbeat increases. All of these changes are controlled by regulatory centers located in the brain.

Although you can hold your breath for a short time, it is impossible to stop breathing voluntarily to the point of severe oxygen deprivation. Eventually, involuntary regulatory

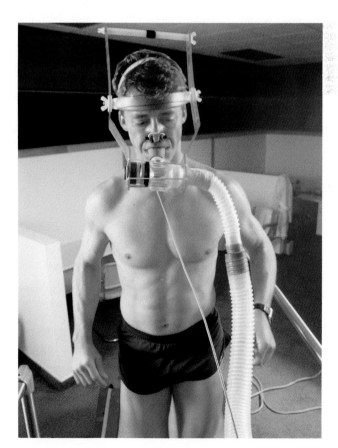

FIGURE 29-7

Measuring the rate of oxygen consumption of a person running on a treadmill.

◁ THE HUMAN PERSPECTIVE ▷
Dying for a Cigarette?

On the average, smoking cigarettes will cut approximately 6 to 8 years off your life, more than 5 minutes for every cigarette smoked! Cigarette smoking is the greatest cause of preventable death in the United States. According to a 1991 report by the Centers for Disease Control (CDC), nearly 450,000 Americans die each year from smoking-related causes. Smoking accounts for 87 percent of all lung-cancer deaths, and smokers are more susceptible to cancer of the esophagus, larynx, mouth, pancreas, and bladder than are nonsmokers. The increased incidence of lung cancer deaths among smokers compared to nonsmokers is shown in Figure 1*a*, and the benefits attained by quitting is shown in Figure 1*b*. The effects of smoking on lung tissue is shown in Figure 2. Atherosclerosis, heart disease, and peptic ulcers also strike smokers with greater frequency than they do nonsmokers. For example, long-term smokers are 3.5 times more likely to suffer from severe arterial disease than are nonsmokers. Emphysema (a condition caused by the destruction of lung tissue, producing severe difficulty in breathing) and bronchitis (an inflammation of the airways) are 20 times more prevalent among smokers.

Smokers also endanger other people. Each year, smokers are responsible for the deaths of thousands of "innocent bystanders," nonsmokers who share the same air with smokers. The risks of passive (involuntary) smoking are well known; secondhand smoke can make you seriously ill (Figure 3). Children of smokers have double the frequency of respiratory infections as do children who are not exposed to tobacco smoke in the home. Being married to a smoker is especially hazardous; 20 percent of lung-cancer deaths among nonsmokers are attributable to inhaling other

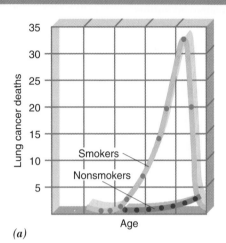

(a)

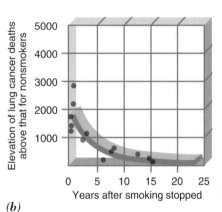

(b)

FIGURE 1 ⎯⎯⎯⎯⎯⎯⎯⎯

people's tobacco smoke. Another "innocent bystander" is a fetus developing in the uterus of a woman who smokes. Smoking increases the incidence of miscarriage and stillbirth and decreases the birthweight of the infant. Once born, these babies suffer twice as many respiratory infections as do babies of nonsmoking mothers.

Why is smoking so bad for your health? The smoke emitted from a burning cigarette contains more than 2,000 identifiable substances, many of which are either irritants or carcinogens. These compounds include carbon monoxide, sulfur dioxide, formaldehyde, nitrosamines, toluene, ammonia, and radioactive isotopes. Autopsies of respiratory tissues from smokers (and from nonsmokers who have lived for long periods with smokers) show widespread cellular changes, including the presence of precancerous cells (cells that may become malignant, given time) and a marked reduction in the number of cilia that play a vital role in the removal of bacteria and debris from the airways.

Of all the compounds found in tobacco (including smokeless varieties), the most important is nicotine, not because it is carcinogenic, but because it is so addictive. Nicotine is addictive because it acts like a neurotransmitter by binding to certain acetylcholine receptors (page 477), stimulating postsynaptic neurons. The physiological effects of this stimulation include the release of epinephrine, an increase in blood sugar, an elevated heart rate, and the constriction of blood vessels, causing elevated blood pressure. A smoker's nervous system becomes "accustomed" to the presence of nicotine and decreases the output of the natural neurotransmitter. As a result, when a person tries to stop smoking, the sudden absence of nicotine, together with the decreased level of the natural transmitter, decreases stimulation of postsynaptic neurons, which creates a craving for a cigarette—a "nicotine fit." Ex-smokers may be so conditioned to the act of smoking that the craving for cigarettes can continue long after the physiological addiction disappears.

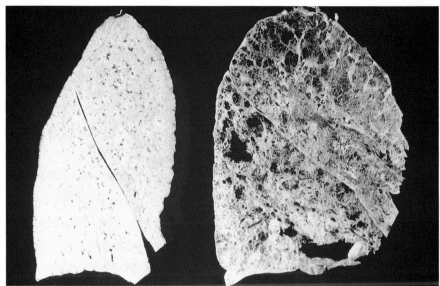

NON SMOKER   SMOKER

FIGURE 2

FIGURE 3

mechanisms restart "automatic" breathing. These same regulatory mechanisms keep you breathing while you sleep or when your attention is on other matters. They also cause you to breathe more often and more deeply as your need for oxygen delivery increases. The breathing control center, or **respiratory center,** is located in the medulla, a portion of the brainstem that regulates automatic activities (page 490). But just how does the respiratory center know how often you need to take a breath and how deep that breath should be?

The respiratory center of the brain receives information from two sources: *peripheral chemoreceptors* located in the walls of the aorta and the carotid arteries of the neck, and *central chemoreceptors* located in the brain itself. Let us consider the role of the more important central chemoreceptors. Recall that an increase in the level of carbon dioxide in the blood leads to a rise in the concentration of hydrogen ions. In other words, the increased carbon dioxide that is generated during more vigorous activity leads to an increased acidity of the blood. Chemoreceptors in the brain are stimulated by even slight increases in acidity, activating the respiratory center of the medulla. This activation, in turn, sends signals to the diaphragm and the intercostal muscles, causing you to breathe more rapidly and deeply. The effect of elevated carbon dioxide levels can be demonstrated in the following ways.

1. If a person is allowed to breathe air that contains a constant level of oxygen but an increasing content of carbon dioxide, the breathing rate will increase markedly, even though the person is not engaged in any physical activity. Since the level of oxygen is held constant, it must be the rising carbon dioxide level that is elevating the respiratory rate.

2. If a person is told to breathe very rapidly and deeply for a period of time and then to breathe in a normal relaxed manner, the breathing rate drops dramatically below that of the normal resting state. This phenomenon is explained on the basis of blood carbon dioxide levels. The period of rapid breathing (known as *hyperventilation*) is not accompanied by increased metabolic activity; thus, there is an increased loss of carbon dioxide from the lungs without a corresponding increase in carbon dioxide production by the tissues. Consequently, blood carbon dioxide concentrations drop. The brain senses the lowered carbon dioxide level—as a drop in hydrogen ion concentration—and interprets the message that further inhalation is not necessary. Underwater divers sometimes use this physiological response to hold their breaths for longer periods of time than would normally be possible. This is actually a very dangerous practice since hyperventilation has no effect on blood oxygen levels. Even though the individual may not feel the need to breathe, the blood oxygen content can become drastically depleted. As a result, the individual may lose consciousness before realizing that he or she is running out of oxygen.

> Increased demands for oxygen create alterations in blood chemistry that trigger an increased uptake of oxygen from the environment. As oxygen-saturated blood circulates through the tissues, its delivery is self-regulating; the greater the tissue's need, the more oxygen that is released to the cells of that tissue. (See CTQ # 5.)

ADAPTATIONS FOR EXTRACTING OXYGEN FROM WATER VERSUS AIR

▐▶ The successful respiratory strategy is one that is adapted to the medium in which the animal lives. Animals that absorb oxygen from water, for example, utilize a different type of respiratory system than do animals that extract oxygen from air.

EXTRACTING OXYGEN FROM WATER

The amount of oxygen dissolved in water is at best 21 times less than that which is dissolved in air. As a result, aquatic animals that acquire their oxygen from the surrounding water have less oxygen available than do their air-breathing counterparts. In addition, water is much heavier than air; thus, water requires considerably more energy in order to be moved over the respiratory surface. We use very little energy to move air in and out of our lungs, whereas a fish uses much of the energy obtained from its food just to pump water through its body to acquire oxygen as needed.

Most animals that extract oxygen from water possess **gills**—outgrowths of the body surface which project into the aqueous environment and are rich with blood vessels. These outgrowths would be ill-adapted in a terrestrial animal since the delicate projections would rapidly dehydrate in air. This is why a fish quickly dies of asphyxiation when it is removed from water even though it is surrounded by a higher concentration of oxygen than that of its normal habitat.

The complexity of an animal's gills correlates with the animal's oxygen requirements. The gills of small or slow-moving aquatic animals, such as sea slugs, sea stars, and aquatic salamanders, are typically simple projections of the body surface (Figure 29-8). Animals with greater oxygen requirements have more complex gills that are capable of harvesting additional oxygen. Such gills are usually covered by a protective flap, such as that found in lobsters or fishes. The form of the complex gills of larger, active animals are well-suited for their function. These gills branch into smaller outgrowths that amplify the surface area for gas exchange without increasing the space required to house them. Gill complexity has reached its current pinnacle in fishes (Figure 29-9a). Each gill consists of fingerlike projections, called **gill filaments,** from which rows of thin, flattened **lamellae** project into the flowing stream of water. The lamellae are the sites of gas exchange and have a rich supply of capillaries into which oxygen molecules diffuse.

Physiologists studying fish gills noted that the flow of blood within the lamellae and the flow of water across the outer surface of the lamellae (over the gills) occurred in opposite directions (Figure 29-9a,b). The movement of the blood and water in opposite directions is imperative to the success of highly active "water breathers" because it maximizes the amount of oxygen that is extracted from water. This principle is called **countercurrent flow,** providing another example of the relationship between form and function. Because the two fluids flow in opposite directions, there will always be a higher concentration of oxygen in the water than in the blood (Figure 29-9b, upper part). This relationship favors the continual diffusion of oxygen from the surrounding water into the bloodstream. Over 80 percent of the oxygen passing through the gills is extracted by this type of respiratory mechanism, much more than that which could be attained if the two currents were flowing in the same direction (Figure 29-9b, lower part).

Not all aquatic animals have gills for extracting oxygen from the water. A variety of aquatic animals, such as porpoises, whales, sea snakes, and some water spiders, have evolved from air-breathing terrestrial ancestors. These animals still rely on air for their oxygen supply, even if they have to bring an external supply of oxygen underwater with them (Figure 29-10).

EXTRACTING OXYGEN FROM AIR

Terrestrial animals live in an environment that is rich in oxygen but low in water content; thus, oxygen is readily available but the loss of some of its precious body water is inevitable. Therefore, the respiratory surfaces of most terrestrial animals—whether a snail, spider, or human—occur as inpocketings of the body surface which are tucked away within the moist, internal environment, thereby minimizing water loss. Two very different types of respiratory systems have evolved in air-breathing animals. Insects, and some of their arthropod relatives, possess *tracheae*, while most other air breathers, including snails, reptiles, birds, and mammals, have *lungs*.

Respiration by Tracheae

If you look at the surface of an insect's abdomen under a microscope, you will find paired rows of openings, called *spiracles* (Figure 29-11a). Each spiracle leads into a tube, which, in turn, branches into a network of finer and finer tubes. These tubes, called **tracheae,** carry air to the most remote recesses of the animal's body (Figure 29-11b). Ultimately, the finer tracheal tubes give rise to microscopic, dead-end, fluid-filled **tracheoles,** which terminate either very close to or actually within the body's cells. Oxygen travels down this complex network of air tubes directly to the cells, without any help from the animal's circulatory system.

While a tracheal respiratory system is very efficient for a small insect, it becomes very limiting in larger species. Even though many insects actively suck fresh air into their larger tracheal tubes (which explains why a grasshopper can sometimes be seen rhythmically contracting its abdomen), the system largely depends on the simple diffusion of air through a network of narrow tubing. This is analogous to a crowd of people trying to breathe at the dead end of a long mine shaft; the available oxygen may be rapidly depleted. The lack of participation of the insect's circulatory system in respiration is one of the key factors that has limited the size

FIGURE 29-8

Simple gills. Extensions of the external surface of this nudibranch (sea slug) increases the animal's oxygen-absorbing efficiency.

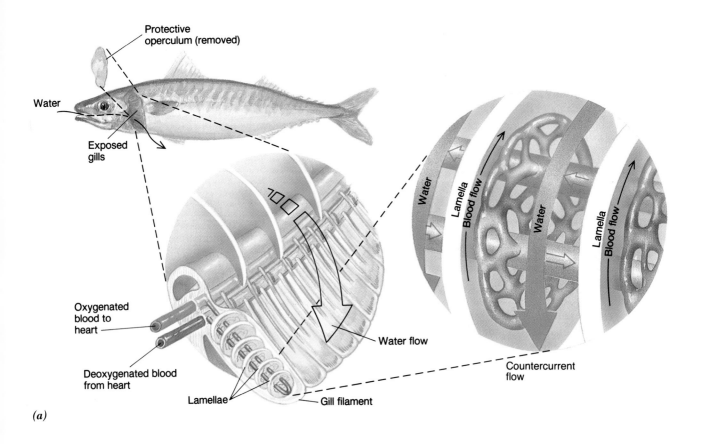

Protective operculum (removed)

Water

Exposed gills

Oxygenated blood to heart

Deoxygenated blood from heart

Lamellae

Gill filament

Water flow

Water

Lamella

Blood flow

Water

Lamella

Blood flow

Countercurrent flow

(a)

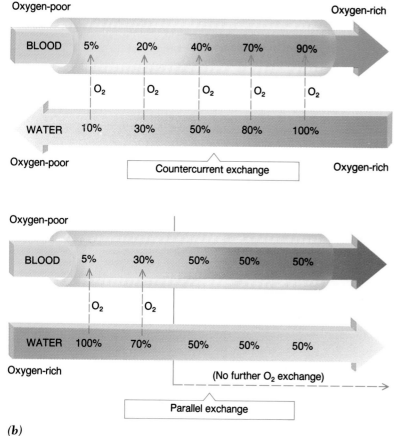

Oxygen-poor

Oxygen-rich

BLOOD   5%   20%   40%   70%   90%

O$_2$ O$_2$ O$_2$ O$_2$ O$_2$

WATER   10%   30%   50%   80%   100%

Oxygen-poor

Oxygen-rich

Countercurrent exchange

Oxygen-poor

BLOOD   5%   30%   50%   50%   50%

O$_2$ O$_2$

WATER   100%   70%   50%   50%   50%

Oxygen-rich

(No further O$_2$ exchange)

Parallel exchange

(b)

FIGURE 29-9

Complex gills. Fish have complex gills that have an enormous surface area compacted into a small space *(a)*. In this diagram, blood flowing through each gill filament changes from blue to red as it acquires oxygen from the passing water. The process is enhanced by a countercurrent exchange system, whereby water is forced past the outer surface of the lamellae in a direction counter to that of blood flow within the lamellae. As the blood acquires oxygen, it moves toward water that is even richer in the gas so that at every point along the exchange surface, the oxygen level in the blood is lower than is that of the surrounding water (upper diagram of *b*). This relationship favors the continuing diffusion of oxygen along the entire lamella. If the flow were in the same direction (lower diagram of *b*), absorption would occur only at one end, and much less oxygen could be obtained.

FIGURE 29-10

Carrying more than dinner. This diving spider carries its own oxygen supply in a glistening bubble of air around its abdomen. The spider is hauling its catch home to the larger, submerged air bubble in which it dines, sleeps, and mates.

of these animals, a condition that is undoubtedly fortunate for the rest of the animal life on this planet.

Respiration by Lungs

As we discussed earlier, the lungs are sac-like invaginations of the body surface that contain an extremely thin respiratory surface and a rich supply of microscopically thin blood capillaries. In less active animals, such as snails, frogs, and snakes, the lungs tend to be relatively simple sacs with little internal surface area. In contrast, the more complex lungs of mammals contain millions of microscopic alveoli, which provide an enormous surface area packed into a relatively small space (see Figure 29-3).

Compared to water, air is dry, light-weight, and rich in oxygen. These environmental differences have selected for different types of respiratory organs in air-breathers versus water-breathers. (See CTQ # 6.)

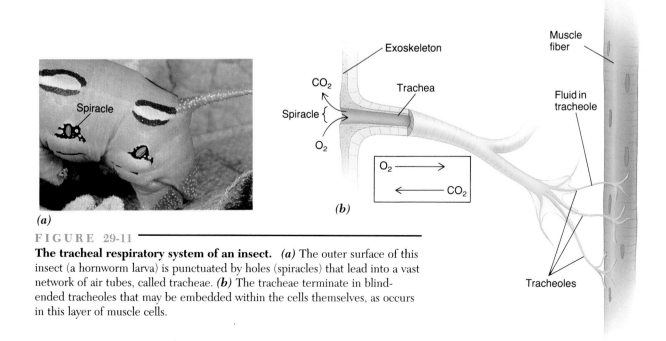

Muscle fiber

Exoskeleton

CO_2

Trachea

Spiracle

Fluid in tracheole

Spiracle {

O_2

$O_2 \longrightarrow$

$\longleftarrow CO_2$

(b)

Spiracle

(a)

Tracheoles

FIGURE 29-11

The tracheal respiratory system of an insect. *(a)* The outer surface of this insect (a hornworm larva) is punctuated by holes (spiracles) that lead into a vast network of air tubes, called tracheae. *(b)* The tracheae terminate in blind-ended tracheoles that may be embedded within the cells themselves, as occurs in this layer of muscle cells.

EVOLUTION OF THE VERTEBRATE RESPIRATORY SYSTEM

▶ Vertebrates have two very different types of respiratory organs: gills and lungs. Even though they serve the same basic function—gas exchange—these two respiratory organs have different structures and have arisen along two separate evolutionary paths.

The ancestors of fishes were thought to have resembled the tiny, modern-day lancelet (pictured in Figure 29-12). The mouth of a lancelet opens into a pharynx that is punctuated with mucus-covered, microscopic slits, causing it to resemble a woven basket. The animal is a filter feeder; it pumps water into its mouth and through the pharyngeal "basket," while suspended food particles become trapped in the mucus lining the slits. While primarily a feeding organ, this type of pharyngeal apparatus is ideally suited for carrying out gas exchange. A constant stream of oxygen-rich water flows through the pharyngeal basket, which contains a large surface area across which gas exchange may occur. While the early fishes may have been filter feeders like their ancestors, most became adapted to feeding on larger food items. The pharyngeal apparatus lost its food-capturing function and evolved into a system of gills that delivered oxygen to the bloodstream.

The evolutionary shift of the pharynx from a feeding to a respiratory organ can be appreciated by following the development of one of the most primitive vertebrates, the lamprey eel (Chapter 39). The larval lamprey lives on the bottom of streams and feeds on particles strained from the water by its pharyngeal basket. Following metamorphosis, the adult lamprey becomes a predator, but its pharyngeal apparatus is retained as its respiratory organ.

The evolutionary origin of vertebrate lungs is less certain. There are no fossil remains of these soft, spongy

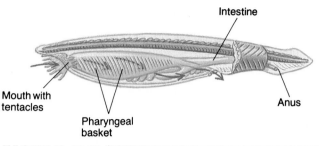

FIGURE 29-12

The pharyngeal feeding apparatus of a lancelet. The pharyngeal basket of these small, aquatic invertebrates is thought to resemble that of the ancestors of fishes. The basket is perforated with openings that strain food from the water that flows through the animal's body. While the pharyngeal basket of the lancelet functions primarily as a feeding organ, a similar organ in an ancient ancestor may have given rise to vertebrate gills.

organs, and little has been learned by studying modern-day representatives. Regardless of how they evolved, lungs were probably present in some of the earliest fish. These primitive vertebrates lived in freshwater ponds, and their lungs are thought to have helped them survive periods of drought when the ponds became shallow and unsuitable for aquatic respiration. These air-breathing fishes include the evolutionary predecessors of all of the land vertebrates.

Both the gills and lungs of modern vertebrates have ancient evolutionary roots. Gills are thought to have evolved from feeding baskets present in filter-feeding prevertebrates, while lungs probably evolved in ancient fishes that became dependent on air as their aquatic environments either stagnated or evaporated. (see CTQ # 7.)

REEXAMINING THE THEMES

Relationship between Form and Function

The form of the human respiratory system is closely correlated with its function. This is evident by the hairs in the nasal cavity, which filter the air; the flaplike epiglottis, which seals the airways from food and liquid; the mucus-secreting cells, which line the respiratory tract and keep the air moist; the cartilage bands in the trachea, which prevent the tube from collapsing as we breathe; and the lungs, which consist of millions of thin-walled pouches, whose combined surface area is equivalent to the size of a badminton court yet is packed into a chest cavity smaller than a badminton racket.

Biological Order, Regulation, and Homeostasis

The supply of oxygen to the body's tissues is maintained by a number of homeostatic mechanisms, some of which are self-regulating and occur automatically, without the intervention of the nervous system. The amount of oxygen delivered to a given tissue is determined largely by the oxygen needs of that tissue. More active tissues utilize more oxygen, producing a steeper oxygen gradient between themselves and the blood, causing the release of more oxygen from the blood. In addition, more active tissues produce more carbon dioxide, which increases the acidity of the blood as it passes through the tissues, causing a change in the shape of the hemoglobin molecules and an additional release of oxygen into these areas. The amount of oxygen that is taken into the lungs is regulated by neural mechanisms. As the level of carbon dioxide in the blood rises during periods of physical activity, chemoreceptors in the brain are stimulated, and the respiratory center in the medulla is activated, leading to an increased rate and depth of breathing.

Acquiring and Using Energy

Respiratory systems have evolved as the means by which animals obtain oxygen. Oxygen serves only one major function in animals: It is the terminal acceptor of the electron transport system of the mitochondria (the system by which over 90 percent of the ATP used by most animals is generated). Without oxygen, animals cannot sustain their energy-requiring reactions. Thus, even though oxygen provides no energy itself, the gas is necessary for most animals to obtain energy from food by oxidation.

Unity within Diversity

Respiratory organs come in many different shapes and sizes, ranging from the outer skin of a frog, to the gills of a fish, to the lungs in your chest. Despite this diversity, all of these organs possess a respiratory surface that is extremely thin, highly permeable, and continually moist. In addition, all respiratory surfaces are underlain by a network of thin-walled capillaries and are characterized by a large surface area. Air breathers, from snails to whales, package their respiratory surface within an inpocketing of their body surface, whereas, water-breathing animals, such as starfish and codfish, utilize delicate filaments that project directly into their aquatic environment.

Evolution and Adaptation

The gills of vertebrates evolved from food-trapping pouches, providing an excellent example of how the function of a structure can change over the course of evolution. Since the currents of water that supply food also carried oxygen into the body, this type of feeding strategy set the stage for the evolutionary conversion of a food-collecting organ into oxygen-collecting gills.

SYNOPSIS

The need for gas exchange. All animals must obtain oxygen for aerobic respiration and must dispose of carbon dioxide waste. Most complex animals have impermeable outer surfaces with specialized gas exchange structures (gills in water-breathing animals, tracheae in most terrestrial arthropods, and lungs in air-breathing vertebrates). These respiratory structures contain respiratory surfaces that are thin, permeable, and moist and are characterized by an extensive surface area.

The human respiratory system consists of a branched passageway that leads into a pair of lungs. The lungs consist of millions of microscopic, thin-walled pouches (alveoli) that are underlain by a network of capillaries. The alveoli provide an extensive surface across which gas exchange can occur. Air enters the lungs when the thoracic cavity is expanded by contraction of the diaphragm and intercostal muscles. Expansion of the chest creates a pressure gradient that sucks air into the lungs. When the muscles relax, the walls of the thoracic cavity recoil, forcing stale air out of the lungs.

Gas exchange in the lungs and tissues is driven by the diffusion of oxygen and carbon dioxide down their respective concentration gradients. Blood entering the lungs from the tissues is relatively high in carbon dioxide and low in oxygen. In the lungs, oxygen diffuses from the alveoli into the blood where over 95 percent of the oxygen molecules bind to hemoglobin in the erythrocytes. The blood becomes completely saturated with oxygen, even when we are undergoing strenuous exercise and the blood is moving rapidly through the lung capillaries. When oxygenated blood is pumped through metabolically active tissues, oxygen molecules diffuse out of the plasma, promoting the release of oxygen molecules bound to hemoglobin. The process of gas exchange in the tissues is self-regulating: The more active the tissue, the lower its oxygen concentration and the steeper the oxygen gradient between the blood and that tissue. This relationship favors the release of additional

oxyggen. More active tissues also contain higher carbon dioxide levels, producing an increased acidity that alters the shape of the hemoglobin molecules, causing the release of additional oxygen molecules.

The rate and depth of our breathing is regulated by our need for oxygen. Breathing is controlled by the respiratory center located in the medulla of the brainstem. The most important factor involved in regulating breathing is the acidity of the blood, which is proportional to the carbon dioxide concentration. As carbon dioxide levels rise, the pH drops, and impulses are sent to the diaphragm and intercostal muscles, increasing the rate and depth of expansion of the thoracic cavity.

Water breathers and air breathers have different types of respiratory organs. Water breathers typically possess gills, which consist of delicate, fingerlike projections that extend into the surrounding water. Air breathers typically possess either lungs — inpocketings of the external surface into the body where the respiratory surface can be kept moist — or tracheae — blind passageways through which air diffuses into the tissues. Tracheae are found in arthropods and carry air to the tissues without the intervention of the circulatory system.

Key Terms

respiratory system (p. 622)
gas exchange system (p. 622)
respiratory surface (p. 622)
cutaneous respiration (p. 622)
respiratory tract (p. 623)
lungs (p. 623)
nasal cavity (p. 624)
pharynx (p. 624)
glottis (p. 624)

larynx (p. 624)
epiglottis (p. 624)
vocal cords (p. 624)
trachea (p. 625)
bronchiole (p. 625)
alveoli (p. 625)
thoracic cavity (p. 626)
pleura (p. 626)
diaphragm (p. 626)

hemoglobin (p. 627)
erythrocyte (p. 627)
respiratory center (p. 632)
gill (p. 632)
gill filament (p. 633)
lamellae (p. 633)
countercurrent flow (p. 633)
tracheae (p. 633)
tracheole (p. 633)

Review Questions

1. Distinguish among aerobic respiration, gas exchange in the lungs and tissues, and breathing.

2. Compare the respiratory strategies of the following: a frog, a snail, a cockroach, a fish, a seal, and a human.

3. Describe the functional interconnection between the human circulatory system and the respiratory system.

4. Describe two mechanisms that help match the amount of oxygen released with a tissue's need for oxygen.

5. Distinguish between the larynx and the trachea; oxygen present in the plasma and in erythrocytes; the pH of blood with a high versus a low carbon dioxide concentration; the mechanism of inhalation and exhalation; the composition of air being inhaled versus air being exhaled; alveoli and gill lamellae; hyperventilation and normal breathing.

Critical Thinking Questions

1. Why do you suppose the diving reflex is more adaptive if it occurs as a response to submersion rather than as a response to metabolic changes resulting from oxygen deprivation?

2. The graph below shows the relationship between the surface area of the alveoli of the lungs of various animals and their rate of oxygen consumption. Explain what the graph shows about this relationship. What characteristics of respiratory surfaces are reflected in this relationship?

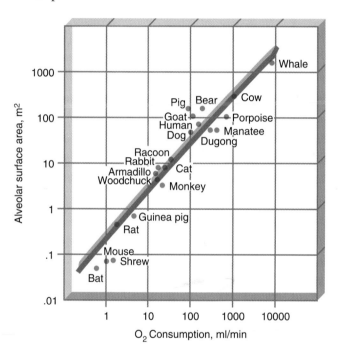

3. Trace the path that a molecule of oxygen would take from the outside atmosphere to a muscle cell in your leg, naming all of the structures through which it would pass.

4. Certain fish living in Antarctic waters have blood that is completely lacking hemoglobin, yet these animals are as active as are related fish that live in warmer waters and have hemoglobin in their blood. Can you suggest any reason that these Antarctic fish can survive without hemoglobin?

5. Suppose you were the subject of a study in which you were exposed to a constant level of carbon dioxide but an increasing level of oxygen. What effect do you think this would have on your rate and depth of breathing? After you have thought about this for a minute, consider the fact that your blood becomes fully saturated with oxygen even when you are exerting yourself and when blood is moving as fast as possible through your lung capillaries.

6. Why is the one-way flow of oxygen-bearing water through the body of a fish better suited for a fish than is the type of movement that occurs in humans, where the oxygen-containing medium is drawn in and out through the same opening?

7. A closed circulatory system would seem to be more efficient than an open circulatory system and, therefore, better adapted to an active way of life. What adaptations enable insects to maintain their high level activity in spite of possessing an open circulatory system?

Additional Readings

American Cancer Society. 1980. *The dangers of smoking; Benefits of quitting.* (Introductory)

Comroe, J. H. 1966. The lung. *Sci. Amer.* Feb:56-71. (Intermediate)

Douville, J. A. 1990. *Active and passive smoking hazards in the workplace.* New York: Van Nostrand Reinhold. (Intermediate)

Powers, S. K., and E. T. Howley. 1990. *Exercise physiology.* Dubuque, Iowa: W. C. Brown. (Intermediate-Advanced)

Sataloff, R.T. 1992. The human voice. *Sci. Amer.* Dec:108–115. (Intermediate)

Scholander, P. F. 1957. The wonderful net. *Sci. Amer.* April:96-110. (Intermediate)

West, J. 1989. *Respiratory physiology: The essentials,* 4th ed. Baltimore: Williams & Wilkins. (Intermediate)

Zapol, W. M. 1987. Diving adaptations of the Weddell seal. *Sci. Amer.* June:100-107. (Intermediate)

Internal Defense: The Immune System

STEPS
TO
DISCOVERY
On the Trail of a Killer: Tracking the AIDS Virus

NONSPECIFIC MECHANISMS: A FIRST LINE OF DEFENSE

Cellular Nonspecific Defenses

Molecular Nonspecific Defenses

THE IMMUNE SYSTEM: MEDIATOR OF SPECIFIC MECHANISMS OF DEFENSE

The Nature of the Immune System

ANTIBODY MOLECULES: STRUCTURE AND FORMATION

Antibody Specificity: A Unique Combination of Polypeptides

Antibody Formation: Discovery of a Unique Genetic Mechanism

IMMUNIZATION

EVOLUTION OF THE IMMUNE SYSTEM

BIOLINE

Treatment of Cancer with Immunotherapy

THE HUMAN PERSPECTIVE

Disorders of the Human Immune System

S TEPS TO DISCOVER Y

On the Trail of a Killer: Tracking the AIDS Virus

*I*n Fall 1980, a resident at the UCLA Medical Center was visited by a 31-year-old male who was suffering from a persistent fever, weight loss, swollen lymph nodes, and a severe yeast infection in his mouth and throat. This latter condition, known as *candidiasis,* is sometimes observed in patients undergoing chemotherapy during cancer treatment or in babies who are born with an immune system deficiency.

The patient came to the attention of Michael Gottlieb, an immunologist who had recently arrived at UCLA. Gottlieb obtained a sample of the patient's lung fluid and blood. The lung sample revealed the presence of a protozoan,

Pneumocystis carinii, that was known to cause a rare type of pneumonia (called PCP), usually in people suffering from lymphoid cancers. In other words, this patient was suffering from two rare types of infections simultaneously. In a healthy person, pathogenic organisms, such as these yeast and protozoa, are readily attacked and eliminated by certain white blood cells, called T cells. This particular patient lacked an entire class of T cells, called helper T cells, leaving Gottlieb and his colleagues dumbfounded.

In February 1981, Gottlieb was confronted by another patient suffering the same combination of infections and depleted content of helper T cells. Like the previous pa-

AIDS can affect anyone. The disease is caused by a virus that infects T-cells, and reproduces by budding from the infected

tient, this young man was also homosexual. By April, four similar cases had come to Gottlieb's attention, all of which involved male homosexuals. Gottlieb wrote a brief report that was submitted to the weekly newsletter of the U.S. Centers for Disease Control (CDC). This was the first publication concerning the disease that would soon be given the name Acquired Immune Deficiency Syndrome, or AIDS.

One of the physicians who received a copy of the newsletter was Willy Rozenbaum of the Claude Bernard Hospital in Paris. Three years earlier, Rozenbaum had treated a man suffering from a combination of rare diseases, including PCP. The man had recently spent several years in Africa. The next year, two African women were referred to Rozenbaum, both suffering from PCP. By the time Gottlieb's paper was published, all three of Rozenbaum's patients had died from their infections. Rozenbaum had little doubt that the disease plaguing homosexual men in Los Angeles (and simultaneously in New York and San Francisco) was the same malady that had killed his patients from Africa. Furthermore, the fact that the disease was appearing in very different populations on two sides of the world made it very unlikely that the disease was caused by some kind of toxic substance in the environment. Rather, Rozenbaum concluded that the affliction was the result of an infectious agent. Several years had passed between the visit of the African victims and the first patients to exhibit the disease in the United States, suggesting that the disease had been in the human population for several years and, furthermore, that the disease had a long *incubation period:* the time between the infection and the appearance of symptoms. The implications of this finding were frightening: The disease could take hold within a population long before any evidence of its presence would be detected.

Since the African cases of the disease predated those in the United States, it seemed likely that the disease had originated in Africa and had been transported to the United States by an infected male homosexual, who unwittingly passed it into the homosexual community. Soon, the disease began to appear in individuals who had received blood transfusions or blood products, most notably hemophiliacs, and intravenous-drug users who shared hypodermic needles.

As more cases were reported, Don Francis, a scientist at the CDC, began to consider the type of agent responsible for the new plague. Francis had received his doctorate working on a feline leukemia that was caused by a virus that ultimately destroyed the animal's immune system. Like this new human disease, feline leukemia resulted in the animal's contracting a host of unusual parasitic infections and was characterized by a long incubation period. It had been determined that feline leukemia was due to a *retrovirus*—a type of virus whose genetic material is RNA, which is transcribed into a molecule of DNA that is inserted into the host cell's chromosomal DNA. Just 1 year earlier, in 1980, a team led by Robert Gallo of the National Institutes of Health had shown that a retrovirus could cause a human disease—a rare type of leukemia that affected a person's T cells. Taken together, these strands of evidence pointed to a retrovirus as the causative agent of this new disease.

In Spring 1982, Gallo's lab began culturing lymphocytes from patients with AIDS. Meanwhile, another expert on retroviruses, Luc Montagnier, of the Pasteur Institute in Paris, was conducting similar research, searching for a retrovirus in infected lymphocytes. In 1983, both the U.S. and French teams reported the isolation of the retrovirus responsible for AIDS, setting the stage for one of the major scientific controversies of the decade: Which lab should be credited with the discovery of the AIDS virus, later called Human Immunodeficiency Virus (HIV)? For a number of reasons, including the fact that Gallo's virus appears to have been isolated from cells donated by the French scientists, Luc Montagnier is credited with the discovery of HIV.

cell surface.

As you read this sentence, you are being attacked. You are, in fact, fighting for your life. The surface of your body is populated by billions of microorganisms: bacteria and other agents that would invade and use your tissues as their next meal if allowed uncontested entry into your body. With every breath you take, you inhale more than a million microorganisms that are suspended in air, many of which would cause fatal infections if you didn't have an arsenal of weapons to protect your respiratory tract against their invasion. Some microorganisms have already entered your bloodstream, perhaps through cuts in your skin so small you were unaware of them. In your internal tissues and fluids, a microorganism has found a bonanza of nutrients and hospitable environmental conditions. The onslaught never stops.

Repelling this invasion is just as important in maintaining the stability of your internal environment as is the removal of waste products or the gain or loss of heat. The system responsible for this aspect of homeostasis is the **immune system;** it is capable of recognizing breaches in security and effectively eliminating invading microorganisms before they can destroy the biological order required for life to continue. The homeostatic activities that protect humans and other vertebrates from foreign agents are divided into two broad categories: nonspecific mechanisms and specific mechanisms. These two types of defenses are distinguished by the types of cells involved and by whether or not the specific identity of the target is recognized before it is attacked.

▼ ▼ ▼

NONSPECIFIC MECHANISMS: A FIRST LINE OF DEFENSE

The body's most evident protective strategy is to keep viruses, bacteria, and parasites from penetrating the living tissue. As we discussed in Chapter 26, the skin forms a

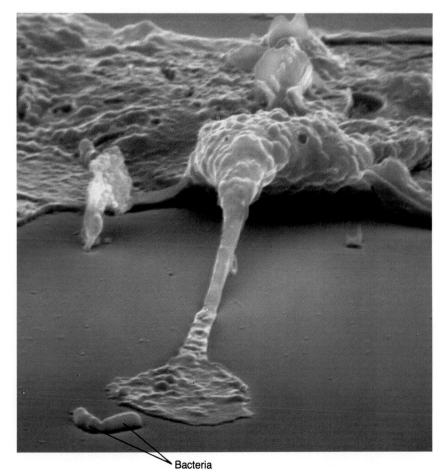

Bacteria

FIGURE 30-1

Death Grip. Among your many weapons against disease are phagocytic cells, which engulf and destroy foreign microbes, such as these doomed bacteria in the clutches of this protective white blood cell.

relatively impregnable outer body layer, as long as it remains undamaged. The mucous membranes that line the respiratory, digestive, and urinary tracts are less formidable barriers than the skin, but they are protected by a sticky layer of mucus, which is continually shed and replenished. The flushing of your urinary tract each time urine is voided is another nonspecific defense.

CELLULAR NONSPECIFIC DEFENSES

If a bacterium or other pathogenic microorganism breaches the body surface and reaches the tissues of the body, it will likely encounter a phagocytic cell that possesses the capacity to ingest and destroy it (see Figure 30-1). These cells are carried to the infected tissues through the blood and lymphatic vessels (Figure 30-2).

Not all potentially dangerous cells arrive from outside the body, however. Cancer cells are spawned within the body and have the same potential for human destruction as do external invading organisms. The transformation of a normal cell into a cancer cell is often accompanied by a change at the cell's surface, which may include the appearance of new or modified membrane proteins. The body contains nonspecific, lymphocyte-like cells, called **natural killer (NK) cells,** which can recognize cells having an altered surface, including cancer cells, and kill them. The precise mechanism by which these cells kill their targets is unclear, but it does not involve phagocytosis.

Inflammation

As important as phagocytic cells are to defending the body, these cells provide little security unless they can be called into the area where protection is needed. The body's overall "strategy" for attracting these and other protective cells to sites of danger, such as an infected wound, is called **inflammation.** Inflammation is initiated when chemicals from cells in the injured tissue are released, causing local blood vessels to dilate, bringing additional blood into the affected region. These chemicals also attract phagocytic leukocytes, such as neutrophils, which escape the vessels by squeezing through the spaces in the vessel wall (Figure 30-2). Protective leukocytes accumulate in the injured or infected region, often forming a yellowish liquid, called *pus.*

The accumulation of fluid and cells in the inflamed area causes painful swelling of the tissues due to stimulation of local neurons. The area grows hot and red from the additional blood. Inflammation creates many of the uncomfortable symptoms we often associate with disease rather than healing. Consequently, inflammation is usually targeted as a problem rather than an ally. Use of drugs to subdue inflammation (such as cortisone) may provide immediate relief from symptoms, but in doing so it suppresses this line of defense.

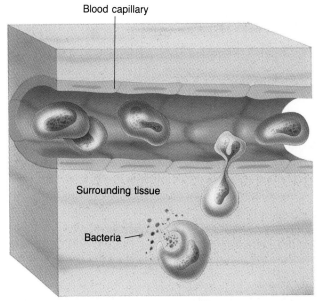

F I G U R E 30-2

Phagocytic leukocytes are able to squeeze through openings between the cells that line the capillaries and enter an inflamed area, where they can carry out phagocytosis.

MOLECULAR NONSPECIFIC DEFENSES

Several *molecular* mechanisms also operate to destroy microbes within the body. Blood contains a group of proteins collectively called **complement.** One of the complement proteins binds to the surface of a bacterium, initiating a series of reactions that poke holes in the plasma membrane of the foreign cell. Fluids and salts can then enter the cell, causing it to burst and die.

The progress of viral infections is blocked by a different type of nonspecific mechanism, spearheaded by the protein *interferon.* Suppose a virus penetrates into the respiratory membranes and infects a cell. Once the virus has entered the cell, the cell responds by secreting interferon into the extracellular fluid. The interferon molecules bind to the surfaces of uninfected, neighboring cells and initiate a series of reactions that block the neighboring cells' ability to manufacture viral proteins. This renders cells incapable of supporting viral reproduction and halts the spread of the infection.

The presence of a foreign cell within the body immediately attracts a nonspecific arsenal of weapons that includes phagocytic cells, blood-borne proteins, and secreted molecules that can attack and destroy the invader. (See CTQ #2.)

THE IMMUNE SYSTEM MEDIATOR OF SPECIFIC MECHANISMS OF DEFENSE

Edward Jenner, an eighteenth-century English physician, was interested in the prevention of smallpox, a deadly and widespread disease characterized by the development of elevated blisters filled with pus. Jenner practiced in the English countryside, where he noticed that the maids who tended the cows were typically spared the ravages of smallpox. Jenner concluded that the milkmaids were somehow "immune" to smallpox because they were infected at an early age with cowpox, a harmless disease they contracted from their cows. Cowpox produces blisters that resemble the pus-filled blisters of smallpox, but the cowpox blisters are localized and disappear, causing nothing more serious than a scar at the site of infection.

In 1796, Jenner performed one of the most famous (and risky) medical experiments of all time. First, he infected an 8-year-old boy with cowpox and gave the boy time to recover. Six weeks later—in an experiment that would be considered unethical today—Jenner intentionally infected the boy with smallpox by injecting pus from a smallpox lesion directly under the boy's skin. The boy remained healthy. Within a few years, thousands of people had become immune to smallpox by intentionally infecting themselves with cowpox. This procedure was termed *vaccination,* after "vacca," the Latin word for cow. Under the leadership of the World Health Organization, vaccination against smallpox has eradicated this once dreaded disease.

You may be wondering how a previous infection with cowpox protects people against a much more serious illness. Infection with cowpox stimulates cells of the immune system to produce **antibodies**—proteins that bind to foreign molecules in a highly specific manner. Any foreign substance that elicits production of an immune response (such as antibody production) is called an **antigen.** Both cowpox and smallpox viruses contain proteins in their outer capsule (Chapter 36) that act as antigens when they enter the body, evoking antibodies that react with these viral proteins. The proteins present in the cowpox and smallpox viruses have a very similar molecular structure—so similar, in fact, that the antibodies produced by the body's cells against the cowpox virus also react and neutralize the smallpox virus. Once infected with cowpox, these antibodies can be produced rapidly at any time during a person's life, protecting the individual against the subsequent development of smallpox.

THE NATURE OF THE IMMUNE SYSTEM

The immune system works by assembling an arsenal of protective cells and molecules that search out a particular

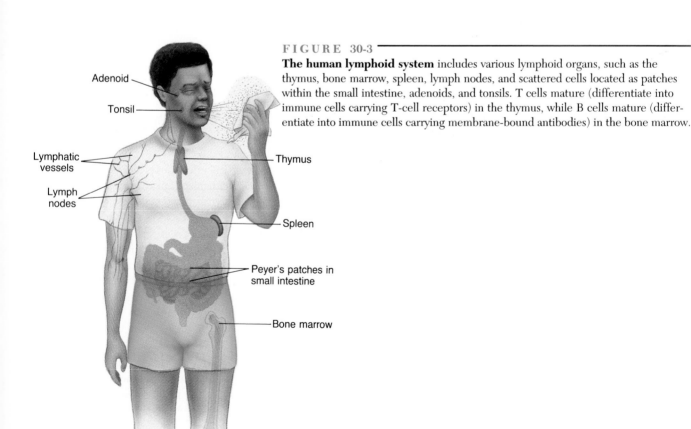

FIGURE 30-3

The human lymphoid system includes various lymphoid organs, such as the thymus, bone marrow, spleen, lymph nodes, and scattered cells located as patches within the small intestine, adenoids, and tonsils. T cells mature (differentiate into immune cells carrying T-cell receptors) in the thymus, while B cells mature (differentiate into immune cells carrying membrane-bound antibodies) in the bone marrow.

Adenoid

Tonsil

Lymphatic vessels

Lymph nodes

Thymus

Spleen

Peyer's patches in small intestine

Bone marrow

antigen and destroy it. The fact that the immune system is not always successful in its endeavors is all too evident by the AIDS virus, which is capable of infecting the very cells that have evolved to destroy such viruses.

The immune system is composed of cells that are scattered throughout the body and particularly concentrated in the lymphoid tissues, which include the thymus, spleen, lymph nodes, bone marrow, and tonsils (Figure 30-3). The most prominent cells of the immune system are lymphocytes, which circulate throughout the blood and lymph. Lymphocytes are aided by *macrophages*, large phagocytic cells that are derived from monocytes (see Figure 28-17*b*). As lymphocytes and macrophages travel throughout the body, they come into contact with foreign materials, which initiates an *immune response.*

The body's immunological arsenal is composed of two major types of lymphocytes: *B lymphocytes* (or simply **B cells**) and *T lymphocytes* (**T cells**). Both types of lymphocytes work in cooperation with each other and with macrophages, but they carry out very different immune responses.

B-Cell Immunity: Protection by Soluble Antibodies

✪ When you come down with a cold or a throat infection, your immune system responds by producing specific antibody molecules that dissolve in the blood, where they circulate throughout the body and bind to the invading pathogenic agents. Once a virus or bacterial cell is covered with these antibody molecules, the pathogen may be inactivated or may become more susceptible to ingestion by a patrolling phagocyte. These soluble, blood-borne antibody molecules are proteins that are secreted by **plasma cells.** Plasma cells are formed from B lymphocytes present within the bone marrow by the process illustrated in Figure 30-4. As we will see below, the B cell recognizes the foreign agent (which acts as an antigen) and responds by transforming into a plasma cell that secretes antibody molecules that can combine with that agent.

Some bacteria (such as those responsible for typhus or meningitis) and all viruses (such as those responsible for colds, measles, or mumps) enter a host cell, reproduce inside the cell, then kill the cell, releasing more bacteria or viruses into the extracellular fluid. It is when these pathogens are *outside* the host cells that they are vulnerable to direct attack by soluble antibodies that are distributed in the body's fluids: tears, milk, nasal and intestinal mucus, interstitial fluid, and blood. In contrast, since soluble antibodies cannot penetrate infected cells, they have little effect on a pathogen while it is inside a host cell. A pathogen growing *inside* the body's cells is eliminated by the other branch of the immune system, T-cell immunity.

T-Cell Immunity: Protection by Intact Cells

While B cells carry out their immune function by secreting antibodies, T cells function by interacting directly with

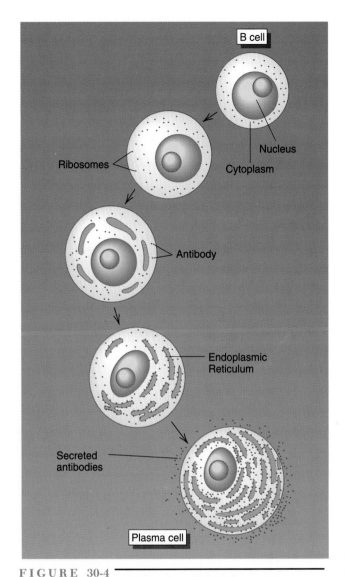

FIGURE 30-4

Differentiation of antibody-secreting plasma cells from B cells. The differentiated plasma cell contains extensive rough ER where antibody proteins are synthesized.

other cells. In other words, T cells function in **cell-mediated immunity.** T cells are able to recognize infected or abnormal target cells by virtue of the antibody-like proteins the T cells carry embedded in their plasma membranes; these membrane proteins are called *T cell receptors.*

There are three functionally distinct subclasses of T cells:

1. **Cytotoxic (killer) T cells** (Figure 30-5) *recognize aged, malignant, or infected cells and release perforins* —proteins that produce holes in the membrane of the target cell, causing its death. By killing infected cells, cytotoxic T cells can eliminate viruses, and intracellular bacteria, yeast, protozoa, and parasites, even after they enter a host cell. Cytotoxic T cells may also play a role in

◁ B I O L I N E ▷
Treatment of Cancer with Immunotherapy

Cancer is currently being treated by three approaches: surgery, radiation, and chemotherapy. Surgery physically removes as much of the malignancy as possible; sometimes all of it. Radiation kills cancer cells in the specific part of the body that contains the malignancy. Chemotherapy poisons cancer cells wherever they might have spread throughout the body. Some forms of cancer, such as childhood leukemias and Hodgkin's disease, have responded well to these treatments, while others, such as lung cancer and pancreatic cancer, have not.

Both radiation and chemotherapy work primarily by killing cells as they divide. Because of the higher rate of division of cancer cells compared to normal cells, cancer cells are more susceptible to the killing action of radiation and chemotherapy. But many normal cells are also killed, and side effects such as nausea and anemia can be debilitating. For several decades, researchers have looked for ways to manipulate the immune system to help in the fight against cancer. Cancer cells often exhibit new proteins at their cell surfaces, which can be recognized by antibodies as foreign antigens. A number of laboratories have been trying to take advantage of the presence of these so-called *tumor-specific antigens* to produce antibodies that can be injected into patients and bring about the death of the tumor cells. While the prospects of antibody therapies have proven disappointing, a new approach has emerged that may prove more promising. This new approach has been pioneered by Stephen Rosenberg of the National Institutes of Health.

Rosenberg's interest in the subject began in 1968 when, as a young surgeon, he removed the gallbladder from a 63-year-old man. It was a routine operation, except for the fact that 12 years earlier this same patient had been diagnosed with ad-

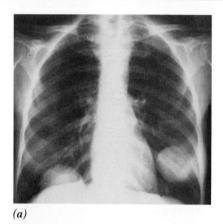
(a)

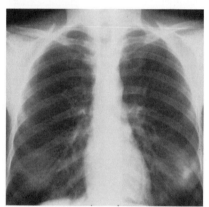
(b)

FIGURE 1 ─────────

One promise of success against cancer utilizes tumor-infiltrating leukocytes (TILs). *(a)* X-ray showing a melanoma that has spread to both lungs. *(b)* Two months after treatment with the patient's own TILs, the tumor masses are greatly reduced in size.

vanced stomach cancer and sent home to die. Within a few months, the cancer had spontaneously disappeared, as had the cancer cells that had spread to other parts of the man's body. Such rare cases of spon-

taneous remission convinced Rosenberg that the immune system has the *potential* to rid the body of its malignant cells.

In the late 1980s, Rosenberg and his colleagues found that human tumors contained cytotoxic T cells that could specifically attack the tumor cells. Under normal circumstances, these T cells are present in too small a quantity and arrive too late to stem the tide of the growing malignancy. Rosenberg devised a plan. What if he were to remove the tumor, isolate the cytotoxic T cells contained in the tumor mass, culture these cells to increase their number, and then reinject them into the same patient? Would these T cells (called **tumor-infiltrating lymphocytes,** or **TILs**) become concentrated in the tumor masses from which they were derived? Would the injected TILs kill the tumor cells? Both of these questions have been answered with an encouraging "yes." When the injected T cells are specially labeled for tracking throughout the body, they are found to accumulate within the tumor. In the first clinical trials on patients with advanced cancer, over half of the treated patients experienced a marked reduction in their tumors (Figure 1); several appeared to have remained in partial or complete remission for years. Moreover, the TILs had no adverse effects on normal tissues, a unique benefit over conventional treatments.

Rosenberg has recently received permission to modify TILs in a way that gives the cells extra killing power. This is done by genetically engineering the cells so that they carry a gene for a protein that is highly toxic to tumor cells. It is hoped that the TILs will be selectively drawn to the tumor and, by delivering the toxic protein, will kill additional tumor cells. If the technique proves effective, within just a few years we may be able to use immunotherapy as a fourth weapon in the war against cancer.

FIGURE 30-5

This large cancer cell is being attacked by numerous, smaller killer T cells. When specific contact is made between the two cells, the T cell releases a protein that kills the target cancer cell by poking holes in its plasma membrane.

inside the sternum. The thymus is quite large in mammals during the period from late fetal life through puberty, after which it gradually shrinks in size and importance. The process by which a T cell matures and produces a particular version of the T-cell receptor occurs within the thymus gland. If the thymus gland fails to develop, T-cell maturation is blocked, and all the various cell-mediated defensive mechanisms fail to appear.

The role of T cells first came to light during studies in the 1960s on mice whose thymus glands had been removed soon after birth. Such mice showed a variety of immunological deficiencies, including the inability to reject tissue and organ grafts from other individuals, which would normally occur. This finding provided one of the first indications that T cells are capable of killing healthy *foreign* cells, which is the basis for the rejection of transplanted human organs, which we will discuss shortly.

More recently, a strain of mice has been developed with a genetic deficiency that leads to the congenital absence of a thymus. These *nude mice* (Figure 30-6), as they are called, are currently playing an important role in cancer research because they accept grafts of malignant human tissue that other mice would reject. Researchers are able to test experimental treatments to stop tumor growth on these mice, which could not be used on human patients. Even more remarkably, when nude mice are given lymphoid tissue that is taken from a human fetus, they actually develop a functioning "human" immune system, complete with both T and B lymphocytes. These "human mice" are currently

destroying cancer cells (see Bioline: Treatment of Cancer with Immunotherapy).

2. **Helper T cells** are regulatory cells, not killers. They regulate immune responses by recognizing and activating other lymphocytes, including both B cells and cytotoxic T cells. Activation is achieved as the helper T cell releases substances that activate the target cell. The best-studied activator substance is *interleukin II,* which is currently being produced by recombinant DNA technology and tested as a treatment for certain types of cancer. Helper T cells are the cells that are hardest hit by AIDS. Ironically, HIV, the virus responsible for AIDS, gains entry to the T cell by binding to the T-cell receptor, the membrane protein normally responsible for detecting the presence of virally infected cells.

3. **Suppressor T cells** also regulate immune responses, but they do so by *inhibiting* the activation of other lymphocytes by a mechanism that remains unknown.

T cells are named for the **thymus gland**—a mass of lymphoid tissue that is situated in the chest cavity, just

FIGURE 30-6

A nude mouse. These homely mice are homozygous for a mutation that leads to the absence of a thymus gland. Since the thymus gland is the site where T cells undergo the process that provides them with a specific T-cell receptor, these animals are lacking immunologically competent T cells; thus, they are unable to reject foreign grafts.

being used as an animal model for AIDS research, since their donated immune systems are susceptible to HIV.

Organ Transplants and Graft Rejection The greatest hurdle that medical research has had to overcome in the transplantation of organs from one person to another is *graft rejection*, which occurs when cells of the transplanted organ are recognized as foreign and are subsequently attacked by the recipient's cytotoxic T cells. Currently, the likelihood of graft rejection is minimized by two factors.

- The first strategy involves matching the donor to the recipient; that is, finding a donor whose cell-surface proteins (called *histocompatibility antigens*) are as closely matched as possible to the recipient. The closer the match, the less likely the rejection. The risk is always present, however, since no two people, other than identical twins, have identical histocompatibility antigens.

- The second strategy requires treating the recipient with drugs (such as the fungal compound cyclosporine) that suppress the person's cell-mediated immunity, thereby reducing the capacity for graft rejection. Even though these drugs are now quite effective, the suppression of the immune system leaves the recipient more vulnerable to infection.

The immune system consists of scattered organs that produce lymphocytes—cells that recognize foreign materials (antigens) and mount a specific response. The immune response may be mediated by soluble antibodies that bind to antigens or by activated T cells that bind to infected cells or to cells with altered surfaces. (See CTQ #3.)

ANTIBODY MOLECULES: STRUCTURE AND FORMATION

B cells are stimulated to produce antibodies by the presence of antigens; that is, by substances that are recognized as foreign. Most antigens are proteins or polysaccharides, such as those present on the surfaces of viruses, bacteria, and parasites. Immune responses are always specific; that is, antigens stimulate production of only those antibodies that can specifically combine with that particular antigen. If you are exposed to the measles virus, for example, the presence of that virus elicits the production of antibodies that will combine only with the measles virus protein (or one with a very similar structure). Therefore, immunity to measles provides no protection against polio, colds, influenza, or any infection other than measles.

You possess the immune capacity to respond to nearly every foreign intruder you will encounter during your lifetime. It is estimated that humans can produce millions of different types of antibodies, which are capable of combin-

ing with virtually any molecule of virtually any shape. To appreciate how antibodies possess such specificity and why there are so many different types of antibodies, it is necessary to examine their structure and formation.

ANTIBODY SPECIFICITY: A UNIQUE COMBINATION OF POLYPEPTIDES

Remarkably, all of the millions of different antibodies your body can produce, with their ability to combine with diverse types of antigens, are constructed according to a similar blueprint, using very similar building materials. All antibody molecules (also called *immunoglobulins*) are composed of two basic types of polypeptide chains, called **heavy chains** and **light chains,** based on their relative sizes. The most common class of immunoglobulins, the IgG (immunoglobulin G) class, is composed of two light chains and two heavy chains, linked together by covalent bonds (Figure 30-7). Early studies of a number of different IgG molecules revealed a consistent pattern: Approximately one half of each light chain was *constant* (C) in amino acid sequence among various antibodies, while the other half was *variable* (V) from antibody to antibody. The heavy chains of these antibodies also contained a variable (V) and a constant (C) portion.

Once the basic structure of a number of IgG molecules was compared, the mechanism of antibody function became evident. Each IgG molecule has two antigen-binding sites located at the ends of a Y-shaped molecule. Each antigen-binding site is formed by the association of the variable portion of a light chain with the variable portion of a heavy chain (Figure 30-7). The amino acids that make up the antigen-binding sites give that portion of the protein a three-dimensional structure that is complementary to that of the corresponding antigen (Figure 30-7). Different antibodies contain different light and heavy chains and thus have different shapes and specificities. Since antibody molecules are made up of two different types of polypeptides, an individual is able to produce a tremendous variety of antibodies from a relatively modest number of different heavy and light chains. If, for example, there are 1,000 different light chains and 1,000 different heavy chains, 1 million (1,000 × 1,000) different antibodies—each with a distinct antigen binding site—can be formed.

ANTIBODY FORMATION: DISCOVERY OF A UNIQUE GENETIC MECHANISM

The discovery that antibodies are composed of constant and variable portions raised several basic questions. We can each produce millions of different IgG molecules, all of which have light (or heavy) chains with identical C portions but highly diverse V portions. But just how can different polypeptides have an identical amino acid sequence in one part, and diverse amino acid sequences in another part?

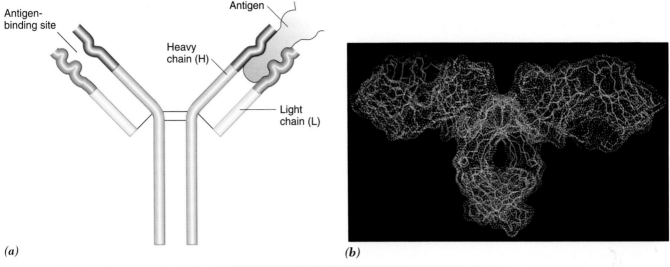

FIGURE 30-7

Structure of an IgG immunoglobulin molecule and its interaction with antigen. *(a)* Each IgG antibody molecule consists of two light chains and two heavy chains, each of which has a constant and a variable region. The chains are held together by covalent (—S—S—) bonds. The antigen-combining sites are present at the forked ends of the Y-shaped molecule. Each combining site is made up of the variable (green) portion of both a heavy and a light chain. In this drawing, one site has combined with a complementary-shaped antigen; the other is unbound. *(b)* A computerized, three-dimensional drawing of an IgG molecule.

In 1965, William Dreyer of the California Institute of Technology, and J. C. Bennett of the University of Alabama put forward the "two gene–one polypeptide" hypothesis to account for antibody structure. In essence, Dreyer and Bennett proposed that each antibody chain is coded by two separate genes—a C gene and a V gene—that somehow combine to form one continuous "gene" that codes for a single light or heavy chain. But there was no evidence that such DNA rearrangement was even *possible,* much less that it occurred.

Over a decade later, in 1976, at a Swiss research institute, Susumu Tonegawa provided clear evidence in favor of the DNA rearrangement hypothesis. Using newly developed techniques, Tonegawa and his colleagues measured the distance in the DNA between the nucleotide sequences coding for the C and V portions of a particular antibody chain. They compared this distance in the DNA isolated from two different types of mouse cells, those of embryos and those of antibody-secreting plasma cells. The group found that, while these two DNA segments were widely separated from each other in DNA obtained from embryos, the segments were very close to each other in the DNA obtained from the plasma cells. Tonegawa had shown that parts of the DNA actually rearranged themselves during the formation of antibody-producing cells, earning him a Nobel Prize in 1987.

Later research revealed that the DNA of a particular chromosome contained a single C gene and a large number of different V genes. During the rearrangement process, which occurred while the B cell was maturing in the bone marrow, that single C gene was moved very close to one of the V genes, enabling a single messenger RNA (mRNA) to form from these combined genes (Figure 30-8). The mRNA was then translated into a single polypeptide, containing both a C and a V amino acid sequence. This mechanism of DNA rearrangement is unique to antibody-forming cells and represents an adaptation by which an individual can produce a large number of different antibodies with a minimal amount of genetic information.

The Clonal Selection Theory of Antibody Formation

When you receive a vaccination that contains an inactivated microorganism or you contract a bacterial or viral infection, your body reacts to the presence of the foreign agent by producing a restricted population of antibody molecules, all of which are able to combine with this particular foreign antigen. How does the body know which antibodies to manufacture in response to the presence of a particular antigen?

In 1957, a far-sighted, Nobel Prize-winning proposal was made by an Australian immunologist, Macfarlane Burnet, to answer this question. Burnett's *clonal selection*

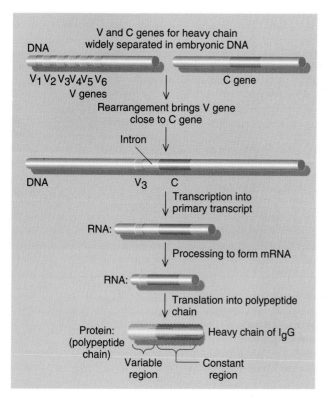

FIGURE 30-8

Antibody genes are formed by DNA rearrangement, which brings the C gene into close proximity with one of the many V genes (in this case, the V_3 DNA segment encoding part of a heavy chain) on the chromosome. The region between the C and V genes after rearrangement represents an intervening sequence (intron). Once DNA rearrangement has occurred, the combined C and V regions are transcribed into one mRNA molecule that codes for a single, combined polypeptide.

theory has gained virtually complete acceptance and remains compatible with all the experimental evidence accumulated since its original proposal. The principles of the clonal selection theory in its present, updated form bring together many different aspects of immune function and are summarized in Figure 30-9. According to this theory, the following events take place during antibody formation.

1. *Each B cell becomes committed to producing one particular antibody.* During embryonic development, B cells are formed from a population of undifferentiated, indistinguishable stem cells. As one of these stem cells becomes a B cell, however, DNA rearrangements occur (see Figure 30-8) which commit the cell to the production of only one particular antibody molecule. Even though a population of B cells remains identical under the microscope, the cells are distinguished by the antibodies they can produce.

2. *B cells become committed to antibody formation in the absence of antigen.* The entire diversity of antibody-producing cells an individual will ever possess is already present within the lymphoid tissues *prior to stimulation by an antigen* (Figure 30-9a) and is independent of the presence of foreign materials. Each B cell "displays" its particular antibody on its surface, with the antigen-reactive portion facing outward; the cell is literally coated with receptors that can fit with one and only one antigen (Figure 30-9a). While most of the cells of the lymphoid tissues will never be called on

to respond during a person's lifetime, the immune system is primed to respond immediately to virtually any type of antigen to which an individual might be exposed.

3. *Antibody production follows selection of B cells by antigen.* Antigens that enter the body trigger the production of complementary antibodies by *selecting* the appropriate antibody-producing cell. When a foreign object, such as a pneumococcus bacterium or the particulate debris from tobacco smoke, enters the body, it is ingested by a phagocytic macrophage. The foreign material is only partially digested, however, generating fragments that make their way to the surface of the macrophage (Figure 30-9b), where they are "presented" to a B cell (Figure 30-9c). This process is called **antigen presentation.** B cells with the appropriate "sample" antibodies at their surface will bind to the antigen on the surface of the antigen-presenting macrophage (Figure 30-10). The binding between the two cells activates the B cell to divide (Figure 30-9d), forming a population, or *clone,* of lymphocytes, all of which are specialized for making the same antibody. Some of these activated cells will then differentiate into plasma cells and begin secreting tens of thousands of antibody molecules per minute.

4. *Immunological memory provides long-term immunity.* Not all of the B lymphocytes that are acti-

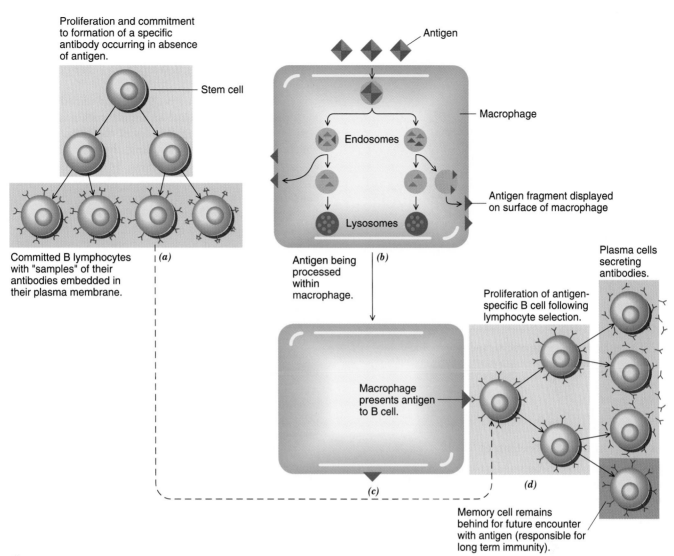

Proliferation and commitment to formation of a specific antibody occurring in absence of antigen.

Stem cell

Committed B lymphocytes with "samples" of their antibodies embedded in their plasma membrane.

(a)

Antigen

Macrophage

Endosomes

Antigen fragment displayed on surface of macrophage

Lysosomes

(b)

Antigen being processed within macrophage.

Plasma cells secreting antibodies.

Proliferation of antigen-specific B cell following lymphocyte selection.

Macrophage presents antigen to B cell.

(c)

(d)

Memory cell remains behind for future encounter with antigen (responsible for long term immunity).

FIGURE 30-9

A modern version of the clonal selection theory. *(a)* Undifferentiated stem cells undergo proliferation, forming a population of B cells that become committed to the formation of a specific antibody. Commitment occurs as a result of DNA rearrangement. Once committed, the B cell carries "sample" antibodies embedded in its plasma membrane. This phase of B-cell development is antigen independent. *(b)* Foreign materials are ingested by wandering macrophages and are packaged into cytoplasmic vesicles (endosomes). Some of the foreign material is completely digested by lysosomal enzymes (the red triangles), while other pieces remain undigested and are moved to the cell surface (the blue triangles), where they can be presented to lymphocytes. *(c)* The activation of a lymphocyte, whose sample antibodies have combining sites that can bind to the processed antigen on the macrophage surface. B cell activation usually requires the participation of "helper" T cells (not shown), which release the soluble factors that stimulate growth and differentiation into antibody producing cells. *(d)* Proliferation of those lymphocytes that have been selected by the antigen form a clone of cells committed to produce antibodies capable of binding to the antigen. Most of these cells go on to differentiate into antibody-secreting plasma cells, but a percentage remain as committed memory cells, which are capable of responding very rapidly should the antigen be reintroduced at a later date.

vated by an antigen differentiate into antibody-secreting plasma cells. Some remain in the lymphoid tissues as **memory cells** (Figure 30-9d), which can respond rapidly at a later date if the antigen reappears in the body. Although plasma cells die off following removal of the antigenic stimulus, memory cells persist, often for years, providing *immunological recall.* When stimulated by the same antigen, some of the memory cells rapidly proliferate into plasma cells, generating a secondary immune response in a matter of hours rather than the days required for the original response. Thanks to this secondary response, measles, mumps, chickenpox, and many other diseases are suffered only once during a lifetime. The first encounter with the intruder leaves survivors immune to subsequent infection. This type of protection is called **active immunity.**

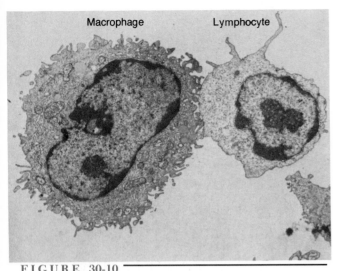

FIGURE 30-10
Interaction between a lymphocyte and macrophage during antigen presentation.

Recognition and Protection of "Self"

Maintaining homeostasis and biological order requires that the immune system be able to distinguish between those substances that have entered the body from the external world and those that "belong" in the body. That is, the system must recognize "self" and refrain from producing antibodies against it, while also recognizing foreign substances ("nonself") and launching an immunological attack against them.

The process by which the immune system becomes *tolerant* of its own tissues is not well understood. During the period of B- or T-cell maturation, if a cell is produced that has antibodies or T-cell receptors capable of reacting with the body's own tissues, that cell is somehow either killed or converted to a nonresponsive state. This process, known as *clonal deletion*, ensures that B cells capable of producing potentially dangerous **autoantibodies** — antibodies against one's own tissues — are rendered incapacitated (see The Human Perspective: Disorders of the Human Immune System). Similarly, formation of cytotoxic T cells capable of interacting with and killing normal, healthy cells is also blocked.

Monoclonal Antibodies: Making Antibodies Available for Use

Antibodies can be very useful proteins. Since they interact so specifically, antibodies can be used to identify or locate particular molecules. For example, the development of prostate cancer in men is associated with the elevation in the blood of a particular protein, called *prostate specific antigen* (PSA), that is shed by prostatic cells. A diagnostic test for prostate cancer measures the level of PSA in the blood by its interaction with a specific antibody. Similar tests are currently being developed for other cancers, including breast cancer.

In earlier decades, antibodies were obtained by injecting a purified substance into an animal, such as a rabbit or goat, and then obtaining samples of the animal's blood, which contained antibodies against the injected, foreign antigen. This approach has severe limitations, however. Today, antibodies are being produced by special cultured cells, called *hybridomas*. Hybridomas are formed by the fusion of two cells: a B cell that is capable of producing a particular antibody, and a malignant cell. Unlike B cells, which have a limited lifespan, hybridomas divide indefinitely in culture, producing an unlimited supply of antibody-producing cells. All of the cells formed from a single fused B cell synthesize and secrete the same antibody molecule into the medium. Different antibodies are obtained by starting with different B cells. Since, in each case, antibody molecules are produced by a clone of identical cells derived from a single fused cell, they are called **monoclonal antibodies**.

The production of monoclonal antibodies has become a billion-dollar biotechnology business, providing antibodies used in a variety of procedures, including cancer screening, pregnancy tests, and basic research. For example, the colorful cells shown in Figure 30-11 were obtained by treating the cells with monoclonal antibodies against specific cytoplasmic proteins. The location of the antibodies — and thus of the cytoplasmic proteins — is made visible by linking the antibody molecules to colored dye molecules.

Antibodies contain sites that are capable of binding molecules that have a complementary shape. The diversity of these binding sites among different antibodies derives from the great variety of polypeptide chains from which antibodies can be constructed. Much of this polypeptide diversity is the result of genetic rearrangements that occur during lymphocyte maturation. Remarkably, an individual possesses the ability to recognize and inactivate millions of different antigens, most of which will never be encountered during his or her lifetime (See CTQ #4.).

IMMUNIZATION

You are probably immune to many diseases from which you have never had to suffer an immunizing infection. Instead, you received a vaccine that "tricked" the immune system into responding and producing immunological memory. Jenner first demonstrated that this was possible when he intentionally stimulated active immunity to smallpox by vaccination with a related cowpox virus. Vaccines have since become one of our most potent weapons against disease. Most vaccines are modified forms of disease-causing agents; these variants have lost the ability to cause the disease but retain some of the same antigens that their danger-

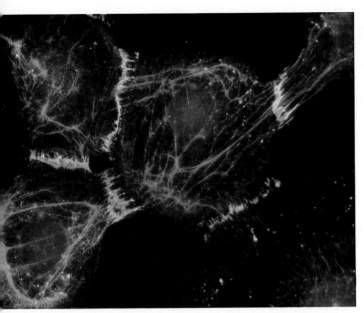

FIGURE 30-11
The use of monoclonal antiodies to locate specific proteins within cells. The antibody molecules are linked to colored dyes to reveal their location.

ous counterparts possess. An immune system exposed to a vaccine, and often several "booster" doses, builds memory cells against a pathogenic agent without the danger of developing the disease. In this fashion, we can now safely promote active immunity against polio, measles, mumps, diphtheria, tetanus, rabies, and many other dangerous diseases.

Sometimes there is no time to wait for an immune system to become sensitized. An attack of diphtheria or tetanus, or the bite of a poisonous snake requires immediate immunity to neutralize the life-threatening antigens. In these cases, antibodies obtained from the blood of an immune individual (or animal) can be injected into the person who needs immediate protection, producing a temporary state of **passive immunity.** For example, the blood of William Haast who, as director of the Miami Serpentarium, had been bitten by a variety of poisonous snakes, has been collected to provide passive immunity to individuals bitten by exotic vipers.

In mammals, passive immunity can be acquired naturally by the transfer of antibodies across the placenta from a mother to her developing fetus. Such passive protection is augmented after birth by the presence of antibodies in breast milk, providing breast-fed babies with more disease resistance—particularly to gastrointestinal infections—than bottle-fed babies. A decline in breast-feeding often results in an increased infant mortality.

Normally, a period of days is required before a person can mount a specific immune response against a toxin or disease-causing microorganism that has entered the body for the first time. A person who has been actively or passively immunized against the foreign intruder can eliminate a foreign agent, before disease develops. (See CTQ #5.)

EVOLUTION OF THE IMMUNE SYSTEM

Dissecting an animal reveals its nerves, vascular channels, digestive tract, and most other systems. It is more difficult to determine whether that same animal has an immune system. The existence of an immune system is determined by the capacity of an animal to respond to foreign materials rather than by the presence of a particular type of organ.
➠ Most invertebrates respond to the injection of foreign material by sending out phagocytic cells in a manner similar to the inflammatory reaction seen in vertebrates. Many invertebrates also appear capable of synthesizing soluble substances that can kill pathogenic organisms, but these substances lack the high degree of specificity that is characteristic of vertebrate antibodies. Most studies indicate that invertebrates lack the ability to reject foreign grafts, suggesting that these animals lack cell-mediated immune systems.

Highly specific, cell-mediated immune mechanisms probably first appeared in vertebrates, as did the ability to produce antibodies. All known species of living vertebrates, from the most primitive, jawless fishes to mammals, have (1) lymphoid tissue that produces lymphocytes, (2) circulating antibodies, and (3) the ability to reject foreign grafts. There is a progression in the complexity of the immune system through the vertebrate classes, however, culminating in the systems found in birds and mammals.

The complexity and effectiveness of immune responses have increased markedly over the course of animal evolution. (See CTQ #6.)

◁ THE HUMAN PERSPECTIVE ▷
Disorders of the Human Immune System

A number of conditions can be traced to a malfunction of the immune system, including allergies, autoimmune diseases, and immunodeficiency disorders.

ALLERGIES

An allergic reaction is triggered when the immune system reacts in an inappropriate manner to a foreign substance, usually one that has been eaten, inhaled, or acquired during injection by an insect sting or a hypodermic syringe. The most common allergic reactions occur in response to exposure to pollen, dust, mold, or certain food products. The allergic individual produces a special class of antibodies (IgE antibodies) against one or more of these *allergens* (Figure 1). The interaction between allergen and antibodies activates various cells to release highly active substances, including prostaglandins (page 532) and *histamine,* which evoke the symptoms of allergy. The

diameter and permeability of blood vessels increase, leading to inflammation; secretion of mucus increases, leading to nasal congestion; and the smooth muscles in the wall of the respiratory airways contract, leading to difficulty in breathing. Allergies are commonly treated with antihistamines, which block the release of histamine. If airways become severely constricted, as can occur during an asthma attack, the quickest relief is obtained by inhaling a mist of epinephrine, which causes a dilation of the airways. Extended periods of asthma may require treatment with antiinflammatory steroids. People with persistent allergies may be treated by a desensitization program, in which increasing amounts of the allergen are administered by injection over a period of time.

The most severe type of allergic reaction is **anaphylaxis;** it occurs if the substance triggering the reaction is introduced directly into the bloodstream, such as fol-

lowing a bee sting or a shot of penicillin. Under these conditions, allergy-producing substances, such as histamine, are released throughout the body, and the severe reaction, known as *anaphylactic shock,* can be life-threatening.

Not all allergic reactions are triggered by soluble antibodies. The skin rash produced by exposure to poison oak or poison ivy occurs following activation of T cells. In these cases, the allergen is an oil from the plant, which is absorbed into the skin and binds to cells. T cells then release substances that cause inflammation at the site of allergen attachment.

AUTOIMMUNE DISEASES

Under normal conditions, the immune system does not mount an immune response against its own tissues. Occasionally, however, the discriminatory ability of the system breaks down, and the lymphoid

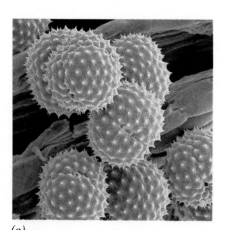

(a)

(b)

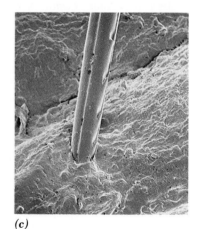

(c)

FIGURE 1

Common allergens. Scanning electron micrographs of selected materials to which many people are allergic: ragweed pollen *(a)* and the house dust mite *(b),* a minute arthropod that lives in dust. *(c)* The sting of a honeybee is, at the worst, painful for most people. People who are allergic to bee venom, however, suffer allergic responses that can be fatal within 10 minutes of the sting.

tissues attack normal components of the body. This phenomenon is responsible for some very serious **autoimmune diseases,** including thyroiditis, rheumatic fever, and systemic lupus erythematosus (often called "lupus"). In all autoimmune diseases, the victim is literally rejecting his or her own tissues.

In the case of thyroiditis, the body produces antibodies that react with tissues of the thyroid gland. The disease probably occurs after some of the protein that is normally stored in the thyroid gland leaks into the bloodstream, stimulating the immune system to produce antibodies against the protein. Rheumatic fever may occur following the body's recovery from infections by the streptococcal bacteria that cause strep throat. These bacteria possess cell-surface molecules that coincidentally are similar in structure to those found on some people's heart valves. In the case of rheumatic fever, antibodies produced against the bacteria *cross react* with the heart tissue after the microbe has been eliminated, causing heart-valve damage. Systemic lupus erythematosus is a debilitating disease in which many different tissues are attacked by antibodies that react with various molecules, including DNA. The reason for the production of these antibodies is unknown. The disease may result in death, usually as a result of kidney damage.

IMMUNODEFICIENCY DISORDERS

Until recently, diseases that left a person with a seriously deficient immune system were very rare and were usually a result of an inherited disorder. AIDS has changed that pattern dramatically, and today immunodeficiency is a growing cause of death. The biology of AIDS is described in Chapter 36. We will restrict the present discussion to the current attempts to halt further spread of the disease and to the treatment of those people already infected.

The offensive against AIDS consists of a three-pronged attack. Education is the first, and currently most effective, weapon. People must be taught how to minimize the risk of contracting AIDS and how to

keep from spreading the disease if infected. The virus (Figure 2) is known to be spread by three avenues: sex, blood, and passive transmission from mother to fetus. The spread of AIDS would be greatly reduced if sexually active individuals followed safe sex practices (such as using a condom) and intravenous drug abusers stopped sharing needles.

The second weapon in the fight against AIDS is research to find a treatment that will kill or control the virus with minimum toxicity to the infected person. Presently, AZT (3'-azido-3'-deoxythymidine) is the primary drug that is widely prescribed for use against AIDS, but other drugs have been approved or are under development. AZT inhibits replication of the AIDS virus (HIV) by competing with thymidine (one of the four nucleotide building blocks of DNA) for the active site of the viral DNA-synthesizing enzyme. AZT has prolonged the survival of patients with AIDS and has delayed the onset of the

FIGURE 2 —————
Human Immunodeficiency Virus (HIV) particles coating the surface of a white blood cell.

disease in infected people who are asymptomatic. Recent research suggests that treatment of patients with a combination of antiviral drugs may prove more effective than do single drug regimes. Effective drugs have also been developed against most of the secondary infectious diseases that strike AIDS patients, including pneumocystis pneumonia (caused by the protist *Pneumocystis*), toxoplasmosis (caused by the protist *Toxoplasma*), and oral yeast infections (caused by the fungus *Candida*). Controlling these secondary infections allows AIDS patients to live much longer with less suffering. Still, finding a cure for AIDS remains a distant goal, despite the enormous efforts of researchers around the world.

The third approach is the development of a vaccine to immunize people against the virus, creating an AIDS-resistant population. Most vaccines against viral diseases, such as polio, measles, and mumps, consist of attenuated viral particles—viruses that are still active and infectious but are no longer capable of injuring the body—to which the body mounts an immune response. But these viruses do not integrate their DNA into human chromosomes as does HIV. The consensus is that it would be too dangerous to use whole HIV as part of a vaccine, no matter how inactivated or altered it might be, since the virus has the potential to become reactivated as a result of interaction with a person's own DNA. Instead, it is hoped that a vaccine can be produced using parts of proteins from the virus's outer coat, rather than the virus itself. A similar approach has been used to develop a vaccine against hepatitis B, and tests for AIDS vaccines using this approach are ongoing. To date, the results of these tests have not been encouraging since the antibody response has been neither effective nor sustained. Instead, researchers have recently turned to techniques whereby genes that code for proteins of HIV are incorporated into other, safer viruses. It is hoped that infection with a "combination virus" may cause mild infections that stimulate immunity to AIDS.

REEXAMINING THE THEMES

Relationship between Form and Function

The hallmark of immune function is specific molecular interaction. The functioning of both soluble antibodies and T cells—the two weapons of the immune system—depends on their ability to interact specifically with a target. This specificity depends on a complementary shape existing between the target and the combining site of the antibody molecule or the T-cell receptor.

Biological Order, Regulation, and Homeostasis

The maintenance of internal order requires that foreign materials, whether they are disease-bearing pathogens or simple debris, must be eliminated from the body. This is the job of the immune system: to keep the "nonself" out of the "self." This is accomplished by first recognizing the "nonself" antigens and then inactivating or destroying them. Foreign agents can usually be eliminated, even if they are hiding within the body's own cells. The importance of the immune system in maintaining homeostasis is readily revealed by the effects of diseases that impair the system, such as AIDS.

Unity Within Diversity

All T cells are basically alike, as are all B cells and all IgG molecules, yet each group is incredibly diverse. Different B cells contain different membrane-bound antibodies; different T cells contain different T-cell receptors; and different IgG molecules contain different antigen-binding sites. Such diversity allows the body to react to virtually any type of foreign agent and stems from subtle differences in the amino acid sequences of a handful of different categories of polypeptide chains.

Evolution and Adaptation

A sophisticated immune system is a relatively recent evolutionary arrival. While invertebrates possess the ability to resist infectious agents, their defenses lack the high degree of specificity that is found in the human system. Immune systems that are capable of producing specific antibodies and cellular responses are found throughout the vertebrates. Even among vertebrates, however, there is a marked progression in complexity of the immune system, reaching its peak in birds and mammals.

SYNOPSIS

Homeostatic mechanisms that afford protection against invading pathogens are divided into nonspecific and specific defenses. Nonspecific mechanisms include external body surfaces that prevent entry to pathogens; cells that either engulf or poison pathogens that have breached the body's surface; and molecules, such as complement and interferon, that attach to the surface of pathogens or infected cells.

The immune system mediates specific mechanisms of defense, whereby foreign substances are specifically recognized and destroyed. The immune system is composed of cells that are scattered throughout the body and concentrated in lymphoid tissues, such as the lymph nodes and bone marrow. Immune responses may be mediated by either (1) soluble antibody molecules secreted by plasma cells that differentiate from B cells, or (2) cell-mediated immunity mediated by T cells. Antibodies attach to specific targets that are present outside of host cells, leading to their destruction, while T cells interact with other cells. Immune

responses are stimulated by foreign substances called antigens.

Antibodies contain binding sites that combine specifically with the antigen responsible for production of that antibody. Antibodies are constructed of both light and heavy polypeptide chains. Part of the polypeptide is constant from chain to chain, while the remainder is highly variable. Both heavy and light antibody chains are encoded by genes that form as a result of DNA rearrangement. Antigen-binding sites are formed by the association of the variable portions of a heavy and light chain.

Antibodies are formed by clonal selection. Lymphoid tissues contain a population of B cells, each of which is committed to forming a particular antibody. Samples of these antibodies are present in the B-cell plasma membrane. When an antigen binds to a cell's membrane-bound antibody, it stimulates the lymphocyte to proliferate, forming a clone of cells that is capable of producing that antibody. Some of the cells differentiate into plasma cells that secrete the antibody; others become memory cells that can mount a rapid response if the antigen reappears at a later time. This process of antigen-mediated lymphocyte stimulation is mimicked by vaccination. In this case, an antigen, such as an inactivated virus, is injected into the body, where it stimulates the proliferation of lymphocytes that are able to produce antibodies against the injected material.

T cells differentiate in the thymus gland by producing a particular membrane-bound T-cell receptor, which mediates the interaction of the T cell with another cell. Cytotoxic T cells recognize infected or altered host cells and kill them. Helper and suppressor T lymphocytes have a regulatory function; they either specifically activate or inhibit other lymphocytes.

The immune system is able to distinguish self from nonself. The body must be able to suppress the formation of T cells and antibodies capable of interacting with the body's own cells and tissue components. The body becomes tolerant of its own tissues by a process whereby cells capable of reacting with "self" are either suppressed or destroyed. When this process breaks down, the body may produce autoantibodies against itself, causing serious disease.

Key Terms

immune system (p. 644)
natural killer cell (p. 645)
inflammation (p. 645)
complement (p. 645)
antibody (p. 646)
antigen (p. 646)
B cell (p. 647)
T cell (p. 647)

plasma cell (p. 647)
cell-mediated immunity (p. 647)
cytotoxic (killer) T cell (p. 647)
helper T cell (p. 649)
suppressor T cell (p. 649)
tumor-infiltrating lymphocyte (TIL) (p. 648)
thymus gland (p. 649)
heavy chain (p. 650)
light chain (p. 650)

antigen presentation (p. 652)
memory cell (p. 653)
active immunity (p. 653)
autoantibody (p. 654)
monoclonal antibody (p. 654)
anaphylaxis (p. 656)
autoimmune disease (p. 657)
passive immunity (p. 655)

Review Questions

1. Distinguish between B cells, plasma cells, and memory cells; complement and interferon; normal mice and nude mice; C genes and V genes; light chains and heavy chains; active and passive immunity; T-cell receptors and histocompatibility antigens.

2. Describe the elements of the clonal selection theory.

3. Describe the steps that occur between the time a cold virus penetrates the nasal epithelium and the time it is eliminated by the immune system.

4. Why does infection with cowpox protect a person from smallpox?

5. How is it that you have only one C gene per haploid set of chromosomes for a light antibody chain, yet you are able to synthesize hundreds of thousands of different antibody molecules?

6. What is the role of the thymus gland? How does this role change following receipt of an organ by transplantation?

Critical Thinking Questions

1. When AIDS first appeared, some clinicians speculated that the condition was due to the male homosexual practice of inhaling amyl nitrate, a stimulant of the heart. What evidence might you seek either to support or to refute this suggestion about the cause of AIDS?

2. Why are the nonspecific defenses of the body insufficient? How is this characteristic illustrated by disorders that affect the immune system, such as AIDS or thyroiditis?

3. How does the immune system illustrate each of the following characteristics, and how does each contribute to the successful defense of the body against disease: decentralization, division of labor, specificity?

4. One might argue that it is wasteful for B cells to produce antibodies by a process that is independent of the presence of antigens since most of these antibodies will never be needed. Do you agree or disagree? Why?

5. If you were to see your doctor after stepping on a rusty nail, you might receive two very different types of shots, both of which would protect you against tetanus. What types of materials do you suspect might be in these different types of shots, and how would the type of protection they provide differ?

6. Mammals have evolved the most complex and effective immune systems among animals, and yet some disease organisms, such as the HIV that cause AIDS and the protozoa that cause sleeping sickness elude the immune system. HIV do so by a rapid rate of mutation; the sleeping sickness protozoa do so by a reshuffling of genes that code for surface proteins. What do you think are the prospects for humans to evolve mechanisms for fighting these diseases successfully? What do you think are the prospects for humans to develop medical responses to these?

Additional Readings

Boon, T. 1993. Teaching the immune system to fight cancer. *Sci. Amer.* March:82–89. (Intermediate)

Cohen, I. R. 1988. The self, the world and autoimmunity. *Sci. Amer.* April:52–60. (Intermediate)

Grey, H. M., A. Sette, and S. Buus. 1989. How T cells see antigen. *Sci. Amer.* Nov:56–64. (Intermediate)

Hoth, D. F., and M. W. Myers. 1991. Current status of HIV, therapy. *Hospital Practice* Jan:174–197, Feb:105–113. (Advanced)

Koff, W. C. 1991. The prospects for AIDS vaccines. *Hospital Practice* April:99–106. (Advanced)

Mills, J., and H. Masur. 1990. AIDS-related infections. *Sci. Amer.* Aug:50–57. (Intermediate)

Radetsky, P. 1993. Magic missiles. *Discover* March: 42–47. (Intermediate)

Rennie, J. 1990. The body against itself. *Sci. Amer.* Dec:106–115. (Intermediate)

Rosenberg, S. A. 1990. Adoptive immunotherapy for cancer. *Sci. Amer.* May:62–69. (Intermediate)

Rothman, D. J., and H. Edgar. 1991. AIDS, activism, and ethics. *Hospital Practice* July:135–142. (Introductory)

Shilts, R. 1987. *And the band played on.* New York: St. Martin's Press. (Introductory—Excellent prize-winning story of the early history of AIDS)

von Boehmer, and P. Kesielow. 1991. How the immune system learns about self. *Sci. Amer.* Oct:74–81. (Intermediate)

Young, J. D., and Z. A. Cohn. 1988. How killer cells kill. *Sci. Amer.* Jan:38–44. (Intermediate)

CHAPTER
◂ 31 ▸

Generating Offspring: The Reproductive System

STEPS
TO
DISCOVERY
The Basis of Human Sexuality

REPRODUCTION: ASEXUAL VERSUS SEXUAL

Types of Asexual Reproduction

The Advantage of Sexual Reproduction

Strategies for Sexual Reproduction

Unity and Diversity Among Reproductive Systems

HUMAN REPRODUCTIVE SYSTEMS

The Male Reproductive System

The Female Reproductive System

CONTROLLING PREGNANCY: CONTRACEPTIVE METHODS

Chemical Intervention: Oral Contraceptives, Hormones, and Spermicides

Mechanical Intervention

Permanent Sterilization

Natural Methods

SEXUALLY TRANSMITTED DISEASES

BIOLINE
Sexual Rituals

THE HUMAN PERSPECTIVE
Overcoming Infertility

BIOETHICS
Frozen Embryos and Compulsive Parenthood

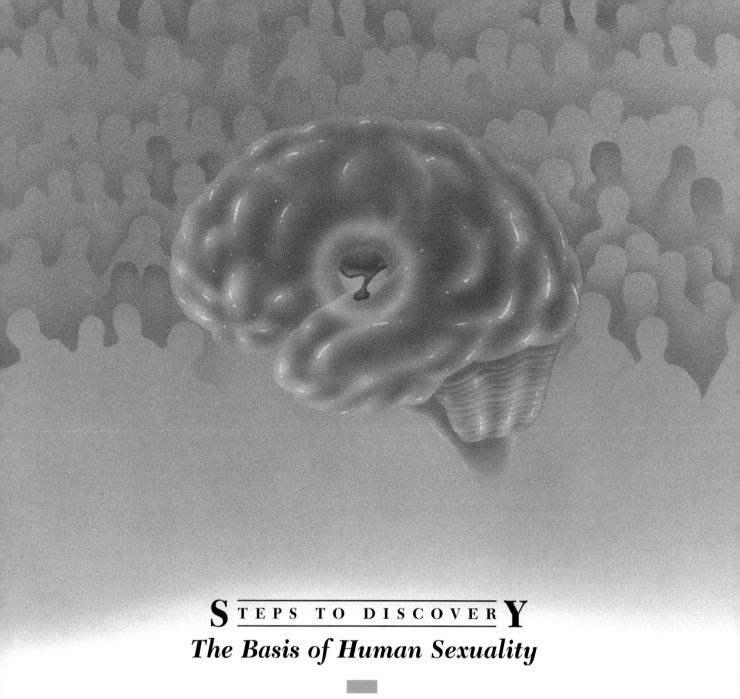

The Basis of Human Sexuality

During the eighteenth and nineteenth centuries, physiologists learned a great deal about the workings of the circulatory system, the pathways of the nerves, and the microscopic structure of the kidneys and lungs. Because of the sensitivity of the subject, however, our knowledge of human reproductive physiology lagged woefully behind. One popular hypothesis of the late 1800s even held that a variety of reproductive system dysfunctions were due to disorders of the epithelium lining the nose. One prominent professor at Johns Hopkins University went so far as to compare the anatomy of the nose to that of the penis, concluding that men who masturbated frequently suffered from various nasal diseases, including chronic nosebleeds.

Interest in human sexuality underwent a major revival following the publication of Charles Darwin's second major work, *The Descent of Man, and Selection in Relation to Sex,*

in 1871. In this work, Darwin applied his basic theory of evolution via natural selection to the evolution of humans; in doing so, he created even more furor. Darwin considered the evolution of the human mind as well as the body. For example, he noted that apes and monkeys showed many of the same emotions as humans, including fear, aggression, surprise, sexual desire, and even jealousy.

Darwin became interested in child behavior and closely followed the behavior of his own son during the child's first few years of life. His observations led Darwin to suggest that very young children exhibit signs of sexual behavior. Darwin's writings on this matter had a great influence on the psychiatric world of the late nineteenth century. Darwin called attention to the presence of the instinctive qualities of human behavior, which had been inherited from our less inhibited ancestors. The Darwinian view of

Anonymous questionnaires uncover the social and psychological aspects of human sexuality, while physiological experimentation

human evolution led many psychotherapists to conclude that human behavior was molded largely by two basic, innate drives: self-preservation and sexual gratification.

Just 6 years after the publication of *The Descent of Man*, a paper appeared in a German journal of biology on the structure of the nervous system of the larval stage of a lamprey eel, one of the most primitive of all living vertebrates. The paper was authored by a young medical student at the University of Vienna, who was interested in becoming a neuroanatomist (a biologist who studies the anatomic organization of the nervous system). The author of the paper was Sigmund Freud. Freud received his medical degree a few years later and attempted to secure a position in neuroanatomy at the Physiology Institute. He was discouraged by his mentor Ernst Brucke, who argued that the prospects there were slim and that Freud would be better off pursuing a career in clinical practice. Eventually, Freud opened his own practice in Vienna.

In 1895, Freud and his friend and colleague Josef Breuer published a book entitled *Studies on Hysteria,* in which the two men evaluated hundreds of hours of conversations with patients who were suffering from various forms of mental disorders *(neuroses).* Freud concluded that the basis of these neuroses could be traced to childhood sexual trauma and repressed sexual desires. He recognized a conflict between the human sexual drive, which is inherited from our prehuman ancestors, and the restrictions placed on this drive by modern civilization. Over the following years, Freud became more and more convinced of the pervasive importance of sexuality, particularly infantile sexuality, in shaping our adult personalities.

During the first half of the twentieth century, Freud and other psychotherapists continued to probe the role of sexuality in shaping individual human personalities, but no attempt was made to study sexuality in the human population as a whole. This would soon be changed by Alfred Kinsey of Indiana University, a prominent *taxonomist*—a biologist who studies the anatomic, physiological, and behavioral characteristics that are used to classify a particular group of organisms. During his extensive studies of a group of tiny insects called gall wasps, Kinsey was struck by the genetic variation that existed among the individual insects, noting, for example, that wing length could vary by a factor of over tenfold from one member of a local population to another. As an insect taxonomist, Kinsey was an unlikely student of human sexuality.

In 1944, students at Indiana University petitioned for a biology course on sexuality and marriage. Kinsey was named to head a seven-member faculty to deliver the lectures. At the end of the course, Kinsey passed out a questionnaire asking students for their evaluation of the course as well as personal questions regarding their sexual activities; he planned to discuss the answers to the questions when the course was next offered. From this unlikely beginning, Kinsey realized that by using interviews and questionnaires he could learn about human sexuality, an important subject about which the scientific community was almost totally ignorant. Within a few years, Kinsey had trained a number of interviewers, and the group had collected several thousand sexual case histories from people of all different ages, economic groups, races, religions, and sexual orientations. The data from the sexual case histories were compiled into two large volumes, entitled *Sexual Behavior in the Human Male* and *Sexual Behavior in the Human Female* and were published in 1948 and 1953, respectively.

Kinsey's books provided a wealth of insights into human sexuality. They revealed the striking variation in sexual practices among different individuals; the importance of socioeconomic factors in sexual habits, particularly among males; the differences and similarities in male and female sexuality; the importance of religious convictions in shaping human sexual behavior; the ages at which males and females reach peak sexual interest; and more. Not surprisingly, the books captured the attention of the public as well as the behavioral scientists, bringing both praise and scathing criticism. Although there were shortcomings in Kinsey's method of selecting the individuals to interview, the methods of interviewing, and the handling of data, the Kinsey reports remain among the most cited of all works in the behavioral sciences to this day.

For the past 2 decades, biologists have been investigating the neural and hormonal basis of human sexuality. Particular attention has been focused on the hypothalamus, a region of the brain known to govern sexual responses and to contain distinct anatomic differences between men and women. For example, one cluster of hypothalmic nerve cells, called INAH-3, typically is more than twice as large in men as it is in women. While the reasons for these gender-related differences are not understood, evidence indicates they are caused by the levels of sex hormones (estrogen and testosterone) that circulate in the fetus and newborn. In 1991, Simon LeVay, a neuroanatomist at the Salk Institute, dropped a bombshell when he reported that the INAH-3 cluster in male homosexuals who had died from AIDS tended to be much smaller than that of heterosexual males; in fact, the cluster in homosexuals was about the same size as that in women. Whether or not LeVay's findings are confirmed remains to be seen; regardless, they have created considerable controversy. For example, the findings affirm the conviction of many homosexuals that their sexual orientation is biologically determined and cannot be altered behaviorally. On the other hand, many homosexuals fear that the information could be used as a basis for screening the population in order to assess an individual's sexual orientation.

revealed the importance of the hypothalamus in regulating sexual function.

*O*f all animals, humans are the only species apparently aware of their own mortality. We all know that we will die, but we also know that our lineage can continue by having children. Consider for a moment that the genes we possess have been passed down, generation to generation, from distant ancestors. A living descendent of Andrew Jackson, for example, may have the same genetic information for Jackson's distinctive eyebrows. Some of the information encoded in our genes may be billions of years old, such as that which directs the formation of semipermeable cell membranes or oxygen-dependent respiratory chains. Some of these genes have been preserved—with changes—for millions of generations.

The transmission of genetic information from one generation to the next is a crucial part of **reproduction:** the process by which new offspring are generated. All organisms, from bacteria to redwood trees to humans, have a finite life span. Reproduction is the process by which the species is perpetuated. But reproduction is more than simply a continuation of life; it is also a process of rejuvenation. As each individual grows older, it shows increasing signs of age. Yet the parents' aging is not passed on to the offspring; each new generation begins life with a "fresh start."

▼ ▼ ▼

REPRODUCTION: ASEXUAL VERSUS SEXUAL

Some animals can reproduce by **asexual reproduction;** that is, they produce more of themselves without the participation of a mate, gametes, or fertilization. As we discussed in Chapter 10, since neither meiosis nor fertilization is part of the process, asexual reproduction produces offspring that are genetically identical to the parent. However, most animals produce offspring by **sexual reproduction,** which requires

1. The formation of two different types of haploid gametes—eggs and sperm—by a process that includes meiosis, and
2. the union of a single egg and sperm at fertilization to form a zygote.

You will recall that gametes are formed in reproductive organs called *gonads.* Sperm are produced in the male gonads (testes), whereas eggs are produced in the female gonads (ovaries).

TYPES OF ASEXUAL REPRODUCTION

Asexual reproduction occurs in various ways, depending on the species. During **fission** (Figure 31-1*a*), an animal splits into two or more parts, each of which becomes a complete individual. Fission is common among sea anemones and various groups of worms. Some animals reproduce asexually by **budding,** whereby offspring develop as an outgrowth of some part of the parent. Budding is common in sponges, hydras, and corals (Figure 31-1*b*), particularly in those species that form large colonies in which individual members remain connected to one another by a common "pipeline." In such species, each colony arises from a single founder that buds repeatedly, generating new individuals that remain physically joined to one another (as in Figure 1-3*a*, on page 8).

One type of reproduction is neither strictly asexual or sexual. Although gametes are produced, **parthenogenesis** requires only one parent—a female—who produces eggs that develop into adults without fertilization. Parthenogenesis commonly occurs among some insects, such as the aphids shown in Figure 31-1*c*, and less commonly in reptiles, amphibians, and fishes.

THE ADVANTAGE OF SEXUAL REPRODUCTION

Why do organisms engage in sexual reproduction? Why don't all organisms simply reproduce asexually, forming an exact, yet younger, version of themselves? After all, asexual reproduction has some obvious adaptive advantages. First, it generates progeny without the greater investment of energy and resources associated with gamete production, mate seeking, and fertilization. Second, asexual reproduction is very efficient; one individual, living by itself, can generate large numbers of offspring very rapidly.

▌▌▶ But sexual reproduction is the predominant mode of reproduction among animals. Even among animal species that reproduce asexually, most do not do so exclusively. Rather, individuals in a population typically reproduce asexually for a period of time; then, in response to some environmental trigger, such as food depletion or overcrowding, these same animals switch to a sexual mode of reproduction. Clearly, there must be some selective disadvantage to a total reliance on asexual reproduction. That disadvantage is presumed to be **genetic monotony;** that is, generation after generation, progeny will be genetically the same. In contrast, sexual reproduction combines traits from two genetically distinct parents in a single individual so that each offspring acquires a unique genetic mix. In addition to gene mixing during fertilization, variation is boosted even further by independent assortment and crossing over during meiosis (Chapter 11).

Asexual reproduction is advantageous only in certain situations. If a parent is successful in surviving and reproducing in a particular habitat, identical offspring are also

FIGURE 31-1

Types of asexual reproduction in animals *(a)* During fission, this sea anemone splits into two individuals. *(b)* In this hydra, a new individual forms as a bud that grows out of the body wall. *(c)* This brood of aphids developed parthenogenetically from unfertilized eggs within the larger female.

likely to succeed. Asexual reproduction is a successful means of reproducing in uniform environments where conditions do not change much from one place to another or from one year to the next. Most of the earth's habitats are not uniform, however. Conditions change over time, as exemplified by the appearance of new diseases, such as AIDS in humans, or the repeated ice ages that have occurred over the past few million years. If an individual member of a species is ill-equipped for a particular change in environmental conditions, then all the identical offspring produced by asexual reproduction will be similarly ill-equipped. The entire species could be wiped out by a single adverse development in the environment. In contrast, sexually reproducing species produce offspring that have different characteristics; consequently, chances are improved that some individuals may be able to survive and reproduce in new habitats, changing environments, or in the face of new diseases. In this way, genetic variation is the foundation of evolutionary change.

STRATEGIES FOR SEXUAL REPRODUCTION

Sexual reproduction requires that eggs and sperm of the same species come together at the same time and place (see Bioline: Sexual Rituals). The simplest strategy, employed by most aquatic animals, is simply to spew sperm and eggs into the surrounding water, which leads to **external fertilization:** fertilization that takes place outside the body (Figure 31-2*a*). With this strategy, the likelihood that a sperm and egg will cross paths is largely a matter of probability; the more gametes produced, the greater the chance of fertilization. Animals that utilize external fertilization typically possess extensive gonads and produce enormous numbers of gametes. A single oyster, for example, releases more than 100 million eggs each season. Only a tiny fraction of these eggs will ever be fertilized, and only a tiny fraction of the resulting zygotes will actually develop into adults. The chance of an egg being fertilized externally is greatly improved when the male discharges his sperm directly onto the eggs, as occurs in many amphibians (Figure 31-2*b*).

External fertilization occurs only in watery environments, where sperm can swim to an egg and fertilize it. Therefore, animals that mate on dry land rely on **internal fertilization** (Figure 31-2*c*), which occurs inside a chamber in the female's body, where a local aquatic environment can be maintained. Internal fertilization has an important selective advantage: It improves the odds that gametes will encounter one another. Animals that utilize internal fertilization typically produce fewer eggs, each of which has a better likelihood of being fertilized.

In many internal fertilizers, sperm is transferred using an intrusive structure called a **penis.** Some internal fertiliz-

(a)

(b)

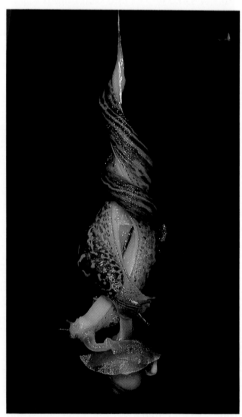

(d)

(c)

FIGURE 31-2

Sexual variety. *(a) External fertilization.* A sponge (lower right) discharging a cloud of gametes into the surrounding sea water. *(b) Coordinated external fertilization.* Among these wood frogs, the female of the species discharges eggs (the black circles surrounded by jelly capsules), which are fertilized by sperm released simultaneously by the male. *(c) Internal fertilization* between male and female damselflies. *(d) Hermaphroditic internal fertilization.* These two entwined slugs exchange sperm that is discharged from the blue penis that extends from each partner's head. This is an example of cross fertilization between individuals that possess both male and female reproductive systems.

ers lack these copulatory organs. For example, the male octopus produces packages of sperm, which it transfers to the female with one of its eight arms. When inserted into the female, part of the arm breaks off and eventually ruptures, showering the eggs with sperm. Afterward, the male regenerates the tip of the lost appendage.

In many invertebrate species, one individual possesses both male and female reproductive systems; such individuals are called **hermaphrodites** (Figure 31-2*d*). When isolated from other members of their species, some hermaphrodites are capable of reproducing by fertilizing their own eggs. This is particularly true of parasites, such as tapeworms, that may find themselves socially isolated inside a host's digestive tract.

UNITY AND DIVERSITY AMONG REPRODUCTIVE SYSTEMS

⬥ Regardless of the type of animal, all sexual reproductive systems have a clearly defined, universal function: the production of eggs and sperm. Animal gametes may be so similar in appearance that the living eggs or sperm of a jellyfish, snail, sea urchin, or human may be nearly indistinguishable when seen through a microscope.

Despite their similar function, reproductive systems are probably more diverse in structure than is any other group of organs. In some species, such as those of parasitic roundworms or marine invertebrates, the reproductive systems are so complex and voluminous that they fill much of

the internal space within the animal (Figure 31-3*a*). For example, the reproductive system of a parasitic flatworm (Figure 31-3*b*) contains a bewildering variety of tubes, chambers, glands, and receptacles, none of which bears any obvious evolutionary relationship to the parts of the reproductive systems of other groups of animals. Rather than attempting to discuss the evolution or diversity of reproductive systems, we will focus on the complex reproductive organs within our own bodies.

Sexual reproduction, which includes meiosis and fertilization, is the predominant mode of reproduction among animals. Although sexual reproduction makes greater demands on animals than does asexual reproduction, it generates genetically diverse offspring, which gives a population a survival advantage. (See CTQ #2.)

HUMAN REPRODUCTIVE SYSTEMS

Both the male and female human reproductive systems consist of a pair of gonads, in which gametes are formed, and a reproductive tract possessing various accessory functions that are discussed later in the chapter. The vastly different structures of the human male and female reproductive systems reflect the difference in their roles during

(a)

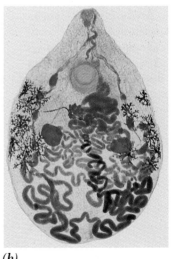

(b)

FIGURE 31-3

Simple animals, complex reproductive systems. *(a)* Huge reproductive organs dominate the anatomy of this jellyfish, testifying to the large number of gametes that must be produced in a species whose eggs are fertilized in the open sea. *(b)* Virtually all of the structures seen in this parasitic flatworm belong to the animal's reproductive system.

◁ B I O L I N E ▷
Sexual Rituals

FIGURE 1

"Timing is the thing; it's true. It was timing that brought me to you."

These words from a popular love song are far more universal than is the "boy-meets-girl" situation the songwriter is describing. Sexual reproduction among virtually all animal species requires some form of timing that increases the likelihood that sperm will meet egg; this is often accomplished by increasing the likelihood that male will meet female. In other words, reproductive timing coordinates the courtship behavior of many animals. The songs of crickets and cicadas, the blinking lights of fireflies, the aromas of prometheus moths, and the nocturnal croaking of frogs are all inherited mechanisms that help mates find and identify one another as members of the same species and opposite genders—in other words, as potential sex partners.

The right smell or the correct approach not only promotes fertilization but can also protect a mate from being mistaken for a potential meal. A male spider, which tends to be smaller than the female of the species, could be consumed by his potential mate before fertilizing her eggs. Not only must the male spider approach the female cautiously, he must also exhibit some behavior that identifies him as a mate, not a meal. Some spiders do so by vibrating the female's web in a rhythm that the female recognizes as an approaching male. Others placate their potentially dangerous mates by providing a "gift," usually a captured meal wrapped in silk (Figure 1). The female spider usually accepts the gift and allows the male to copulate with her. Males occasionally try to trick a female into accepting an empty package. When their deceit is discovered, the "empty-handed" male may become the gift himself.

↻ Younger members of a species, whether a floating sea urchin larva or a newborn colt, are less hardy than are their adult counterparts. For this and other reasons (such as the availability of food), reproduction usually occurs at a particular season of the year, a time that is most favorable for offspring survival. The onset of seasonal reproductive behavior is often coordinated by some specific environmental change that informs the animal's nervous system that the mating season has arrived. For many animals, this environmental factor is *photoperiod*, the length of daylight or darkness. Spring is the most common season for reproduction; the physiological and behavioral changes that lead up to reproduction are often triggered by the increasing length of daylight at this time of year.

In many mammals, the increasing spring daylight stimulates hormonal changes that bring about the onset of *estrus*, or sexual "heat," a period of sexual receptivity. For example, in dogs, as in many other mammals, females in estrus produce a sexual attractant, the odor of which stimulates males to begin courtship activities at a time when eggs are present to be fertilized. The odds of fertilization can be further increased by mechanisms that guarantee that eggs will be available when sperm are deposited. In rabbits, for example, the egg is retained in the ovary until after mating, at which time it is released into the reproductive tract for fertilization.

Reproductive behavior in humans and other primates, such as monkeys and apes, is not limited by such estrus restraints, and sexual arousal follows no seasonal pattern. In fact, the songwriter quoted earlier might find it ironic that the human mating behavior is one of the few behaviors *not* governed by biologically determined timing mechanisms.

reproduction. The male produces huge numbers (trillions during a lifetime) of tiny motile gametes, called **spermatozoa** (singular *spermatozoon*), commonly referred to as sperm, and delivers these cells into the female's reproductive tract. In contrast, the female produces a small number (a few hundred over a lifetime) of relatively large gametes, called **ova** (singular *ovum*), commonly referred to as eggs. The female is responsible for more than just gamete production. She also provides an environment in which the eggs can be fertilized and another environment in which the zygote can develop into a fully formed human infant.

The reproductive activities of both sexes are regulated by a battery of hormones. These hormones are responsible for stimulating the initial development of the reproductive system in the embryo; for causing its maturation during puberty; and for maintaining its day-to-day operation throughout the reproductive years.

THE MALE REPRODUCTIVE SYSTEM

In human males, the paired testes produce sperm throughout adult life at the average rate of 30 million sperm per day. The male reproductive tract is primarily a hollow conduit, equipped with accessory glands, that moves sperm out of the body and into the reproductive tract of the female.

Male External Genitals: From and Function

The male external genitals (Figure 31-4) consist of the penis and a sac, called the **scrotum,** which houses the testes, or testicles. The temperature within the scrotum is several degrees cooler than is the rest of the body; this cooler temperature is necessary for the formation of sperm. If the testes do not descend into the scrotum during fetal development, as occurs on rare occasion, the elevated gonadal temperature results in sterility (the inability to produce offspring) after about 4 years inside the body. Surgically maneuvering the testes from the body cavity into the scrotum prevents such loss of fertility. A passageway, called the **urethra,** runs through the length of the penis, conveying both sperm and urine, although not simultaneously.

🔖 In order to deposit sperm into the vagina, the penis must have penetrating capacity, which is accomplished when the organ is engorged with blood. The interior of the penis contains three long cylinders of spongy *erectile tissue.* During periods of sexual excitement, impulses travel along parasympathetic nerves from the brain to the smooth muscle cells that line the arterioles of the penis. Relaxation of the smooth muscles leads to vasodilation of the arterioles, causing blood to engorge the spaces of the erectile tissue, which expands as it fills. As a result, the penis enlarges and becomes rigid enough for vaginal penetration; this is known as an erection.

The failure to develop an erection (known as *impotence*) is a common type of male sexual dysfunction. In younger men, the cause is usually psychological; in older men, it is often a failure of the organ to receive sufficient blood, which may be due to diabetes, atherosclerosis, or as a side effect of medication taken for treatment of high blood pressure. Whether the root cause is psychological or physiological, the condition can, in most cases, be corrected by counseling or medical treatment.

The surface of the penis, especially the tip, or *glans* (Figure 31-4), is richly endowed with sensory receptors that increase tactile sensitivity. In the flaccid (nonerect) penis, the glans is surrounded by the *prepuce*, or "foreskin," which is often removed shortly after birth by a process called **circumcision.** Circumcision is usually performed for religious or hygenic reasons, but it is also used to treat a painful condition called *phimosis*, in which the prepuce cannot retract over the glans as normally occurs during erection.

The Formation of Sperm

A cross section through a testis reveals that the structure is composed almost entirely of tubular elements, called **seminiferous tubules** (Figure 31-5), whose combined length is equivalent to about seven times the length of a football field. Within these tubules, **spermatogenesis,** the formation of the male gametes, takes place. Between the tubules are scattered clusters of *interstitial cells:* the endocrine cells responsible for the production of testosterone, the male sex hormone.

Spermatogenesis takes place in stages, which are revealed in an examination of a cross section of a seminiferous tubule (Figure 31-5). Cells enter spermatogenesis at the outer edge of the tubule and are gradually moved toward its inner lumen as the process continues. The outer edge of the tubule contains a self-perpetuating layer of **spermatogonia,** germ cells that have not yet begun meiosis. Some of these spermatogonia are about ready to enter meiosis and develop into sperm. Others continue to divide by *mitosis* (Figure 31-6), producing more spermatogonia, replacing those that have gone on to form gametes. In the first step of spermatogenesis, a spermatogonium grows in size and enters meiosis, forming a **primary spermatocyte.** Each diploid primary spermatocyte subsequently undergoes the first meiotic division to form two **secondary spermatocytes,** each of which undergoes the second meiotic division to form a total of four haploid **spermatids.** Each spermatid undergoes a dramatic transformation as it is converted into a sperm cell, one of the most specialized cells in the body.

The central lumen of a seminiferous tubule contains germ cells that have almost completed spermatogenesis and resemble, in outward appearance, mature sperm (inset, Figure 31-5). These cells are not yet ready to fertilize an egg, however. Sperm complete their maturation only after they are pushed out of the seminiferous tubule and into the

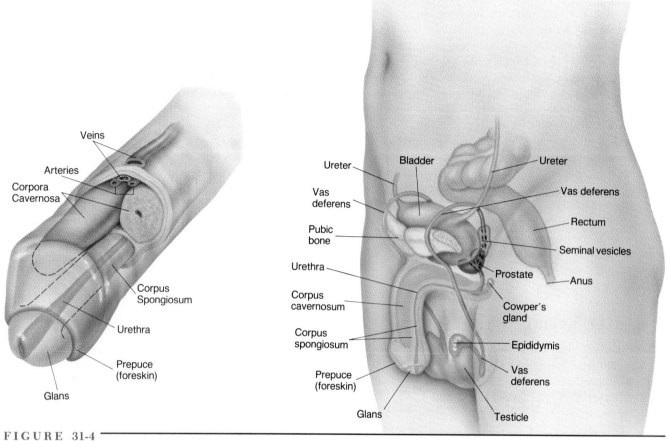

FIGURE 31-4

The male human reproductive system. The purple structures trace the route sperm travel from the testes to the penis. When the arterioles of the penis become dilated, and the erectile tissues (composed of the *corpus spongiosum* and paired *corpora cavernosa*) become engorged with blood, the penis becomes erect.

epididymis. Each **epididymis** (Figure 31-5) is a tubule, approximately 6 meters (20 feet) long; it is coiled into a compact mass attached to each testis. Sperm spend several weeks in the epididymis and then move into the **vas deferens,** a tubule that transports the male gametes to the urethra.

Form and Function of Sperm

A sperm is a compact, streamlined cell (Figure 31-5), whose function it is to move up the female reproductive tract and fuse with an egg. The entire structure of a sperm can be correlated with this function. As it differentiates, the sperm loses nearly all of its cytoplasmic and nuclear fluid, which accounts for more than 90 percent of the volume of the spermatid. The head of a sperm consists of two parts: the nucleus and the acrosome. The nucleus of a sperm is the ultimate in chromosome compactness; the chromosomal material is condensed to a virtual crystalline state. The

acrosome, which forms a cap over the nucleus, contains digestive enzymes that are released as the sperm "digests" its way through the protective layers that surround an egg. The middle portion of the sperm contains tightly packed mitochondria that, by virtue of the ATP they produce, will power the sperm's movements (see Figure 5-14*d*). The tail of a sperm contains a flagellum that whips against the surrounding fluid, driving the sperm toward and into the egg. Sperm are launched like self-propelled torpedoes with a limited range. If they do not reach the egg within the allotted time (24 to 48 hours in humans), the sperm exhaust their fuel supply and die.

The Role of the Male Accessory Glands

Before their release from the penis, sperm are mixed with secretions from several glands (see Figure 31-4) to form **semen,** the sperm-containing fluid that is expelled from the body as the result of strong muscle contractions that occur

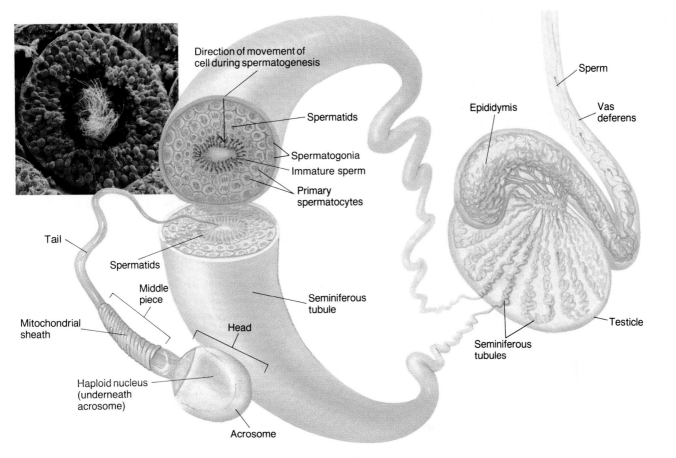

Direction of movement of
cell during spermatogenesis

Spermatids

Spermatogonia

Immature sperm

Primary
spermatocytes

Sperm

Epididymis

Vas
deferens

Tail

Spermatids

Middle
piece

Mitochondrial
sheath

Head

Seminiferous
tubule

Testicle

Seminiferous
tubules

Haploid nucleus
(underneath
acrosome)

Acrosome

FIGURE 31-5

Spermatogenesis. The cross section through a seminiferous
tubule shows cells in different stages of spermatogenesis. The
outer layer contains spermatogonia that have not yet begun
spermatogenesis. The next layer contains primary spermatocytes that have entered meiosis. Further inward are the secondary spermatocytes and the spermatids—the products of meiosis
—and within the lumen are nearly differentiated sperm. The
structure of a single sperm is also shown, illustrating several
major features such as the head, the mitochondria-rich middle
piece, and the tail. The inset shows a scanning electron micrograph of a cross section through a seminiferous tubule; tails of
the sperm are clearly seen extending into the tubule's lumen.

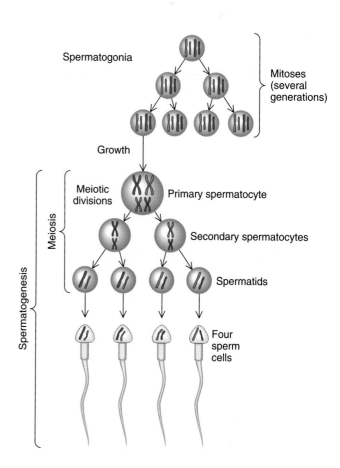

Spermatogonia

Mitoses
(several
generations)

Growth

Meiotic
divisions

Primary spermatocyte

Meiosis

Secondary spermatocytes

Spermatogenesis

Spermatids

Four
sperm
cells

FIGURE 31-6

The stages of spermatogenesis.

during **ejaculation.** Sperm are only a small percentage of the semen's final volume; 5 billion sperm—a quantity large enough to repopulate the world with humans—would fit into a space the size of an aspirin tablet.

Together, the **prostate gland** and the paired **seminal vesicles** and **Cowper's glands** produce most of the ejaculatory fluid. These secretions are rich in fructose, a sugar that serves as an energy source for the sperm. The fluid also contains prostaglandins (page 532), which stimulate the contraction of the smooth muscles of the female reproductive tract, helping to propel sperm cells toward their encounter with an egg. The complete semen is ejaculated from the penis by strong muscular contractions that occur during **orgasm,** a brief period of peak excitement that constitutes the sexual climax. The ejaculatory fluid nourishes and protects the sperm; it provides a liquid medium needed for sperm motility; and, because of its alkaline pH, it temporarily neutralizes vaginal acidity that might otherwise impair sperm-cell activity.

Hormonal Control of Male Reproductive Function

Proper functioning of the testes is maintained by the presence of two hormones that are secreted by the anterior pituitary: **follicle-stimulating hormone (FSH)** and **luteinizing hormone (LH).** These hormones were originally named because of their effects on the female reproductive system, as we will discuss later in the chapter. The fact that the identical hormones are present in both genders but elicit different responses in males and females illustrates an important principle of endocrine function: The nature of the *target cell* determines the type of response, while the *hormone* itself acts simply as a stimulus or trigger.

FSH and LH are described as *gonadotropins* because they stimulate the activities of the gonads (Figure 31-7). LH acts primarily on the interstitial cells of the testes, stimulating the production and secretion of testosterone. FSH is required for spermatogenesis. The secretion of both LH and FSH by the pituitary is, in turn, regulated by the level of **gonadotropin-releasing hormone (GnRH)** that is secreted by the hypothalamus.

While the gonadotropins act on the gonads, testosterone, the hormone produced by the gonads, acts on the other tissues associated with male sexuality. Testosterone secretion stimulates the differentiation of the male reproductive tract in the embryo, the descent of the testes into the scrotum, the further development of the reproductive tract and penis during puberty, and the development of male *second-*

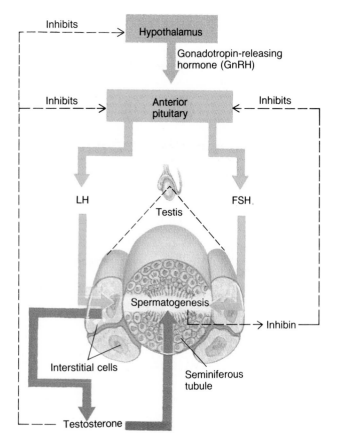

FIGURE 31-7

Production and regulation of male sex hormones. As testosterone concentrations in the bloodstream increase, the release of GnRH by the hypothalamus and LH by the anterior pituitary is inhibited. The reduction in LH concentration, in turn, causes a decrease in testosterone secretion by the interstitial cells of the testes, which, in turn, releases the inhibition on the secretion of GnRH and LH. As a result of this negative-feedback mechanism, a balanced level of all hormones is achieved. FSH production is also believed to be regulated by a negative-feedback mechanism involving a protein hormone called *inhibin,* which is produced by the seminiferous tubules and acts on the anterior pituitary. Both testosterone and FSH are needed for spermatogenesis to occur within the seminiferous tubules.

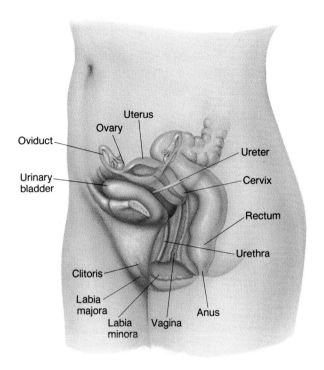

FIGURE 31-8

The female human reproductive system. Eggs are formed in the ovary and swept into the oviducts. Meanwhile, sperm enter the body in the vagina and travel through an opening in the cervix up to the oviducts, where fertilization takes place. The resulting embryo passes down the oviduct and into the uterus, where it is implanted into the uterine wall and develops.

ary sex characteristics, including a beard and chest hair, enlargement of the larynx, and increased muscle mass. Testosterone also plays a role in the development and maintenance of the male *libido,* or sexual desire.

THE FEMALE REPRODUCTIVE SYSTEM

A woman's body performs many activities that are essential to reproductive success: Her ovaries are the sites of **oogenesis,** or formation of ova, and her reproductive tract nourishes, houses, and protects the developing fetus. After birth, a woman's body provides breast milk, which nourishes the developing infant. As in the male, each aspect of reproductive activity is under the complex control of hormones.

Female External Genitals and Reproductive Tract: Form and Function

The female's external genitals (Figure 31-8) are collectively known as the **vulva.** The most prominent features of the vulva are an outer and inner pair of lips, the **labia majora** and **labia minora,** respectively. The **clitoris** protrudes from the point where the labia minora merge. Rich in sensory neurons and erectile tissue, the clitoris resembles the penis in its sexual sensitivity and erectile capacity. Unlike the penis, however, the clitoris has no urinary or ejaculatory function; its sole function is to receive sexual stimulation,

which it transmits to the central nervous system. Thus, the clitoris is the only human structure dedicated exclusively to the enhancement of sexual pleasure.

The urethra, which, in the female, is involved solely with the release of urine, opens between the labia minora. Also located between the labia minora is the opening to the **vagina,** an elastic channel that receives the sperm that are ejaculated from the penis and forms the birth canal through which an infant leaves its mother's body during childbirth. The vagina leads to the remainder of the female reproductive tract (Figure 31-8), which consists of the uterus and paired oviducts. The vagina is separated from the uterus by the **cervix,** which contains the opening through which sperm must pass on their way to fertilize an egg. The **uterus** (or *womb*) is a pear-shaped, thick-walled chamber that houses the embryo and fetus during pregnancy. The **oviducts** (or *fallopian tubes*) emerge from the uterus and serve as the sites where sperm and egg become united during fertilization.

The Formation of Eggs

A cross section through an ovary (Figure 31-9) shows an appearance very different from that of a testis. There are no tubules in an ovary. Instead, the **oocytes**—the germ cells that have the potential to form eggs—are housed within spherical compartments, called **follicles,** which are scattered throughout the tissue of the ovary. Each follicle con-

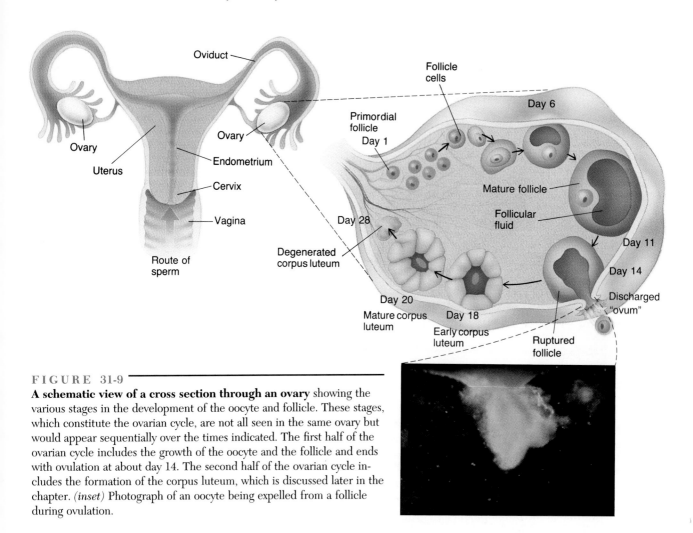

FIGURE 31-9

A schematic view of a cross section through an ovary showing the various stages in the development of the oocyte and follicle. These stages, which constitute the ovarian cycle, are not all seen in the same ovary but would appear sequentially over the times indicated. The first half of the ovarian cycle includes the growth of the oocyte and the follicle and ends with ovulation at about day 14. The second half of the ovarian cycle includes the formation of the corpus luteum, which is discussed later in the chapter. *(inset)* Photograph of an oocyte being expelled from a follicle during ovulation.

tains a single oocyte, surrounded by one or more layers of **follicle cells,** which provide the materials that support the growth and differentiation of the enclosed germ cell. Unlike the male gonad, the ovary of a woman contains no **oogonia,** or premeiotic germ cells. All of the oogonia that are produced during embryonic development have already entered prophase of meiosis I (page 216) by the time of birth. Oocytes will remain suspended in meiotic prophase for years, some for several decades.

The vast majority of the follicles of the adult ovary are small and consist of an undifferentiated oocyte surrounded by a single layer of follicle cells. These *primordial follicles* are storehouses of oocytes that will provide the eggs produced during the reproductive life of the woman. A pair of human ovaries contains approximately 400,000 primordial follicles at the time of birth. Once a female reaches reproductive maturity, a few oocytes will undergo oogenesis during each monthly **ovarian cycle.** During oogenesis, which occurs in the first half of the ovarian cycle, the oocyte increases in size from about 25 to 100 micrometers in diameter. Like the sperm, the architecture of the egg is correlated

with its function. Whereas the sperm is shaped for movement and activation, the egg is packed with nutrients that will be utilized by the embryo during the first days following fertilization. Nutrients take up considerable space and, for the most part, account for the large size of the egg.

The changes in the oocyte that occur during oogenesis are accompanied by a dramatic transformation of the entire follicle. By the time it completes its growth, a *mature follicle* is large enough to be seen as an obvious bulge at the surface of the ovary. Although several follicles typically undergo maturation within each ovarian cycle, one usually outpaces the others. Ultimately, the wall of this follicle suddenly ruptures, and the enclosed oocyte is released from the ovary (inset, Figure 31-9). This is **ovulation.** The ovulated oocyte is swept into the broad, funnel-shaped opening of the oviduct, where fertilization occurs.

Meiotic Divisions of the Female Germ Cell As in the male, the reduction of the chromosome number by meiosis is necessary during the formation of the female gamete. Unlike spermatogenesis, where meiosis occurs before the differentiation of the sperm, meiotic divisions in the female occur after the entire process of growth and differentiation of the ovum has essentially been completed (Figure 31-10).

The first meiotic division is completed in the oocyte just before ovulation. Unlike spermatogenesis, where meiosis I produces two equal-sized cells, the meiotic division in the female produces one large cell and one tiny cell, called a **polar body,** that eventually deteriorates. The second meiotic division begins while the oocyte is in the oviduct, but it does not run to completion. Instead, the meiotic process stops after the chromosomes have lined up for metaphase II. At this stage, the human egg is fertilized. The second meiotic division is completed after fertilization. Meiosis II is also a highly unequal cell division, producing a single, large cell that goes on to develop into a new individual, and another polar body that disintegrates. Thus, unlike meiosis in the male, which produces four equal-sized gametes, meiosis in the female produces only one large egg, which conserves all the cytoplasmic material in one package.

Hormonal Control of Female Reproductive Function

As in the male, the development, maturation, and function of the female reproductive system is under hormonal control. The maturation of the female system during puberty is thought to be initiated within the brain by the production of the hypothalamic hormone GnRH, which stimulates the secretion of LH and FSH by the anterior pituitary. In the female, the gonadotropins LH and FSH act cooperatively on the follicle cells of the ovary, stimulating them to produce the primary female sex hormone, estrogen, which has numerous target tissues. Increased estrogen levels at puberty stimulate maturation of external genitals and the development of secondary sex characteristics, such as the enlargement of breasts and the growth of hair in the armpits and pubic regions. Internally, elevated estrogen levels induce the maturation of the tissues of the uterus, enabling this organ to house and nurture a developing fetus.

Puberty is also the time when the ovary begins to produce mature ova on a cyclical basis, a process that will continue until *menopause,* the cessation of a woman's reproductive cycle. Menopause typically occurs in a woman's late forties; it is probably the result of a drop in GnRH secretion rather than an aging ovary. This is suggested by an experiment carried out by Terry Parkening and co-workers at the University of Texas. If the ovary from an aging mouse is transplanted into the body of a young mouse, the ovary continues to produce viable eggs. This suggests that the brain and pituitary of the young mouse are still able to stimulate oogenesis in an ovary that would ordinarily have stopped functioning if left in the older mouse. In contrast, a "young" ovary transplanted into an older mouse's body stops producing and ovulating oocytes.

The complex relationship among changing levels of hormones, the ovarian cycle, and the menstrual cycle is shown in Figure 31-11. The first phase of the ovarian cycle is characterized by the growth and differentiation of an oocyte. This maturation process is stimulated by both LH and FSH as well as by estrogen, which is produced within the follicle itself. As the growth of one or a few ovarian

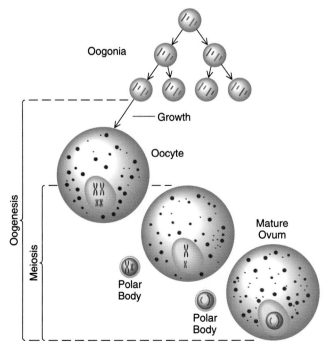

Oogonia

Growth

Oocyte

Oogenesis

Meiosis

Polar Body

Polar Body

Mature Ovum

FIGURE 31-10

Gamete formation in the female. Oogonia are premeiotic cells that produce more of their own kind by mitosis. All of these oogonia enter oogenesis during fetal development, generating a large number of primary oocytes. Each month, a few of these primary oocytes (and their surrounding follicles, not shown) undergo growth and differentiation, after which meiosis takes place. Meiosis in the female produces only one viable egg containing the nutrient reserves (yolk) that support early embryonic development. The other products of meiosis are polar bodies that disintegrate. This series of events can be contrasted to that occurring in the male (31-6).

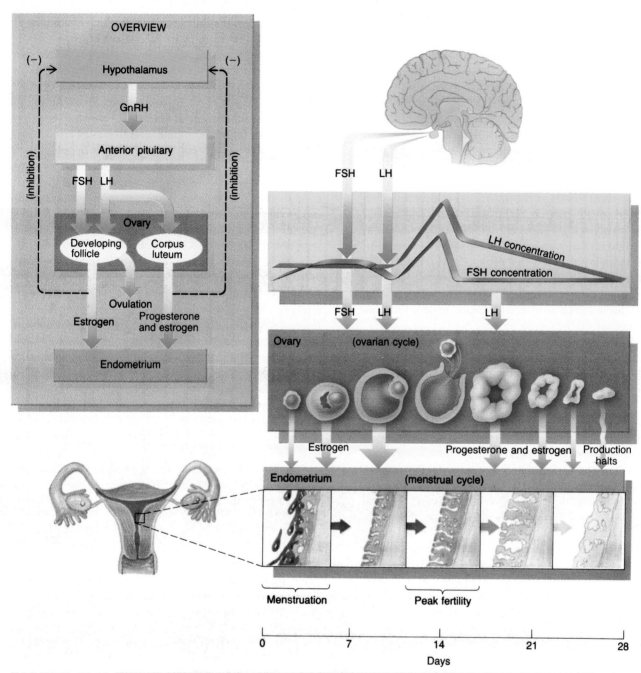

FIGURE 31-11

Synchronizing the ovarian and menstural (uterine) cycles. FSH is released from the anterior pituitary and promotes development of a follicle and the oocyte it contains. The maturing follicle secretes estrogen which stimulates endometrium development in the uterus. Presumably, when estrogen reaches a critical concentration, it stimulates the hypothalamus and anterior pituitary to flood the body with LH. This surge of LH triggers ovulation. LH also transforms the follicle into a corpus luteum. Progesterone (and some estrogen) produced by the corpus luteum maintain the enriched endometrium, so that it is receptive to implantation. Progesterone from the corpus luteum also inhibits GnRH release, so LH secretion stops, and the gonadotropin declines over the next 14 days. In the absence of implantation, the corpus luteum deteriorates (due to declining LH concentrations) until progesterone and estrogen production halts. Without these two hormones, the extra endometrial tissue cannot be maintained and is sloughed as menstrual fluid. Menstruation signals the beginning of a new cycle. With the disappearance of progesterone, inhibition of GnRH is relaxed and FSH is again released and another follicle begins to mature.

follicles nears completion, there is a marked surge in the pituitary's secretion of LH, which triggers ovulation.

The Menstrual (Uterine) Cycle

As the follicles are growing during each ovarian cycle, the uterus undergoes cyclical changes in form that are related to its function as a potential residence for the embryo. These dramatic changes in the uterus constitute the **menstrual cycle,** which takes an average of 28 days to complete (Figure 31-11). The first day of the menstrual cycle (which is defined as the day on which menstruation begins) is characterized by the flow of blood and discarded tissue from the uterus through the vagina. Menstruation takes place when the body "becomes aware" chemically that no fertilization or pregnancy has occurred following the last ovulation. As a result, the interior lining of the uterus, which had been prepared to receive a developing embryo, is broken down and rebuilt. The destruction of the bulk of the uterine wall is initiated by the constriction of arterioles, thereby cutting off the blood supply to the thickened *endometrium*, the inner lining of the uterus. The dead and dying tissue is then expelled from the uterus during menstruation by the contraction of the *myometrium* (uterine muscles).

After the period of menstruation is over, the rebuilding of the uterus begins, preparing the endometrium for the possible reception of a fertilized egg in the upcoming ovarian cycle. The first phase of uterine reconstruction produces the dramatic thickening of the inner glandular endometrium (Figure 31-11). This period is marked by the growth of the endometrial blood vessels and the formation of many new secretory glands. These changes are induced by rising levels of estrogen, produced by the follicle.

Following ovulation, on about day 14 of the menstrual cycle, the ruptured follicle remaining in the ovary is rapidly converted into a yellowish, glandular structure, called the **corpus luteum** (*corpus* = body, *luteum* = yellow). The corpus luteum (see Figure 31-9) secretes estrogen and large quantities of a second ovarian steroid hormone, progesterone. The increasing levels of estrogen and progesterone in the second half of the menstrual cycle act on the uterus to further its growth and development. If the ovulated ovum is fertilized, the resulting embryo will implant itself in the uterus by the eighth day following fertilization. The implanted embryo produces a gonadotropic hormone, called *human chorionic gonadotropin (HCG),* that acts on the ovary to maintain the activity of the corpus luteum. Biochemical tests for detecting HCG in a woman's blood or urine provide a reliable means of determining pregnancy.

If implantation occurs, the sustained corpus luteum continues to produce estrogen and progesterone, which maintain the endometrium where the embryo is developing. Since progesterone also inhibits the release of FSH, follicles cannot mature during pregnancy, making it highly unlikely that ovulation will occur while a woman is pregnant. Some birth control pills prevent ovulation by artificially increasing the concentrations of these hormones.

What happens if the ovulated ovum is not fertilized? Since the egg is not fertilized, there is no embryo available to secrete HCG. In the absence of HCG production, there is a rapid deterioration of the corpus luteum, which quickly loses its ability to produce estrogen and progesterone. When these hormones drop below a critical level, the uterus sloughs its extra endometrial tissue, and menstruation begins.

Women who exercise strenuously, such as marathon runners and gymnasts, often stop having their periods and become temporarily infertile. This response is thought to be due to a shutdown in the secretion of GnRH. Other causes of female infertility are discussed in The Human Perspective: Overcoming Infertility. The Bioethics box entitled "Frozen Embryos and Compulsive Parenthood" discusses the ethical implications of one method of overcoming infertility.

Gamete formation in the male and female are very different processes; each process is suited to the functions of the particular gamete. The role of the ovary is to produce a small number of large cells that contain the nutrients and storage materials needed to carry the fertilized egg through its early stages of development. In contrast, the role of the testis is to produce a large number of small, motile cells, one of which will fuse with the egg and contribute half of the genetic material (See CTQ #3.)

CONTROLLING PREGNANCY: CONTRACEPTIVE METHODS

Throughout history, people have devised many ways to prevent reproduction. *Contraceptive* methods have ranged from inserting elephant and crocodile manure into the vagina in order to block the entrance of sperm into the uterus to "fumigating" the vagina over a charcoal burner after intercourse to kill the sperm. Though primitive, these methods employed the same strategies as do some of the most effective modern contraceptive approaches; that is, physically blocking the uterine entrance or killing the sperm in the vagina.

Modern methods of avoiding pregnancy prevent ovulation, halt spermatogenesis, trap gametes in the oviducts or vas deferens, or interfere with implantation of the fertilized ovum in the uterus. A number of pharmaceutical companies are currently working on contraceptive *vaccines* for both men and women, whereby a person is stimulated to produce antibodies against proteins present on the surfaces of sperm or eggs. Questions have been raised as to the safety of inducing these types of autoimmune responses (page 657) and to the reversibility of the procedure. In the following discussion, *effectiveness* is expressed as the percentage of sexually active women who do not become pregnant in

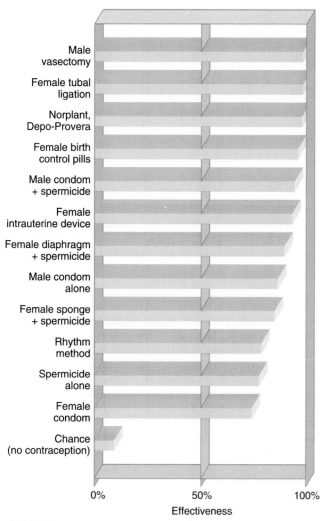

FIGURE 31-12
Effectiveness of various forms of contraception. Effectiveness is expressed as the percentage of sexually active women who fail to become pregnant in one year when a particular form of contraception is practiced diligently.

FIGURE 31-13
Norplant: an implantable contraceptive. These capsules, which are implanted under the skin, contain a synthetic progesteronelike hormone called levonorgestrel. Six of these capsules provide contraception for about 5 years.

the first year of practicing that form of birth control. In the absence of birth control, "effectiveness" would be approximately 10 percent. The effectiveness of various methods is summarized in Figure 31-12.

CHEMICAL INTERVENTION: ORAL CONTRACEPTIVES, HORMONES, AND SPERMICIDES

Synthetic estrogens and progesterones inhibit the release of FSH and LH which, in turn, prevents ovulation. Taken orally, either alone or in combination, these two hormones constitute *birth control pills*, the most effective (97 percent) *temporary* contraceptive method known today. Birth control pills must be taken daily to be effective. In 1990, the FDA approved the use of a new contraceptive called Nor-

plant, whereby thin, rubber capsules (Figure 31-13) are implanted beneath the skin of a woman's arm. The capsules release a synthetic hormone for a period of 5 years, blocking ovulation and preventing pregnancy. Another new hormonal contraceptive that provides extended protection is Depo-Provera, which consists of hormonal injections every 3 months.

To date, hormone-based contraceptives have been limited to women. Use of hormones to switch off sperm production in men also inhibits testosterone production, causing feminization and a reduction in libido. The World Health Organization has been testing weekly shots of testosterone as a spermatogenesis inhibitor, but excess levels of this hormone may carry unacceptable risks (page 528).

The most controversial chemical contraceptive is a drug called RU486, which is produced in France and interferes with the production of progesterone. Because RU486

◁ THE HUMAN PERSPECTIVE ▷
Overcoming Infertility

Approximately 10 percent of couples who try to have children are unable to do so because of physiological problems. This condition, known as **infertility,** may be due to a number of causes. Infertility in females often results from an inability to ovulate or from a blockage of the reproductive tract, while infertility in males typically results from either the production of abnormal sperm or a low sperm count. On the average, a male's ejaculate contains about 100 million sperm per milliliter of semen. When this number drops below about 20 million sperm per milliliter, fertility is markedly decreased.

Failure of a woman to ovulate is often due to an insufficient surge of LH secretion in the middle of the menstrual cycle. Women with this problem can be administered the hormone prior to the time ovulation would be expected to occur. Since the clinical procedure is not as "fine-tuned" as is the natural surge of LH, several ova are frequently released from enlarged follicles. If more than one of these cells should

be fertilized, as is often the case in such situations, multiple births can occur.

When the cause of infertility is a blockage in the female reproductive tract, a couple may try to conceive by *in vitro fertilization*, a more invasive and expensive procedure. In this procedure, oocytes are removed by suction from the enlarged follicles of the ovary prior to ovulation, and the isolated oocytes are fertilized *in vitro* —outside the body in a laboratory dish; babies conceived in this manner are commonly referred to as "test-tube babies." Eggs fertilized *in vitro* are allowed to develop for a few days in culture medium. One of the embryos is then transferred to the woman's uterus, where it has about a 20 percent chance of implanting and developing to term. In recent years, eggs that have been fertilized *in vitro* have been kept alive in a frozen state so that subsequent attempts at implantation can be made without having to operate again to remove additional oocytes.

In a small percentage of cases when the wife is infertile, couples have turned to a *surrogate:* a woman who is willing to be artificially inseminated with the husband's sperm, carry the resulting offspring to term, and give the baby to the couple after birth. Ethical questions concerning the practice of surrogacy were raised in the late 1980s when a surrogate named Mary Beth Whitehead decided she wanted to keep the baby to whom she had given birth, despite having signed a contract in which she agreed to give up the baby in exchange for payment. In 1987, the New Jersey courts awarded custody to the Sterns, the couple who had paid Whitehead for bearing the child; Whitehead was granted surrogate visitation rights. The state court also outlawed the practice of surrogacy, ruling that it amounted to the sale of babies. Since the ruling, a number of other states have passed laws governing surrogacy contracts. Adoption remains the most common choice among couples who are not able to bear children of their own.

---

renders the uterus incapable of receiving an embryo for implantation, the drug has been used in Europe as a "morning after" pill; that is, a pill that blocks pregnancy (which, by most definitions, begins at implantation) for several days after unprotected sex. RU486 also acts as an "abortion" pill by causing the uterus to reject an implanted embryo. The drug has been prescribed in Europe to terminate pregnancies through about the seventh week. Banned in the United States by the FDA, RU486 became the center of a storm of controversy in 1992 when a woman attempted to bring the drug into the United States aboard a flight from Paris. The incident reopened discussion as to whether or not RU486 should be approved for use in the United States.

Spermicides are sperm-killing chemicals that are inserted into the vagina before intercourse. When used alone, spermicides are not very reliable (79 percent effective) and thus are often used in combination with a diaphragm or condom.

MECHANICAL INTERVENTION

There are various forms of mechanical intervention to prevent pregnancy. The most popular include the IUD, diaphragm, condoms, and the sponge.

Intrauterine Device (IUD)

The *intrauterine device,* or *IUD,* is a plastic or metal device that is inserted by a physician into the uterus, where it prevents implantation of a developing embryo. In spite of its effectiveness (94 percent) and convenience, once inserted, the IUD may cause serious side effects that negate its usefuless for some women. Uterine puncture, a potentially fatal complication, occurs in about 1 in every 1,000 women. The IUD may also aggravate existing low-grade infections of the uterus and lead to a serious and painful condition called *pelvic inflammatory disease.* The most common side effects are painful cramping and irregular

◁ B I O E T H I C S ▷
Frozen Embryos and Compulsive Parenthood
By ARTHUR CAPLAN
Director of the Center for Biomedical Ethics at the University of Minnesota

The Tennessee Supreme Court issued a ruling on June 1, 1992, concerning the fate of seven embryos frozen in a liquid nitrogen tank in a Knoxville, Tennessee, fertility clinic. The ruling has important implications not only for the fate of the more than 22,000 frozen embryos at clinics around the nation but also for the national debate about abortion.

Mary Sue and Junior Lewis Davis were married on April 26, 1980. They wanted children. Mary Sue got pregnant five times, but each time she suffered a tubal pregnancy. She could conceive, but the embryos kept getting stuck in her oviducts instead of implanting in her uterus. Finally, Mary Sue had to have her tubes removed, leaving her able to make eggs but unable to bear a child.

In 1985, the Davises went through six attempts at in vitro fertilization, but they had no luck. They decided after their last attempt, in December 1988, that instead of giving up they would freeze the seven embryos and try again at some future time. Junior Davis filed for divorce in February 1989, however, claiming that he had known his marriage was falling apart for some time, but he had hoped that the birth of a baby might have saved it. The Davises were divorced, but there was one unre-

solved issue: who would get custody of the frozen embryos? Mary Sue wanted custody, initially so that she could try to have a baby. Junior Lewis objected. Mary Sue went to court to obtain sole custody of the embryos.

The first court to hear the case awarded Mary Sue custody of the embryos on the grounds that the embryos were "human beings" from the moment of fertilization and ought to have the "opportunity . . . to be brought to term through implantation." Junior Lewis appealed the decision, and a Tennessee appellate court reversed the trial court, assigning joint custody over the embryos on the grounds that Junior had a "constitutionally protected right not to beget a child" without his consent.

Mary Sue decided to appeal this ruling to the Tennessee Supreme Court. She no longer wanted to use the embryos herself, however; she wanted to donate them to another childless couple, rather than have them destroyed. Junior Lewis remained adamantly opposed, preferring that the embryos be destroyed.

In its ruling, the Tennessee Supreme Court scolded fertility clinics for not making the disposition of unused or unwanted embryos part of the consent process. The

court rejected the view that legal or moral rights begin at the point of conception. It found no legal basis for assigning personhood to an eight-cell embryo or for giving an embryo the same legal standing as a child. Instead, the court held that there is a fundamental right to privacy, according to the laws of Tennessee and the U.S. Constitution, that forbids compelling Junior Lewis Davis to become a parent against his will. The court concluded that "Mary Sue Davis' interest in donating [the embryos] is not as significant as the interest Junior Davis has in avoiding parenthood" and forbade the implantation of the seven embryos into any woman unless Junior Lewis agrees. In 1993, the U.S. Supreme Court upheld the Tennessee ruling.

The significance of the case goes further than the court ruling. The Tennessee Supreme Court has now dismissed the claim that rights begin at conception. It has also asserted the view that no man should be forced to parent against his will. These two rulings have obvious implications for the future of abortion policy in this country. For example, the U.S. Supreme Court will find it very difficult to ignore the Davis case in deliberating the fate of *Roe* v. *Wade,* the case that legalized abortion.

bleeding, especially immediately after the device is inserted. Although infrequent, these complications created enough concern to cause IUDs to be taken off the market in the United States in 1987. A few types of IUDs are again being made available, notably those that have few negative side effects.

Diaphragm with Spermicide

Legend has it that the famous lover Casanova used half of a hollowed out lemon to cover the cervix of a lover in an attempt to reduce the number of new little Casanovas he sired. The modern *diaphragm*, a thin rubber dome, works in the same way, although with considerably more success (90 percent effectiveness) because it is precisely fitted to the woman's cervix and is used in conjunction with a spermicide. The diaphragm is inserted just before intercourse and must remain in place for 8 hours following intercourse to be fully effective.

Condoms

Condoms are thin sheaths that are worn over the penis and trap sperm, preventing fertilization. Condoms have the additional advantage of preventing sexually transmitted diseases, which is why they are also called "prophylactics," meaning "disease preventing." Condoms have become an important line of defense in the war against AIDS. Used alone, condoms are 88 percent effective as contraceptives; their effectiveness increases to 95 percent when used with a spermicide. Inconvenience and diminished sensations are two drawbacks to condoms.

Recently, a female "condom" has become available, which fits into the vagina, forming a plastic lining that cannot be penetrated by sperm. The sheath is held in place by a ring that remains outside the vagina. The effectiveness of the female condom remains to be determined.

Vaginal Sponge with Spermicide

The spermicide-saturated, cup-shaped *sponge* fits over the cervix prior to intercourse, much like a diaphragm but without the need for fitting by a physician. This advantage may be offset by a serious concern: a slight increase in the incidence of *toxic shock syndrome*. This potentially fatal disease is usually due to excessive growth of a particular strain of toxin-producing bacteria in the vagina. When the sponge is used in conjunction with spermicide, the effectiveness is around 85 percent.

PERMANENT STERILIZATION

In a *vasectomy*, a surgical procedure for men, a portion of each vas deferens is removed, and the cut ends of the tubes are tied. Although spermatogenesis continues, sperm are blocked from reaching the penis, so the man's ejaculate contains no sperm cells, making the procedure 99.85 per-

cent effective. The volume of semen and the sensations of orgasm are unaffected by vasectomy.

In a *tubal ligation*, the woman's oviducts are surgically cut and sealed, preventing an egg from reaching the uterus or from even coming in contact with sperm but allowing ovulation to continue. Tubal ligation is 99.6 percent effective.

NATURAL METHODS

There are two natural methods of birth control: rhythm and coitus interruptus. The *rhythm method* requires abstinence from intercourse during a woman's fertile period, about 12 hours before and 48 hours after ovulation. A major problem with the rhythm method is the difficulty in predicting when ovulation occurs, especially in women with irregular ovarian cycles. (Ovulation occurs 14 days *before* the end of the cycle, not 14 days after the beginning of the cycle, making the prediction difficult.) Even when practiced diligently, failure rates range from 15 to 35 percent; thus, effectiveness is variable, at best. Effectiveness can be improved by keeping a daily record of the woman's body temperature, which rises about half a degree at the onset of her peak fertility days. Another clue is the change in consistency of cervical mucus from thick and sticky to thin and clear just before ovulation.

During *coitus interruptus*, also known as withdrawal, the man removes his penis from the vagina before ejaculation. Because it requires willpower and reduces gratification, failures are very likely.

Prevention of pregnancy can be accomplished by interfering with either the process of sperm–egg union or implantation. There are diverse ways of preventing pregnancy, each with varying degrees of effectiveness and safety. (See CTQ #5.)

SEXUALLY TRANSMITTED DISEASES

No discussion of human sexual reproduction is complete without considering those diseases that are transmitted from one sexual partner to another, a group of disorders called **sexually transmitted diseases (STDs).** These diseases include AIDS, syphilis, gonorrhea, genital herpes and more than a dozen others. To understand STDs, it is necessary to be familiar with the properties of the viruses and bacteria that cause most of these disorders. For this reason, we have deferred the discussion of STDs until Chapter 36, the chapter in which the structure and function of viruses and bacteria are considered (see The Human Perspective, page 794).

REEXAMINING THE THEMES

Relationship between Form and Function

⬚ The architecture of a sperm cell is tailored by evolution to accomplish the following tasks: swimming to the egg, penetrating the egg's surface barriers, and delivering a nucleus that contains the paternal genes. These functions are evident in the sperm's streamlined shape, its virtual lack of cytoplasmic and nuclear fluid, its cap of hydrolytic enzymes that are capable of digesting their way through the egg's outer layers, its compact sheath of mitochondria that provide the chemical energy needed to fuel the sperm's journey, its single flagellum that provides the motile force for locomotion, and its condensed nucleus that contains the full haploid number of chromosomes packed into the least possible volume. While the egg lacks such elaborate adaptations, its large size and nutrient content make the egg an ideal package for supporting the needs of embryonic development.

Biological Order, Regulation, and Homeostasis

⬚ Virtually every aspect of reproductive activity in mammals—from the development of the system in the embryo, to the maturation of the system at puberty, to the continuing day-to-day functioning of the system in the adult—is under tight regulation by the body's endocrine system. The master controls are located in the hypothalamus. The neuroendocrine cells located in this part of the brain secrete hormones that control the activity of the pituitary gland which, in turn, secretes hormones that control the activities of the gonads. The gonads secrete hormones of their own, particularly steroids, such as estrogen, progesterone, and testosterone. The timely secretion of these steroids determines the ability of the individual to reproduce successfully.

Unity within Diversity

⬚ The process of sexual reproduction is basically the same in all animals: Males produce very small, motile sperm, while females produce much larger, immotile eggs. Both types of gametes form in conjunction with meiosis, which reduces their chromosome number in half. These characteristics of sexual reproduction must have appeared at an early stage in the evolution of eukaryotes and have been retained ever since. Despite the similarities in gamete formation and structure, there is a great variety in the structure of reproductive systems and in the strategies used to bring the gametes together. Sex itself provided the genetic variety that allowed this diversity to develop.

Evolution and Adaptation

⬚ Sexual reproduction is one of life's most important evolutionary developments. Mixing genes from two parents has generated a vast diversity of organisms, introducing combinations of traits that never would have existed had asexual reproduction remained the only option. Asexual reproduction creates sameness, whereas sexual reproduction generates diversity, creating new types of organisms subject to natural selection; those with favorable adaptations proliferated. Today's rich diversity of life would not have been possible without sexual reproduction.

SYNOPSIS

Both asexual and sexual reproduction have advantages. Asexual reproduction is a biologically economical way to generate identical copies of oneself rapidly. In contrast, sexual reproduction is more costly energetically, but it produces offspring with variable characteristics, which makes it much more likely that a species population will be able to survive changing environmental conditions.

Animals employ several strategies that increase the likelihood of fertilization. Animals that depend on external fertilization release large numbers of sperm and eggs at about the same time, in the same location. External fertilizers use various mechanisms to help coordinate the release of gametes. Internal fertilizers produce fewer eggs, each with a much higher probability of developing to adulthood. Some animals are hermaphrodites; individuals that possess both male and female reproductive systems.

Gametes are produced by gonads. In humans, spermatozoa are generated in the seminiferous tubules of the male's testes, and ova are generated in the follicles of the female's ovaries. Sperm are highly differentiated cells that derive from spermatids that have formed from primary spermatocytes by meiosis. Primary spermatocytes are derived from spermatogonia, whose mitotic divisions provide a continuous source of germ cells throughout life. Sperm produced in the testes move into the adjacent tubules of the epididymis and, ultimately, into the vas deferens, where they are mixed with fluids from the seminal vesicles, Cowper's glands, and prostate gland to form semen, which can be ejaculated during orgasm.

In the female, all germ cells have entered meiosis by the time of birth. Most of these oocytes are located in small, primordial follicles. During each ovarian cycle, a number of follicles enlarge, and their contained oocyte undergoes enlargement and differentiation to form a cell that contains the nutrients needed to support embryonic development. One of these oocytes is released into an oviduct during each cycle. The oocyte then continues to the metaphase of the second meiotic division and awaits fertilization. Meiosis in the female includes highly unequal divisions, producing one cell that retains virtually all of the cytoplasmic material and forms the egg, while the others contain little more than nuclei and eventually disintegrate.

Sexual development and gamete formation in both sexes are regulated by hormones. In both males and females, the hypothalamus produces gonadotropin-releasing hormone (GnRH), which controls the production and release of two gonadotropins, LH and FSH, by the anterior pituitary. In the male, FSH promotes spermatogenesis, while LH stimulates the interstitial cells of the testes to produce testosterone, which maintains male reproductive function. In the female, both FSH and LH control the ovarian cycle and stimulate the production of estrogen. These three hormones, in addition to the progesterone that is produced by the corpus luteum (the follicle from which the oocyte is ovulated) control the menstrual (uterine) cycle. During each cycle, the lining of the uterus is thickened and vascularized in anticipation of an implanted embryo. If the embryo implants itself, HCG is produced and the corpus luteum and uterus are maintained. If the ovulated egg is not fertilized, the failure to produce an implanted embryo leads to the deterioration of the corpus luteum, a drop in progesterone and estrogen production, and the breakdown and sloughing of the uterine wall.

Key Terms

reproduction (p. 664)
asexual reproduction (p. 664)
sexual reproduction (p. 664)
fission (p. 664)
budding (p. 664)
parthenogenesis (p. 664)
external fertilization (p. 665)
internal fertilization (p. 665)
penis (p. 665)
hermaphrodite (p. 667)
spermatozoa (p. 669)
ova (p. 669)
scrotum (p. 669)
urethra (p. 669)
circumcision (p. 669)
seminiferous tubule (p. 669)
spermatogenesis (p. 669)
spermatogonia (p. 669)

primary spermatocyte (p. 669)
secondary spermatocyte (p. 669)
spermatid (p. 669)
epididymis (p. 670)
vas deferens (p. 670)
acrosome (p. 670)
semen (p. 670)
ejaculation (p. 672)
prostate gland (p. 672)
seminal vesicle (p. 672)
Cowper's gland (p. 672)
orgasm (p. 672)
follicle-stimulating hormone (FSH) (p. 672)
luteinizing hormone (LH) (p. 672)
gonadotropin-releasing hormone (GnRH) (p. 672)
oogenesis (p. 673)

vulva (p. 673)
labia majora (p. 673)
labia minora (p. 673)
clitoris (p. 673)
vagina (p. 673)
cervix (p. 673)
uterus (p. 673)
oviduct (p. 673)
oocyte (p. 673)
follicle (p. 673)
follicle cell (p. 674)
oogonia (p. 674)
ovarian cycle (p. 674)
ovulation (p. 674)
polar body (p. 675)
menstrual cycle (p. 677)
corpus luteum (p. 677)
infertility (p. 679)

Review Questions

1. Describe three differences in gamete production by men and women.

2. Trace the odyssey of a sperm cell from the spermatogonial stage to fertilization.

3. Explain why LH and FSH have such different effects in human males and females. Describe these effects and how the levels of these gonadotropins are controlled in males and females.

4. Compare and contrast the vas deferens and oviducts; penis and clitoris; testes and ovaries; testosterone and estrogen; parthenogenesis and fission; interstitial cells and follicle cells; ovarian cycle and menstrual cycle; uterus and cervix.

Critical Thinking Questions

1. Speculate on the differences in the use and value of scientific information obtained from the studies of individuals (Freud's method) and the studies of populations (Kinsey's method).

2. Compare the advantages and disadvantages of external fertilization versus internal fertilization and asexual reproduction versus sexual reproduction.

3. What, if any, is the difference in chromosome composition between a spermatogonium and a primary spermatocyte; a spermatid and a sperm; an oocyte in a primordial follicle, an ovulated egg, and an egg awaiting fertilization?

4. Explain why a woman who is taking birth control pills must stop taking them for several days every month to allow menstruation to occur.

5. The United States has one of the highest teenage pregnancy rates in the world. What do you think could be done to reduce the number of unwanted pregnancies among teens? (Obviously, there is no right or wrong answer to this question.)

Additional Readings

Barinaga, M. 1991. Is homosexuality biological? *Science* 253:956–957. (Intermediate)

Bullough, V. L., and B. Bullough. 1990. *Contraception: A guide to birth control methods.* Prometheus Books. Buffalo, NY: (Introductory)

Christenson, C. V. 1971. *Kinsey: A biography.* Indiana University Press. (Introductory)

Daly, M., and M. Wilson. 1983. *Sex, evolution, and behavior.* New York: Wadsworth. (Intermediate)

Duellman, W. E. 1992. Reproductive strategies of frogs. *Sci. Amer.* July:80–87. (Intermediate)

Knobil, E., and J. D. Neill. 1988. *The physiology of sex.* New York: Raven. (Advanced)

Marx, J. L. 1988. Sexual responses are—almost—all in the brain. *Science* 241:903–904. (Intermediate)

Silber, S. J. 1987. *How not to get pregnant.* New York: Scribners. (Introductory)

Symons, D. 1979. *The evolution of human sexuality.* New York: Oxford University Press. (Intermediate)

Sulloway, F. J. 1979. *Freud, biologist of the mind.* New York: Basic Books. (Introductory)

Witters, W., and P. Witters. 1980. *Human sexuality, a biological perspective.* New York: Van Nostrand. (Introductory–Intermediate).

Animal Growth and Development: Acquiring Form and Function

STEPS
TO
DISCOVERY
Genes that Control Development

THE HUMAN PERSPECTIVE

The Dangerous World of a Fetus

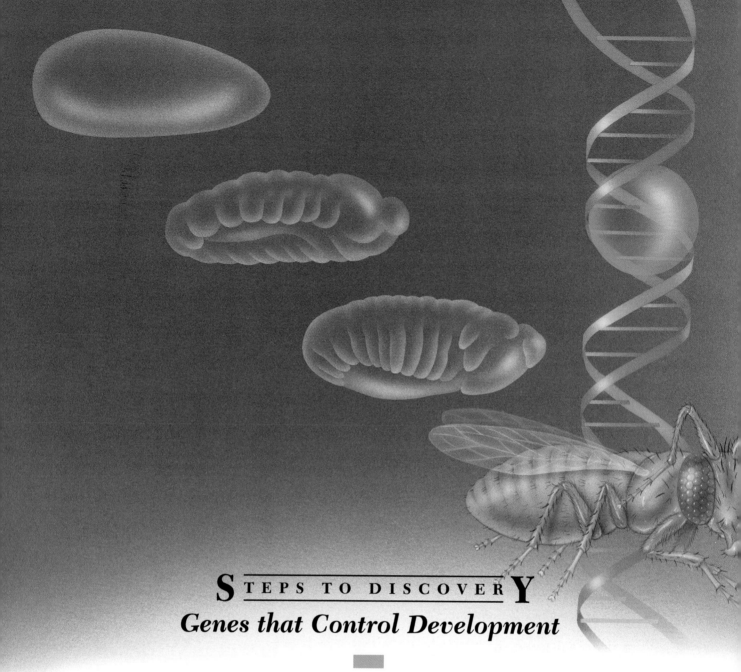

STEPS TO DISCOVERY
Genes that Control Development

During the 1940s, Edward B. Lewis, a geneticist at the California Institute of Technology, studied a mutant fruit fly with an abnormal body organization. An insect is composed of a head, thorax, and abdomen, each part containing a defined number of segments. The last segment of the thorax of a fruit fly is usually wingless, but Lewis's mutant fly had a second pair of wings on this segment; thus, the mutant was named *bithorax*.

In following years, other mutant fruit flies were isolated that showed even more profound disturbances in body organization. For example, the mutant *antennapedia*, studied

by John Postlethwait and Howard Schneiderman of Case Western Reserve University in the late 1960s, has a pair of legs growing out of its head in the place where antennae are normally found. Genes such as these, that affect the spatial arrangement of the body parts, are called **homeotic genes.** While humans are not known to suffer such drastic homeotic mutations as bithorax and antennapedia, there are many examples of serious developmental malformations that could be the result of mutations in homeotic genes. In addition, human embryos with drastic developmental defects tend to abort spontaneously, so the existence of mu-

Mutations in a crucial gene can "throw a switch" in the stages of development of a fruit fly that causes a pair of legs to develop

tant homeotic alleles might easily go undetected.

Homeotic genes are thought to play a role in the basic process by which each part of an embryo becomes committed to developing along a particular pathway—toward forming a leg rather than an antenna, for example. One way homeotic genes might exert such profound influence on the course of development is by acting as a type of "master" gene. As such, homeotic genes would control the transcriptional activity of other genes, whose products actually form the various tissues. The antennapedia gene, for example, might code for a protein that normally switches on the genes required for antenna formation in the appropriate cells of the developing head. If the antennapedia gene becomes defective, a different cluster of genes may be switched on, and the cells of that part of the body differentiate into a leg instead of an antenna.

In 1983, Walter Gehring, a Swiss biologist, discovered that a number of homeotic genes in the fruit fly contained a common sequence of about 180 nucleotides; this sequence was named the **homeobox.** Once the homeobox DNA from the fruit fly had been isolated, Gehring and others were able to search the DNAs of other organisms to see if they contained similar DNA segments. The homeobox sequence was soon found to exist within the DNA of many different animals, from worms to humans; in fact, it was even present in plants. This finding suggested that similar types of genetic processes take place during the development of very different types of organisms, but it left an important question unanswered: What was the function of the homeobox?

Recall from Chapter 17 that deciphering the amino acid sequence encoded by a gene is a relatively simple matter, once the nucleotide sequence of the DNA has been determined. When the amino acid sequence of the homeobox was deciphered and compared to the sequence of other known proteins, it was found to be very similar to a gene regulatory protein found in yeast that was known to bind to DNA. This correlation suggested that homeotic genes encoded DNA-binding proteins. By binding to a specific portion of a particular chromosome, the product of a homeotic gene could activate or repress transcription of nearby genes, much like a repressor protein in bacterial operons or a steroid receptor protein in eukaryotes (Chapter 15). In this way, homeotic genes might control the course of development. This concept was confirmed in 1988, when Patrick O'Farrell and his colleagues at the University of California demonstrated that products of two of the homeobox-containing genes in the fruit fly actually bound to DNA and altered the rate of transcription of nearby genes.

In the past few years, a number of investigators studying homeotic genes have turned their attention from fruit flies to mice. In doing so, they have invited speculation on the role of these genes in human development. In 1991, for example, Osamu Chisaka and Mario Capecchi of the University of Utah produced transgenic mice (page 319) that carried a genetically engineered version of a homeotic gene in place of the normal gene. The mice that were born with this altered homeotic gene exhibited a variety of severe abnormalities, including deformations of the throat and heart. Interestingly, a similar complex of abnormalities occurs in DiGeorge's syndrome, a rare human disorder that usually causes the death of the affected infant within the first few months of life. It is possible that this human condition is a result of a mutation in a homeotic gene whose expression is required during human development.

where a pair of antennae would normally be located.

*T*he newborn humpback whale being pushed by his mother to the ocean's surface for his first breath of air bears little resemblance to his appearance just 12 months earlier. At that time, he was a single, microscopic cell—a zygote—more potential than whale (Figure 32-1). Yet, packaged in these few micrograms of cellular material are all the blueprints and metabolic machinery needed to direct the zygote's transformation into one of the largest animals on earth. The body of an adult mammal—whether whale, mouse, or human—is an intricate composite of nerve cells, bone cells, muscle fibers, and a few hundred other cell types, all organized into complex tissues, organs, and organ systems. How does this complex structure emerge from what appears in a microscope to be a relatively simple fertilized egg? The answers to this question are found in the study of **embryos**—developing organisms during those stages in the early life of an animal that follow fertilization.

▼ ▼ ▼

THE COURSE OF EMBRYONIC DEVELOPMENT

Embryonic development is a programmed course of events that carries an organism along a path that leads from a zygote to an increasingly more complex, ordered structure.

The program for development is concealed within the zygote in two forms. (1) Most of the program is encoded within the genes the zygote inherits from its parents. As the zygote develops into an embryo, cells are formed by mitosis, ensuring that each daughter cell receives two full sets of homologous chromosomes. Consequently, all the cells in an animal—even highly differentiated cells, such as those of the skin—retain all the genetic information (page 297). Development is the result of each cell selectively expressing the genes relevant for its specialization. (2) The remainder of the program for development resides in the cytoplasm of the egg, which includes a store of messenger RNAs that help direct the course of early development. Specific mRNAs are located in specific parts of the egg, influencing the development of that particular region of the embryo.

◐ The development of a biologically complex individual is driven by the input of energy. In most animals, the energy required to fuel the construction of the embryo is present in the form of **yolk,** a mixture of lipids, proteins, and polysaccharides that are stored within the egg. Yolk provides the embryo with nutrients until the developing animal can obtain food for itself. Some animals, such as sharks, reptiles, and birds, have large yolk supplies that provide nutrients throughout development. For example, a bird hatches from the egg at a relatively advanced anatomic stage, without having had to find its own food prior to that time. Some animals, including sea urchins, produce microscopic-sized eggs that have very little yolk. A sea urchin hatches at an early stage; within a matter of hours, it develops into a **larva,** a self-feeding immature form of an animal. Mammals also develop from very small eggs with very little yolk, but, unlike sea urchins, they receive the required nutrients directly from their mother through an interfacing of the maternal and embryonic bloodstreams.

(a)

FIGURE 32-1 ━━━━━━
The course of embryonic development.
(a) Micrograph of a mammalian egg showing the cytoplasm, with its undistinguished appearance, and the blue-stained chromosomes. The egg reveals no visible evidence of its potential to form an incredibly complex animal, such as the humpback whale shown in *(b).* In fact, the zygote has a less complex structure than virtually all of the cells that descend from it. The smaller cell associated with the egg is a polar body (see Figure 31-10).

(b)

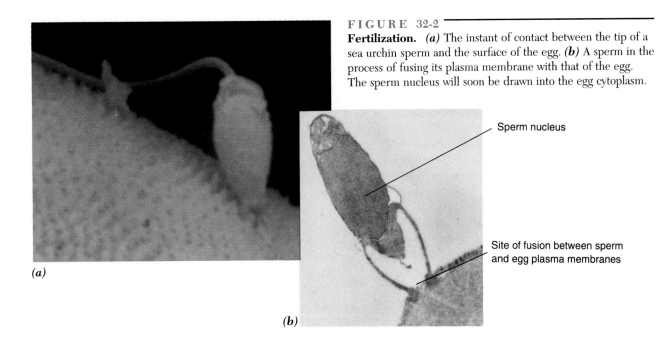

FIGURE 32-2
Fertilization. *(a)* The instant of contact between the tip of a sea urchin sperm and the surface of the egg. *(b)* A sperm in the process of fusing its plasma membrane with that of the egg. The sperm nucleus will soon be drawn into the egg cytoplasm.

Sperm nucleus

Site of fusion between sperm and egg plasma membranes

(a)

(b)

All eggs undergo similar processes of early development, passing through the stages of fertilization, cleavage, blastulation, and gastrulation. We will begin our discussion as development begins: with an unfertilized egg awaiting a chance encounter with a sperm.

FERTILIZATION: ACTIVATING THE EGG

An unfertilized egg is a cell that is primed and ready to begin development. The egg must first be *activated*, however, a function of the fertilizing sperm. Protruding from the plasma membranes of the egg and sperm of the same species are proteins with complementary molecular structures. As the tip of the sperm contacts the plasma membrane of the egg (Figure 32-2*a*), these interacting proteins bind to one another, fusing the two membranes (Figure 32-2*b*) and forming a single cell that contains both maternal and paternal sets of chromosomes. Providing the egg with this second, homologous set of chromosomes is another important function of the fertilizing sperm.

The eggs of most species initially respond to sperm contact by forming barriers that prevent entry of additional sperm. In most animals, the entry of more than one sperm causes the zygote to contain an excess number of chromosomes, which is so disruptive that it leads to early embryonic death. In those species that have been studied —such as the sea urchin, frog, and mouse—a wave of electrical activity sweeps around the surface of the egg within a second or two of sperm contact, much as an action potential moves along a neuron. This wave makes the egg's plasma membrane instantly unresponsive to the advances of other sperm in the neighborhood besides the one that has already made contact.

The stimulus provided by sperm contact is transmitted from the egg surface to the interior of the egg by the release of calcium ions from storage sites in the cytoplasm. One of the functions of the calcium ions is to trigger the secretion of materials from the cytoplasm into the region surrounding the egg's surface. These materials form an extracellular *fertilization membrane* that protects the egg as it develops and prevents penetration by additional sperm (Figure 32-3). Another indication of activation is the sharp increase in oxygen consumption that occurs in most zygotes in the period following fertilization.

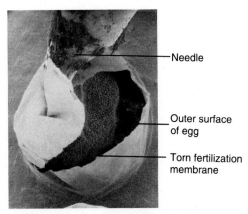

Needle

Outer surface of egg

Torn fertilization membrane

FIGURE 32-3
A fertilized sea urchin egg surrounded by an extracellular fertilization membrane that was formed by materials expelled from the egg cytoplasm very soon after sperm contact. The outer surface of the egg can be seen after tearing the fertilization membrane with a fine needle.

CLEAVAGE AND BLASTULATION: DIVIDING A LARGE ZYGOTE INTO SMALLER CELLS

A fertilized egg is an unbalanced cell; it has a huge amount of cytoplasm but only two sets of homologous chromosomes. This situation changes very rapidly during early development, as the egg undergoes a succession of mitotic divisions, called **cleavage.** Cleavage is not a time of growth but a period when the oversized egg is divided into a large number of smaller cells, known as **blastomeres** (Figure 32-4). Cleavage ends with **blastulation,** the formation of a ball of cells, called a **blastula.** In most animals, the blastula contains an internal, fluid-filled chamber, called the **blastocoel,** whose relative size and location depends primarily on the amount of yolk in the egg (Figure 32-5). For example, both sea urchin and mammalian embryos are essen- tially devoid of yolk; both develop via a blastula stage that has a large, central blastocoel. In contrast, frogs and birds produce larger eggs with considerably more yolk. Conse- quently, the relative size of the blastocoel in these animals is dramatically reduced.

The Developmental Potential of Cleaving Cells

During cleavage, the fertilized egg divides into many cells, each with a specific developmental "fate." A cell may be- come part of the liver, the brain, or some other part of the body. One of the first questions developmental biologists asked was: Can the cells of an embryo be experimentally "tricked" into developing into structures other than those they would normally form? In other words, how rigidly is the fate of a cell determined? The answer depends on sev-

First cleavage furrow

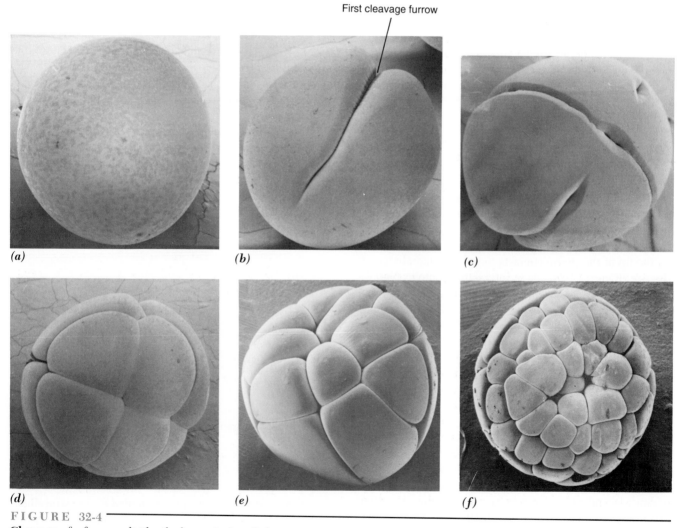

(a)　　　　　*(b)*　　　　　*(c)*

(d)　　　　　*(e)*　　　　　*(f)*

FIGURE 32-4

Cleavage of a frog egg divides the large, single-celled zygote into a number of smaller cells, called blastomeres. Although the egg contains more cells following cleavage, the total volume of cellular ma- terial is the same as is that of the original zygote.

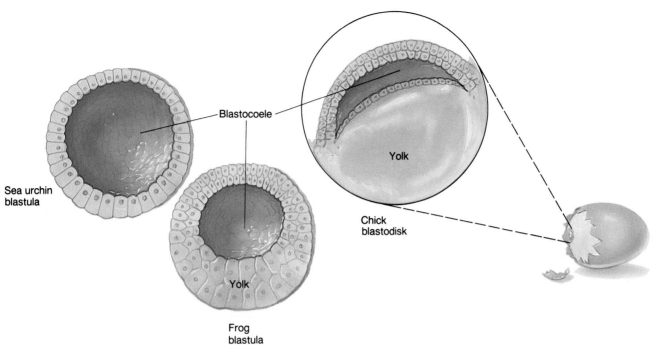

FIGURE 32-5

The blastula stage of a sea urchin, frog, and chick.   A sea urchin develops rapidly from a small, relatively yolk-free egg into a feeding larva. The sea urchin's blastula has a large blastocoel, surrounded by a single layer of cells. Frogs and other amphibians have eggs that contain a relatively larger content of yolk, which leaves relatively less room inside the blastocoel. Reptiles and birds— animals whose embryos develop on dry land—must supply their developing offspring with enough food and water to last until they hatch from the shells that enclose them. The enormous yolk and albumen (egg white) provide these resources and occupy most of the egg space. Because of the presence of so much yolk, the growth of early embryonic cells is restricted to a small area on the yolk surface, where an embryonic disk, or *blastodisk,* is formed.

eral factors, particularly the species and the stage of development being studied. We will examine the question at an early stage in the development of a frog.

In most cases, the first cleavage furrow in the frog (see Figure 32-4) divides the egg into two cells that form the left and right halves of the animal. This can be demonstrated by injecting a fluorescent dye into one of the first two blastomeres; only the cells on one half of the body receive the dye. When the first two cells of a frog embryo are separated from each other with a fine needle, each cell is able to develop *independently* into a complete embryo and larva (Figure 32-6, upper). We can conclude that, when separated from its neighbor, each cell somehow responds to the fact that it is now alone and forms a complete individual rather than the half of an individual it would have formed if left attached to the other cell. This is a remarkable observation. How can a part of an embryo have a "sense" of the whole, "knowing" it is now alone and must form the entire organism? How can a cell that is to form a part of an embryo suddenly regulate

its development and form additional parts not in its normal repertoire? These remain among the most basic, unanswered questions in developmental biology.

The Roles of the Genes and the Cytoplasm

Genes and cytoplasmic materials work together to generate the biological complexity that characterizes embryonic development. The importance of both components—genes and cytoplasm—is readily demonstrated. For example, if a fertilized sea urchin, frog, or mouse egg is treated with a drug that inhibits the transcription of genes, the egg will develop relatively normally up to the blastula stage, at which time development stops. This simple experiment illustrates two important aspects of development:

1. *An embryo doesn't need its genes for the very first stages of embryonic development*   Whatever new proteins are needed to carry an embryo from the form of a fertilized egg to the blastula stage are synthesized using

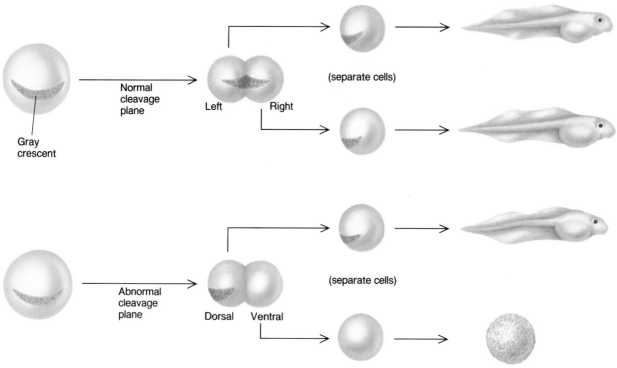

FIGURE 32-6

The developmental fate of isolated frog blastomeres. In the upper panel, the first cleavage plane divides the right and left halves of the egg, as indicated by the bisection of the *gray crescent;* each cell is able to develop independently into an intact tadpole. In the lower panel, the first division is altered so it doesn't bisect the gray crescent; only the dorsal cell containing the gray crescent develops normally.

mRNAs that were stored in the egg at the time it was released from the ovary.

2. *An embryo must use its genes to get past the blastula stage* Before an embryo can begin the next stage of development, its genes must spring into action, producing new mRNAs that direct the sweeping changes that are to occur during later development.

Further experiments provide evidence of the importance of the egg's cytoplasm. Occasionally, the first cleavage plane of an amphibian egg fails to divide an egg into left and right halves. Instead, the plane divides the cell into blastomeres that will give rise to the dorsal and ventral halves of the animal (Figure 32-6, lower). (The dorsal side of the egg will become the dorsal side of the animal—the side containing the animal's backbone. The opposite, or ventral, side will become the belly of the animal.) The division of the egg into dorsal and ventral blastomeres, rather than left and right blastomeres, can be detected by the failure of the cleavage plane to pass through the **gray crescent,** a distinctive, crescent-shaped, cytoplasmic landmark (shown in Figure 32-6) that always forms after fertilization on the *dorsal* surface of a frog egg. One might expect that a dorsal or ventral blastomere would have the same developmental

potential as a right or left blastomere; that is, the potential to form an entire embryo. This is not the case, however. Whereas, the isolated dorsal blastomere will develop into a normal embryo, the isolated ventral blastomere forms little more than a ball of cells. Only the blastomere that contains the gray crescent has all the cytoplasmic components needed for development, even though both cells contain identical genetic information. The important role of the gray crescent in frog development is discussed further below.

GASTRULATION: REORGANIZING THE EMBRYO

Gastrulation is the process whereby the undistinguished blastula is transformed into a more complex stage of development, called a **gastrula.** Gastrulation is characterized by an extensive series of coordinated cellular movements, whereby regions of the blastula are displaced into radically different locations. The dramatic events that take place during gastrulation in a frog (Figure 32-7) provide an overview of the process.

⬢ Even though gastrulation takes place in markedly different ways in different types of embryos, in principle, the

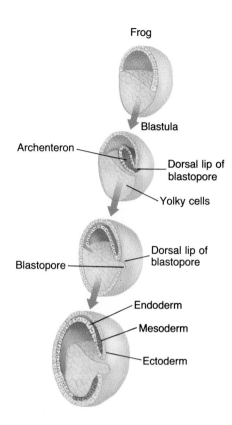

Frog

Blastula

Archenteron

Dorsal lip of
blastopore

Yolky cells

Blastopore

Dorsal lip of
blastopore

Endoderm

Mesoderm

Ectoderm

FIGURE 32-7

Gastrulation in the frog. The first indication of the onset of gastrulation in the frog is the appearance of a groove on the dorsal side of the embryo, just below the gray crescent. The opening into the interior of the embryo is the *blastopore,* and the fold above the groove is the *dorsal lip* of the blastopore. During gastrulation, cells at the rim of the blastopore migrate into the interior of the embryo and are replaced by new cells that move over the surface toward the blastoporal lip. Once inside the embryo, the cells move deeper into the interior, away from the blastopore, forming interior walls of an increasingly spacious cavity called the *archenteron.* The walls of the archenteron consist of endodermal cells (in purple) that will give rise to the digestive tract. The archenteron remains open to the outside through the blastopore, which corresponds in position to the future anus. Between the ectoderm and the endoderm and above the archenteron, a third group of cells develop into the mesoderm (in green), which will give rise to the skeleton, muscles, and other mesodermal derivatives. Once gastrulation is complete, the entire external layer of the embryo is composed of ectodermal cells. (Note that gastrulation in other animals, such as the sea urchin or chick, occurs by a very different pathway.)

process achieves a similar end result. Regardless of the animal, by the time gastrulation has been completed, the embryo (gastrula) can be divided into an inner, middle, and outer layer, which correspond to the three embryonic *germ layers:* the endoderm, mesoderm, and ectoderm. In vertebrates (such as the frog of Figure 32-7), the inner layer of the gastrula, or **endoderm,** gives rise to the digestive tract and its derivatives, including the lungs, liver, and pancreas; the outer layer, or **ectoderm,** gives rise to the epidermal layer of the skin and to the entire nervous system; and the middle layer, or **mesoderm,** gives rise to the remaining body components, including the dermal layer of the skin, bones and cartilage, blood vessels, kidneys, gonads, muscles, and the inner linings of the body cavities (Figure 32-8).

FIGURE 32-8

A schematic illustration of the fate of the three germ layers in vertebrates.

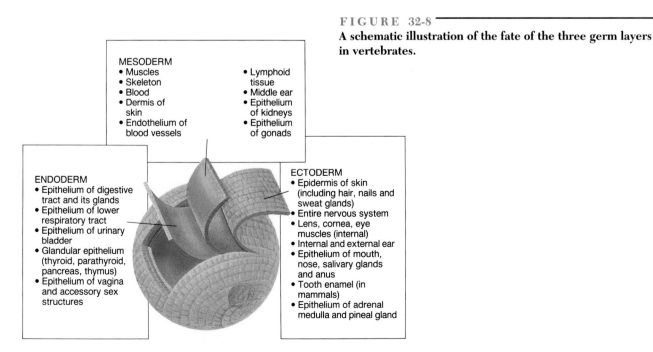

MESODERM
• Muscles
• Skeleton
• Blood
• Dermis of
  skin
• Endothelium of
  blood vessels
• Lymphoid
  tissue
• Middle ear
• Epithelium
  of kidneys
• Epithelium
  of gonads

ENDODERM
• Epithelium of digestive
  tract and its glands
• Epithelium of lower
  respiratory tract
• Epithelium of urinary
  bladder
• Glandular epithelium
  (thyroid, parathyroid,
  pancreas, thymus)
• Epithelium of vagina
  and accessory sex
  structures

ECTODERM
• Epidermis of skin
  (including hair, nails and
  sweat glands)
• Entire nervous system
• Lens, cornea, eye
  muscles (internal)
• Internal and external ear
• Epithelium of mouth,
  nose, salivary glands
  and anus
• Tooth enamel (in
  mammals)
• Epithelium of adrenal
  medulla and pineal gland

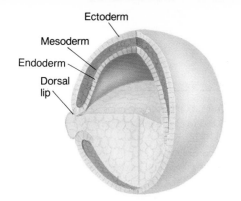

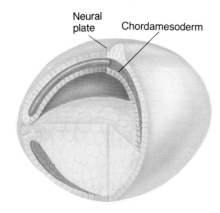

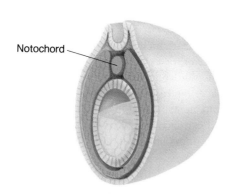

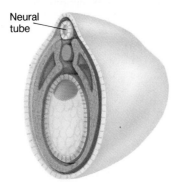

Successive steps in the formation of the neural tube in a frog. The first sign of the development of the future nervous system is the thickening of the ectoderm along the dorsal midline, which forms the neural plate. The edges of the neural plate roll upward and fuse together, forming a hollow neural tube along the midline. The formation of the neural tube is induced by the underlying chordamesoderm, which moved into this position during gastrulation.

NEURULATION: LAYING THE FOUNDATION OF THE NERVOUS SYSTEM

Toward the end of gastrulation in vertebrates, the ectodermal cells situated along the embryo's dorsal surface become elongated, forming a tall epithelial layer, called the **neural plate** (Figure 32-9). This single layer of cells will develop into the entire nervous system in the following manner. First, the neural plate thickens. Then, it rolls upward and fuses with itself to form a cylindrical, hollow **neural tube.** The neural tube is wider in the anterior portion of the embryo, where it will differentiate into the brain, and narrower in the posterior portion, where it gives rise to the spinal cord. On occasion, the neural tube fails to close completely during human development and a baby is born with *spina bifida occulta,* a condition in which a portion of the spinal cord remains open to the outside. A person born with this type of spina bifida may suffer partial paralysis and is usually confined to a wheelchair.

Formation of the Neural Tube

The ectodermal cells that lie in the dorsal midline of the gastrula cannot develop into a neural tube and then into a brain and spinal cord without a little help from underlying tissues. Beneath the ectoderm that will give rise to the neural tube is a layer of mesodermal cells, called the **chordamesoderm** (Figure 32-9). The chordamesoderm will form the **notochord,** a central dorsal rod in the embryo, which is later replaced by the backbone, or vertebral column. If the chordamesoderm is surgically removed from an embryo, that portion of the dorsal ectoderm above the mesodermal "hole" does not form a neural plate or neural tube. The role of the chordamesoderm in the formation of the nervous system was revealed in a classic experiment published in 1924 by a graduate student, Hilde Mangold, and her mentor and Nobel Laureate, Hans Spemann (Figure 32-10).

The cells that form the chordamesoderm are derived from that part of the frog egg that contains the gray crescent. At the beginning of gastrulation, these cells are located at the dorsal lip of the blastopore (see Figure 32-7). In their experiment, Mangold and Spemann surgically removed the dorsal lip of the blastopore from an early gastrula and transplanted it beneath the *ventral* ectoderm of a different gastrula of a similar age (Figure 32-10). Two different, but closely related, species were used in the experiment so the investigators could follow which cells were derived from the "host" embryo that received the transplant and which from the "donor" embryo that provided the transplanted cells.

Normally, the ventral portion of a frog embryo forms part of the belly, and the ventral ectoderm gives rise to belly skin. After transplantation, however, a remarkable transformation occurred. A secondary embryo developed in the ventral portion of the embryo at the site of transplantation,

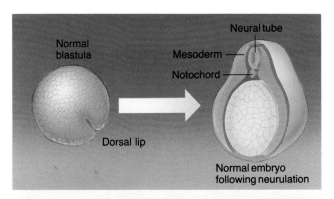

(b)

F I G U R E 32-10

Demonstration of the role of the chordamesoderm in neural induction. *(a)* In this classic experiment by Mangold and Spemann, the dorsal lip of the blastopore (which normally becomes the chordamesoderm) was removed from an early gastrula and transplanted beneath the ventral ectoderm of a gastrula of a recognizably different species. As expected, the donor tissue gave rise to the notochord and adjacent mesodermal tissue (shown in red). What was *not* expected was the differentiation of the ventral ectoderm of the host embryo into a second neural tube rather than the belly skin that it would normally form. Together, the donor mesoderm and the host tissues that are induced constitute a secondary embryo, complete with archenteron, neural tube, notochord, and other mesodermal structures. *(b)* The formation of a "double embryo" following transplantation of a dorsal lip.

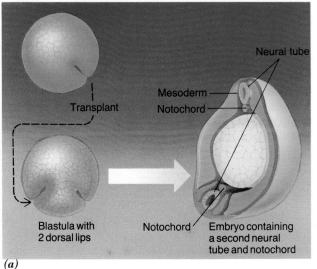

(a)

joined like a Siamese twin to the primary embryo. Analysis of sections of the embryos showed that the transplanted dorsal lip cells, which could be distinguished by their lighter color, differentiated into the notochord and other mesodermal tissues of the secondary embryo. But the neural tube of the secondary embryo and the entire nervous system it produced were clearly formed from the recipient's tissue; specifically, from the ventral ectoderm on the underside of the embryo.

This experiment reveals important properties of the ventral ectoderm and the chordamesoderm (dorsal lip), both of which participate in the formation of the secondary nervous system. First, it tells us that the chordamesoderm, which is derived from the gray crescent of the fertilized egg, has special properties that allow it to *induce* the ventral ectoderm to differentiate along a pathway it would not normally have taken. This process is known as **induction.** Secondly, it reveals that the the fate of the ventral ectoderm is not rigidly *determined* (committed) to form belly skin *at the early gastrula stage;* otherwise, it would not have been influenced by the chordamesoderm to differentiate into something besides skin tissue. When this same type of transplantation experiment is performed on a late gastrula

rather than an early gastrula, however, the transplanted dorsal lip is no longer able to alter the course of differentiation of ventral ectoderm; the ventral ectoderm develops into belly skin, regardless of the presence of nearby chordamesoderm. We can conclude that during gastrulation an important change occurs in the ventral ectoderm (and other parts of the embryo), whereby the cells become determined to form the structure dictated by their location and can no longer be influenced to go in other developmental directions. This is just one of many irreversible events that occur as embryos progress down the path of development.

The chordamesoderm is so important in the development of a vertebrate that it is often called the *primary organizer.* Biologists have spent over 50 years trying to determine precisely what it is that the chordamesoderm passes to the ectoderm to induce it to develop into nervous tissue, but the question remains unanswered. The mechanism by which the ectoderm responds to the inductive signal also remains a mystery. One possible answer is that the chordamesoderm releases a substance that activates homeotic genes within the adjacent ectoderm. The products of these genes would then direct the differentiation of these ectodermal cells into the nervous system.

DEVELOPMENT BEYOND NEURULATION

By the time neurulation has been completed, the ectoderm, mesoderm, and endoderm are in position to begin forming the body's organs. But the cells of the embryo remain undifferentiated at this stage, revealing little of the structural complexity that will soon emerge. The conversion of this organized mass of cells into an organism that contains complex, functional organs is one of the least understood aspects of embyronic development. Organ formation, or **organogenesis,** is a complex process whereby two or more specialized tissues develop in precise relationship with one another. The mechanisms that underlie organogenesis are remarkably similar among different organs in different animals, even though the organs produced are remarkably diverse in structure and function.

Positional Information

↻ What determines the spatial order of the parts that make up an animal's body? Why, for example, does a hand form at the end of a forelimb, and a foot at the end of a hindlimb? Why does a single, long bone form in the upper arm, and a pair of long bones in the forearm? These questions concern **positional information:** information that determines the location or spatial organization of parts of the body. In order for a cell to "know" what to form, it must receive some type of chemical or physical signal from its surroundings which inform it of its relative position within the whole.

The importance of positional information is readily demonstrated by a simple experiment performed on the developing limb of a chick embryo in 1959 by John Saunders of Marquette University. Although wings and legs have a very different structure and function in the adult bird, both body parts arise from similar-looking *limb buds* in the early embryo. The inner portion of the bud is destined to develop into either the thigh of the leg or the inner portion of the wing, while the outer portion of the bud is destined to develop into either the foot or the wing tip. Consider what might happen if a small block of undifferentiated tissue is removed from the inner portion of a *leg* bud and transplanted to the *tip* of the wing bud (Figure 32-11). You might expect that this block of transplanted tissue would develop into either (1) thigh tissue, if it ignored the influences from its new surroundings and developed as it would have in its original location, or (2) wing tip tissue, if it developed in harmony with its new environment. In fact, the transplanted tissue does neither; instead, it develops into a toe. The transplanted tissue "remembers" its previous position as part of a leg bud rather than that of a wing. At the same time, the tissue responds to its new location in the outer portion of a limb bud rather than at its base so it forms a toe rather than a thigh. It would appear that the block of transplanted tissue has received two distinct positional signals: An early signal tells the cells that they are part of a leg, while a later signal tells them they are in the outer portion of the bud and should develop into the outer portion of the leg (a toe).

Morphogenesis

Each part of the body has a characteristic shape and internal architecture that can be correlated with its particular function. The spinal cord is basically a hollow tube; the kidney consists of microscopic tubules; the salivary gland consists of groups of secretory cells clustered around a duct; the lungs consist of microscopic air spaces lined by thin cellular sheets; and so forth. The development of form and internal architecture within the embryo is called **morphogenesis** and is the product of several distinct cellular activities. These activities include increased or decreased rates of cell division; changes in the adhesion of cells to their neighbors; changes in cell shape; deposition of extracellular materials; passage of inductive signals from one group of cells to another; the programmed death of cells in certain locations; and movement of cells from one place to another. The ways in which some of these processes can shape embryonic

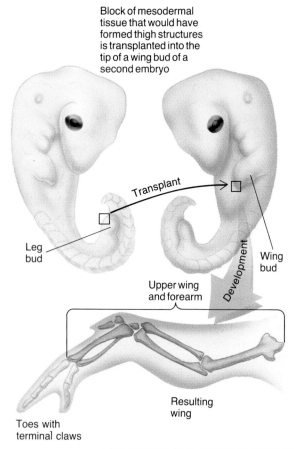

Block of mesodermal tissue that would have formed thigh structures is transplanted into the tip of a wing bud of a second embryo

Transplant

Leg bud

Wing bud

Development

Upper wing and forearm

Toes with terminal claws

Resulting wing

FIGURE 32-11

When a block of prospective thigh mesoderm is grafted into the tip of the wing, it differentiates into a toe.

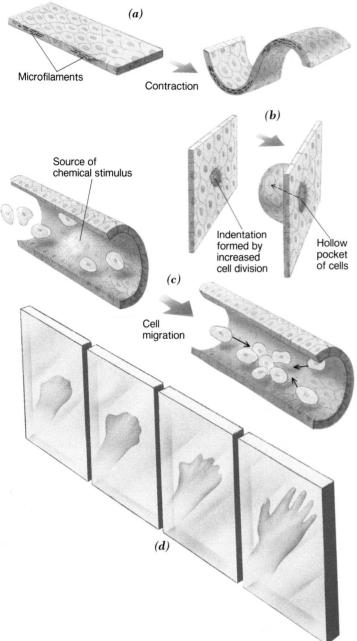

(a)

Microfilaments

Contraction

(b)

Source of
chemical stimulus

Indentation
formed by
increased
cell division

Hollow
pocket
of cells

(c)

Cell
migration

(d)

FIGURE 32-12

Morphogenesis: Shaping the embryo. *(a) Contraction of microfilaments* causes the curvature of a sheet of cells. *(b) Increased cell division* by a group of cells in a sheet can lead to a hollow outgrowth. *(c) Cell migration*, in this case in response to a chemical, leads to the movement of cells from one site to another. *(d) Cell death* leads to the elimination of cells from specific parts of a developing limb.

development is illustrated in Figure 32-12. Similar processes occur during morphogenesis in all embyos.

The formation of the human hand provides an example of morphogenesis that results from cell death. The fingers of your hand were "carved" out of a paddle-shaped structure by the *programmed* cell death of the tissue between the digits (Figure 32-12*d*). Cell death is described as "programmed" because studies on chick embryos indicate that the cells that will die receive a "death message" over 20 hours before they show any signs of ill health. If parts of the chick limb are removed at this earlier stage and transplanted to a nonlimb region of another embryo or placed in culture the cells die at approximately the same time as they

would have had they been left in place. It is as if the cells have a "death clock" that tells them that the time to self-destruct has arrived.

Formation of the eyeball exemplifies how organogenesis can occur as the result of an orderly, highly regulated series of changes in form. In the eyeball, these changes are driven by the induction of one group of cells by another. The eye begins its development as an outpocketing of the rudimentary embryonic brain, called the *optic vesicle* (Figure 32-13), which grows outward toward the overlying ectoderm. When the optic vesicle reaches its target, it induces the ectodermal cells to form a disc-shaped thickening, called the *lens placode*. In subsequent stages, the outer wall

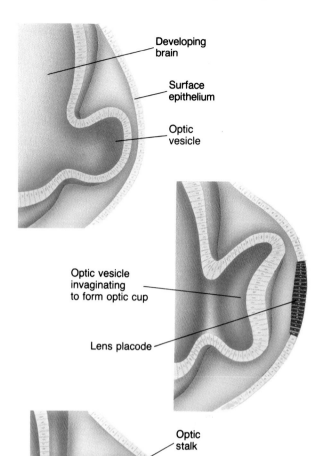

of the optic vesicle invaginates to form the *optic cup,* the forerunner of the retina, while the lens placode invaginates to form the lens. This invagination within an invagination ensures that the lens will be tightly enclosed by the surrounding layers of the eye. The invagination of each cellular sheet is driven largely by changes in the shape of the component cells.

Meanwhile, the developing lens induces the overlying ectoderm to develop into the *cornea*—the transparent covering of the eyeball. Each of the changes illustrated in Figure 32-13 assures the proper positioning of the retina, lens, and cornea of the eye, allowing light to enter the eye and become focused on the retina.

Cell Differentiation

During **cell differentiation,** the internal contents of a cell become assembled into a structure that allows the cell to carry out a specific set of activities. As a result, the cytoplasm of each type of cell becomes visibly different from

---

FIGURE 32-13

Steps in the formation of the eye. *(a)* The optic vesicle forms as an outgrowth of the wall of the brain and then induces the overlying ectoderm to form the lens placode. The outer layer of the optic vesicle invaginates to form the optic cup, which becomes the retina. The optic cup remains connected to the brain by the optic stalk through which the optic nerve later will pass. The lens placode invaginates within the optic cup to form the lens. *(b)* Scanning electron micrograph showing the arrangement of the parts of the embryonic eye.

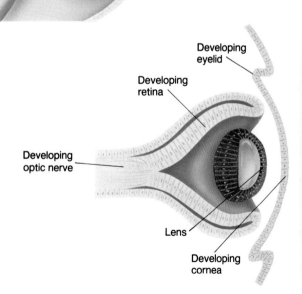

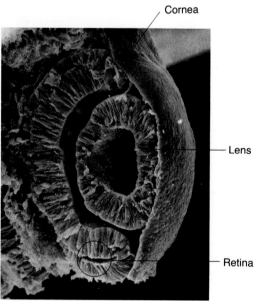

(a)   *(b)*

that of other cell types. A salivary gland cell, for example, develops an extensive rough endoplasmic reticulum and Golgi apparatus, two specializations necessary for the production and secretion of mucus; a muscle cell develops an extensive array of cytoplasmic filaments, which are required for the generation of tension; and a cell of the adrenal cortex develops an extensive smooth endoplasmic reticulum in which steroid hormones are synthesized.

While all cells have the same genetic information, different types of cells utilize that inheritance differently. For example, the cells of the salivary gland transcribe those genes that code for the proteins of mucus; the cells of a muscle utilize the genes that code for contractile proteins, such as muscle myosin and actin; and the cells of a differentiating red blood cell utilize the genes that code for hemoglobin. Each of the several hundred different types of cells in your body presumably contains a unique set of gene regulatory proteins that promote the activation of a unique set of genes. As a result, each cell transcribes only those genes whose products are required for the function of that particular cell type.

Embryonic development is a programmed course of events that transform a structurally unspecialized cell — the fertilized egg—into a complex animal containing functional organs that consist of highly differentiated cells. The events of embryonic development are governed primarily by gene transcription; fueled by nutrients that may be present in the egg or delivered by the mother; and brought about by a variety of mechanisms, including changes in cell shape, movement of cells from one place to another, cell death, and changes in the internal architecture of differentiating cells. (See CTQ #2.)

BEYOND EMBRYONIC DEVELOPMENT: CONTINUING CHANGES IN BODY FORM

In most animals, the end of embryonic development marks the end of the formation of new structures. This is particularly true for animals that have roughly acquired their adult form by the time they are born or hatch. In these organisms, postembryonic change generally constitutes an increase in size, although not all parts grow at equal rates. For example, the head of a human embryo is disproportionately larger than are other body parts. After birth, however, the arms, legs, and trunk grow much faster than does the head. If you make a circle with your arms, touching your fingertips above your head, you can see how large your head would be if it had grown at the same rate as the rest of your body. The development of reproductive capacity is another example of a postembryonic change, which is often accompanied by the appearance of secondary sex characteristics.

METAMORPHOSIS: EXTREMES IN POSTEMBRYONIC CHANGES

Postembryonic changes are most striking in those animals that develop through a larval stage that bears little resemblance to the adult. For these animals, the path to the adult passes through a dramatic transformation called **metamorphosis.** The degree to which metamorphosis can alter the body form of an animal is illustrated by the familiar example of the transformation of a water-bound, tailed tadpole into an air-breathing, tailless frog (Figure 32-14). The entire conversion takes place over a period of weeks. Among many invertebrates, including sea urchins, such drastic changes in

(a)   *(b)*

FIGURE 32-14

Metamorphosis. This larval wood frog *(a)* is in the process of transforming into an adult *(b).*

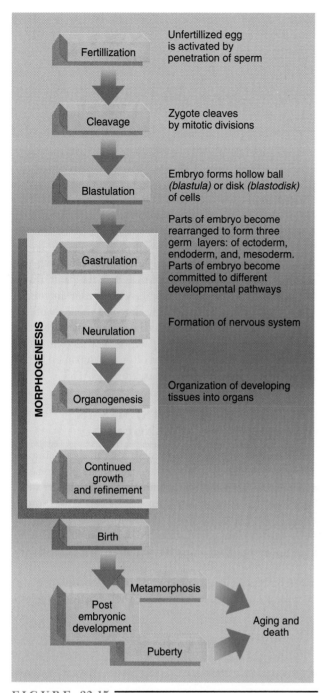

FIGURE 32-15
Summary of the events of growth and development in animals.

form can occur in a matter of minutes. A generalized summary of the stages of animal development is illustrated in Figure 32-15.

Developmental changes do not stop with the formation of an embryo. Animals continue to develop as they achieve a larger body form and undergo sexual maturation. Most animals achieve their basic adult body form as the result of gradual changes; others do so by a rapid, dramatic metamorphosis. (See CTQ #3.)

HUMAN DEVELOPMENT

The development of a human embryo begins with fertilization; 6 to 8 days later, the embryo "burrows" into the prepared endometrium of the uterus (Figure 32-16), rupturing uterine blood vessels as it penetrates and implanting itself in the lining. By the time the entry site reseals itself, the embryo is surrounded by a pool of maternal blood, which is continually replenished by fresh blood from maternal vessels. Branching projections, called *villi*, sprout from the **chorion,** the embryo's outer membrane covering. These chorionic villi provide a large surface for the exchange of respiratory gases, nutrients, and wastes between the embryo and its mother, until a **placenta,** constructed from the embryo's chorion and the mother's uterine lining, takes over the exchange. Veins and arteries develop in the villi, connecting the embryo's early circulatory system with the blood-filled space, serving as a temporary life support system. These connecting vessels soon become channeled within an **umbilical cord.** One end of the cord is attached to the embryo's belly, the other end to the placenta. In the placenta, the embryo's vascularized chorionic villi are bathed in maternal blood. The two blood supplies are separated by a thin layer of cells, permeable to nutrients, gases, and wastes.

THE STAGES OF HUMAN DEVELOPMENT

During the first 2 months following fertilization, all the major organs appear, many in a functioning capacity. By the end of the eighth week, the embryo has acquired a distinctively human form. Up to this point, the developing offspring is still called an embryo. During the last 7 months in the uterus, however, it is referred to as a **fetus.** In the fetal stage, organ refinement accompanies overall growth, as the fetus develops the ability to survive outside the uterus.

The 266 or so days between conception and birth are traditionally grouped into three stages, each called a *trimester*. The most dramatic changes occur during the first of these periods (Figure 32-17*a–d*).

The First Trimester

During the first 3 months of development, the embryo is transformed from a single-celled zygote into an ensemble of organ systems that are only slightly less complex than are those found in a full-term baby. Cleavage, blastulation, and implantation take place during the first week or so following fertilization. At the time of implantation, the embryo is called a **blastocyst,** and it contains two different groups of cells (Figure 32-16) with very different fates. Inside the spacious cavity of the blastocyst is a clump of cells called the **inner cell mass,** which will give rise to the tissues of the embryo itself. The outer wall of the blastocyst is called the **trophoblast;** it secretes the enzymes that allow the blasto-

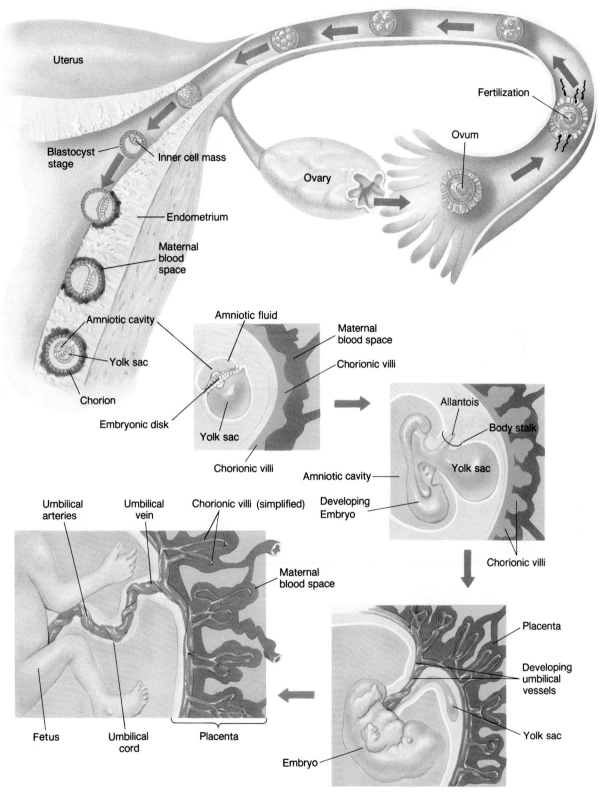

FIGURE 32-16

Implantation and placenta formation. Following fertilization, cleavage, and blastulation a blastocyst is formed that consists of a single outer layer of cells—the trophoblast—and an inner cell mass.
Soon after the embryo has implanted in the uterine wall, the inner cell mass gives rise to the embryonic disk, from which the embryo will develop. The amnion contains the liquid in which the embryo
floats as it develops, while the chorion and allantois of the body stalk help form the placenta and umbilical cord. Notice that fetal and maternal blood do not mix but are separated by the selectively permeable membranes of the chorion and the capillaries.

(a) *3 weeks:* Organogenesis begins with neural tube formation, seen here as enlarged crests that delineate the neural groove from which the spinal cord and brain develop.
Length = 0.25 cm (1/10 in).

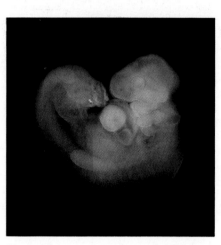

(b) *4 weeks:* A pumping heart, developing eye, and arm and leg buds are evident. The embryo has a long tail, gill arches, and a relatively enormous head.
Length = 0.7 cm (1/3 in).

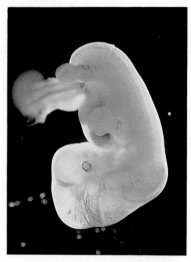

(c) *5 weeks:* Internal organ development is well under way. Fingers are faintly suggested. The fetal circulatory system is evident, and the body stalk is now an umbilical cord.
Length = 1.2 cm (1/2 in).

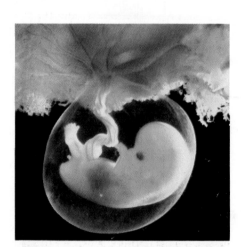

(d) *2.5 months:* The fetus, seen here floating in the amniotic cavity, now has all its major organ systems. The umbilical blood vessels and placenta are well defined, and the tail and gill arches have disappeared.
Length = 3 cm (1-1/2 in).

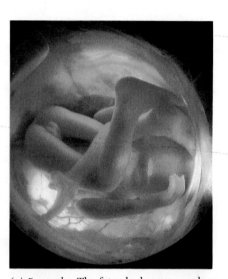

(e) *5 months:* The fetus looks very much as she will at birth. Bone marrow is assuming more of the blood-producing duties. The mother begins to feel fetal movements, even hiccuping.
Length = 25 cm (10 in).

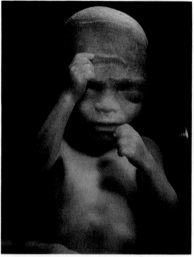

(f) *5.5 months:* The vernix caseosa is accentuated by a groove the fetus has scraped away with its thumb. Internal organs now occupy their permanent positions (except for the testes in males).
Length = 30 cm (12 in).

FIGURE 32-17

Some stages in human embryonic and fetal development.

cyst to penetrate the uterine wall and then differentiates into the **chorion,** the outermost of four *extraembryonic membranes.* These membranes are termed "extraembryonic" because they do not become part of the embryo and are eventually discarded prior to birth. In addition, the chorion secretes human chorionic gonadotropin (HCG), the hormone that stimulates the corpus luteum to continue to produce progesterone (page 677), which maintains the uterine lining, where the embryo is developing.

▣▶ The innermost extraembryonic membrane, the **amnion,** envelops the young embryo and encloses the *amniotic fluid* that suspends and cushions the developing body throughout its duration in the uterus. A third membrane forms the **yolk sac** which, in humans, simply contains fluid (in birds and reptiles, the yolk sac is filled with nutrient-containing yolk). The wall of the human yolk sac is a source of the embryo's blood cells and germ cells, the latter of which will migrate from this extraembryonic tissue into the gonads of the developing embryo. The fourth membrane, the **allantois,** is rich in blood vessels and eventually helps form the vascular connections between mother and fetus. These same four extraembryonic membranes (chorion, amnion, yolk sac, and allantois) are also present in the eggs of birds and reptiles, revealing the evolutionary relationship among the three classes of "higher" vertebrates: reptiles, birds, and mammals.

A few days after implantation, the embryo begins to gastrulate. The cells of the inner cell mass become rearranged to form a double layer of cells, called an *embryonic disk,* which is separated from the overlying chorion by a newly formed space, called the *amniotic cavity* (Figure 32-16). Your entire body, with the exception of the germ cells that migrate in from the yolk sac, is derived from the cells that make up the embryonic disk. By the end of the first month, the embryo, though still less than half a centimeter (³⁄₁₆ of an inch) long, has begun forming its nervous system, lungs, liver, and several other internal organs. At this stage, the heart is a four-chambered pump; the first signs of eyes and a nose appear; and four buds protrude from the side of the "**C**-shaped" embryo. By the end of the following week, these buds will be vaguely recognizable as developing arms and legs.

During the second month, the liver takes over its temporary job as the main blood-producing organ of the embryo, while an intricate cartilage-containing skeleton begins to change into bone. The brain develops cerebral hemispheres, and spinal nerves grow out from the spinal cord. The face acquires a distinctively human look, with its slate-colored eyes and the beginnings of eyelids. Muscles begin to form and assume their permanent relationships. Fingers and toes become evident. The embryo has become a fetus.

The third month is devoted primarily to the growth and development of existing structures, with the exception of the formation of genitals, which reveals the sex of the fetus. The limbs are clearly recognizable, with well-sculpted fingers and toes, complete with nails. The fetus begins reflex movements that go undetected by the mother.

The first trimester is unquestionably the most dangerous time for the developing embryo. The magnitude of the developmental changes that occur during this time renders the embryo particularly vulnerable to pathogens and chemicals that would be relatively harmless to an older fetus or a newborn human (see The Human Perspective: The Dangerous World of a Fetus).

The Second Trimester

As the 7.5-centimeter fetus enters its fourth month, it develops sucking and swallowing reflexes and soon starts kicking. The mother does not usually feel these movements until the fifth month. The developing skeleton is clearly distinguishable by X-ray examination, and the bone marrow begins blood-cell production. The fetus acquires a downy coat of soft body hair, called *lanugo,* that covers the skin. Lanugo is believed to help the fat droplets that are secreted by fetal sebaceous glands stay on the skin. These droplets form a "greasy" protective coating, the *vernix caseosa,* which conditions the fetal skin during its aquatic tenure in the amniotic fluid. Midway through the pregnancy, two heartbeats become discernible in the mother: her own (about 72 beats per minute) and that of the fetus (up to 150 beats per minute). Convolutions appear in the cerebral cortex of the brain, and sense organs begin supplying the fetus with limited information about its environment.

Although most of the organ systems are at least partially functional by the end of the second trimester, the fetus has little chance of survival if it is born before the seventh month. Even with expert medical care, a 23-week-old fetus has only a 10 percent chance of surviving outside the uterus. If born a month later, its chances of survival jump to 50 percent.

The Third Trimester

During the last 3 months of pregnancy, the fetus increases its body weight by 500 to 600 percent. The brain and peripheral nervous system enlarge and mature at an especially rapid rate during the final trimester. Neurological performance later in life, including intelligence, depends on the availability of protein needed for fetal brain development. Women who suffer protein deficiency during this time tend to have babies that are mentally slower throughout life. Furthermore, this condition cannot be reversed by providing the child with a protein-rich diet. By the end of the third trimester, the fetus is capable of regulating its temperature and can control its own breathing. This latter ability, together with the degree of lung maturation, generally determines a premature infant's chances of survival.

The formerly lean fetus changes in appearance, as fat deposits form under the skin, imparting the rounded, chubby shape characteristic of many babies. In males, the testes descend into the scrotum. Most fetuses change posi-

◁ THE HUMAN PERSPECTIVE ▷
The Dangerous World of a Fetus

The "womb," or uterus, is often used as a metaphor for warmth, protection, and security—a world free of fear and danger. Yet, a fetus may be exposed to a great many threats while residing in its mother's uterus, including nutrient deficiency, infectious microbes, radiation, alcohol, and the chemicals found in drugs, cigarettes, and environmental pollutants. The placenta screens out most harmful substances, but some dangerous agents may pass through the placental barrier and enter the fetal circulation. For example, some radiation passes through *all* living barriers, bombarding the fetus with potential *teratogenic* (embryo-deforming) consequences. Large doses of X-rays during the first trimester of pregnancy may cause mental retardation, skeletal malformations, and reduced head size, and may predispose the unborn child to the development of cancer later in life. When a cell is genetically impaired during the first trimester, all the millions of cells that develop from the affected cell inherit the impairment, amplifying the effect. These cells may fail to differentiate normally and, consequently, produce organs that are malformed, some so severely that the fetus or newborn child has little chance of survival.

The virus that causes *rubella*, or "German measles," illustrates the vulnerability of the embryo to infectious microbes during its first trimester. The disease is so mild that a pregnant woman with rubella may be unaware that she has the disease. The virus can pass through the placenta, however, and infect the emerging embryonic organs, interfering with their normal development. The result can be physical malformations, mental retardation, deafness, or death in about half of the babies exposed during their first 6 weeks of development. Yet, second- and third-trimester fetuses exposed to rubella suffer no impairment. Routine vaccination against rubella has dramatically reduced these tragic occurrences. In recent years, the virus presenting the greatest epidemiologic threat to newborns is HIV, the virus responsible for AIDS. Roughly half of all newborns delivered by infected mothers will be infected with HIV; most of these babies will develop the disease and die within the first few years of life. It is not clear why some babies resist becoming infected in the womb, while others succumb.

Another microbe that can reach the fetus is the streptococcus bacterium, which may infect as many as 12,000 babies in the United States each year. While the bacteria may cause no symptoms in the mother, in the infant, the infection can cause permanent brain damage, cerebral palsy, lung and kidney damage, blindness, and even death. All pregnant women are urged to be tested for Group B streptococcus, the chief cause of preventable newborn infections. Fortunately, most microbes cannot cross the placenta, so the fetus usually remains healthy while the mother is combating a cold, the flu, or most other microbial onslaughts.

Many chemicals to which the mother may be intentionally or accidentally exposed can also compromise fetal development. Some of these chemicals may be found in medications. In the 1950s, thousands of women who took the mild tranquilizer thalidomide during early pregnancy delivered babies whose arms and legs had failed to develop. Although stricter drug regulations resulted from this tragic episode, many pharmaceutical products that are potentially dangerous to fetuses are still available on the market today. These include some vitamins, which can be dangerous when taken in large doses (especially D, K, and C), cortisone, antibiotics, birth control pills, tranquilizers, anticoagulants, and thyroid drugs. Drugs such as LSD, marijuana, cocaine, and alcohol expose the fetus to proven toxins or teratogenic agents. Alcohol consumption by a pregnant woman during a critical period of fetal development can lead to *fetal alcohol syndrome,* a collection of childhood deformities that includes mental retardation, reduced head size, facial irregularities, and learning disabilities later in life. Children born to mothers addicted to crack cocaine also experience severe emotional and learning disabilities.

Tobacco smoke constitutes yet another serious danger to the developing fetus and may contribute to infant mortality. Cigarette smoking during pregnancy tends to impair fetal growth and reduce resistance to such respiratory infections as bronchitis and pneumonia during the first year of life. The carbon monoxide in tobacco smoke competes with oxygen for hemoglobin binding sites, reducing the amount of oxygen available to the fetus. Nicotine constricts blood vessels, further reducing the blood supply. Tobacco smoke may also contain teratogens and may contribute to the development of heart and brain defects, a cleft palate, or sudden infant death syndrome (SIDS).

tion in the uterus, becoming aligned for a head-first delivery through the birth canal.

One system that does not mature by the time of birth is the immune system; human babies are born immunologically deficient. During the final fetal month, however, antibodies from the mother's blood cross the placenta and temporarily fortify the baby's defenses against infectious diseases until its own immune arsenal becomes competent. In breast-fed babies, this immunologic gift is supplemented by the maternal antibodies and immune cells found in breast milk.

BIRTH

After 9 months of development, a human fetus is 6 billion times larger than was the original zygote. Yet, this astonishing size increase probably plays no direct role in inducing **birth,** or *parturition.* In fact, the fetus is believed to have little to do with initiating the uterine contractions of birth. Although the birth-triggering factor(s) remains a mystery, the onset of labor is preceded in most women by a drop in the concentration of progesterone, the hormone that inhibits contractions of the smooth uterine muscles during pregnancy. Once labor begins, a cascade of events promote uterine muscle contractions of increasing strength and frequency. The posterior pituitary releases oxytocin, which stimulates muscle contraction in the uterus. This response then triggers a reflex release of even more oxytocin, establishing a positive feedback loop that intensifies contractions. Prostaglandins (page 532) are also believed to participate in labor; this contention is supported by the ability of aspirin, a prostaglandin antagonist, to delay birth. Both oxytocin and prostaglandins may be administered by injection to induce labor when its delay threatens the health of the mother or fetus.

Birth occurs in three distinct stages. During the first stage, the mucous plug that blocks the cervical canal and prevents microbial invasion of the intrauterine environment is expelled. The amniotic sac ("bag of waters") ruptures during this stage, and its fluid is discharged through the vagina. Contractions during this stage force the cervical canal to dilate until its diameter enlarges to about 10 centimeters (3.9 inches), marking the onset of the second stage of labor: the expulsion of the fetus. Powerful contractions move the fetus out of the uterus and through the vagina, usually head first. Fewer than 4 percent of all deliveries are "breech births," whereby the buttocks or legs come out first. Contractions during the final stage expel the detached placenta, called the *afterbirth,* from the uterus. Additional contractions help stop maternal bleeding by closing vessels that were severed when the placenta detached.

↻ A newborn infant quickly acclimates to its terrestrial environment. With the umbilical cord clamped and severed, the baby rapidly depletes its source of maternal oxygen; still unable to breathe, the baby's respiratory waste (carbon dioxide) accumulates in the blood. The brain's respiratory center responds to this increase by activating the breathing process, and the lungs inflate with air for the first time. But before the lungs can oxygenate the blood, a circulatory modification is required because the unneeded fetal lungs were bypassed by the bloodstream in the uterus. The septum between the right and left atria of the *fetal* heart is perforated by a large hole, called the *foramen ovale,* that allows blood to flow directly from one side of the heart to the other, rather than traveling through the pulmonary circulation. At birth, the oval window is immediately closed by a hinged flap, forcing blood to travel through the lungs to get to the other side of the heart. At the same time, blood vessels to and from the now useless umbilical cord are sealed by muscular contractions, as are vessels that bypassed the fetal lungs and liver.

Humans progress through the same recognizable stages of early development—fertilization, blastulation, gastrulation, and neurulation—as do embryos of other species of vertebrates. As in reptiles, birds, and other mammals, only a small part of a human egg's contents is converted into embryonic tissues; most is used in the formation of extraembryonic membranes which play a key role in providing the human embryo and fetus with oxygen, nutrients, and waste removal. Most human organs are formed within the first 2 months of development; the remaining 7 months are devoted primarily to growth and refinement of structure. (See CTQ #4.)

EMBRYONIC DEVELOPMENT AND EVOLUTION

▥▶ It would be hard to mistake a fish for a bird, or a turtle for a human, yet the embryos of these vertebrates are remarkably similar (Figure 32-18). Embryos tend to change much more slowly over evolution than do the corresponding adults. For this reason, similarities in embryonic development are often used as evidence for evolutionary relatedness. For example, mollusks (such as snails) and flatworms (such as tapeworms) are so different as adults that there is no reason to think the two groups are more closely related to one another than are mollusks and vertebrates. Yet, the pattern of cleavage of certain mollusks and flatworms is so similar that evolutionary schemes typically show mollusks as descendants of flatworm-like ancestors.

The various parts of the embryos depicted in Figure 32-18 appear so similar because they are *homologous* structures; that is, they are derived from the same structure that was present in a common ancestor (Chapter 34). We have seen in this chapter how the resemblance between the four extraembryonic membranes of reptiles, birds, and mam-

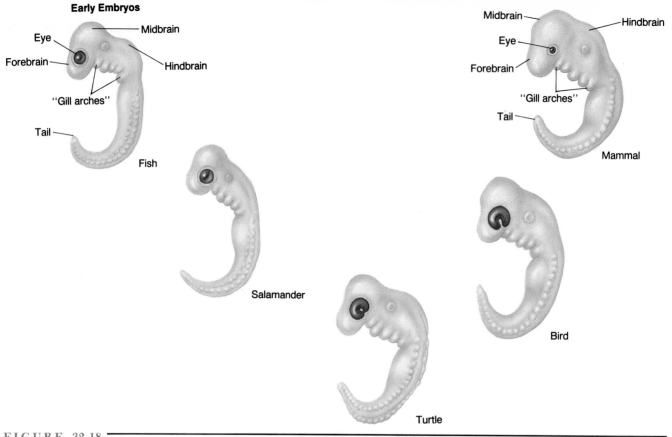

Early Embryos

Eye — Midbrain
Forebrain —
"Gill arches"
Tail —
Fish

Midbrain — Hindbrain
Eye —
Forebrain —
"Gill arches"
Tail —
Mammal

Salamander

Bird

Turtle

FIGURE 32-18 ━━━━━━━━━━━━━━━━━━━━━━━━━━
The striking similarities in the structure of the early embryos of vertebrates reveals their common ancestry.

mals testifies to the common ancestry of all three groups. Even when the original function of a membrane is no longer needed by a particular species, the membrane remains as a vestigial reminder of the organism's evolutionary origins. For example, humans and other mammals have no yolk in their embryo, yet they still develop a yolk sac—a vestigial remnant left over from an early "yolked" ancestor. Similarly, all vertebrate embryos, including those of humans and other air-breathing mammals, develop pharyngeal gill slits (page 636), even though these structures never become functional, even in the embryo. Instead, the gill slits of the human embryo are only transient structures, appearing

rapidly in an early stage and then giving rise to totally different structures, including parts of the jaw, ear, and thyroid gland. The entire jaw apparatus of vertebrates—from fishes to humans—is derived from part of the gills that appear in the embryo.

Embryos of diverse animal species tend to be more similar than are the corresponding adults. As a result, the study of development often provides insights into evolutionary relationships among animals and into the evolutionary pathway by which particular structures arose. (See CTQ #6.)

REEXAMINING THE THEMES

The Relationship between Form and Function

Each of the several hundred cell types found in a human or other mammal has a specialized function supplied by that cell's peculiar structure and metabolic machinery. During embryonic development, each type of cell acquires its specialized morphology. Muscle cells synthe-

size contractile proteins, such as actin and myosin, which become organized into cytoplasmic filaments; salivary gland cells develop an elaborate, interconnected, membranous network that promotes the synthesis and export of secretory proteins; nerve cells develop highly elongated axonal processes, which facilitate cell-to-cell communication; and so forth.

Biological Order, Regulation, and Homeostasis

◑ Embryonic development is characterized by a striking increase in biological order. A fertilized egg, whether a sea urchin or a human, has one of the least specialized internal structures of any animal cell. Yet, from this relatively simple beginning emerges an animal of striking complexity. Even though a fertilized egg may not show obvious evidence of its awesome potential, the cell has a developmental program that unfolds according to a predetermined plan. This program is encoded within the DNA of the chromosomes and as mRNAs and other cytoplasmic materials. The expression of this program requires such a high degree of regulation that we are now only *beginning* to understand the underlying mechanisms.

Acquiring and Utilizing Energy

☀ The increasing complexity that characterizes embryonic development is driven by the chemical energy that is consumed by the embryo. In some cases, as in birds, the energy is provided in the form of yolk, which is packaged into the egg as it forms in the ovary. In other cases, as in sea urchins, the individual begins life with very little energy reserves and rapidly develops into a free-swimming larva that must obtain food for itself. Mammals produce eggs with very little yolk, but the growth of the embryo and fetus is fueled by the nutrients delivered by the maternal bloodstream.

Unity within Diversity

◭ Although adult animals show great diversity, there is an undeniable similarity in the underlying mechanisms of embryonic development in all species. Virtually all embryos pass through stages of cleavage, blastulation, and gastrulation, even though the embryos themselves may bear little superficial resemblance to one another. The existence of homeotic genes that contain virtually identical homeobox sequences suggests a similarity in the types of genes that control development. Changes in the diverse form of embryos are accomplished by similar morphogenetic mechanisms, including changes in cell shape, cell growth and division, cell death, and cellular migrations.

Evolution and Adaptation

▮▶ Since the embryos of animals tend to be much more similar morphologically to one another than are the corresponding adults, the study of embryonic development provides one of the best tools to understanding the evolutionary relatedness among animals. In addition, the pathway by which a particular structure develops often reveals insights into the evolutionary path by which the structure arose. For example, the origin of the vertebrate jaw from part of the pharyngeal gills is hardly evident by examining an adult vertebrate but becomes immediately apparent by observing the formation of jaws in an early vertebrate embryo.

S Y N O P S I S

Embryonic development is a programmed course of events that carries an animal from fertilization through cleavage, blastulation, gastrulation, and organ formation. This program is encoded within the fertilized egg in both the genes and the organization of the cytoplasm.

A fertilizing sperm activates an unfertilized egg and donates a set of homologous chromosomes. Unfertilized eggs respond to sperm contact by surface changes that prevent penetration by additional sperm and by the release of calcium, which triggers various responses, including the formation of an extracellular membrane.

Cleavage divides the unbalanced egg into a large number of smaller blastomeres. Cleavage leads to the formation of a blastula, a stage that contains an internal chamber, or blastocoel, whose size and location depend primarily on the amount of yolk in the egg. Cleavage can occur in the absence of transcription, indicating that newly synthesized proteins are formed using stored mRNAs. De-

velopment beyond the blastula requires the activation of embryonic genes. Some blastomeres may be able to form an entire embryo in isolation, while others have a much more restricted potential. In frogs, only those blastomeres possessing a portion of the gray crescent can develop normally.

During gastrulation the various parts of the blastula become rearranged to form an embryo with three defined germ layers. The outer ectoderm gives rise to the skin and nervous system; the inner endoderm to the digestive tract and related organs; and the middle mesoderm to the remainder of the embryo. After gastrulation in vertebrates, the dorsal strip of ectoderm becomes thickened into the neural plate, which rolls into a tube that ultimately gives rise to the entire nervous system. This transformation requires induction from the underlying chordamesoderm, those cells that will form the notochord.

The formation of organs depends on a number of processes. Cells receive chemical and physical signals from their surroundings which inform them of their relative postion within the embryo. The shape of the organ formed by developing cells depends on morphogenetic processes, including changes in the rate of cell division, changes in cell shape, changes in cell adhesion, and programmed cell death. During formation of an organ, the internal architecture of the cells assumes a differentiated state, characteristic of that cell type. Each type of cell transcribes a restricted set of genes, forming a characteristic set of proteins.

Embryonic development provides a window to evolution. Embryos are typically much more similar to one another than are the corresponding adults; thus, embryos are useful in establishing evolutionary relatedness. Following the course of development of a particular structure may reveal information about the evolution of that structure.

Human development occurs in the uterus. The blastocyst implants itself in the endometrium and gastrulates to form an embryonic disk. Of the four extraembryonic membranes, the chorion and allantois contribute to placenta formation; the amnion protects the fetus; and the yolk sac temporarily manufactures blood cells. By the end of the second month, the formation of virtually all embryonic organs has begun. During the remaining 7 months, the fetus refines these structures and grows in size until uterine contractions expel it through the vagina.

Key Terms

homeotic gene (p. 686)
homeobox (p. 687)
embryo (p. 688)
yolk (p. 688)
larva (p. 688)
cleavage (p. 690)
blastomere (p. 690)
blastulation (p. 690)
blastula (p. 690)
blastocoel (p. 690)
gray crescent (p. 692)
gastrulation (p. 692)
gastrula (p. 692)

endoderm (p. 693)
ectoderm (p. 693)
mesoderm (p. 693)
neural plate (p. 694)
neural tube (p. 694)
chordamesoderm (p. 694)
notochord (p. 694)
induction (p. 695)
organogenesis (p. 696)
positional information (p. 696)
morphogenesis (p. 696)
cell differentiation (p. 698)
metamorphosis (p. 699)

chorion (p. 700)
placenta (p. 700)
umbilical cord (p. 700)
fetus (p. 700)
blastocyst (p. 700)
inner cell mass (p. 700)
trophoblast (p. 700)
amnion (p. 703)
yolk sac (p. 703)
allantois (p. 703)
birth (p. 705)

Review Questions

1. Arrange these terms in the order of development and briefly describe the role of each process in embryonic development: neural tube formation, birth, blastulation, fertilization, gastrulation, gametogenesis (meiosis).

2. What is the relationship between the amount of yolk in an egg and the relative size of the blastocoel?

3. Discuss the role of each of the four extraembryonic membranes in human development and indicate their evolutionary significance.

4. Distinguish between inner cell mass and trophoblast; primary and secondary induction; neural plate and neural tube; optic vesicle and optic cup; homeotic and hemoglobin genes; trophoblast and inner cell mass.

Critical Thinking Questions

1. Different vertebrae along the backbone can be distinguished by their shape. Occasionally, an infant is born with vertebrae in his or her lower back shaped like those normally found in the neck. In other words, the infant possesses cervical-type vertebrae in the lumbar region of the spine. Do you think this condition could be due to a homeotic mutation? Why or why not?

2. How does embryonic induction help determine the sequence of organogenesis? Why is sequence so important in embryonic development?

3. The graph (right) shows the growth of the body, heart, and brain of a person from birth to age 30. What does this show about the relative growth of the parts of the body? Why is the brain not much larger at age 30 than at age 5, but the heart is?

4. Prepare a timeline, to scale, of the embryonic development of a human. Your timeline should go from 0 to 266 days and should indicate the approximate times of significant events in development. You can indicate these events with words, drawings, or photographs from magazines.

5. Suppose you treated a fertilized snail egg with a drug that is assumed to inhibit transcription and found that the egg developed normally into a snail. What conclusion might you draw? Are there any experiments you might run to be sure of your conclusion? (Consider

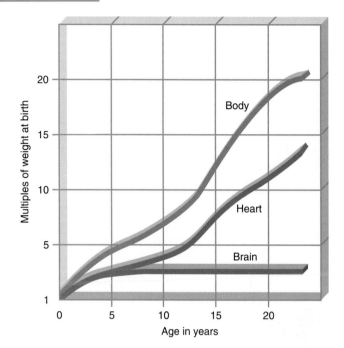

drug permeability and the actual ability of the drug to block transcription in your eggs.)

6. How does the study of comparative embryology contribute to the understanding of evolution? Give examples.

Additional Readings

Beardsley, T. 1991. Smart genes. *Sci. Amer.* Aug:86–95. (Intermediate)

Alberts, B. M. 1989. *Molecular biology of the cell,* 2d. ed. New York: Garland. (Advanced)

Browder, L., C. Erickson, and W. Jeffrey. 1991. *Developmental biology,* 3d ed. Philadelphia: Saunders. (Advanced)

Carlson, B. 1988. *Patten's foundations of embryology,* 5th ed. New York: McGraw-Hill. (Advanced)

Cherfas, J. 1990. Embryology gets down to the molecular level. *Science* 250:33–34. (Advanced)

DeRobertis, E. M., G. Oliver, and C. V. E. Wright. 1990. Homeobox genes and the vertebrate body plan. *Sci. Amer.* July:46–52. (Intermediate)

Gilbert, S. F. 1991. *Developmental biology,* 3d ed. Sunderland, MA: Sinauer. (Advanced)

Hoffman, M. 1990. The embryo takes its vitamins. *Science* 250:372–373. (Intermediate)

Melton, D. A. 1991. Pattern formation during animal development. *Science* 252:234–241. (Advanced)

Steinmetz, G. 1992. Fetal alcohol syndrome. *Nat'l Geog.* Feb:36–39. (Introductory)

Wassarman, P. M. 1988. Fertilization in mammals. *Sci. Amer.* Feb:78–84. (Intermediate)

APPENDIX

◀ A ▶

Metric and Temperature Conversion Charts

Metric Unit (symbol)		Metric to English	English to Metric
Length			
kilometer (km)	= 1,000 (10^3) meters	1 km = 0.62 mile	1 mile = 1.609 km
meter (m)	= 100 centimeters	1 m = 1.09 yards	1 yard = 0.914 m
		= 3.28 feet	1 foot = 0.305 m
centimeter (cm)	= 0.01 (10^{-2}) meter	1 cm = 0.394 inch	1 inch = 2.54 cm
millimeter (mm)	= 0.001 (10^{-3}) meter	1 mm = 0.039 inch	1 inch = 25.4 mm
micrometer (μm)	= 0.000001 (10^{-6}) meter		
nanometer (nm)	= 0.000000001 (10^{-9}) meter		
angstrom (Å)	= 0.0000000001 (10^{-10}) meter		
Area			
square kilometer (km^2)	= 100 hectares	1 km^2 = 0.386 square mile	1 square mile = 2.590 km^2
hectare (ha)	= 10,000 square meters	1 ha = 2.471 acres	1 acre = 0.405 ha
square meter (m^2)	= 10,000 square centimeters	1 m^2 = 1.196 square yards	1 square yard = 0.836 m^2
		= 10.764 square feet	1 square foot = 0.093 m^2
square centimeter (cm^2)	= 100 square millimeters	1 cm^2 = 0.155 square inch	1 square inch = 6.452 cm^2
Mass			
metric ton (t)	= 1,000 kilograms	1 t = 1.103 tons	1 ton = 0.907 t
	= 1,000,000 grams		
kilogram (kg)	= 1,000 grams	1 kg = 2.205 pounds	1 pound = 0.454 kg
gram (g)	= 1,000 milligrams	1 g = 0.035 ounce	1 ounce = 28.35 g
milligram (mg)	= 0.001 gram		
microgram (μg)	= 0.000001 gram		
Volume Solids			
1 cubic meter (m^3)	= 1,000,000 cubic centimeters	1 m^3 = 1.308 cubic yards	1 cubic yard = 0.765 m^3
		= 35.315 cubic feet	1 cubic foot = 0.028 m^3
1 cubic centimeter (cm^3)	= 1,000 cubic millimeters	1 cm^3 = 0.061 cubic inch	1 cubic inch = 16.387 cm^3
Volume Liquids			
kiloliter (kl)	= 1,000 liters	1 kl = 264.17 gallons	
liter (l)	= 1,000 milliliters	1 l = 1.06 quarts	1 gal = 3.785 l
			1 qt = 0.94 l
			1 pt = 0.47 l
milliliter (ml)	= 0.001 liter	1 ml = 0.034 fluid ounce	1 fluid ounce = 29.57 ml
microliter (μl)	= 0.000001 liter		

TEMPERATURE

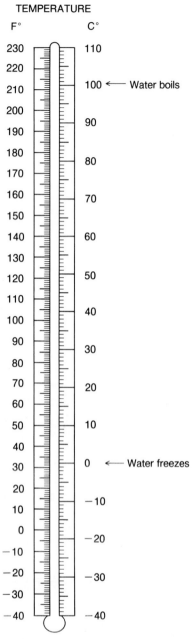

Fahrenheit to Centigrade: $^\circ C = \frac{5}{9}(^\circ F - 32)$
Centigrade to Fahrenheit: $^\circ F = \frac{9}{5}(^\circ C + 32)$

APPENDIX
◂ B ▸

Microscopes: Exploring the Details of Life

Microscopes are the instruments that have allowed biologists to visualize objects that are vastly smaller than anything visible with the naked eye. There are broadly two types of specimens viewed in a microscope: whole mounts which consist of an intact subject, such as a hair, a living cell, or even a DNA molecule, and thin sections of a specimen, such as a cell or piece of tissue.

THE LIGHT MICROSCOPE

A light microscope consists of a series of glass lenses that bend (refract) the light coming from an illuminated specimen so as to form a visual image of the specimen that is larger than the specimen itself (a). The specimen is often stained with a colored dye to increase its visibility. A special phase contrast light microscope is best suited for observing unstained, living cells because it converts differences in the density of cell organelles, which are normally invisible to the eye, into differences in light intensity which can be seen.

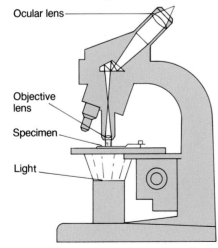

Ocular lens

Objective lens

Specimen

Light

(a)

All light microscopes have limited *resolving power*—the ability to distinguish two very close objects as being separate from each other. The resolving power of the light microscope is about 0.2 μm (about 1,000 times that of the naked eye), a property determined by the wave length of visible light. Consequently, objects closer to each other than 0.2 μm, which includes many of the smaller cell organ-

elles, will be seen as a single, blurred object through a light microscope.

THE TRANSMISSION ELECTRON MICROSCOPE

Appreciation of the wondrous complexity of cellular organization awaited the development of the transmission electron microscope (or TEM), which can deliver resolving powers 1000 times greater than the light microscope. Suddenly, biologists could see strange new structures, whose function was totally unknown—a breakthrough that has kept cell biologists busy for the past 50 years. The TEM (b) works by shooting a beam of electrons through very thinly sliced specimens that have been stained with heavy metals,

(b)

such as uranium, capable of deflecting electrons in the beam. The electrons that pass through the specimen undeflected are focused by powerful electromagnets (the lenses of a TEM) onto either a phosphorescent screen or high-contrast photographic film. The resolution of the TEM is so great—sufficient to allow us to see individual DNA molecules—because the wavelength of an electron beam is so small (about 0.0005 μm).

THE SCANNING ELECTRON MICROSCOPE

Specimens examined in the scanning electron microscope (SEM) are whole mounts whose surfaces have been coated with a thin layer of heavy metals. In the SEM, a fine beam of electrons scans back and forth across the specimen and the image is formed from electrons bouncing off the hills and valleys of its surface. The SEM produces a three-dimensional image of the surface of the specimen—which can

(c)   *(d)*

range in size from a virus to an insect head (c,d)—with remarkable depth and clarity. The SEM produces black and white images; the colors seen in many of the micrographs in the text have been added to enhance their visual quality. Note that the insect head (d) is that of an antennatedia mutant as described on p 687.

APPENDIX
C

The Hardy-Weinberg Principle

If the allele for brown hair is dominant over that for blond hair, and curly hair is dominant over straight hair, then why don't all people by now have brown, curly hair? The **Hardy-Weinberg Principle** (developed independently by English mathematician G. H. Hardy and German physician W. Weinberg) demonstrates that the frequency of alleles remains the same from generation to generation unless influenced by outside factors. The outside factors that would cause allele frequencies to change are mutation, immigration and emigration (movement of individuals into and out of a breeding population, respectively), natural selection of particular traits, and breeding between members of a small population. In other words, unless one or more of these forces influence hair color and hair curl, the relative number of people with brown and curly hair will not increase over those with blond and straight hair.

To illustrate the Hardy-Weinberg Principle, consider a single gene locus with two alleles, *A* and *a*, in a breeding population. (If you wish, consider *A* to be the allele for brown hair and *a* to be the allele for blond hair.) Because there are only two alleles for the gene, the sum of the frequencies of *A* and *a* will equal 1.0. (By convention, allele frequencies are given in decimals instead of percentages.) Translating this into mathematical terms, if

p = the frequency of allele *A*, and
q = the frequency of allele *a*,

then $p + q = 1$.

If *A* represented 80 percent of the alleles in the breeding population ($p = 0.8$), then according to this formula the frequency of *a* must be 0.2 ($p + q = 0.8 + 0.2 = 1.0$).

After determining the allele frequency in a starting population, the predicted frequencies of alleles and genotypes in the next generation can be calculated. Setting up a Punnett square with starting allele frequencies of $p = 0.8$ and $q = 0.2$:

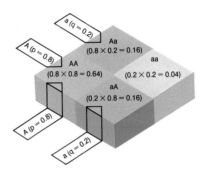

The chances of each offspring receiving any combination of the two alleles is the product of the probability of receiving one of the two alleles alone. In this example, the chances of an offspring receiving two *A* alleles is $p \times p = p^2$, or $0.8 \times 0.8 = 0.64$. A frequency of 0.64 means that 64 percent of the next generation will be homozygous dominant (*AA*). The chances of an offspring receiving two *a* alleles is $q^2 = 0.2 \times 0.2 = 0.04$, meaning 4 percent of the next generation is predicted to be *aa*. The predicted frequency of heterozygotes (*Aa* or *aA*) is 0.32 or $2pq$, the sum of the probability of an individual being *Aa*($p \times q = 0.8 \times 0.2 = 0.16$) plus the probability of an individual being *aA*($q \times p = 0.2 \times 0.8 = 0.16$). Just as all of the allele frequencies for a particular gene must add up to 1, so must all of the possible genotypes for a particular gene locus add up to 1. Thus, the Hardy-Weinberg Principle is

$$p^2 + 2pq + q^2 = 1$$
$$(0.64 + 0.32 + 0.04 = 1)$$

So after one generation, the frequency of possible genotypes is

$$AA = p^2 = 0.64$$
$$Aa = 2pq = 0.32$$
$$aa = q^2 = 0.04$$

Now let's determine the actual allele frequencies for *A* and *a* in the new generation. (Remember the original allele frequencies were 0.8 for allele *A* and 0.2 for allele *a*. If the Hardy-Weinberg Principle is right, there will be no change in the frequency of either allele.) To do this we sum the frequencies for each genotype containing the allele. Since heterozygotes carry both alleles, the genotype frequency must be divided in half to determine the frequency of each allele. (In our example, heterozygote *Aa* has a frequency of 0.32, 0.16 for allele *A*, plus 0.16 for allele *a*.) Summarizing then:

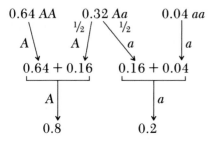

As predicted by the Hardy-Weinberg Principle, the frequency of allele *A* remained 0.8 and the frequency of allele *a* remained 0.2 in the new generation. Future generations can be calculated in exactly the same way, over and over again. As long as there are no mutations, no gene flow between populations, completely random mating, no natural selection, and no genetic drift, there will be no change in allele frequency, and therefore no evolution.

Population geneticists use the Hardy-Weinberg Principle to calculate a starting point allele frequency, a reference that can be compared to frequencies measured at some future time. The amount of deviation between observed allele frequencies and those predicted by the Hardy-Weinberg Principle indicates the degree of evolutionary change. Thus, this principle enables population geneticists to measure the rate of evolutionary change and identify the forces that cause changes in allele frequency.

APPENDIX
◄ **D** ►

Careers in Biology

Although many of you are enrolled in biology as a requirement for another major, some of you will become interested enough to investigate the career opportunities in life sciences. This interest in biology can grow into a satisfying livelihood. Here are some facts to consider:

- Biology is a field that offers a very wide range of possible science careers

- Biology offers high job security since many aspects of it deal with the most vital human needs: health and food

- Each year in the United States, nearly 40,000 people obtain bachelor's degrees in biology. But the number of newly created and vacated positions for biologists is increasing at a rate that exceeds the number of new graduates. Many of these jobs will be in the newer areas of biotechnology and bioservices.

Biologists not only enjoy job satisfaction, their work often changes the future for the better. Careers in medical biology help combat diseases and promote health. Biologists have been instrumental in preserving the earth's life-supporting capacity. Biotechnologists are engineering organisms that promise dramatic breakthroughs in medicine, food production, pest management, and environmental protection. Even the economic vitality of modern society will be increasingly linked to biology.

Biology also combines well with other fields of expertise. There is an increasing demand for people with backgrounds or majors in biology complexed with such areas as business, art, law, or engineering. Such a distinct blend of expertise gives a person a special advantage.

The average starting salary for all biologists with a Bachelor's degree is $22,000. A recent survey of California State University graduates in biology revealed that most were earning salaries between $20,000 and $50,000. But as important as salary is, most biologists stress job satisfaction, job security, work with sophisticated tools and scientific equipment, travel opportunities (either to the field or to scientific conferences), and opportunities to be creative in their job as the reasons they are happy in their career.

Here is a list of just a few of the careers for people with degrees in biology. For more resources, such as lists of current openings, career guides, and job banks, write to Biology Career Information, John Wiley and Sons, 605 Third Avenue, New York, NY 10158.

A SAMPLER OF JOBS THAT GRADUATES HAVE SECURED IN THE FIELD OF BIOLOGY°

Agricultural Biologist	Bioanalytical Chemist	Brain Function	Environmental Center
Agricultural Economist	Biochemical/Endocrine	Researcher	Director
Agricultural Extension	Toxicologist	Cancer Biologist	Environmental Engineer
Officer	Biochemical Engineer	Cardiovascular Biologist	Environmental Geographer
Agronomist	Pharmacology Distributor	Cardiovascular/Computer	Environmental Law Specialist
Amino-acid Analyst	Pharmacology Technician	Specialist	Farmer
Analytical Biochemist	Biochemist	Chemical Ecologist	Fetal Physiologist
Anatomist	Biogeochemist	Chromatographer	Flavorist
Animal Behavior	Biogeographer	Clinical Pharmacologist	Food Processing Technologist
Specialist	Biological Engineer	Coagulation Biochemist	Food Production Manager
Anticancer Drug Research	Biologist	Cognitive Neuroscientist	Food Quality Control
Technician	Biomedical	Computer Scientist	Inspector
Antiviral Therapist	Communication Biologist	Dental Assistant	Flower Grower
Arid Soils Technician	Biometerologist	Ecological Biochemist	Forest Ecologist
Audio-neurobiologist	Biophysicist	Electrophysiology/	Forest Economist
Author, Magazines & Books	Biotechnologist	Cardiovascular Technician	Forest Engineer
Behavioral Biologist	Blood Analyst	Energy Regulation Officer	Forest Geneticist
Bioanalyst	Botanist	Environmental Biochemist	Forest Manager

Forest Pathologist
Forest Plantation Manager
Forest Products Technologist
Forest Protection Expert
Forest Soils Analyst
Forester
Forestry Information Specialist
Freeze-Dry Engineer
Fresh Water Biologist
Grant Proposal Writer
Health Administrator
Health Inspector
Health Scientist
Hospital Administrator
Hydrologist
Illustrator
Immunochemist
Immunodiagnostic
   Assay Developer
Inflammation Technologist
Landscape Architect
Landscape Designer
Legislative Aid
Lepidopterist
Liaison Scientist,
   Library of Medicine
Computer Biologist
Life Science Computer
   Technologist
Lipid Biochemist
Livestock Inspector
Lumber Inspector

Medical Assistant
Medical Imaging Technician
Medical Officer
Medical Products Developer
Medical Writer
Microbial Physiologist
Microbiologist
Mine Reclamation Scientist
Molecular Endocrinologist
Molecular Neurobiologist
Molecular Parasitologist
Molecular Toxicologist
Molecular Virologist
Morphologist
Natural Products Chemist
Natural Resources Manager
Nature Writer
Nematode Control Biologist
Nematode Specialist
Nematologist
Neuroanatomist
Neurobiologist
Neurophysiologist
Neuroscientist
Nucleic Acids Chemist
Nursing Aid
Nutritionist
Occupational Health Officer
Ornamental Horticulturist
Paleontologist
Paper Chemist
Parasitologist

Pathologist
Peptide Biochemist
Pharmaceutical Writer
Pharmaceutical Sales
Pharmacologist
Physiologist
Planning Consultant
Plant Pathologist
Plant Physiologist
Production Agronomist
Protein Biochemist
Protein Structure & Design
   Technician
Purification Biochemist
Quantitative Geneticist
Radiation Biologist
Radiological Scientist
Regional Planner
Regulatory Biologist
Renal Physiologist
Renal Toxicologist
Reproductive Toxicologist
Research and Development
   Director
Research Technician
Research Liaison Scientist
Research Products Designer
Research Proposal Writer
Safety Assessment Sanitarian
Scientific Illustrator
Scientific Photographer
Scientific Reference Librarian

Scientific Writer
Soil Microbiologist
Space Station Life Support
   Technician
Spectroscopist
Sports Product Designer
Steroid Health Assessor
Taxonomic Biologist
Teacher
Technical Analyst
Technical Science Project
   Writer
Textbook Editor
Theoretical Ecologist
Timber Harvester
Toxicologist
Toxic Waste Treatment
   Specialist
Urban Planner
Water Chemist
Water Resources Biologist
Wood Chemist
Wood Fuel Technician
Zoning and Planning
   Manager
Zoologist
Zoo Animal Breeder
Zoo Animal Behaviorist
Zoo Designer
Zoo Inspector

°Results of one survey of California State University graduates. Some careers may require advanced degrees

Glossary

Abiotic Environment   Components of ecosystems that include all nonliving factors. (41)

Abscisic Acid (ABA)   A plant hormone that inhibits growth and causes stomata to close. ABA may not be commonly involved in leaf drop. (21)

Abscission   Separation of leaves, fruit, and flowers from the stem. (21)

Acclimation   A physiological adjustment to environmental stress. (28)

Acetyl CoA   Acetyl coenzyme A. A complex formed when acetic acid binds to a carrier coenzyme forming a bridge between the end products of glycolysis and the Krebs cycle in respiration. (9)

Acetylcholine   Neurotransmitter released by motor neurons at neuromuscular junctions and by some interneurons. (23)

Acid Rain   Occurring in polluted air, rain that has a lower pH than rain from areas with unpolluted air. (40)

Acids   Substances that release hydrogen ions (H^+) when dissolved in water. (3)

Acid Snow   Occurring in polluted air, snow that has a lower pH than snow from areas with unpolluted air. (40)

Acoelomates   Animals that lack a body cavity between the digestive cavity and body wall. (39)

Acquired Immune Deficiency Syndrome (AIDS)   Disease caused by infection with HIV (Human Immunodeficiency Virus) that destroys the body's ability to mount an immune response due to destruction of its helper T cells. (30, 36)

Actin   A contractile protein that makes up the major component of the thin filaments of a muscle cell and the microfilaments of nonmuscle cells. (26)

Action Potential   A sudden, dramatic reversal of the voltage (potential difference) across the plasma membrane of a nerve or muscle cell due to the opening of the sodium channels. The basis of a nerve impulse. (23)

Activation Energy   Energy required to initiate chemical reaction. (6)

Active Site   Region on an enzyme that binds its specific substrates, making them more reactive. (6)

Active Transport   Movement of substances into or out of cells against a concentration gradient, i.e., from a region of lower concentration to a region of higher concentration. The process requires an expenditure of energy by the cell. (7)

Adaptation   A hereditary trait that improves an organism's chances of survival and/or reproduction. (33)

Adaptive Radiation   The divergence of many species from a single ancestral line. (33)

Adenosine Triphosphate (ATP)   The molecule present in all living organisms that provides energy for cellular reactions in its phosphate bonds. ATP is the universal energy currency of cells. (6)

Adenylate Cyclase   An enzyme activated by hormones that converts ATP to cyclic AMP, a molecule that activates resting enzymes. (25)

Adrenal Cortex   Outer layer of the adrenal glands. It secretes steroid hormones in response to ACTH. (25)

Adrenal Medulla   An endocrine gland that controls metabolism, cardiovascular function, and stress responses. (25)

Adrenocorticotropic Hormone (ACTH)   An anterior pituitary hormone that stimulates the cortex of the adrenal glands to secrete cortisol and other steroid hormones. (25)

Adventitious Root System   Secondary roots that develop from stem or leaf tissues. (18)

Aerobe   An organism that requires oxygen to release energy from food molecules. (9)

Aerobic Respiration   Pathway by which glucose is completely oxidized to CO_2 and H_2O, requiring oxygen and an electron transport system. (9)

Afferent (Sensory) Neurons   Neurons that conduct impulses from the sense organs to the central nervous system. (23)

Age-Sex Structure   The number of individuals of a certain age and sex within a population. (43)

Aggregate Fruits   Fruits that develop from many pistils in a single flower. (20)

AIDS   See Acquired Immune Deficiency Syndrome.

Albinism   A genetic condition characterized by an absence of epidermal pigmentation that can result from a deficiency of any of a variety of enzymes involved in pigment formation. (12)

Alcoholic Fermentation   The process in which electrons removed during glycolysis are transferred from NADH to form alcohol as an end product. Used by yeast during the commercial process of ethyl alcohol production. (9)

Aldosterone   A hormone secreted by the adrenal cortex that stimulates reabsorption of sodium from the distal tubules and collecting ducts of the kidneys. (23)

Algae   Any unicellular or simple colonial photosynthetic eukaryote. (37,38)

Algin   A substance produced by brown algae harvested for human application because of its ability to regulate texture and consistency of products. Found in ice cream, cosmetics, marshmellows, paints, and dozens of other products. (38)

Allantois   Extraembryonic membrane that serves as a repository for nitrogenous wastes. In placental mammals, it helps form the vascular connections between mother and fetus. (32)

Allele   Alternative form of a gene at a particular site, or locus, on the chromosome. (12)

Allele Frequency   The relative occurrence of a certain allele in individuals of a population. (33)

Allelochemicals   Chemicals released by some plants and animals that deter or kill a predator or competitor. (42)

Allelopathy   A type of interaction in which one organism releases allelochemicals that harm another organism. (42)

Allergy   An inappropriate response by the immune system to a harmless foreign substance leading to symptoms such as itchy eyes, runny nose, and congested airways. If the reaction occurs throughout the body (anaphylaxis) it can be life threatening. (30)

Allopatric Speciation   Formation of new species when gene flow between parts of a population is stopped by geographic isolation. (33)

Alpha Helix   Portion of a polypeptide chain organized into a defined spiral conformation. (4)

Alternation of Generations   Sequential change during the life cycle of a plant in which a haploid (1N) multicellular stage (gametophyte) alternates with a diploid (2N) multicellular stage (sporophyte). (38)

Alternative Processing   When a primary RNA transcript can be processed to form more than one mRNA depending on conditions. (15)

Altruism   The performance of a behavior that benefits another member of the species at some cost to the one who does the deed. (44)

Alveolus   A tiny pouch in the lung where gas is exchanged between the blood and the air; the functional unit of the lung where CO_2 and O_2 are exchanged. (29)

Alzheimer's Disease   A degenerative disease of the human brain, particularly affecting acetylcholine-releasing neurons and the hippocampus, characterized by the presence of tangled fibrils within the cytoplasm of neurons and amyloid plaques outside the cells. (23)

Amino Acids   Molecules containing an amino group ($-NH_2$) and a carboxyl group ($-COOH$) attached to a central carbon atom. Amino acids are the subunits from which proteins are constructed. (4)

Amniocentesis   A procedure for obtaining fetal cells by withdrawing a sample of the fluid that surrounds a developing fetus (amniotic fluid) using a hypodermic needle and syringe. (17)

Amnion Extraembryonic membrane that envelops the young embryo and encloses the amniotic fluid that suspends and cushions it. (32)

Amoeba A protozoan that employs pseudopods for motility. (37)

Amphibia A vertebrate class grouped into three orders: Caudata (tailed amphibians); Anura (tail-less amphibians); Apoda (rare worm-like, burrowing amphibians). (39)

Anabolic Steroids Steroid hormones, such as testosterone, which promote biosynthesis (anabolism), especially protein synthesis. (25)

Anabolism Biosynthesis of complex molecules from simpler compounds. Anabolic pathways are endergonic, i.e., require energy. (6)

Anaerobe Organism that does not require oxygen to release energy from food molecules. (9)

Anaerobic Respiration Pathway by which glucose is completely oxidized, using an electron transport system but requiring a terminal electron acceptor other than oxygen. (Compare with fermentation.) (9)

Analogous Structures (Homoplasies) Structures that perform a similar function, such as wings in birds and insects, but did not originate from the same structure in a common ancestor. (33)

Anaphase Stage of mitosis when the kinetochores split and the sister chromatids (now termed chromosomes) move to opposite poles of the spindle. (10)

Anatomy Study of the structural characteristics of an organism. (18)

Angiosperm (Anthophyta) Any plant having its seeds surrounded by fruit tissue formed from the mature ovary of the flowers. (38)

Animal A mobile, heterotrophic, multicellular organism, classified in the Animal kingdom. (39)

Anion A negatively charged ion. (3)

Annelida The phylum which contains segmented worms (earthworms, leeches, and bristleworms). (39)

Annuals Plants that live for one year or less. (18)

Annulus A row of specialized cells encircling each sporangium on the fern frond; facilitates rupture of the sporangium and dispersal of spors. (38)

Antagonistic Muscles Pairs of muscles whose contraction bring about opposite actions as illustrated by the biceps and triceps, which bends or straightens the arm at the elbow, respectively. (26)

Antenna Pigments Components of photosystems that gather light energy of different wavelengths and then channel the absorbed energy to a reaction center. (8)

Anterior In anatomy, at or near the front of an animal; the opposite of posterior. (39)

Anterior Pituitary A true endocrine gland manufacturing and releasing six hormones

when stimulated by releasing factors from the hypothalamus. (25)

Anther The swollen end of the stamen (male reproductive organ) of a flowering plant. Pollen grains are produced inside the anther lobes in pollen sacs. (20)

Antibiotic A substance produced by a fungus or bacterium that is capable of preventing the growth of bacteria. (2)

Antibodies Proteins produced by plasma cells. They react specifically with the antigen that stimulated their formation. (30)

Anticodon Triplet of nucleotides in tRNA that recognizes and base pairs with a particular codon in mRNA. (14)

Antidiuretic Hormone (ADH) One of the two hormones released by the posterior pituitary. ADH increases water reabsorption in the kidney, which then produces a more concentrated urine. (25)

Antigen Specific foreign agent that triggers an immune response. (30)

Aorta Largest blood vessel in the body through which blood leaves the heart and enters the systemic circulation. (28)

Apical Dominance The growth pattern in plants in which axillary bud growth is inhibited by the hormone auxin, present in high concentrations in terminal buds. (21)

Apical Meristems Centers of growth located at the tips of shoots, axillary buds, and roots. Their cells divide by mitosis to produce new cells for primary growth in plants. (18)

Aposematic Coloring Warning coloration which makes an organism stand out from its surroundings. (42)

Appendicular Skeleton The bones of the appendages and of the pectoral and pelvic girdles. (26)

Aquatic Living in water. (40)

Archaebacteria Members of the kingdom Monera that differ from typical bacteria in the structure of their membrane lipids, their cell walls, and some characteristics that resemble those of eukaryotes. Their lack of a true nucleus, however, accounts for their assignment to the Moneran kingdom. (36)

Archenteron In gastrulation, the hollow core of the gastrula that becomes an animal's digestive tract. (32)

Arteries Large, thick-walled vessels that carry blood away from the heart. (28)

Arterioles The smallest arteries, which carry blood toward capillary beds. (28)

Arthropoda The most diverse phylum on earth, so called from the presence of jointed limbs. Includes insects, crabs, spiders, centipedes. (39)

Ascospores Sexual fungal spore borne in a sac. Produced by the sac fungi, Ascomycota. (37)

Asexual Reproduction Reproduction without the union of male and female gametes. (31)

Association In ecological communities, a major organization characterized by uniformity and two or more dominant species. (41)

Asymmetric Referring to a body form that cannot be divided to produce mirror images. (39)

Atherosclerosis Condition in which the inner walls of arteries contain a buildup of cholesterol-containing plaque that tends to occlude the channel and act as a site for the formation of a blood clot (thrombus). (7)

Atmosphere The layer of air surrounding the Earth. (40)

Atom The fundamental unit of matter that can enter into chemical reactions; the smallest unit of matter that possesses the qualities of an element. (3)

Atomic Mass Combined number of protons and neutrons in the nucleus of an atom. (3)

Atomic Number The number of protons in the nucleus of an atom. (3)

ATP (see **Adenosine Triphosphate**)

ATPase An enzyme that catalyzes a reaction in which ATP is hydrolyzed. These enzymes are typically involved in reactions where energy stored in ATP is used to drive an energy-requiring reaction, such as active transport or muscle contractility. (7, 26)

ATP Synthase A large protein complex present in the plasma membrane of bacteria, the inner membrane of mitochondria, and the thylakoid membrane of chloroplasts. This complex consists of a baseplate in the membrane, a channel across the membrane through which protons can pass, and a spherical head (F_1 particle) which contains the site where ATP is synthesized from ADP and P_i (8, 9)

Atrioventricular (AV) Node A neurological center of the heart, located at the top of the ventricles. (28)

Atrium A contracting chamber of the heart which forces blood into the ventricle. There are two atria in the hearts of all vertebrates, except fish which have one atrium. (28)

Atrophy The shrinkage in size of structure, such as a bone or muscle, usually as a result of disuse. (26)

Autoantibodies Antibodies produced against the body's own tissue. (30)

Autoimmune Disease Damage to a body tissue due to an attack by autoantibodies. Examples include thyroiditis, multiple sclerosis, and rheumatic fever. (30)

Autonomic Nervous System The nerves that control the involuntary activities of the internal organs. It is composed of the parasympathetic system, which functions during normal activity, and the sympathetic system, which operates in times of emergency or prolonged exertion. (23)

Autosome Any chromosome that is not a sex chromosome. (13)

Autotrophs Organisms that satisfy their own nutritional needs by building organic molecules photosynthetically or chemosynthetically from inorganic substances. (8)

Auxins Plant growth hormones that promote cell elongation by softening cell walls. (21)

Axial Skeleton The bones aligned along the long axis of the body, including the skull, vertebral column, and ribcage. (26)

Axillary Bud A bud that is directly above each leaf on the stem. It can develop into a new stem or a flower. (18)

Axon The long, sometimes branched extension of a neuron which conducts impulses from the cell body to the synaptic knobs. (23)

◀ B ▶

Bacteriophage A virus attacking specific bacteria that multiplies in the bacterial host cell and usually destroys the bacterium as it reproduces. (36)

Balanced Polymorphism The maintenance of two or more alleles for a single trait at fairly high frequencies. (33)

Bark Common term for the periderm. A collective term for all plant tissues outside the secondary xylem. (18)

Base Substance that removes hydrogen ions (H^+) from solutions. (3)

Basidiospores Sexual spores produced by basidiomycete fungi. Often found by the millions on gills in mushrooms. (37)

Basophil A phagocytic leukocyte which also releases substances, such as histamine, that trigger an inflammatory response. (28)

Batesian Mimicry The resemblance of a good-tasting or harmless species to a species with unpleasant traits. (42)

Bathypelagic Zone The ocean zone beneath the mesopelagic zone, characterized by no light; inhabited by heterotrophic bacteria and benthic scavengers. (40)

B Cell A lymphocyte that becomes a plasma cell and produces antibodies when stimulated by an antigen. (30)

Benthic Zone The deepest ocean zone; the ocean floor, inhabitated by bottom dwelling organisms. (40)

Bicarbonate Ion HCO_{-3}. (3, 29)

Biennials Plants that live for two years. (18)

Bilateral Symmetry The quality possessed by organisms whose body can be divided into mirror images by only one median plane. (39)

Bile Salts Detergentlike molecules produced by the liver and stored by the gallbladder that function in lipid emulsification in the small intestine. (27)

Binomial A term meaning "two names" or "two words". Applied to the system of nomenclature for categorizing living things with a

genus and species name that is unique for each type of organism. (1)

Biochemicals Organic molecules produced by living cells. (4)

Bioconcentration The ability of an organism to accumulate substances within its' body or specific cells. (41)

Biodiversity Biological diversity of species, including species diversity, genetic diversity, and ecological diversity. (43)

Biogeochemical Cycles The exchanging of chemical elements between organisms and the abiotic environment. (41)

Biological Control Pest control through the use of naturally occurring organisms such as predators, parasites, bacteria, and viruses. (41)

Biological Magnification An increase in concentration of slowly degradable chemicals in organisms at successively higher trophic levels; for example, DDT or PCB's. (41)

Bioluminescence The capability of certain organisms to utilize chemical energy to produce light in a reaction catalyzed by the enzyme luciferase. (9)

Biomass The weight of organic material present in an ecosystem at any one time. (41)

Biome Broad geographic region with a characteristic array of organisms. (40)

Biosphere Zone of the earth's soil, water, and air in which living organisms are found. (40)

Biosynthesis Construction of molecular components in the growing cell and the replacement of these compounds as they deteriorate. (6)

Biotechnology A new field of genetic engineering; more generally, any practical application of biological knowledge. (16)

Biotic Environment Living components of the environment. (40)

Biotic Potential The innate capacity of a population to increase tremendously in size were it not for curbs on growth; maximum population growth rate. (43)

Blade Large, flattened area of a leaf; effective in collecting sunlight for photosynthesis. (18)

Blastocoel The hollow fluid-filled space in a blastula. (32)

Blastocyst Early stage of a mammalian embryo, consisting of a mass of cells enclosed in a hollow ball of cells called the trophoblast. (32)

Blastodisk In bird and reptile development, the stage equivalent to a blastula. Because of the large amount of yolk, cleavage produces two flattened layers of cells with a blastocoel between them. (32)

Blastomeres The cells produced during embryonic cleavage. (32)

Blastopore The opening of the archenteron that is the embryonic predecessor of the anus in vertebrates and some other animals. (32)

Blastula An early developmental stage in many animals. It is a ball of cells that encloses a cavity, the blastocoel. (32)

Blood A type of connective tissue consisting of red blood cells, white blood cells, platelets, and plasma. (28)

Blood Pressure Positive pressure within the cardiovascular system that propels blood through the vessels. (28)

Blooms are massive growths of algae that occur when conditions are optimal for algae proliferation. (37)

Body Plan The general layout of a plant's or animal's major body parts. (39)

Bohr Effect Increased release of O_2 from hemoglobin molecules at lower pH. (29)

Bone A tissue composed of collagen fibers, calcium, and phosphate that serves as a means of support, a reserve of calcium and phosphate, and an attachment site for muscles. (26)

Botany Branch of biology that studies the life cycles, structure, growth, and classification of plants. (18)

Bottleneck A situation in which the size of a species' population drops to a very small number of individuals, which has a major impact on the likelihood of the population recovering its earlier genetic diversity. As occurred in the cheetah population. (33)

Bowman's Capsule A double-layered container that is an invagination of the proximal end of the renal tubule that collects molecules and wastes from the blood. (28)

Brain Mass of nerve tissue composing the main part of the central nervous system. (23)

Brainstem The central core of the brain, which coordinates the automatic, involuntary body processes. (23)

Bronchi The two divisions of the trachea through which air enters each of the two lungs. (29)

Bronchioles The smallest tubules of the respiratory tract that lead into the alveoli of the lungs where gas exchange occurs. (29)

Bryophyta Division of non-vascular terrestrial plants that include liverworts, mosses, and hornworts. (38)

Budding Asexual process by which offspring develop as an outgrowth of a parent. (39)

Buffers Chemicals that couple with free hydrogen and hydroxide ions thereby resisting changes in pH. (3)

Bundle Sheath Parenchyma cells that surround a leaf vein which regulate the uptake and release of materials between the vascular tissue and the mesophyll cells. (18)

◀ C ▶

C_3 Synthesis The most common pathway for fixing CO_2 in the synthesis reactions of photosynthesis. It`is so named because the first detectable organic molecule into which CO_2 is incorporated is a 3-carbon molecule, phosphoglycerate (PGA). (8)

C_4 Synthesis Pathway for fixing CO_2 during the light-independent reactions of photosynthesis. It is so named because the first detectable organic molecule into which CO_2 is incorporated is a 4-carbon molecule. (8)

Calcitonin A thyroid hormone which regulates blood calcium levels by inhibiting its release from bone. (25)

Calorie Energy (heat) necessary to elevate the temperature of one gram of water by one degree Centigrade ($1°$ C). (6)

Calvin Cycle The cyclical pathway in which CO_2 is incorporated into carbohydrate. See C_3 synthesis. (8)

Calyx The outermost whorl of a flower, formed by the sepals. (20)

CAM Crassulacean acid metabolism. A variation of the photosynthetic reactions in plants, biochemically identical to C_4 synthesis except that all reactions occur in the same cell and are separated by time. Because CAM plants open their stomates at night, they have a competitive advantage in hot, dry climates. (8)

Cambium A ring or cluster of meristematic cells that increase the width of stems and roots when they divide to produce secondary tissues. (18)

Camouflage Adaptations of color, shape and behavior that make an organism more difficult to detect. (42)

Cancer A disease resulting from uncontrolled cell divisions. (10,13)

Capillaries The tiniest blood vessels consisting of a single layer of flattened cells. (28)

Capillary Action Tendency of water to be pulled into a small-diameter tube. (3)

Carbohydrates A group of compounds that includes simple sugars and all larger molecules constructed of sugar subunits, e.g. polysaccharides. (4)

Carbon Cycle The cycling of carbon in different chemical forms, from the environment to organisms and back to the environment. (41)

Carbon Dioxide Fixation In photosynthesis, the combination of CO_2 with carbon-accepting molecules to form organic compounds. (8)

Carcinogen A cancer-causing agent. (13)

Cardiac Muscle One of the three types of muscle tissue; it forms the muscle of the heart. (26)

Cardiovascular System The organ system consisting of the heart and the vessels through which blood flows. (28)

Carnivore An animal that feeds exclusively on other animals. (42)

Carotenoid A red, yellow, or orange plant pigment that absorbs light in 400-500 nm wavelengths. (8)

Carpels Central whorl of a flower containing the female reproductive organs. Each separate carpel, or each unit of fused carpels, is called a pistil. (20)

Carrier Proteins Proteins within the plasma membrane that bind specific substances and facilitate their movement across the membrane. (7)

Carrying Capacity The size of a population that can be supported indefinitely in a given environment. (43)

Cartilage A firm but flexible connective tissue. In the human, most cartilage originally present in the embryo is transformed into bones. (26)

Casparian Strip The band of waxy suberin that surrounds each endodermal cell of a plant's root tissue. (18)

Catabolism Metabolic pathways that degrade complex compounds into simpler molecules, usually with the release of the chemical energy that held the atoms of the larger molecule together. (6)

Catalyst A chemical substance that accelerates a reaction or causes a reaction to occur but remains unchanged by the reaction. Enzymes are biological catalysts. (6)

Cation A positively charged ion. (3)

Cecum A closed-ended sac extending from the intestine in grazing animals lacking a rumen (e.g., horses) that enables them to digest cellulose. (27)

Cell The basic structural unit of all organisms. (5)

Cell Body Region of a neuron that contains most of the cytoplasm, the nucleus, and other organelles. It relays impulses from the dendrites to the axon. (23)

Cell Cycle Complete sequence of stages from one cell division to the next. The stages are denoted G_1, S, G_2, and M phase. (10)

Cell Differentiation The process by which the internal contents of a cell become assembled into a structure that allows the cell to carry out a specific set of activities, such as secretion of enzymes or contraction. (32)

Cell Division The process by which one cell divides into two. (10)

Cell Fusion Technique whereby cells are caused to fuse with one another producing a large cell with a common cytoplasm and plasma membrane. (5, 10)

Cell Plate In plants, the cell wall material deposited midway between the daughter cells during cytokinesis. Plate material is deposited by small Golgi vesicles. (5, 10)

Cell Sap Solution that fills a plant vacuole. In addition to water, it may contain pigments, salts, and even toxic chemicals. (5)

Cell Theory The fundamental theory of biology that states: 1) all organisms are composed of one or more cells, 2) the cell is the basic organizational unit of life, 3) all cells arise from pre-existing cells. (5)

Cellular Respiration (See **Aerobic respiration**)

Cellulose The structural polysaccharide comprising the bulk of the plant cell wall. It is the most abundant polysaccharide in nature. (4, 5)

Cell Wall Rigid outer-casing of cells in plants and other organisms which gives support, slows dehydration, and prevents a cell from bursting when internal pressure builds due to an influx of water. (5)

Central Nervous System In vertebrates, the brain and spinal cord. (23)

Centriole A pinwheel-shaped structure at each pole of a dividing animal cell. (10)

Centromere Indented region of a mitotic chromosome containing the kinetochore. (10)

Cephalization The clustering of neural tissues and sense organs at the anterior (leading) end of the animal. (39)

Cerebellum A bulbous portion of the vertebrate brain involved in motor coordination. Its prominence varies greatly among different vertebrates . (23)

Cerebral Cortex The outer, highly convoluted layer of the cerebrum. In the human, this is the center of higher brain functions, such as speech and reasoning. (23)

Cerebrospinal Fluid Fluid present within the ventricles of the brain, central canal of the spinal cord, and which surrounds and cushions the central nervous system. (23)

Cerebrum The most dominant part of the human forebrain, composed of two cerebral hemispheres, generally associated with higher brain functions. (23)

Cervix The lower tip of the uterus. (31)

Chapparal A type of shrubland in California, characterized by drought-tolerant and fire-adapted plants. (40)

Character Displacement Divergence of a physical trait in closely related species in response to competition. (42)

Chemical Bonds Linkage between atoms as a result of electrons being shared or donated. (3)

Chemical Evolution Spontaneous synthesis of increasingly complex organic compounds from simpler molecules. (35)

Chemical Reaction Interaction between chemical reactants. (6)

Chemiosmosis The process by which a pH gradient drives the formation of ATP. (8, 9)

Chemoreceptors Sensory receptors that respond to the presence of specific chemicals. (24)

Chemosynthesis An energy conversion process in which inorganic substances (H, N, Fe, or S) provide energized electrons and hydrogen for carbohydrate formation (9, 36)

Chiasmata Cross-shaped regions within a tetrad, occurring at points of crossing over or genetic exchange. (11)

Chitin Structural polysaccharide that forms the hard, strong external skeleton of many arthropods and the cell walls of fungi. (4)

Chlamydia Obligate intracellular parasitic bacteria that lack a functional ATP-generating system. (36)

Chlorophyll Pigments Major light-absorbing pigments of photosynthesis. (8)

Chlorophyta Green algae, the largest group of algae; members of this group were very likely the ancestors of the modern plant kingdom. (38)

Chloroplasts An organelle containing chlorophyll found in plant cells in which photosynthesis occurs. (5, 8)

Cholecystokinin (CCK) Hormone secreted by endocrine cells in the wall of the small intestine that stimulates the release of digestive products by the pancreas. (27)

Chondrocytes Living cartilage cells embedded within the protein-polysaccharide matrix they manufacture. (26)

Chordamesoderm In vertebrates, the block of mesoderm that underlies the dorsal ectoderm of the gastrula, induces the formation of the nervous system, and gives rise to the notochord. (32)

Chordate A member of the phylum Chordata possessing a skeletal rod of tissue called a notochord, a dorsal hollow nerve cord, gill slits, and a post-anal tail at some stage of its development. (39)

Chorion The outermost of the four extraembryonic membranes. In placental mammals, it forms the embryonic portion of the placenta. (32)

Chorionic Villus Sampling (CVS) A procedure for obtaining fetal cells by removing a small sample of tissue from the developing placenta of a pregnant woman. (17)

Chromatid Each of the two identical subunits of a replicated chromosome. (10)

Chromatin DNA-protein fibers which, during prophase, condense to form the visible chromosomes. (5, 10)

Chromatography A technique for separating different molecules on the basis of their solubility in a particular solvent. The mixture of substances is spotted on a piece of paper or other material, one end of which is then placed in the solvent. As the solvent moves up the paper by capillary action, each substance in the mixture is carried a particular distance depending on its solubility in the moving solvent. (8)

Chromosomes Dark-staining structures in which the organism's genetic material (DNA) is organized. Each species has a characteristic number of chromosomes. (5, 10)

Chromosome Aberrations Alteration in the structure of a chromosome from the normal state. Includes chromosome deletions, duplications, inversions, and translocations. (13)

Chromosome Puff A site on an insect polytene chromosome where the DNA has unraveled and is being transcribed. (15)

Cilia Short, hairlike structures projecting from the surfaces of some cells. They beat in coordinated ways, are usually found in large numbers, and are densely packed. (5)

Ciliated Mucosa Layer of ciliated epithelial cells lining the respiratory tract. The beating of cilia propels an associated mucous layer and trapped foreign particles. (29)

Circadian Rhythm Behavioral patterns that cycle during approximately 24 hour intervals.

Circulatory System The system that circulates internal fluids throughout an organism to deliver oxygen and nutrients to cells and to remove metabolic wastes. (28)

Class (Taxonomic) A level of the taxonomic hierarchy that groups together members of related orders. (1)

Classical Conditioning A form of learning in which an animal develops a response to a new stimulus by repeatedly associating the new stimulus with a stimulus that normally elicits the response. (44)

Cleavage Successive mitotic divisions in the early embryo. There is no cell growth between divisions. (32)

Cleavage Furow Constriction around the middle of a dividing cell caused by constriction of microfilaments. (10)

Climate The general pattern of average weather conditions over a long period of time in a specific region, including precipitation, temperature, solar radiation, and humidity. (40)

Climax Final or stable community of successional stages, that is more or less in equilibrium with existing environmental conditions for a long period of time. (41)

Climax Community Community that remains essentially the same over long periods of time; final stage of ecological succession. (41)

Clitoris A protrusion at the point where the labia minora merge; rich in sensory neurons and erectile tissue. (31)

Clonal Selection Mechanism The mechanism by which the body can synthesize antibodies specific for the foreign substance (antigen) that stimulated their production. (30)

Clones Offspring identical to the parent, produced by asexual processes. (15)

Closed Circulatory System Circulatory system in which blood travels throughout the body in a continuous network of closed tubes. (Compare with open circulatory system). (28)

Clumped Pattern Distribution of individuals of a population into groups, such as flocks or herds. (43)

Cnidaria A phylum that consists of radial symmetrical animals that have two cell layers. There are three classes: 1) Hydrozoa (hydra), 2) Scyphozoa (jellyfish), 3) Anthozoa (sea anemones, corals). Most are marine forms that live in warm, shallow water. (39)

Cnidocytes Specialized stinging cells found in the members of the phylum Cnidaria. (39)

Coastal Waters Relatively warm, nutrient-rich shallow water extending from the high-tide mark on land to the sloping continental shelf. The greatest concentration of marine life are found in coastal waters. (40)

Coated Pits Indentations at the surfaces of cells that contain a layer of bristly protein (called clathrin) on the inner surface of the plasma membrane. Coated pits are sites where cell receptors become clustered. (7)

Cochlea Organ within the inner ear of mammals involved in sound reception. (24)

Codominance The simultaneous expression of both alleles at a genetic locus in a heterozygous individual. (12)

Codon Linear array of three nucleotides in mRNA. Each triplet specifies a particular amino acid during the process of translation. (14)

Coelomates Animals in which the body cavity is completely lined by mesodermally-derived tissues. (39)

Coenzyme An organic cofactor, typically a vitamin or a substance derived from a vitamin. (6)

Coevolution Evolutionary changes that result from reciprocal interactions between two species, e.g., flowering plants and their insect pollinators. (33)

Cofactor A non-protein component that is linked covalently or noncovalently to an enzyme and is required by the enzyme to catalyze the reaction. Cofactors may be organic molecules (coenzymes) or metals. (6)

Cohesion The tendency of different parts of a substance to hold together because of forces acting between its molecules. (3)

Coitus Sexual union in mammals. (31)

Coleoptile Sheath surrounding the tip of the monocot seedling, protecting the young stem and leaves as they emerge from the soil. (21)

Collagen The most abundant protein in the human body. It is present primarily in the extracellular space of connective tissues such as bone, cartilage, and tendons. (26)

Collenchyma Living plant cells with irregularly thickened primary cell walls. A supportive cell type often found inside the epidermis of stems with primary growth. Angular, lacunar and laminar are different types of collenchyma cells. (18)

Commensalism A form of symbiosis in which one organism benefits from the union while the other member neither gains nor loses. (42)

Community The populations of all species living in a given area. (41)

Compact Bone The solid, hard outer regions of a bone surrounding the honey-combed mass of spongy bone. (26)

Companion Cell Specialized parenchyma cell associated with a sieve-tube member in phloem. (18)

Competition Interaction among organisms that require the same resource. It is of two types: 1) intraspecific (between members of the same species); 2) interspecific (between members of different species). (42)

Competitive Exclusion Principle (Gause's Principle) Competition in which a winner species captures a greater share of resources, increasing its survival and reproductive capacity. The other species is gradually displaced. (42)

Competitive Inhibition Prevention of normal binding of a substrate to its enzyme by the presence of an inhibitory compound that competes with the substrate for the active site on the enzyme. (6)

Complement Blood proteins with which some antibodies combine following attachment to antigen (the surface of microorganisms). The bound complement punches the tiny holes in the plasma membrane of the foreign cell, causing it to burst. (28)

Complementarity The relationship between the two strands of a DNA molecule determined by the base pairing of nucleotides on the two strands of the helix. A nucleotide with guanine on one strand always pairs with a nucleotide having cytosine on the other strand; similarly with adenine and thymine. (14)

Complete Digestive Systems Systems that have a digestive tract with openings at both ends—a mouth for entry and an anus for exit. (27)

Complete Flower A flower containing all four whorls of modified leaves—sepels, petals, stamen, and carpels. (20)

Compound Chemical substances composed of atoms of more than one element. (3)

Compound Leaf A leaf that is divided into leaflets, with two or more leaflets attached to the petiole. (18)

Concentration Gradient Regions in a system of differing concentration representing potential energy, such as exist in a cell and its environment, that cause molecules to move from areas of higher concentration to lower concentration. (7)

Conditioned Reflex A reflex ("automatic") response to a stimulus that would not normally have elicited the response. Conditioned reflexes develop by repeated association of a new stimulus with an old stimulus that normally elicits the response. (44)

Conformation The three-dimensional shape of a molecule as determined by the spatial arrangement of its atoms. (4)

Conformational Change Change in molecular shape (as occurs, for example, in an en-

zyme as it catalyzes a reaction, or a myosin molecule during contraction). (6)

Conjugation A method of reproduction in single-celled organisms in which two cells link and exchange nuclear material. (11)

Connective Tissues Tissues that protect, support, and hold together the internal organs and other structures of animals. Includes bone, cartilage, tendons, and other tissues, all of which have large amounts of extracellular material. (22)

Consumers Heterotrophs in a biotic environment that feed on other organisms or organic waste. (41)

Continental Drift The continuous shifting of the earth's land masses explained by the theory of plate tectonics. (35)

Continuous Variation An inheritance pattern in which there is graded change between the two extremes in a phenotype (compare with discontinuous variation). (12)

Contraception The prevention of pregnancy. (31)

Contractile Proteins Actin and myosin, the protein filaments that comprise the bulk of the muscle mass. During contraction of skeletal muscle, these filaments form a temporary association and slide past each other, generating the contractile force. (26)

Control (Experimental) A duplicate of the experiment identical in every way except for the one variable being tested. Use of a control is necessary to demonstrate cause and effect. (2)

Convergent Evolution The evolution of similar structures in distantly related organisms in response to similar environments. (33)

Cork Cambium In stems and roots of perennials, a secondary meristem that produces the outer protective layer of the bark. (18)

Coronary Arteries Large arteries that branch immediately from the aorta, providing oxygen-rich blood to the cardiac muscle. (28)

Corpus Callosum A thick cable composed of hundreds of millions of neurons that connect the right and left cerebral hemispheres of the mammalian brain. (23)

Corpus Luteum In the mammalian ovary, the structure that develops from the follicle after release of the egg. It secretes hormones that prepare the uterine endometrium to receive the developing embryo. (31)

Cortex In the stem or root of plants, the region between the epidermis and the vascular tissues. Composed of ground tissue. In animals, the outermost portion of some organs. (18)

Cotyledon The seed leaf of a dicot embryo containing stored nutrients required for the germinated seed to grow and develop, or a food digesting seed leaf in a monocot embryo. (20)

Countercurrent Flow Mechanism for increasing the exchange of substances or heat from one stream of fluid to another by having the two fluids flow in opposite directions. (29)

Covalent Bonds Linkage between two atoms which share the same electrons in their outermost shells. (3)

Cranial Nerves Paired nerves which emerge from the central stalk of the vertebrate brain and innervate the body. Humans have 12 pairs of cranial nerves. (23)

Cranium The bony casing which surrounds and protects the vertebrate brain. (23)

Cristae The convolutions of the inner membrane of the mitochondrion. Embedded within them are the components of the electron transport system and proton channels for chemiosmosis. (9)

Crossing Over During synapsis, the process by which homologues exchange segments with each other. (11)

Cryptic Coloration A form of camouflage wherein an organism's color or patterning helps it resemble its background. (42)

Cutaneous Respiration The uptake of oxygen across virtually the entire outer body surface. (29)

Cuticle 1) Waxy layer covering the outer cell walls of plant epidermal cells. It retards water vapor loss and helps prevent dehydration. (18) 2) Outer protective, nonliving covering of some animals, such as the exoskeleton of anthropods. (26, 39)

Cyanobacteria A type of prokaryote capable of photosynthesis using water as a source of electrons. Cyanobacteria were responsible for initially creating an O_2-containing atmosphere on earth. (35, 36)

Cyclic AMP (Cyclic adenosine monophosphate) A ring-shaped molecular version of an ATP minus two phosphates. A regulatory molecule formed by the enzyme adenylate cyclase which converts ATP to cAMP. A second messenger. (25)

Cyclic Pathways Metabolic pathways in which the intermediates of the reaction are regenerated while assisting the conversion of the substrate to product. (9)

Cyclic Photophosphorylation A pathway that produces ATP, but not NADPH, in the light reactions of photosynthesis. Energized electrons are shuttled from a reaction center, along a molecular pathway, back to the original reaction center, generating ATP en route. (8)

Cysts Protective, dormant structure formed by some protozoa. (37)

Cytochrome Oxidase A complex of proteins that serves as the final electron carrier in the mitochondrial electron transport system, transferring its electrons to O_2 to form water. (9)

Cytokinesis Final event in eukaryotic cell division in which the cell's cytoplasm and the new nuclei are partitioned into separate daughter cells. (10)

Cytokinins Growth-producing plant hormones which stimulate rapid cell division. (21)

Cytoplasm General term that includes all parts of the cell, except the plasma membrane and the nucleus. (5)

Cytoskeleton Interconnecting network of microfilaments, microtubules, and intermediate filaments that serves as a cell scaffold and provides the machinery for intracellular movements and cell motility. (5)

Cytotoxic (Killer) T Cells A class of T cells capable of recognizing and destroying foreign or infected cells. (30)

◄ D ►

Day Neutral Plants Plants that flower at any time of the year, independent of the relative lengths of daylight and darkness. (21)

Deciduous Trees or shrubs that shed their leaves in a particular season, usually autumn, before entering a period of dormancy. (40)

Deciduous Forest Forests characterized by trees that drop their leaves during unfavorable conditions, and leaf out during warm, wet seasons. Less dense than tropical rain forests. (40)

Decomposers (Saprophytes) Organisms that obtain nutrients by breaking down organic compounds in wastes and dead organisms. Includes fungi, bacteria, and some insects. (41)

Deletion Loss of a portion of a chromosome, following breakage of DNA. (13)

Denaturation Change in the normal folding of a protein as a result of heat, acidity, or alkalinity. Such changes result in a loss of enzyme functioning. (4)

Dendrites Cytoplasmic extensions of the cell body of a neuron. They carry impulses from the area of stimulation to the cell body. (23)

Denitrification The conversion by denitrifying bacteria of nitrites and nitrates into nitrogen gas. (41)

Denitrifying Bacteria Bacteria which take soil nitrogen, usable to plants, and convert it to unusable nitrogen gas. (41)

Density-Dependent Factors Factors that control population growth which are influenced by population size. (43)

Density-Independent Factors Factors that control population growth which are not affected by population size. (43)

Deoxyribonucleic Acid (DNA) Double-stranded polynucleotide comprised of deoxyribose (a sugar), phosphate, and four bases (adenine, guanine, cytosine, and thymine). Encoded in the sequence of nucleotides are the instructions for making proteins. DNA is the genetic material in all organisms except certain viruses. (14)

Depolarization A decrease in the potential difference (voltage) across the plasma membrane of a cell typically due to an increase in the movement of sodium ions into the cell. Acts to excite a target cell. (23)

Dermal Bone Bones of vertebrates that form within the dermal layer of the skin, such as the scales of fishes and certain bones of the skull. (26)

Dermal Tissue System In plants, the epidermis in primary growth, or the periderm in secondary growth. (18)

Dermis In animals, layer of cells below the epidermis in which connective tissue predominates. Embedded within it are vessels, various glands, smooth muscle, nerves, and follicles. (26)

Desert Biome characterized by intense solar radiation, very little rainfall, and high winds. (40)

Detrivore Organism that feeds on detritus, dead organisms or their parts, and living organisms' waste. (41)

Deuterostome One path of development exhibited by coelomate animals (e.g., echinoderms and chordates). (39)

Diabetes Mellitus A disease caused by a deficiency of insulin or its receptor, preventing glucose from being absorbed by the cells. (25)

Diaphragm A sheet of muscle that separates the thoracic cavity from the abdominal wall. (29)

Diastolic Pressure The second number of a blood pressure reading; the lowest pressure in the arteries just prior to the next heart contraction. (28)

Diatoms are golden-brown algae that are distinguished most dramatically by their intricate silica shells. (37)

Dicotyledonae (Dicots) One of the two classes of flowering plants, characterized by having seeds with two cotyledons, flower parts in 4s or 5s, net-veined leaves, one main root, and vascular bundles in a circular array within the stem. (Compare with Monocotylenodonae). (18)

Diffusion Tendency of molecules to move from a region of higher concentration to a region of lower concentration, until they are uniformly dispersed. (7)

Digestion The process by which food particles are disassembled into molecules small enough to be absorbed into the organism's cells and tissues. (27)

Digestive System System of specialized organs that ingests food, converts nutrients to a form that can be distributed throughout the animal's body, and eliminates undigested residues. (27)

Digestive Tract Tubelike channel through which food matter passes from its point of ingestion at the mouth to the elimination of indigestible residues from the anus. (27)

Dihybrid Cross A mating between two individuals that differ in two genetically-determined traits. (12)

Dimorphism Presence of two forms of a trait within a population, resulting from diversifying selection. (33)

Dinoflagellates Single-celled photosynthesizers that have two flagella. They are members of the pyrophyta, phosphorescent algae that sometimes cause red tide, often synthesizing a neurotoxin that accumulates in plankton eaters, causing paralytic shellfish poisoning in people who eat the shellfish. (37)

Dioecious Plants that produce either male or female reproductive structures but never both. (38)

Diploid Having two sets of homologous chromosomes. Often written 2N. (10, 13)

Directional Selection The steady shift of phenotypes toward one extreme. (33)

Discontinuous Variation An inheritance pattern in which the phenomenon of all possible phenotypes fall into distinct categories. (Compare with continuous variation). (12)

Displays The signals that form the language by which animals communicate. These signals are species specific and stereotyped and may be visual, auditory, chemical, or tactile. (44)

Disruptive Coloration Coloration that disguises the shape of an organism by breaking up its outline. (42)

Disruptive Selection The steady shift toward more than one extreme phenotype due to the elimination of intermediate phenotypes as has occurred among African swallowtail butterflies whose members resemble more than one species of distasteful butterfly. (33)

Divergent Evolution The emergence of new species as branches from a single ancestral lineage. (33)

Diversifying Selection The increasing frequency of extreme phenotypes because individuals with average phenotypes die off. (33)

Diving Reflex Physiological response that alters the flow of blood in the body of diving mammals that allows the animal to maintain high levels of activity without having to breathe. (29)

Division (or Phylum) A level of the taxonomic hierarchy that groups together members or related classes. (1)

DNA (see **Deoxyribonucleic Acid**)

DNA Cloning The amplification of a particular DNA by use of a growing population of bacteria. The DNA is initially taken up by a bacterial cell—usually as a plasmid—and then replicated along with the bacteria's own DNA. (16)

DNA Fingerprint The pattern of DNA fragments produced after treating a sample of DNA with a particular restriction enzyme and separating the fragments by gel electrophoresis. Since different members of a population have DNA with a different nucleotide sequence, the pattern of DNA fragments produced by this method can be used to identify a particular individual. (16)

DNA Ligase The enzyme that covalently joins DNA fragments into a continuous DNA strand. The enzyme is used in a cell during replication to seal newly-synthesized fragments and by biotechnologists to form recombinant DNA molecules from separate fragments. (14, 16)

DNA Polymerase Enzyme responsible for replication of DNA. It assembles free nucleotides, aligning them with the complementary ones in the unpaired region of a single strand of DNA template. (14)

Dominant The form of an allele that masks the presence of other alleles for the same trait. (12)

Dormancy A resting period, such as seed dormancy in plants or hibernation in animals, in which organisms maintain reduced metabolic rates. (21)

Dorsal In anatomy, the back of an animal. (39)

Double Blind Test A clinical trial of a drug in which neither the human subjects or the researchers know who is receiving the drug or placebo. (2)

Down Syndrome Genetic disorder in humans characterized by distinct facial appearance and mental retardation, resulting from an extra copy of chromosome number 21 (trisomy 21) in each cell. (11, 17)

Duodenum First part of the human small intestine in which most digestion of food occurs. (27)

Duplication The repetition of a segment of a chromosome. (13)

◄ **E** ►

Ecdysis Molting process by which an arthropod periodically discards its exoskeleton and replaces it with a larger version. The process is controlled by the hormone ecydysone. (39)

Ecdysone An insect steroid hormone that triggers molting and metamorphosis. (15)

Echinodermata A phylum composed of animals having an internal skeleton made of many small calcium carbonate plates which have jutting spines. Includes sea stars, sea urchins, etc. (39)

Echolocation The use of reflected sound waves to help guide an animal through its environment and/or locate objects. (24)

Ecological Equivalent Organisms that occupy similar ecological niches in different regions or ecosystems of the world. (41)

Ecological Niche The habitat, functional role(s), requirements for environmental resources and tolerance ranges for each abiotic condition in relation to an organism. (41)

Ecological Pyramid Illustration showing the energy content, numbers of organisms, or biomass at each trophic level. (41)

Ecology The branch of biology that studies interactions among organisms as well as the interactions of organisms and their physical environment. (40)

Ecosystem Unit comprised of organisms interacting among themselves and with their physical environment. (41)

Ecotypes Populations of a single species with different, genetically fixed tolerance ranges. (41)

Ectoderm In animals, the outer germ cell layer of the gastrula. It gives rise to the nervous system and integument. (32)

Ectotherms Animals that lack an internal mechanism for regulating body temperature. "Cold-blooded" animals. (28)

Edema Swelling of a tissue as the result of an accumulation of fluid that has moved out of the blood vessels. (28)

Effectors Muscle fibers and glands that are activated by neural stimulation. (23)

Efferent (Motor) Nerves The nerves that carry messages from the central nervous system to the effectors, the muscles, and glands. They are divided into two systems: somatic and autonomic. (23)

Egg Female gamete, also called an ovum. A fertilized egg is the product of the union of female and male gametes (egg and sperm cells). (32)

Electrocardiogram (EKG) Recording of the electrical activity of the heart, which is used to diagnose various types of heart problems. (28)

Electron Acceptor Substances that are capable of accepting electrons transferred from an electron donor. For example, molecular oxygen (O_2) is the terminal electron acceptor during respiration. Electron acceptors also receive electrons from chlorophyll during photosynthesis. Electron acceptors may act as part of an electron transport system by transferring the electrons they receive to another substance. (8, 9)

Electron Carrier Substances (such as NAD^+ and FAD) that transport electrons from one step of a metabolic pathway to the next or from metabolic reactions to biosynthetic reactions. (8, 9)

Electrons Negatively charged particles that orbit the atomic nucleus. (3)

Electron Transport System Highly organized assembly of cytochromes and other proteins which transfer electrons. During transport, which occurs within the inner membranes of mitochondria and chloroplasts, the energy extracted from the electrons is used to make ATP. (8, 9)

Electrophoresis A technique for separating different molecules on the basis of their size and/or electric charge. There are various ways the technique is used. In gel electrophoresis, proteins or DNA fragments are driven through a porous gel by their charge, but become separated according to size; the larger the molecule, the slower it can work its way through the pores in the gel, and the less distance it travels along the gel. (16)

Element Substance composed of only one type of atom. (3)

Embryo An organism in the early stages of development, beginning with the first division of the zygote. (32)

Embryo Sac The fully developed female gametophyte within the ovule of the flower. (20)

Emigration Individuals permanently leaving an area or population. (43)

Endergonic Reactions Chemical reactions that require energy input from another source in order to occur. (6)

Endocrine Glands Ductless glands, which secrete hormones directly into surrounding tissue fluids and blood vessels for distribution to the rest of the body by the circulatory system. (25)

Endocytosis A type of active transport that imports particles or small cells into a cell. There are two types of endocytic processes: phagocytosis, where large particles are ingested by the cell, and pinocytosis, where small droplets are taken in. (7)

Endoderm In animals, the inner germ cell layer of the gastrula. It gives rise to the digestive tract and associated organs and to the lungs. (32)

Endodermis The innermost cylindrical layer of cortex surrounding the vascular tissues of the root. The closely pressed cells of the endodermis have a waxy band, forming a waterproof layer, the Casparian strip. (18)

Endogenous Plant responses that are controlled internally, such as biological clocks controlling flower opening. (21)

Endometrium The inner epithelial layer of the uterus that changes markedly with the uterine (menstrual) cycle in preparation for implantation of an embryo. (31)

Endoplasmic Reticulum (ER) An elaborate system of folded, stacked and tubular membranes contained in the cytoplasm of eukaryotic cells. (5)

Endorphins (Endogenous Morphinelike Substances) A class of peptides released from nerve cells of the limbic system of the brain that can block perceptions of pain and produce a feeling of euphoria. (23)

Endoskeleton The internal support structure found in all vertebrates and a few invertebrates (sponges and sea stars). (26)

Endosperm Nutritive tissue in plant embryos and seeds. (20)

Endosperm Mother Cell A binucleate cell in the embryo sac of the female gametophyte, occurring in the ovule of the ovary in angiosperms. Each nucleus is haploid; after fertilization, nutritive endosperm develops. (20)

Endosymbiosis Theory A theory to explain the development of complex eukaryotic cells by proposing that some organelles once were free-living prokaryotic cells that then moved into another larger such cell, forming a beneficial union with it. (5)

Endotherms Animals that utilize metabolically produced heat to maintain a constant, elevated body temperature. "Warm-blooded" animals. (28)

End Product The last product in a metabolic pathway. Typically a substance, such as an amino acid or a nucleotide, that will be used as a monomer in the formation of macromolecules. (6)

Energy The ability to do work. (6)

Entropy Energy that is not available for doing work; measure of disorganization or randomness. (6)

Environmental Resistance The factors that eventually limit the size of a population. (43)

Enzyme Biological catalyst; a protein molecule that accelerates the rate of a chemical reaction. (6)

Eosiniphil A type of phagocytic white blood cell. (28)

Epicotyl The portion of the embryo of a dicot plant above the cotyledons. The epicotyl gives rise to the shoot. (20)

Epidermis In vertebrates, the outer layer of the skin, containing superficial layers of dead cells produced by the underlying living epithelial cells. In plants, the outer layer of cells covering leaves, primary stem, and primary root. (26, 18)

Epididymis Mass of convoluted tubules attached to each testis in mammals. After leaving the testis, sperm enter the tubules where they finish maturing and acquire motility. (31)

Epiglottis A flap of tissue that covers the glottis during swallowing to prevent food and liquids from entering the lower respiratory tract. (29)

Epinephrine (Adrenalin) Substance that serves both as an excitatory neurotransmitter released by certain neurons of the CNS and as a hormone released by the adrenal medulla that increases the body's ability to combat a stressful situation. (25)

Epipelagic Zone The lighted upper ocean zone, where photosynthesis occurs; large populations of phytoplankton occur in this zone. (40)

Epiphyseal Plates The action centers for ossification (bone formation). (26)

Epistasis A type of gene interaction in which a particular gene blocks the expression of another gene at another locus. (12)

Epithelial Tissue Continuous sheets of tightly packed cells that cover the body and line its tracts and chambers. Epithelium is a fundamental tissue type in animals. (22)

Erythrocytes Red blood cells. (28)

Erythropoietin A hormone secreted by the kidney which stimulates the formation of erythrocytes by the bone marrow. (28)

Essential Amino Acids Eight amino acids that must be acquired from dietary protein. If even one is missing from the human diet, the synthesis of proteins is prevented. (27)

Essential Fatty Acids Linolenic and linoleic acids, which are required for phospholipid construction and must be acquired from a dietary source. (27)

Essential Nutrients The 16 minerals essential for plant growth, divided into two groups: macronutrients, which are required in large quantities, and micronutrients, which are needed in small amounts. (19)

Estrogen A female sex hormone secreted by the ovaries when stimulated by pituitary gonadotrophins. (31)

Estuaries Areas found where rivers and streams empty into oceans, mixing fresh water with salt water. (40)

Ethology The study of animal behavior. (44)

Ethylene Gas A plant hormone that stimulates fruit ripening. (21)

Etiolation The condition of rapid shoot elongation, small underdeveloped leaves, bent shoot-hook, and lack of chlorophyll, all due to lack of light. (21)

Eubacteria Typical procaryotic bacteria with peptidoglycan in their cell walls. The majority of monerans are eubacteria. (36)

Eukaryotic Referring to organisms whose cellular anatomy includes a true nucleus with a nuclear envelope, as well as other membrane-bound organelles. (5)

Eusocial Species Social species that have sterile workers, cooperative care of the young, and an overlap of generations so that the colony labor is a family affair. (44)

Eutrophication The natural aging process of lakes and ponds, whereby they become marshes and, eventually, terrestrial environments.

Evolution A process whereby the characteristics of a species change over time, eventually leading to the formation of new species that go about life in new ways. (33)

Evolutionarily Stable Strategy (ESS) A behavioral strategy or course of action that depends on what other members of the population are doing. By definition, an ESS cannot be replaced by any other strategy when most of the members of the population have adopted it. (44)

Excitatory Neurons Neurons that stimulate their target cells into activity. (23)

Excretion Removal of metabolic wastes from an organism. (28)

Excretory System The organ system that eliminates metabolic wastes from the body. (28)

Exergonic Reactions Chemical reactions that occur spontaneously with the release of energy. (6)

Exocrine Glands Glands which secrete their products through ducts directly to their sites of action, e.g., tear glands. (26)

Exocytosis A form of active transport used by cells to move molecules, particles, or other cells contained in vesicles across the plasma membrane to the cell's environment. (5)

Exogenous Plant responses that are controlled externally, or by environmental conditions. (21)

Exons Structural gene segments that are transcribed and whose genetic information is subsequently translated into protein. (15)

Exoskeletons Hard external coverings found in some animals (e.g., lobsters, insects) for protection, support, or both. Such organisms grow by the process of molting. (26)

Exploitative Competition A competition in which one species manages to get more of a resource, thereby reducing supplies for a competitor. (42)

Exponential Growth An increase by a fixed percentage in a given time period; such as population growth per year. (43)

Extensor Muscle A muscle which, when contracted, causes a part of the body to straighten at a joint. (26)

External Fertilization Fertilization of an egg outside the body of the female parent. (31)

Extinction The loss of a species. (33)

Extracellular Digestion Digestion occurring outside the cell; occurs in bacteria, fungi, and multicellular animals. (27)

Extracellular Matrix Layer of extracellular material residing just outside a cell. (5)

◀ **F** ▶

F₁ First filial generation. The first generation of offspring in a genetic cross. (12)

F₂ Second filial generation. The offspring of an F₁ cross. (12)

Facilitated Diffusion The transport of molecules into cells with the aid of "carrier" proteins embedded in the plasma membrane. This carrier-assisted transport does not require the expenditure of energy by the cell. (7)

FAD Flavin adenine dinucleotide. A coenzyme that functions as an electron carrier in metabolic reactions. When it is reduced to FADH₂, this molecule becomes a cellular energy source. (9)

Family A level of the taxonomic hierarchy that groups together members of related genera. (1)

Fast-Twitch Fibers Skeletal muscle fibers that depend on anaerobic metabolism to produce ATP rapidly, but only for short periods of time before the onset of fatigue. Fast-twitch fibers generate greater forces for shorter periods than slow-twitch fibers. (9)

Fat A triglyceride consisting of three fatty acids joined to a glycerol. (4)

Fatty Acid A long unbranched hydrocarbon chain with a carboxyl group at one end. Fatty acids lacking a double bond are said to be saturated. (4)

Fauna   The animals in a particular region.

Feedback Inhibition (Negative Feedback) A mechanism for regulating enzyme activity by temporarily inactivating a key enzyme in a biosynthetic pathway when the concentration of the end product is elevated. (6)

Fermentation   The direct donation of the electrons of NADH to an organic compound without their passing through an electron transport system. (9)

Fertility Rate   In humans, the average number of children born to each woman between 15 and 44 years of age. (43)

Fertilization   The process in which two haploid nuclei fuse to form a zygote. (32)

Fetus   The term used for the human embryo during the last seven months in the uterus. During the fetal stage, organ refinement accompanies overall growth. (32)

Fibrinogen   A rod-shaped plasma protein that, converted to fibrin, generates a tangled net of fibers that binds a wound and stops blood loss until new cells replace the damaged tissue. (28)

Fibroblasts   Cells found in connective tissues that secrete the extracellular materials of the connective tissue matrix. These cells are easily isolated from connective tissues and are widely used in cell culture. (22)

Fibrous Root System   Many approximately equal-sized roots; monocots are characterized by a fibrous root system. Also called diffuse root system. (18)

Filament   The stalk of a stamen of angiosperms, with the anther at its tip. Also, the threadlike chain of cells in some algae and fungi. (20)

Filamentous Fungus   Multicellular members of the fungus kingdom comprised mostly of living threads (hyphae) that grow by division of cells at their tips (see molds). (37)

Filter Feeders   Aquatic animals that feed by straining small food particles from the surrounding water. (27, 39)

Fitness   The relative degree to which an individual in a population is likely to survive to reproductive age and to reproduce. (33)

Fixed Action Patterns   Motor responses that may be triggered by some environmental stimulus, but once started can continue to completion without external stimuli. (44)

Flagella   Cellular extensions that are longer than cilia but fewer in number. Their undulations propel cells like sperm and many protozoans, through their aqueous environment. (5)

Flexor Muscle   A muscle which, when contracted, causes a part of the body to bend at a joint. (26)

Flora   The plants in a particular region. (21)

Florigen Proposed   A chemical hormone that is produced in the leaves and stimulates flowering. (21)

Fluid Mosaic Model   The model proposes that the phospholipid bilayer has a viscosity similar to that of light household oil and that globular proteins float like icebergs within this bilayer. The now favored explanation for the architecture of the plasma membrane. (5)

Follicle (Ovarian)   A chamber of cells housing the developing oocytes. (31)

Food Chain   Transfers of food energy from organism to organism, in a linear fashion. (41)

Food Web   The map of all interconnections between food chains for an ecosystem. (41)

Forest Biomes   Broad geographic regions, each with characteristic tree vegetation: 1) tropical rain forests (lush forests in a broad band around the equator), 2) deciduous forests (trees and shrubs drop their leaves during unfavorable seasons), 3) coniferous forest (evergreen conifers). (40)

Fossil Record   An entire collection of remains from which paleontologists attempt to reconstruct the phylogeny, anatomy, and ecology of the preserved organisms. (34)

Fossils   The preserved remains of organisms from a former geologic age. (34)

Fossorial   Living underground.

Founder Effect   The potentially dramatic difference in allele frequency of a small founding population as compared to the original population. (33)

Founder Population   The individuals, usually few, that colonize a new habitat. (33)

Frameshift Mutation   The insertion or deletion of nucleotides in a gene that throws off the reading frame. (14)

Free Radical   Atom or molecule containing an unpaired electron, which makes it highly reactive. (3)

Freeze-Fracture   Technique in which cells are frozen into a block which is then struck with a knife blade that fractures the block in two. Fracture planes tend to expose the center of membranes for EM examination. (5)

Fronds   The large leaf-like structures of ferns. Unlike true leaves, fronds have an apical meristem and clusters of sporangia called sori. (38)

Fruit   A mature plant ovary (flower) containing seeds with plant embryos. Fruits protect seeds and aid in their dispersal. (20)

Fruiting Body   A spore-producing structure that extends upward in an elevated position from the main mass of a mold or slime mold. (37)

FSH   Follicle stimulating hormone. A hormone secreted by the anterior pituitary that prepares a female for ovulation by stimulating the primary follicle to ripen or stimulates spermatogenesis in males. (31)

Functional Groups   Accessory chemical entities (e.g., —OH, —NH₂, —CH₃), which help determine the identity and chemical properties of a compound. (4)

Fundamental Niche   The potential ecological niche of a species, including all factors affecting that species. The fundamental niche is usually never fully utilized. (41)

Fungus   Yeast, mold, or large filamentous mass forming macroscopic fruiting bodies, such as mushrooms. All fungi are eukaryotic nonphotosynthetic heterotrophics with cell walls. (37)

◀ **G** ▶

G₁ Stage   The first of three consecutive stages of interphase. During G₁, cell growth and normal functions occur. The duration of this stage is most variable. (10)

G₂ Stage   The final stage of interphase in which the final preparations for mitosis occur. (10)

Gallbladder   A small saclike structure that stores bile salts produced by the liver. (27)

Gamete   A haploid reproductive cell--either a sperm or an egg. (10)

Gas Exchange Surface   Surface through which gases must pass in order to enter or leave the body of an animal. It may be the plasma membrane of a protistan or the complex tissues of the gills or the lungs in multicellular animals. (29)

Gastrovascular Cavity   In cnidarians and flatworms, the branched cavity with only one opening. It functions in both digestion and transport of nutrients. (39)

Gastrula   The embryonic stage formed by the inward migration of cells in the blastula. (32)

Gastrulation   The process by which the blastula is converted into a gastrula having three germ layers (ectoderm, mesoderm, and endoderm). (32)

Gated Ion Channels   Most passageways through a plasma membrane that allow ions to pass contain "gates" that can occur in either an open or a closed conformation. (7, 23)

Gel Electrophoresis   (See **Electrophoresis**)

Gene Pool   All the genes in all the individuals of a population. (33)

Gene Regulatory Proteins   Proteins that bind to specific sites in the DNA and control the transcription of nearby genes. (15)

Genes   Discrete units of inheritance which determine hereditary traits. (12, 14)

Gene Therapy   Treatment of a disease by alteration of the person's genotype, or the genotype of particular affected cells. (17)

Genetic Carrier   A heterozygous individual who shows no evidence of a genetic disorder but, because they possess a recessive allele for a disorder, can pass the mutant gene on to their offspring. (17)

Genetic Code   The correspondence between the various mRNA triplets (codons, e.g., UGC) and the amino acid that the triplet specifies (e.g., cysteine). The genetic code includes 64 possible three-letter words that constitute the genetic language for protein synthesis. (14)

Genetic Drift   Random changes in allele frequency that occur by chance alone. Occurs primarily in small populations. (33)

Genetic Engineering   The modification of a cell or organism's genetic composition according to human design. (16)

Genetic Equilibrium   A state in which allele frequencies in a population remain constant from generation to generation. (33)

Genetic Mapping   Determining the locations of specific genes or genetic markers along particular chromosomes. This is typically accomplished using crossover frequencies; the more often alleles of two genes are separated during crossing over, the greater the distance separating the genes. (13)

Genetic Recombination   The reshuffling of genes on a chromosome caused by breakage of DNA and its reunion with the DNA of a homologoue. (11)

Genome   The information stored in all the DNA of a single set of chromosomes. (17)

Genotype   An individual's genetic makeup. (12)

Genus   Taxonomic group containing related species. (1)

Geologic Time Scale   The division of the earth's 4.5 billion-year history into eras, periods, and epochs based on memorable geologic and biological events. (35)

Germ Cells   Cells that are in the process of or have the potential to undergo meiosis and form gametes. (11, 31)

Germination   The sprouting of a seed, beginning with the radicle of the embryo breaking through the seed coat. (21)

Germ Layers   Collective name for the endoderm, ectoderm, and mesoderm, from which all the structures of the mature animal develop. (32)

Gibberellins   More than 50 compounds that promote growth by stimulating both cell elongation and cell division. (21)

Gills   Respiratory organs of aquatic animals. (29)

Globin   The type of polypeptide chains that make up a hemoglobin molecule.

Glomerular Filtration   The process by which fluid is filtered out of the capillaries of the glomerulus into the proximal end of the nephron. Proteins and blood cells remain behind in the bloodstream. (28)

Glomerulus   A capillary bundle embedded in the double-membraned Bowman's capsule, through which blood for the kidney first passes. (28)

Glottis   Opening leading to the larynx and lower respiratory tract. (29)

Glucagon   A hormone secreted by the Islets of Langerhans that promotes glycogen breakdown to glucose. (25)

Glucocorticoids   Steroid hormones which regulate sugar and protein metabolism. They are secreted by the adrenal cortex. (25)

Glycogen   A highly branched polysaccharide consisting of glucose monomers that serves as a storage of chemical energy in animals. (4)

Glycolysis   Cleavage, releasing energy, of the six-carbon glucose molecule into two molecules of pyruvic acid, each containing three carbons. (9)

Glycoproteins   Proteins with covalently-attached chains of sugars. (5)

Glycosidic Bond   The covalent bond between individual molecules in carbohydrates. (4)

Golgi Complex   A system of flattened membranous sacs, which package substances for secretion from the cell. (5)

Gonadotropin-Releasing Hormone (GnRH)   Hypothalmic hormone that controls the secretion of the gonadotropins FSH and LH. (31)

Gonadotropins   Two anterior pituitary hormones which act on the gonads. Both FSH (follicle-stimulating hormone) and LH (luteinizing hormone) promote gamete development and stimulate the gonads to produce sex hormones. (25)

Gonads   Gamete-producing structures in animals: ovaries in females, testes in males. (31)

Grasslands   Areas of densely packed grasses and herbaceous plants. (40)

Gravitropisms (Geotropisms)   Changes in plant growth caused by gravity. Growth away from gravitational force is called negative gravitropism; growth toward it is positive. (21)

Gray Matter   Gray-colored neural tissue in the cerebral cortex of the brain and in the butterfly-shaped interior of the spinal cord. Composed of nonmyelinated cell bodies and dendrites of neurons. (23)

Greenhouse Effect   The trapping of heat in the Earth's troposphere, caused by increased levels of carbon dioxide near the Earth's surface; the carbon dioxide is believed to act like glass in a greenhouse, allowing light to reach the Earth, but not allowing heat to escape. (41)

Ground Tissue System   All plant tissues except those in the dermal and vascular tissues. (18)

Growth   An increase in size, resulting from cell division and/or an increase in the volume of individual cells. (10)

Growth Hormone (GH)   Hormone produced by the anterior pituitary; stimulates protein synthesis and bone elongation. (25)

Growth Ring   In plants with secondary growth, a ring formed by tracheids and/or vessels with small lumens (late wood) during periods of unfavorable conditions; apparent in cross section. (18)

Guard Cells   Specialized epidermal plant cells that flank each stomated pore of a leaf. They regulate the rate of gas diffusion and transpiration. (18)

Guild   Group of species with similar ecological niches. (41)

Guttation   The forcing of water and mineral completely out to the tips of leaves as a result of positive root pressure. (19)

Gymnosperms   The earliest seed plants, bearing naked seeds. Includes the pines, hemlocks, and firs. (38)

◀ **H** ▶

Habitat   The place or region where an organism lives. (41)

Habituation   The phenomenon in which an animal ceases to respond to a repetitive stimulus. (23, 44)

Hair Cells   Sensory receptors of the inner ear that respond to sound vibration and bodily movement. (24)

Half-Life   The time required for half the mass of a radioactive element to decay into its stable, non-radioactive form. (3)

Haplodiploidy   A genetic pattern of sex determination in which fertilized eggs develop into females and non-fertilized eggs develop into males (as occurs among bees and wasps). (44)

Haploid   Having one set of chromosomes per cell. Often written as 1N. (10)

Hardy-Weinberg Law   The maintenance of constant allele frequencies in a population from one generation to the next when certain conditions are met. These conditions are the absence of mutation and migration, random mating, a large population, and an equal chance of survival for all individuals. (33)

Haversian Canals   A system of microscopic canals in compact bone that transport nutrients to and remove wastes from osteocytes. (26)

Heart   An organ that pumps blood (or hemolymph in arthropods) through the vessels of the circulatory system. (28)

Helper T Cells   A class of T cells that regulate immune responses by recognizing and activating B cells and other T cells. (30)

Hemocoel   In arthropods, the unlined spaces into which fluid (hemolymph) flows when it leaves the blood vessels and bathes the internal organs. (28)

Hemoglobin   The iron-containing blood protein that temporarily binds O_2 and releases it into the tissues. (4, 29)

Hemophilia   A genetic disorder determined by a gene on the X chromosome (an X-linked trait) that results from the failure of the blood to form clots. (13)

Herbaceous   Plants having only primary growth and thus composed entirely of primary tissue. (18)

Herbivore   An organism, usually an animal, that eats primary producers (plants). (42)

Herbivory   The term for the relationship of a secondary consumer, usually an animal, eating primary producers (plants). (42)

Heredity The passage of genetic traits to off-spring which consequently are similar or identical to the parent(s). (12)

Hermaphrodites Animals that possess gonads of both the male and the female. (31)

Heterosporous Higher vascular plants producing two types of spores, a megaspore which grows into a female gametophyte and a microspore which grows into a male gametophyte. (38)

Heterozygous A term applied to organisms that possess two different alleles for a trait. Often, one allele (A) is dominant, masking the presence of the other (a), the recessive. (12)

High Intertidal Zone In the intertidal zone, the region from mean high tide to around just below sea level. Organisms are submerged about 10% of the time. (40)

Histones Small basic proteins that are complexed with DNA to form nucleosomes, the basic structural components of the chromatin fiber. (14)

Homeobox That part of the DNA sequence of homeotic genes that is similar (homologous) among diverse animal species. (32)

Homeostasis Maintenance of fairly constant internal conditions (e.g., blood glucose level, pH, body temperature, etc.) (22)

Homeotic Genes Genes whose products act during embryonic development to affect the spatial arrangement of the body parts. (32)

Hominids Humans and the various groups of extinct, erect-walking primates that were either our direct ancestors or their relatives. Includes the various species of *Homo* and *Australopithecus*. (34)

Homo the genus that contains modern and extinct species of humans. (34)

Homologous Structures Anatomical structures that may have different functions but develop from the same embryonic tissues, suggesting a common evolutionary origin. (34)

Homologues Members of a chromosome pair, which have a similar shape and the same sequence of genes along their length. (10)

Homoplasy (see **Analogous Structures**)

Homosporous Plants that manufacture only one type of spore, which develops into a gametophyte containing both male and female reproductive structures. (38)

Homozygous A term applied to an organism that has two identical alles for a particular trait. (12)

Hormones Chemical messengers secreted by ductless glands into the blood that direct tissues to change their activities and correct imbalances in body chemistry. (25)

Host The organism that a parasite lives on and uses for food. (42)

Human Chorionic Gonadotropin (HCG) A hormone that prevents the corpus luteum from degenerating, thereby maintaining an ade-quate level of progesterone during pregnancy. It is produced by cells of the early embryo. (25)

Human Immunodeficiency Virus (HIV) The infectious agent that causes AIDS, a disease in which the immune system is seriously disabled. (30, 36)

Hybrid An individual whose parents possess different genetic traits in a breeding experiment or are members of different species. (12)

Hybridization Occurs when two distinct species mate and produce hybrid offspring. (33)

Hybridoma A cell formed by the fusion of a malignant cell (a myeloma) and an antibody-producing lymphocyte. These cells proliferate indefinitely and produce monoclonal antibodies. (30)

Hydrogen Bonds Relatively weak chemical bonds formed when two molecules share an atom of hydrogen. (3)

Hydrologic Cycle The cycling of water, in various forms, through the environment, from Earth to atmosphere and back to Earth again. (41)

Hydrolysis Splitting of a covalent bond by donating the H^+ or OH^- of a water molecule to the two components. (4)

Hydrophilic Molecules Polar molecules that are attracted to water molecules and readily dissolve in water. (3)

Hydrophobic Interaction When nonpolar molecules are "forced" together in the presence of a polar solvent, such as water. (3)

Hydrophobic Molecules Nonpolar substances, insoluble in water, which form aggregates to minimize exposure to their polar surroundings. (3)

Hydroponics The science of growing plants in liquid nutrient solutions, without a solid medium such as soil. (19)

Hydrosphere That portion of the Earth composed of water. (40)

Hydrostatic Skeletons Body support systems found usually in underwater animals (e.g., marine worms). Body shape is protected against gravity and other physical forces by internal hydrostatic pressure produced by contracting muscles encircling their closed, fluid-filled chambers. (26)

Hydrothermal Vents Fissures in the ocean floor where sea water becomes superheated. Chemosynthetic bacteria that live in these vents serve as the autotrophs that support a diverse community of ocean-dwelling organisms. (8)

Hyperpolarization An increase in the potential difference (voltage) across the plasma membrane of a cell typically due to an increase in the movement of potassium ions out of the cell. Acts to inhibit a target cell. (23)

Hypertension High blood pressure (above about 130/90). (28)

Hypertonic Solutions Solutions with higher solute concentrations than found inside the cell. These cause a cell to lose water and shrink. (7)

Hypervolume In ecology, a multidimensional area which includes all factors in an organism's ecological niche, or its' potential niche. (41)

Hypocotyl Portion of the plant embryo below the cotyledons. The hypocotyl gives rise to the root and, very often, to the lower part of the stem. (20)

Hypothalamus The area of the brain below the thalamus that regulates body temperature, blood pressure, etc. (25)

Hypothesis A tentative explanation for an observation or a phenomenon, phrased so that it can be tested by experimentation. (2)

Hypotonic Solutions Solutions with lower solute concentrations than found inside the cell. These cause a cell to accumulate water and swell. (7)

◄ **I** ►

Immigration Individuals permanently moving into a new area or population. (43)

Immune System A system in vertebrates for the surveillance and destruction of disease-causing microorganisms and cancer cells. Composed of lymphocytes, particularly B cells and T cells, and triggered by the introduction of antigens into the body which makes the body, upon their destruction, resistant to a recurrence of the same disease. (30)

Immunoglobulins (IGs) Antibody molecules. (30)

Imperfect Flowers Flowers that contain either stamens or carpels, making them male or female flowers, respectively. (20)

Imprinting A type of learning in which an animal develops an association with an object after exposure to the object during a critical period early in its life. (44)

Inbreeding When individuals mate with close relatives, such as brothers and sisters. May occur when population sizes drastically shrink and results in a decrease in genetic diversity. (33)

Incomplete Digestive Tract A digestive tract with only one opening through which food is taken in and residues are expelled. (27)

Incomplete (Partial) Dominance A phenomenon in which heterozygous individuals are phenotypically distinguishable from either homozygous type. (12)

Incomplete Flower Flowers lacking one or more whorls of sepals, petals, stamen, or pistils. (20)

Independent Assortment The shuffling of members of homologous chromosome pairs in meiosis I. As a result, there are new chromosome combinations in the daughter cells, which later produce offspring with random mixtures of traits from both parents. (11, 12)

Indoleatic Acid (IAA) An auxin responsible for many plant growth responses including apical dominance, a growth pattern in which shoot tips prevent axillary buds from sprouting. (21)

Induction The process in which one embryonic tissue induces another tissue to differentiate along a pathway that it would not otherwise have taken. (32) Stimulation of transcription of a gene in an operon. Occurs when the repressor protein is unable to bind to the operator. (15)

Inflammation A body strategy initiated by the release of chemicals following injury or infection which brings additional blood with its protective cells to the injured area. (30)

Inhibitory Neurons Neurons that oppose a response in the target cells. (23)

Inhibitory Neurotransmitters Substances released from inhibitory neurons where they synapse with the target cell. (23)

Innate Behavior Actions that are under fairly precise genetic control, typically species-specific, highly stereotyped, and that occur in a complete form the first time the stimulus is encountered. (44)

Insight Learning The sudden solution to a problem without obvious trial-and-error procedures. (44)

Insulin One of the two hormones secreted by endocrine centers called Islets of Langerhans; promotes glucose absorption, utilization, and storage. Insulin is secreted by them when the concentration of glucose in the blood begins to exceed the normal level. (25)

Integumentary System The body's protective external covering, consisting of skin and subcutaneous tissue. (26)

Integuments Protective covering of the ovule. (20)

Intercellular Junctions Specialized regions of cell-cell contact between animal cells. (5)

Intercostal Muscles Muscles that lie between the ribs in humans whose contraction expands the thoracic cavity during breathing. (29)

Interference Competition One species' direct interference by another species for the same limited resource; such as aggressive animal behavior. (42)

Internal Fertilization Fertilization of an egg within the body of the female. (31)

Interneurons Neurons situated entirely within the central nervous system. (23)

Internodes The portion of a stem between two nodes. (18)

Interphase Usually the longest stage of the cell cycle during which the cell grows, carries out normal metabolic functions, and replicates its DNA in preparation for cell division. (10)

Interstitial Cells Cells in the testes that produce testosterone, the major male sex hormone. (31)

Interstitial Fluid The fluid between and surrounding the cells of an animal; the extracellular fluid. (28)

Intertidal Zone The region of beach exposed to air between low and high tides. (40)

Intracellular Digestion Digestion occurring inside cells within food vacuoles. The mode of

digestion found in protists and some filter-feeding animals (such as sponges and clams). (27)

Intraspecific Competition Individual organisms of one species competing for the same limited resources in the same habitat, or with overlapping niches. (42)

Intrinsic Rate of Increase (r_m) the maximum growth rate of a population under conditions of maximum birth rate and minimum death rate. (43)

Introns Intervening sequences of DNA in the middle of structural genes, separating exons. (15)

Invertebrates Animals that lack a vertebral column, or backbone. (39)

Ion An electrically charged atom created by the gain or loss of electrons. (3)

Ionic Bond The noncovalent linkage formed by the attraction of oppositely charged groups. (3)

Islets of Langerhans Clusters of endocrine cells in the pancreas that produce insulin and glucagon. (25)

Isolating Mechanisms Barriers that prevent gene flow between populations or among segments of a single population. (33)

Isotopes Atoms of the same element having a different number of neutrons in their nucleus. (3)

Isotonic Solutions Solutions in which the solute concentration outside the cell is the same as that inside the cell. (7)

◀ **J** ▶

Joints Structures where two pieces of a skeleton are joined. Joints may be flexible, such as the knee joint of the human leg or the joints between segments of the exoskeleton of the leg of an insect, or inflexible, such as the joints (sutures) between the bones of the skull. (26)

J-Shaped Curve A curve resulting from exponential growth of a population. (43)

◀ **K** ▶

Karyotype A visual display of an individual's chromosomes. (10)

Kidneys Paired excretory organs which, in humans, are fist-sized and attached to the lower spine. In vertebrates, the kidneys remove nitrogenous wastes from the blood and regulate ion and water levels in the body. (28)

Killer T Cells A type of lymphocyte that functions in the destruction of virus-infected cells and cancer cells. (30)

Kinases Enzymes that catalyze reactions in which phosphate groups are transferred from ATP to another molecule. (6)

Kinetic Energy Energy in motion. (6)

Kinetochore Part of a mitotic (or meiotic) chromosome that is situated within the centromere and to which the spindle fibers attach. (10)

Kingdom A level of the taxonomic hierarchy that groups together members of related phyla or divisions. Modern taxonomy divides all organisms into five Kingdoms: Monera, Protista, Fungi, Plantae, and Animalia. (1)

Klinefelter Syndrome A male whose cells have an extra X chromosome (XXY). The syndrome is characterized by underdeveloped male genitalia and feminine secondary sex characteristics. (17)

Krebs Cycle A circular pathway in aerobic respiration that completely oxidizes the two pyruvic acids from glycolysis. (9)

K-Selected Species Species that produce one or a few well-cared for individuals at a time. (43)

◀ **L** ▶

Lacteal Blind lymphatic vessel in the intestinal villi that receives the absorbed products of lipid digestion. (27)

Lactic Acid Fermentation The process in which electrons removed during glycolysis are transferred from NADH to pyruvic acid to form lactic acid. Used by various prokaryotic cells under oxygen-deficient conditions and by muscle cells during strenuous activity. (9)

Lake Large body of standing fresh water, formed in natural depressions in the Earth. Lakes are larger than ponds. (40)

Lamella In bone, concentric cylinders of calcified collagen deposited by the osteocytes. The laminated layers produce a greatly strengthened structure. (26)

Large Intestine Portion of the intestine in which water and salts are reabsorbed. It is so named because of its large diameter. The large instestine, except for the rectum, is called the colon. (27)

Larva A self-feeding, sexually, and developmentally immature form of an animal. (32)

Larynx The short passageway connecting the pharynx with the lower airways. (29)

Latent (hidden) Infection Infection by a microorganism that causes no symptoms but the microbe is well-established in the body. (36)

Lateral Roots Roots that arise from the pericycle of older roots; also called branch roots or secondary roots. (18)

Law of Independent Assortment Alleles on nonhomologous chromosomes segregate independently of one another. (12)

Law of Segregation During gamete formation, pairs of alleles separate so that each sperm or egg cell has only one gene for a trait. (12)

Law of the Minimum The ecological principle that a species' distribution will be limited by whichever abiotic factor is most deficient in the environment. (41)

Laws of Thermodynamics Physical laws that describe the relationship of heat and mechanical energy. The first law states that energy cannot be created or destroyed, but one form can change into another. The second law states that the total energy the universe decreases as energy conversions occur and some energy is lost as heat. (6)

Leak Channels Passageways through a plasma membrane that do not contain gates and, therefore, are always open for the limited diffusion of a specific substance (ion) through the membrane. (7, 23)

Learning A process in which an animal benefits from experience so that its behavior is better suited to environmental conditions. (44)

Lenticels Loosely packed cells in the periderm of the stem that create air channels for transferring CO_2, H_2O, and O_2. (18)

Leukocytes White blood cells. (28)

LH Luteinizing hormone. A hormone secreted by the anterior pituitary that stimulates testosterone production in males and triggers ovulation and the transformation of the follicle into the corpus luteum in females. (31)

Lichen Symbiotic associations between certain fungi and algae. (37)

Life Cycle The sequence of events during the lifetime of an organism from zygote to reproduction. (39)

Ligaments Strong straps of connective tissue that hold together the bones in articulating joints or support an organ in place. (26)

Light-Dependent Reactions First stage of photosynthesis in which light energy is converted to chemical energy in the form of energy-rich ATP and NADPH. (8)

Light-Independent Reactions Second stage of photosynthesis in which the energy stored in ATP and NADPH formed in the light reactions is used to drive the reactions in which carbon dioxide is converted to carbohydrate. (8)

Limb Bud A portion of an embryo that will develop into either a forelimb or hindlimb. (32)

Limbic System A series of an interconnected group of brain structures, including the thalamus and hypothalamus, controlling memory and emotions. (23)

Limiting Factors The critical factors which impose restraints of the distribution, health, or activities of an organism. (41)

Limnetic Zone Open water of lakes, through which sunlight penetrates and photosynthesis occurs. (40)

Linkage The tendency of genes of the same chromosome to stay together rather than to assort independently. (13)

Linkage Groups Groups of genes located on the same chromosome. The genes of each linkage group assort independently of the genes of other linkage groups. In all eukaryotic organisms, the number of linkage groups is equal to the haploid number of chromosomes. (13)

Lipids A diverse group of biomolecules that are insoluble in water. (4)

Lithosphere The solid outer zone of the Earth; composed of the crust and outermost portion of the mantle. (40)

Littoral Zone Shallow, nutrient-rich waters of a lake, where sunlight reaches the bottom; also the lakeshore. Rooted vegetation occurs in this zone. (40)

Locomotion The movement of an organism from one place to another. (26)

Locus The chromosomal location of a gene. (13)

Logistic Growth Population growth producing a sigmoid, or S-shaped, growth curve. (43)

Long-Day Plants Plants that flower when the length of daylight exceeds some critical period. (21)

Longitudinal Fission The division pattern in flagellated protozoans, where division is along the length of the cell.

Loop of Henle An elongated section of the renal tubule that dips down into the kidney's medulla and then ascends back out to the cortex. It separates the proximal and distal convoluted tubules and is responsible for forming the salt gradient on which water reabsorption in the kidney depends. (28)

Low Density Lipoprotein (LDL) Particles that transport cholesterol in the blood. Each particle consists of about 1,500 cholesterol molecules surrounded by a film of phospholipids and protein. LDLs are taken into cells following their binding to cell surface LDL receptors. (7)

Low Intertidal Zone In the intertidal zone, the region which is uncovered by "minus" tides only. Organisms are submerged about 90% of the time. (40)

Lumen A space within an hollow organ or tube. (28)

Luminescence (see **Bioluminescence**)

Lungs The organs of terrestrial animals where gas exchange occurs. (29)

Lymph The colorless fluid in lymphatic vessels. (28)

Lymphatic System Network of fluid-carrying vessels and associated organs that participate in immunity and in the return of tissue fluid to the main circulation. (28)

Lymphocytes A group of non-phagocytic white blood cells which combat microbial invasion, fight cancer, and neutralize toxic chemicals. The two classes of lymphocytes, B cells and T cells, are the heart of the immune system. (28, 30)

Lymphoid Organs Organs associated with production of blood cells and the lymphatic system, including the thymus, spleen, appendix, bone marrow, and lymph nodes. (30)

Lysis (1) To split or dissolve. (2) Cell bursting.

Lysomes A type of storage vesicle produced by the Golgi complex, containing hydrolytic (digestive) enzymes capable of digesting many kinds of macromolecules in the cell. The membrane around them keeps them sequestered. (5)

 M ▶

M Phase That portion of the cell cycle during which mitosis (nuclear division) and cytokinesis (cytoplasmic division) takes place. (10)

Macroevolution Evolutionary changes that lead to the appearance of new species. (33)

Macrofungus Filamentous fungus so named for the large size of its fleshy sexual structures; a mushroom, for example. (37)

Macromolecules Large polymers, such as proteins, nucleic acids, and polysaccharides. (4)

Macronutrients Nutrients required by plants in large amounts: carbon, oxygen, hydrogen, nitrogen, potassium, calcium, phosphorus, magnesium, and sulfur. (19)

Macrophages Phagocytic cells that develop from monocytes and present antigen to lymphocytes. (30)

Macroscopic Referring to biological observations made with the naked eye or a hand lens.

Mammals A class of vertebrates that possesses skin covered with hair and that nourishes their young with milk from mammary glands. (39)

Mammary Glands Glands contained in the breasts of mammalian mothers that produce breast milk. (39)

Marsupials Mammals with a cloaca whose young are born immature and complete their development in an external pouch in the mother's skin. (39)

Mass Extinction The simultaneous extinction of a multitude of species as the result of a drastic change in the environment. (33, 35)

Maternal Chromosomes The set of chromosomes in an individual that were inherited from the mother. (11)

Mechanoreceptors Sensory receptors that respond to mechanical pressure and detect motion, touch, pressure, and sound. (24)

Medulla The center-most portion of some organs. (23)

Medusa The motile, umbrella-shaped body form of some members of the phylum Cnidaria, with mouth and tentacles on the lower, concave service. (Compare with polyp.) (39)

Megaspores Spores that divide by mitosis to produce female gametophytes that produce the egg gamete. (20)

Meiosis The division process that produces cells with one-half the number of chromosomes in each somatic cell. Each resulting daughter cell is haploid (1N) (11)

Meiosis I A process of reductional division in which homologous chromosomes pair and then segregate. Homologues are partitioned into separate daughter cells. (11)

Meiosis II Second meiotic division. A division process resembling mitosis, except that the haploid number of chromosomes is present. After the chromosomes line up at the meta-phase plate, the two sister chromatids separate. (11)

Melanin A brown pigment that gives skin and hair its color (12)

Melanoma A deadly form of skin cancer that develops from pigment cells in the skin and is promoted by exposure to the sun. (14)

Memory Cells Lymphocytes responsible for active immunity. They recall a previous exposure to an antigen and, on subsequent exposure to the same antigen, proliferate rapidly into plasma cells and produce large quantities of antibodies in a short time. This protection typically lasts for many years. (30)

Mendelian Inheritance Transmission of genetic traits in a manner consistent with the principles discovered by Gregor Mendel. Includes traits controlled by simple dominant or recessive alleles; more complex patterns of transmission are referred to as Nonmendelian inheritance. (12)

Meninges The thick connective tissue sheath which surrounds and protects the vertebrate brain and spinal cord. (23)

Menstrual Cycle The repetitive monthly changes in the uterus that prepare the endometrium for receiving and supporting an embryo. (31)

Meristematic Region New cells arise from this undifferentiated plant tissue; found at root or shoot apical meristems, or lateral meristems. (18)

Meristems In plants, clusters of cells that retain their ability to divide, thereby producing new cells. One of the four basic tissues in plants. (18)

Mesoderm In animals, the middle germ cell layer of the gastrula. It gives rise to muscle, bone, connective tissue, gonads, and kidney. (32)

Mesopelagic Zone The dimly lit ocean zone beneath the epipelagic zone; large fishes, whales and squid occupy this zone; no phytoplankton occur in this zone. (40)

Mesophyll Layers of cells in a leaf between the upper and lower epidermis; produced by the ground meristem. (18)

Messenger RNA (mRNA) The RNA that carries genetic information from the DNA in the nucleus to the ribosomes in the cytoplasm, where the sequence of bases in the mRNA is translated into a sequence of amino acids. (14)

Metabolic Intermediates Compounds produced as a substrate are converted to end product in a series of enzymatic reactions. (6)

Metabolic Pathways Set of enzymatic reactions involved in either building or dismantling complex molecules. (6)

Metabolic Rate A measure of the level of activity of an organism usually determined by measuring the amount of oxygen consumed by an individual per gram body weight per hour. (22)

Metabolic Water Water produced as a product of metabolic reactions. (28)

Metabolism The sum of all the chemical reactions in an organism; includes all anabolic and catabolic reactions. (6)

Metamorphosis Transformation from one form into another form during development. (32)

Metaphase The stage of mitosis when the chromosomes line-up along the metaphase plate, a plate that usually lies midway between the spindle poles. (10)

Metaphase Plate Imaginary plane within a dividing cell in which the duplicated chromosomes become aligned during metaphase. (10)

Microbes Microscopic organisms. (36)

Microbiology The branch of biology that studies microorganisms. (36)

Microevolution Changes in allele frequency of a species' gene pool which has not generated new species. Exemplified by changes in the pigmentation of the peppered moth and by the acquisition of pesticide resistance in insects. (33)

Microfibrils Bundles formed from the intertwining of cellulose molecules, i.e., long chains of glucose molecules in the cell walls of plants. (5)

Microfilaments Thin actin-containing protein fibers that are responsible for maintenance of cell shape, muscle contraction and cyclosis. (5)

Micrometer One millionth (1/1,000,000) of a meter.

Micronutrients Nutrients required by plants in small amounts: iron, chlorine, copper, manganese, zinc, molybdenum, and boron. (19)

Micropyle A small opening in the integuments of the ovule through which the pollen tube grows to deliver sperm. (21)

Microspores Spores within anthers of flowers. They divide by mitosis to form pollen grains, the male gametophytes that produce the plant's sperm. (20)

Microtubules Thin, hollow tubes in cells; built from repeating protein units of tubulin. Microtubules are components of cilia, flagella, and the cytoskeleton. (5)

Microvilli The small projections on the cells that comprise each villus of the intestinal wall, further increasing the absorption surface area of the small intestine. (27)

Middle Intertidal Zone In the intertidal zone, the region which is covered and uncovered twice a day, the zero of tide tables. Organisms are submerged about 50% of the time. (40)

Migration Movements of a population into or out of an area. (44)

Mimicry A defense mechanism where one species resembles another in color, shape, behavior, or sound. (42)

Mineralocorticoids Steroid hormones which regulate the level of sodium and potassium in the blood. (25)

Mitochondria Organelles that contain the biochemical machinery for the Krebs cycle and the electron transport system of aerobic respiration. They are composed of two membranes, the inner one forming folds, or cristae. (9)

Mitosis The process of nuclear division producing daughter cells with exactly the same number of chromosomes as in the mother cell. (10)

Mitosis Promoting Factor (MPF) A protein that appears to be a universal trigger of cell division in eukaryotic cells. (10)

Mitotic Chromosomes Chromosomes whose DNA-protein threads have become coiled into microscopically visible chromosomes, each containing duplicated chromatids ready to be separated during mitosis. (10)

Molds Filamentous fungi that exist as colonies of threadlike cells but produce no macroscopic fruiting bodies. (37)

Molecule Chemical substance formed when two or more atoms bond together; the smallest unit of matter that possesses the qualities of a compound. (3)

Mollusca A phylum, second only to Arthropoda in diversity. Composed of three main classes: 1) Gastropoda (spiral-shelled), 2) Bivalvia (hinged shells), 3) Cephalopoda (with tentacles or arms and no, or very reduced shells). (39)

Molting (Ecdysis) Shedding process by which certain arthropods lose their exoskeletons as their bodies grow larger. (39)

Monera The taxonomic kingdom comprised of single-celled prokaryotes such as bacteria, cyanobacteria, and archebacteria. (36)

Monoclonal Antibodies Antibodies produced by a clone of hybridoma cells, all of which descended from one cell. (30)

Monocotyledae (Monocots) One of the two divisions of flowering plants, characterized by seeds with a single cotyledon, flower parts in 3s, parallel veins in leaves, many roots of approximately equal size, scattered vascular bundles in its stem anatomy, pith in its root anatomy, and no secondary growth capacity. (18)

Monocytes A type of leukocyte that gives rise to macrophages. (28)

Monoecious Both male and female reproductive structures are produced on the same sporophyte individual. (20, 38)

Monohybrid Cross A mating between two individuals that differ only in one genetically-determined trait. (12)

Monomers Small molecular subunits which are the building blocks of macromolecules. The macromolecules in living systems are constructed of some 40 different monomers. (4)

Monotremes A group of mammals that lay eggs from which the young are hatched. (39)

Morphogenesis The formation of form and internal architecture within the embryo brought about by such processes as programmed cell death, cell adhesion, and cell movement. (32)

Morphology The branch of biology that studies form and structure of organisms.

Mortality Deathrate in a population or area. (43)

Motile Capable of independent movement.

Motor Neurons Nerve cells which carry outgoing impulses to their effectors, either glands or muscles. (23)

Mucosa The cell layer that lines the digestive tract and secretes a lubricating layer of mucus. (27)

Mullerian Mimicry Resemblance of different species, each of which is equally obnoxious to predators. (42)

Multicellular Consisting of many cells. (35)

Multichannel Food Chain Where the same primary producer supplies the energy for more than one food chain. (41)

Multiple Allele System Three or more possible alleles for a given trait, such as ABO blood groups in humans. (12)

Multiple Fission Division of the cell's nucleus without a corresponding division of cytoplasm.

Multiple Fruits Fruits that develop from pistils of separate flowers. (20)

Muscle Fiber A muiltinucleated skeletal muscle cell that results from the fusion of several pre-muscle cells during embryonic development. (26)

Muscle Tissue Bundles and sheets of contractile cells that shorten when stimulated, providing force for controlled movement. (26)

Mutagens Chemical or physical agents that induce genetic change. (14)

Mutation Random heritable changes in DNA that introduce new alleles into the gene pool. (14)

Mutualism The symbiotic interaction in which both participants benefit. (42)

Mycology The branch of biology that studies fungi. (37)

Mycorrhizae An association between soil fungi and the roots of vascular plants, increasing the plant's ability to extract water and minerals from the soil. (19)

Myelin Sheath In vertebrates, a jacket which covers the axons of high-velocity neurons, thereby increasing the speed of a neurological impulse. (23)

Myofibrils In striated muscle, the banded fibrils that lie parallel to each other, constituting the bulk of the muscle fiber's interior and powering contraction. (26)

Myosin A contractile protein that makes up the major component of the thick filaments of a muscle cell and is also present in nonmuscle cells. (26)

◀ **N** ▶

NADPH Nicotinamide adenine dinucleotide phosphate. NADPH is formed by reduction of $NADP^+$, and serves as a store of electrons for use in metabolism (see Reducing Power). (9)

NAD⁺ Nicotinamide adenine dinucleotide. A coenzyme that functions as an electron carrier in metabolic reactions. When reduced to NADH, the molecule becomes a cellular energy source. (9)

Natality Birthrate in a population or area. (43)

Natural Killer (NK) Cells Nonspecific, lymphocytelike cells which destroy foreign cells and cancer cells. (30)

Natural Selection Differential survival and reproduction of organisms with a resultant increase in the frequency of those best adapted to the environment. (33)

Neanderthals A subspecies of Homo sapiens different from that of modern humans that were characterized by heavy bony skeletons and thick bony ridges over the eyes. They disappeared about 35,000 years ago. (34)

Nectary Secretory gland in flowering plants containing sugary fluid that attracts pollinators as a food source. Usually located at the base of the flower. (20)

Negative Feedback Any regulatory mechanism in which the increased level of a substance inhibits further production of that substance, thereby preventing harmful accumulation. A type of homeostatic mechanism. (22, 25)

Negative Gravitropism In plants, growth against gravitational forces, or shoot growth upward. (21)

Nematocyst Within the stinging cell (cnidocyte) of cnidarians, a capsule that contains a coiled thread which, when triggered, harpoons prey and injects powerful toxins. (39)

Nematoda The widespread and abundant animal phylum containing the roundworms. (39)

Nephridium A tube surrounded by capillaries found in an organism's excretory organs that removes nitrogenous wastes and regulates the water and chemical balance of body fluids. (28)

Nephron The functional unit of the vertebrate kidney, consisting of the glomerulus, Bowman's capsule, proximal and distal convoluted tubules, and loop of Henle. (28)

Nerve Parallel bundles of neurons and their supporting cells. (23)

Nerve Impulse A propagated action potential. (23)

Nervous Tissue Excitable cells that receive stimuli and, in response, transmit an impulse to another part of the animal. (23)

Neural Plate In vertebrates, the flattened plate of dorsal ectoderm of the late gastrula that gives rise to the nervous system. (32)

Neuroglial Cells Those cells of a vertebrate nervous system that are not neurons. Includes a variety of cell types including Schwann cells. (23)

Neuron A nerve cell. (23)

Neurosecretory Cells Nervelike cells that secrete hormones rather than neurotransmitter substances when a nerve impulse reaches the distal end of the cell. In vertebrates, these cells arise from the hypothalamus. (25)

Neurotoxins Substances, such as curare and tetanus toxin, that interfere with the transmission of neural impulses. (23)

Neurotransmitters Chemicals released by neurons into the synaptic cleft, stimulating or inhibiting the post-synaptic target cell. (23)

Neurulation Formation by embryonic induction of the neural tube in a developing vertebrate embryo. (32)

Neutrons Electrically neutral (uncharged) particles contained within the nucleus of the atom. (3)

Neutrophil Phagocytic leukocyte, most numerous in the human body. (28)

Niche An organism's habitat, role, resource requirements, and tolerance ranges for each abiotic condition. (42)

Niche Breadth Relative size and dimension of ecological niches; for example, broad or narrow niches. (41)

Niche Overlap Organisms that have the same habitat, role, environmental requirements, or needs. (41)

Nitrogen Fixation The conversion of atmospheric nitrogen gas N_2 into ammonia (NH_3) by certain bacteria and cyanobacteria. (19)

Nitrogenous Wastes Nitrogen-containing metabolic waste products, such as ammonia or urea, that are produced by the breakdown of proteins and nucleic acids. (28)

Nodes The attachment points of leaves to a stem. (18)

Nodes of Ranvier Uninsulated (nonmyelinated) gaps along the axon of a neuron. (23)

Noncovalent Bonds Linkages between two atoms that depend on an attraction between positive and negative charges between molecules or ions. Includes ionic and hydrogen bonds. (3)

Non-Cyclic Photophosphorylation The pathway in the light reactions of photosynthesis in which electrons pass from water, through two photosystems, and then ultimately to NADP⁺. During the process, both ATP and NADPH are produced. It is so named because the electrons do not return to their reaction center. (8)

Nondisjunction Failure of chromosomes to separate properly at meiosis I or II. The result is that one daughter will receive an extra chromosome and the other gets one less. (11, 13)

Nonpolar Molecules Molecules which have an equal charge distribution throughout their structure and thus lack regions with a localized positive or negative charge. (3)

Notochord A flexible rod that is below the dorsal surface of the chordate embryo, beneath the nerve cord. In most chordates, it is replaced by the vertebral column. (32)

Nuclear Envelope A double membrane pierced by pores that separates the contents of the nucleus from the rest of the eukaryotic cell. (5)

Nucleic Acids DNA and RNA; linear polymers of nucleotides, responsible for the storage and expression of genetic information. (4, 14)

Nucleoid A region in the prokaryotic cell that contains the genetic material (DNA). It is unbounded by a nuclear membrane. (36)

Nucleoplasm The semifluid substance of the nucleus in which the particulate structures are suspended. (5)

Nucleosomes Nuclear protein complex consisting of a length of DNA wrapped around a central cluster of 8 histones. (14)

Nucleotides Monomers of which DNA and RNA are built. Each consists of a 5-carbon sugar, phosphate, and a nitrogenous base. (4)

Nucleous (pl. nucleoli) One or more darker regions of a nucleus where each ribosomal subunit is assembled from RNA and protein. (5)

Nucleus The large membrane-enclosed organelle that contains the DNA of eukaryotic cells. (5)

Nucleus, Atomic The center of an atom containing protons and neutrons. (3)

◀ **O** ▶

Obligate Symbiosis A symbiotic relationship between two organisms that is necessary for the survival or both organisms. (42)

Olfaction The sense of smell. (24)

Oligotrophic Little nourished, as a young lake that has few nutrients and supports little life. (40)

Omnivore An animal that obtains its nutritional needs by consuming plants and other animals. (42)

Oncogene A gene that causes cancer, perhaps activated by mutation or a change in its chromosomal location. (10)

Oocyte A female germ cell during any of the stages of meiosis. (31)

Oogenesis The process of egg production. (31)

Oogonia Female germ cells that have not yet begun meiosis. (31)

Open Circulatory System Circulatory system in which blood travels from vessels to tissue spaces, through which it percolates prior to returning to the main vessel (compare with closed circulatory system). (28)

Operator A regulatory gene in the operon of bacteria. It is the short DNA segment to which the repressor binds, thus preventing RNA polymerase from attaching to the promoter. (15)

Operon A regulatory unit in prokaryotic cells that controls the expression of structural genes. The operon consists of structural genes that produce enzymes for a particular metabolic pathway, a regulator region composed of a promoter and an operator, and R (regulator) gene that produces a repressor. (15)

Order A level of the taxonomic hierarchy that groups together members of related families. (1)

Organ Body part composed of several tissues that performs specialized functions. (22)

Organelle A specialized part of a cell having some particular function. (5)

Organic Compounds Chemical compounds that contain carbon. (4)

Organism A living entity able to maintain its organization, obtain and use energy, reproduce, grow, respond to stimuli, and display homeostatis. (1)

Organogenesis Organ formation in which two or more specialized tissue types develop in a precise temporal and spatial relationship to each other. (32)

Organ System Group of functionally related organs. (22)

Osmoregulation The maintenance of the proper salt and water balance in the body's fluids. (28)

Osmosis The diffusion of water through a differentially permeable membrane into a hypertonic compartment. (7)

Ossification Synthesis of a new bone. (26)

Osteoclast A type of bone cell which breaks down the bone, thereby releasing calcium into the bloodstream for use by the body. Osteoclasts are activated by hormones released by the parathyroid glands. (26)

Osteocytes Living bone cells embedded within the calcified matrix they manufacture. (26)

Osteoporosis A condition present predominantly in postmenopausal women where the bones are weakened due to an increased rate of bone resorption compared to bone formation. (26)

Ovarian Cycle The cycle of egg production within the mammalian ovary. (31)

Ovarian Follicle In a mammalian ovary, a chamber of cells in which the oocyte develops. (31)

Ovary In animals, the egg-producing gonad of the female. In flowering plants, the enlarged base of the pistil, in which seeds develop. (20)

Oviduct (Fallopian Tube) The tube in the female reproductive organ that connects the ovaries and uterus and where fertilization takes place. (31)

Ovulation The release of an egg (ovum) from the ovarian follicle. (31)

Ovule In seed plants, the structure containing the female gametophyte, nucellus, and integuments. After fertilization, the ovule develops into a seed. (20, 38)

Ovum An unfertilized egg cell; a female gamete. (31)

Oxidation The removal of electrons from a compound during a chemical reaction. For a carbon atom, the fewer hydrogens bonded to a carbon, the greater the oxidation state of the atom. (6)

Oxidative Phosphorylation The formation of ATP from ADP and inorganic phosphate that occurs in the electron-transport chain of cellular respiration. (8, 9)

Oxyhemoglobin A complex of oxygen and hemoglobin, formed when blood passes through the lungs and is dissociated in body tissues, where oxygen is released. (29)

Oxytocin A female hormone released by the posterior pituitary which triggers uterine contractions during childbirth and the release of milk during nursing. (25)

◀ **P** ▶

P680 Reaction Center (P = Pigment) Special chlorophyll molecule in Photosystem II that traps the energy absorbed by the other pigment molecules. It absorbs light energy maximally at 680 nm. (8)

Palisade Parenchyma In dicot leaves, densely packed, columnar shaped cells functioning in photosynthesis. Found just beneath the upper epidermis. (18)

Pancreas In vertebrates, a large gland that produces digestive enzymes and hormones. (27)

Parallel Evolution When two species that have descended from the same ancestor independently acquire the same evolutionary adaptations. (33)

Parapatric Speciation The splitting of a population into two species' populations under conditions where the members of each population reside in adjacent areas. (33)

Parasite An organism that lives on or inside another called a host, on which it feeds. (39, 42)

Parasitism A relationship between two organisms where one organism benefits, and the other is harmed. (42)

Parasitoid Parasitic organisms, such as some insect larvae, which kill their host. (42)

Parasympathetic Nervous System Part of the autonomic nervous system active during relaxed activity. (23)

Parathyroid Glands Four glands attached to the thyroid gland which secrete parathyroid hormone (PTH). When blood calcium levels are low, PTH is secreted, causing calcium to be released from bone. (25)

Parenchyma The most prevalent cell type in herbaceous plants. These thin-walled, polygonal-shaped cells function in photosynthesis and storage. (18)

Parthenogenesis Process by which offspring are produced without egg fertilization. (31)

Passive Immunity Immunity achieved by receiving antibodies from another source, as occurs with a newborn infant during nursing. (30)

Paternal Chromosomes The set of chromosomes in an individual that were inherited from the father. (11)

Pathogen A disease-causing microorganism. (36)

Pectoral Girdle In humans, the two scapulae (shoulder blades) and two clavicles (collarbones) which support and articulate with the bones of the upper arm. (26)

Pedicel A shortened stem carrying a flower. (20)

Pedigree A diagram showing the inheritance of a particular trait among the members of the family. (13)

Pelagic Zone The open oceans, divided into three layers: 1) photo- or epipelagic (sunlit), 2) mesopelagic (dim light), 3) aphotic or bathypelagic (always dark). (40)

Pelvic Girdle The complex of bones that connect a vertebrate's legs with its backbone. (26)

Penis An intrusive structure in the male animal which releases male gametes into the female's sex receptacle. (31)

Peptide Bond The covalent bond between the amino group of one amino acid and the carboxyl group of another. (4)

Peptidoglycan A chemical component of the prokaryotic cell wall. (36)

Percent Annual Increase A measure of population increase; the number of individuals (people) added to the population per 100 individuals. (43)

Perennials Plants that live longer than two years. (18)

Perfect Flower Flowers that contain both stamens and pistils. (20)

Perforation Plate In plants, that portion of the wall of vessel members that is perforated, and contains an area with neither primary nor secondary cell wall; a "hole" in the cell wall. (18)

Pericycle One or more layers of cells found in roots, with phloem or xylem to its' inside, and the endodermis to its' outside. Functions in producing lateral roots and formation of the vascular cambium in roots with secondary growth. (18)

Periderm Secondary tissue that replaces the epidermis of stems and roots. Consists of cork, cork cambium, and an internal layer of parenchyma cells. (18)

Peripheral Nervous System Neurons, excluding those of the brain and spinal cord, that permeate the rest of the body. (23)

Peristalsis Sequential waves of muscle contractions that propel a substance through a tube. (27)

Peritoneum The connective tissue that lines the coelomic cavities. (39)

Permeability The ability to be penetrable, such as a membrane allowing molecules to pass freely across it. (7)

Petal The second whorl of a flower, often brightly colored to attract pollinators; collectively called the corolla. (20)

Petiole The stalk leading to the blade of a leaf. (18)

pH A scale that measures the concentration of hydrogen ions in a solution. The pH scale extends from 0 to 14. Acidic solutions have a pH of less than 7; alkaline solutions have a pH above 7; neutral solutions have a pH equal to 7. (3)

Phagocytosis Engulfing of food particles and foreign cells by amoebae and white blood cells. A type of endocytosis. (5)

Pharyngeal Pouches In the vertebrate embryo, outgrowths from the walls of the pharynx that break through the body surface to form gill slits. (32)

Pharynx The throat; a portion of both the digestive and respiratory system just behind the oral cavity. (29)

Phenotype An individual's observable characteristics that are the expression of its genotype. (12)

Pheromones Chemicals that, when released by an animal, elicit a particular behavior in other animals of the same species. (44)

Phloem The vascular tissue that transports sugars and other organic molecules from sites of photosynthesis and storage to the rest of the plant. (18)

Phloem Loading The transfer of assimilates to phloem conducting cells, from photosynthesizing source cells. (19)

Phloem Unloading The transfer of assimilates to storage (sink) cells, from phloem conducting cells. (19)

Phospholipids Lipids that contain a phosphate and a variable organic group that form polar, hydrophilic regions on an otherwise nonpolar, hydrophobic molecule. They are the major structural components of membranes. (4)

Phosphorylation A chemical reaction in which a phosphate group is added to a molecule or atom. (6)

Photoexcitation Absorption of light energy by pigments, causing their electrons to be raised to a higher energy level. (8)

Photolysis The splitting of water during photosynthesis. The electrons from water pass to Photosystem II, the protons enter the lumen of the thylakoid and contribute to the proton gradient across the thylakoid membrane, and the oxygen is released into the atmosphere. (8)

Photon A particle of light energy. (8)

Photoperiod Specific lengths of day and night which control certain plant growth responses to light, such as flowering or germination. (21)

Photoperiodism Changes in the behavior and physiology of an organism in response to the relative lengths of daylight and darkness, i.e., the photoperiod. (21)

Photoreceptors Sensory receptors that respond to light. (24)

Photorespiration The phenomenon in which oxygen binds to the active site of a CO_2-fixing enzyme, thereby competing with CO_2 fixation, and lowering the rate of photosynthesis. (8)

Photosynthesis The conversion by plants of light energy into chemical energy stored in carbohydrate. (8)

Photosystems Highly organized clusters of photosynthetic pigments and electron/hydrogen carriers embedded in the thylakoid membranes of chloroplasts. There are two photosystems, which together carry out the light reactions of photosynthesis. (8)

Photosystem I Photosystem with a P700 reaction center; participates in cyclic photophosphorylation as well as in noncyclic photophosphorylation. (8)

Photosystem II Photosystem activated by a P680 reaction center; participates only in noncyclic photophosphorylation and is associated with photolysis of water. (8)

Phototropism The growth responses of a plant to light. (21)

Phyletic Evolution The gradual evolution of one species into another. (33)

Phylogeny Evolutionary history of a species. (35)

Phylum The major taxonomic divisions in the Animal kingdom. Members of a phylum share common, basic features. The Animal kingdom is divided into approximately 35 phyla. (39)

Physiology The branch of biology that studies how living things function. (22)

Phytochrome A light-absorbing pigment in plants which controls many plant responses, including photoperiodism. (21)

Phytoplankton Microscopic photosynthesizers that live near the surface of seas and bodies of fresh water. (37)

Pineal Gland An endocrine gland embedded within the brain that secretes the hormone melatonin. Hormone secretion is dependent on levels of environmental light. In amphibians and reptiles, melatonin controls skin coloration. In humans, pineal secretions control sexual maturation and daily rhythms. (25)

Pinocytosis Uptake of small droplets and dissolved solutes by cells. A type of endocytosis. (5)

Pistil The female reproductive part and central portion of a flower, consisting of the ovary, style and stigma. May contain one carpel, or one or more fused carpels. (20)

Pith A plant tissue composed of parenchyma cells, found in the central portion of primary growth stems of dicots, and monocot roots. (18)

Pith Ray Region between vascular bundles in vascular plants. (18)

Pituitary Gland (see **Posterior and Anterior Pituitary**).

Placenta In mammals (exclusive of marsupials and monotremes), the structure through which nutrients and wastes are exchanged between the mother and embryo/fetus. Develops from both embryonic and uterine tissues. (32)

Plant Multicellular, autotrophic organism able to manufacture food through photosynthesis. (38)

Plasma In vertebrates, the liquid portion of the blood, containing water, proteins (including fibrinogen), salts, and nutrients. (28)

Plasma Cells Differentiated antibody-secreting cells derived from B lymphocytes. (30)

Plasma Membrane The selectively permeable, molecular boundary that separates the cytoplasm of a cell from the external environment. (5)

Plasmid A small circle of DNA in bacteria in addition to its own chromosome. (16)

Plasmodesmata Openings between plant cell walls, through which adjacent cells are connected via cytoplasmic threads. (19)

Plasmodium Genus of protozoa that causes malaria. (37)

Plasmodium A huge multinucleated "cell" stage of a plasmodial slime mold that feeds on dead organic matter. (37)

Plasmolysis The shrinking of a plant cell away from its cell wall when the cell is placed in a hypertonic solution. (7)

Platelets Small, cell-like fragments derived from special white blood cells. They function in clotting. (28)

Plate Tectonics The theory that the earth's crust consists of a number of rigid plates that rest on an underlying layer of semimolten rock. The movement of the earth's plates results from the upward movement of molten rock into the solidified crust along ridges within the ocean floor. (35)

Platyhelminthes The phylum containing simple, bilaterally symmetrical animals, the flatworms. (39)

Pleiotropy Where a single mutant gene produces two or more phenotypic effects. (12)

Pleura The double-layered sac which surrounds the lungs of a mammal. (29, 39)

Pneumocytis Pneumonia (PCP) A disease of the respiratory tract caused by a protozoan that strikes persons with immunodeficiency diseases, such as AIDS. (30)

Point Mutations Changes that occur at one point within a gene, often involving one nucleotide in the DNA. (14)

Polar Body A haploid product of meiosis of a female germ cell that has very little cytoplasm and disintegrates without further function. (31)

Polar Molecule A molecule with an unequal charge distribution that creates distinct positive and negative regions or poles. (3)

Pollen The male gametophyte of seed plants, comprised of a generative nucleus and a tube nucleus surrounded by a tough wall. (20)

Pollen Grain The male gametophyte of conifers and angiosperms, containing male gametes. In angiosperms, pollen grains are contained in the pollen sacs of the anther of a flower. (20)

Pollination The transfer of pollen grains from the anther of one flower to the stigma of another. The transfer is mediated by wind, water, insects, and other animals. (20)

Polygenic Inheritance An inheritance pattern in which a phenotype is determined by two or more genes at different loci. In humans, examples include height and pigmentation. (12)

Polymer A macromolecule formed of monomers joined by covalent bonds.. Includes proteins, polysaccharides, and nucleic acids. (4)

Polymerase Chain Reaction (PCR) Technique to amplify a specific DNA molecule using a temperature-sensitive DNA polymerase obtained from a heat-resistant bacterium. Large numbers of copies of the initial DNA molecule can be obtained in a short period of time, even when the starting material is present in vanishingly small amounts, as for example from a blood stain left at the scene of a crime. (16)

Polymorphic Property of some protozoa to produce more than one stage of organism as they complete their life cycles. (37)

Polymorphic Genes Genes for which several different alleles are known, such as those that code for human blood type. (17)

Polyp Stationary body form of some members of the phylum Cnidaria, with mouth and tentacles facing upward. (Compare with medusa.) (39)

Polypeptide An unbranched chain of amino acids covalently linked together and assembled on a ribosome during translation. (4)

Polyploidy An organism or cell containing three or more complete sets of chromosomes. Polyploidy is rare in animals but common in plants. (33)

Polysaccharide A carbohydrate molecule consisting of monosaccharide units. (4)

Polysome A complex of ribosomes found in chains, linked by mRNA. Polysomes contain the ribosomes that are actively assembling proteins. (14)

Polytene Chromosomes Giant banded chromosomes found in certain insects that form by the repeated duplication of DNA. Because of the multiple copies of each gene in a cell, polytene chromosomes can generate large amounts of a gene product in a short time. Transcription occurs at sites of chromosome puffs. (13)

Pond Body of standing fresh water, formed in natural depressions in the Earth. Ponds are smaller than lakes. (40)

Population Individuals of the same species inhabiting the same area. (43)

Population Density The number of individual species living in a given area. (43)

Positive Gravitropism In plants, growth with gravitational forces, or root growth downward. (21)

Posterior Pituitary A gland which manufactures no hormones but receives and later releases hormones produced by the cell bodies of neurons in the hyopthalamus. (25)

Potential Energy Stored energy, such as occurs in chemical bonds. (6)

Preadaptation A characteristic (adaptation) that evolved to meet the needs of an organism in one type of habitat, but fortuitously allows the organism to exploit a new habitat. For example, lobed fins and lungs evolved in ancient fishes to help them live in shallow, stagnant ponds, but also facilitated the evolution of terrestrial amphibians. (33, 39)

Precells Simple forerunners of cells that, presumably, were able to concentrate organic molecules, allowing for more frequent molecular reactions. (35)

Predation Ingestion of prey by a predator for energy and nutrients. (42)

Predator An organism that captures and feeds on another organism (prey). (42)

Pressure Flow In the process of phloem loading and unloading, pressure differences resulting from solute increases in phloem conducting cells and neighboring xylem cells cause the flow of water to phloem. A concentration gradient is created between xylem and phloem cells. (19)

Prey An organism that is captured and eaten by another organism (predator). (42)

Primary Consumer Organism that feeds exclusively on producers (plants). Herbivores are primary consumers. (41)

Primary Follicle In the mammalian ovary, a structure composed of an oocyte and its surrounding layer of follicle cells. (31)

Primary Growth Growth from apical meristems, resulting in an increase in the lengths of shoots and roots in plants. (18)

Primary Immune Response Process of antibody production following the first exposure to an antigen. There is a lag time from exposure until the appearance in the blood of protective levels of antibodies. (30)

Primary Oocyte Female germ cell that is either in the process of or has completed the first meiotic division. In humans, germ cells may remain in this stage in the ovary for decades. (31)

Primary Producers All autotrophs in a biotic environment that use sunlight or chemical energy to manufacture food from inorganic substances. (41)

Primary Sexual Characteristics Gonads, reproductive tracts, and external genitals. (31)

Primary Spermatocyte Male germ cell that is either in the process of or has completed the first meiotic division. (31)

Primary Succession The development of a community in an area previously unoccupied by any community; for example, a "bare" area such as rock, volcanic material, or dunes. (41)

Primary Tissues Tissues produced by primary meristems of a plant, which arise from the shoot and root apical meristems. In general, primary tissues are a result of an increase in plant length. (18)

Primary Transcript An RNA molecule that has been transcribed but not yet subjected to any type of processing. The primary transcript corresponds to the entire stretch of DNA that was transcribed. (15)

Primates Order of mammals that includes humans, apes, monkeys, and lemurs. (39)

Primitive An evolutionary early condition. Primitive features are those that were also present in an early ancestor, such as five digits on the feet of terrestrial vertebrates. (34)

Prions An infectious particle that contains protein but no nucleic acid. It causes slow diseases of animals, including neurological disease of humans. (36)

Processing-Level Control Control of gene expression by regulating the pathway by which a primary RNA transcript is processed into an mRNA. (15)

Products In a chemical reaction, the compounds into which the reactants are transformed. (6)

Profundal Zone Deep, open water of lakes, where it is too dark for photosynthesis to occur. (40)

Progesterone A hormone produced by the corpus luteum within the ovary. It prepares and maintains the uterus for pregnancy, participates in milk production, and prevents the ovary from releasing additional eggs late in the cycle or during pregnancy. (25)

Prokaryotic Referring to single-celled organisms that have no membrane separating the DNA from the cytoplasm and lack membrane-enclosed organelles. Prokaryotes are confined to the kingdom Monera; they are all bacteria. (36)

Prokaryotic Fission The most common type of cell division in bacteria (prokaryotes). Duplicated DNA strands are attached to the plasma membrane and become separated into two cells following membrane growth and cell wall formation. (10, 36)

Prolactin A hormone produced by the anterior pituitary, stimulating milk production by mammary glands. (25)

Promoter A short segment of DNA to which RNA polymerase attaches at the start of transcription. (15)

Prophase Longest phase of mitosis, involving the formation of a spindle, coiling of chromatin fibers into condensed chromosomes, and movement of the chromosomes to the center of the cell. (10)

Prostaglandins Hormones secreted by endocrine cells scattered throughout the body responsible for such diverse functions as contraction of uterine muscles, triggering the inflammatory response, and blood clotting. (25)

Prostate Gland A muscular gland which produces and releases fluids that make up a substantial portion of the semen. (31)

Proteins Long chains of amino acids, linked together by peptide bonds. They are folded into specific shapes essential to their functions. (4)

Prothallus The small, heart-shaped gametophyte of a fern. (38)

Protists A member of the kingdom Protista; simple eukaryotic organisms that share broad taxonomic similarities. (36, 37)

Protocooperation Non-compulsory interactions that benefit two organisms, e.g., lichens. (42)

Proton Gradient A difference in hydrogen ion (proton) concentration on opposite sides of a membrane. Proton gradients are formed during photosynthesis and respiration and serve as a store of energy used to drive ATP formation. (8, 9)

Protons Positively charged particles within the nucleus of an atom. (3)

Protostomes One path of development exhibited by coelomate animals (e.g., mollusks, annelids, and arthropods). (39)

Protozoa Member of protist kingdom that is unicellular and eukaryotic; vary greatly in size, motility, nutrition and life cycle. (37)

Provirus DNA copy of a virus' nucleic acid that becomes integrated into the host cell's chromosome. (36)

Pseudocoelamates Animals in which the body cavity is not lined by cells derived from mesoderm. (39)

Pseudopodia (psuedo = false, pod = foot). Pseudopodia are fingerlike extensions of cytoplasm that flow forward from the "body" of an amoeba; the rest of the cell then follows. (37)

Puberty Development of reproductive capacity, often accompanied by the appearance of secondary sexual characteristics. (31)

Pulmonary Circulation The loop of the circulatory system that channels blood to the lungs for oxygenation. (28)

Punctuated Equilibrium Theory A theory to explain the phenomenon of the relatively sudden appearance of new species, followed by long periods of little or no change. (33)

Punnett Square Method A visual method for predicting the possible genotypes and their expected ratios from a cross. (12)

Pupa In insects, the stage in metamorphosis between the larva and the adult. Within the pupal case, there is dramatic transformation in body form as some larval tissues die and others differentiate into those of the adult. (32)

Purine A nitrogenous base found in DNA and RNA having a double ring structure. Adenine and guanine are purines. (14)

Pyloric Sphincter Muscular valve between the human stomach and small intestine. (27)

Pyrimidine A nitrogenous base found in DNA and RNA having a single ring structure. Cytosine, thymine, and uracil are pyrimidines. (14)

Pyramid of Biomass Diagrammatic representation of the total dry weight of organisms at each trophic level in a food chain or food web. (41)

Pyramid of Energy Diagrammatic representation of the flow of energy through trophic levels in a food chain or food web. (41)

Pyramid of Numbers Similar to a pyramid of energy, but with numbers of producers and consumers given at each trophic level in a food chain or food web. (41)

◄ **Q** ►

Quiescent Center The region in the apical meristem of a root containing relatively inactive cells. (18)

◄ R ►

R-Group The variable portion of a molecule. (4)

r-Selected Species Species that possess adaptive strategies to produce numerous off-spring at once. (43)

Radial Symmetry The quality possessed by animals whose bodies can be divided into mirror images by more than one median plane. (39)

Radicle In the plant embryo, the tip of the hypocotyl that eventually develops into the root system. (20)

Radioactivity A property of atoms whose nucleus contains an unstable combination of particles. Breakdown of the nucleus causes the emission of particles and a resulting change in structure of the atom. Biologists use this property to track labeled molecules and to determine the age of fossils. (3)

Radiodating The use of known rates of radioactive decay to date a fossil or other ancient object. (3, 34)

Radioisotope An isotope of an element that is radioactive. (3)

Radiolarian A prozoan member of the protistan group Sarcodina that secretes silicon shells through which it captures food.

Rainshadow The arid, leeward (downwind) side of a mountain range. (40)

Random Distribution Distribution of individuals of a population in a random manner; environmental conditions must be similar and individuals do not affect each other's location in the population. (43)

Reactants Molecules or atoms that are changed to products during a chemical reaction. (6)

Reaction A chemical change in which starting molecules (reactants) are transformed into new molecules (products). (6)

Reaction Center A special chlorophyll molecule in a photosystem (P_{700} in Photosystem I, P_{680} in Photosystem II). (8)

Realized Niche Part of the fundamental niche of an organism that is actually utilized. (41)

Receptacle The base of a flower where the flower parts are attached; usually a widened area of the pedicel. (20)

Receptor-Mediated Endocytosis The up-take of materials within a cytoplasmic vesicle (endocytosis) following their binding to a cell surface receptor. (7)

Receptor Site A site on a cell's plasma membrane to which a chemical such as a hormone binds. Each surface site permits the attachment of only one kind of hormone. (5)

Recessive An allele whose expression is masked by the dominant allele for the same trait. (12)

Recombinant DNA A DNA molecule that contains DNA sequences derived from different biological sources that have been joined together in the laboratory. (16)

Recombination The rejoining of DNA pieces with those of a different strand or with the same strand at a point different from where the break occurred. (11, 13)

Red Marrow The soft tissue in the interior of bones that produces red blood cells. (26)

Red Tide Growth of one of several species of reddish brown dinoflagellate algae so extensive that it tints the coastal waters and inland lakes a distinctive red color. Often associated with paralytic shellfish poisoning (see dinoflagellates). (37)

Reducing Power A measure of the cell's ability to transfer electrons to substrates to create molecules of higher energy content. Usually determined by the available store of NADPH, the molecule from which electrons are transferred in anabolic (synthetic) pathways. (6)

Reduction The addition of electrons to a compound during a chemical reaction. For a carbon atom, the more hydrogens that are bonded to the carbon, the more reduced the atom. (6)

Reduction Division The first meiotic division during which a cell's chromosome number is reduced in half. (11)

Reflex An involuntary response to a stimulus. (23)

Reflex Arc The simplest example of central nervous system control, involving a sensory neuron, motor neuron, and usually an inter-neuron. (23)

Regeneration Ability of certain animals to replace injured or lost limbs parts by growth and differentiation of undifferentiated stem cells. (15)

Region of Elongation In root tips, the region just above the region of cell division, where cells elongate and the root length increases. (18)

Region of Maturation In root tips, the region above the region of elongation; cells differentiate and root hairs occur in this region. (18)

Regulatory Genes Genes whose sole function is to control the expression of structural genes. (15)

Releaser A sign stimulus that is given by an individual to another member of the same species, eliciting a specific innate behavior. (44)

Releasing Factors Hormones secreted by the tips of hypothalmic neurosecretory cells that stimulate the anterior pituitary to release its hormones. GnRH, for example, stimulates the release of gonadotropins. (25)

Renal Referring to the kidney. (28)

Replication Duplication of DNA, usually prior to cell division. (14)

Replication Fork The site where the two strands of a DNA helix are unwinding during replication. (14)

Repression Inhibition of transcription of a gene which, in an operon, occurs when repressor protein binds to the operator. (15)

Repressor Protein encoded by a bacterial regulatory gene that binds to an operator site of an operon and inhibits transcription. (15)

Reproduction The process by which an organism produces offspring. (31)

Reproductive Isolation Phenomenon in which members of a single population become split into two populations that no longer interbreed. (33)

Reproductive System System of specialized organs that are utilized for the production of gametes and, in some cases, the fertilization and/or development of an egg. (31)

Reptiles Members of class Reptilia, scaly, air-breathing, egg-laying vertebrates such as lizards, snakes, turtles, and crocodiles. (39)

Resolving Power The ability of an optical instrument (eye, microscopes) to discern whether two very close objects are separate from each other. (APP.)

Resource Partitioning Temporal or spatial sharing of a resource by different species. (42)

Respiration Process used by organisms to exchange gases with the environment; the source of oxygen required for metabolism. The process organisms use to oxidize glucose to CO_2 and H_2O using an electron transport system to extract energy from electrons and store it in the high-energy bonds of ATP. (29)

Respiratory System The specialized set of organs that function in the uptake of oxygen from the environment. (29)

Resting Potential The electrical potential (voltage) across the plasma membrane of a neuron when the cell is not carrying an impulse. Results from a difference in charge across the membrane. (23)

Restriction Enzyme A DNA-cutting enzyme found in bacteria. (16)

Restriction Fragment Length Polymorphism (RFLP) Certain sites in the DNA tend to have a highly variable sequence from one individual to another. Because of these differences, restriction enzymes cut the DNA from different individuals into fragments of different length. Variations in the length of particular fragments (RFLPs) can be used as genetic signposts for the identification of nearby genes of interest. (17)

Restriction Fragments The DNA fragments generated when purified DNA is treated with a particular restriction enzyme. (16)

Reticular Formation A series of interconnected sites in the core of the brain (brainstem) that selectively arouse conscious activity. (23)

Retroviruses RNA viruses that reverse the typical flow of genetic information; within the infected cell, the viral DNA serves as a template for synthesis of a DNA copy. Examples include HIV, which causes AIDS, and certain cancer viruses. (36)

Reverse Genetics Determining the amino acid sequence and function of a polypeptide from the nucleotide sequence of the gene that codes for that polypeptide. (17)

Reverse Transcriptase An enzyme present in retroviruses that transcriibes a strand of DNA, using viral RNA as the template. (36)

Rhizoids Slender cells that resemble roots but do not absorb water or minerals. (36)

Rhodophyta Red algae; seaweeds that can absorb deeper penetrating light rays than most aquatic photosynthesizers. (36)

Rhyniophytes Ancient plants having vascular tissue which thrived in marshy areas during the Silurian period.

Ribonucleic Acid (RNA) Single-stranded chain of nucleotides each comprised of ribose (a sugar), phosphate, and one of four bases (adenine, guanine, cytosine, and uracil). The sequence of nucleotides in RNA is dictated by DNA, from which it is transcribed. There are three classes of RNA: mRNA, tRNA, and rRNA, all required for protein synthesis. (4, 14)

Ribosomal RNA (rRNA) RNA molecules that form part of the ribosome. Included among the rRNAs is one that is thought to catalyze peptide bond formation. (14)

Ribosomes Organelles involved in protein synthesis in the cytoplasm of the cell. (14)

Ribozymes RNAs capable of catalyzing a chemical reaction, such as peptide bond formation or RNA cutting and splicing. (15)

Rickettsias A group of obligate intracellular parasites, smaller than the typical prokaryote. They cause serious diseases such as typhus. (36)

River Flowing body of surface fresh water; rivers are formed from the convergence of streams. (40)

RNA Polymerase The enzyme that directs transcription and assembling RNA nucleotides in the growing chain. (14)

RNA Processing The process by which the intervening (noncoding) portions of a primary RNA transcript are removed and the remaining (coding) portions are spliced together to form an mRNA. (15)

Root Cap A protective cellular helmet at the tip of a root that surrounds delicate meristematic cells and shields them from abrasion and directs the growth downward. (18)

Root Hairs Elongated surface cells near the tip of each root for the absorption of water and minerals. (18)

Root Nodules Knobby structures on the roots of certain plants. They house nitrogen-fixing bacteria which supply nitrogen in a form that can be used by the plant. (19)

Root Pressure A positive pressure as a result of continuous water supply to plant roots that assists (along with transpirational pull) the pushing of water and nutrients up through the xylem. (19)

Root System The below-ground portion of a plant, consisting of main roots, lateral roots, root hairs, and associated structures and systems such as root nodules or mycorrhizae. (18)

Rough ER (RER) Endoplasmic reticulum with many ribosomes attached. As a result, they appear rough in electron micrographs. (5)

Ruminant Grazing mammals that possess an additional stomach chamber called rumen which is heavily fortified with cellulose-digesting microorganisms. (27)

◄ S ►

S Phase The second stage of interphase in which the materials needed for cell division are synthesized and an exact copy of cell's DNA is made by DNA replication. (10)

Sac Body The body plan of simple animals, like cnidarians, where there is a single opening leading to and from a digestion chamber.

Saltatory Conduction The "hopping" movement of an impulse along a myelinated neuron from one Node of Ranvier to the next one. (23)

Sap Fluid found in xylem or sieve of phloem. (20)

Saprophyte Organisms, mainly fungi and bacteria, that get their nutrition by breaking down organic wastes and dead organisms, also called decomposers. (42)

Saprobe Organism that obtains its nutrients by decomposing dead organisms. (37)

Sarcolemma The plasma membrane of a muscle fiber. (26)

Sarcomere The contractile unit of a myofibril in skeletal muscle. (26)

Sarcoplasmic Reticulum (SR) In skeletal muscle, modified version of the endoplasmic reticulum that stores calcium ions. (26)

Savanna A grassland biome with alternating dry and rainy seasons. The grasses and scattered trees support large numbers of grazing animals. (40)

Scaling Effect A property that changes disproportionately as the size of organisms increase. (22)

Scanning Electron Microscope (SEM) A microscope which operates by showering electrons back and forth across the surface of a specimen prepared with a thin metal coating. The resultant image shows three-dimensional features of the specimen's surface. (APP.)

Schwann Cells Cells which wrap themselves around the axons of neurons forming an insulating myelin sheath composed of many layers of plasma membrane. (23)

Sclereids Irregularly-shaped sclerenchyma cells, all having thick cell walls; a component of seed coats and nuts. (18)

Sclerenchyma Component of the ground tissue system of plants. They are thick walled cells of various shapes and sizes, providing support or protection. They continue to function after the cell dies. (18)

Sclerenchyma Fibers Non-living elongated plant cells with tapering ends and thick secondary walls. A supportive cell type found in various plant tissues. (18)

Sebaceous Glands Exocrine glands of the skin that produce a mixture of lipids (sebum) that oil the hair and skin. (26)

Secondary Cell Wall An additional cell wall that improves the strength and resiliency of specialized plant cells, particularly those cells found in stems that support leaves, flowers, and fruit. (5)

Secondary Consumer Organism that feeds exclusively on primary consumers; mostly animals, but some plants. (41)

Secondary Growth Growth from cambia in perennials; results in an increase in the diameter of stems and roots. (18)

Secondary Meristems (vascular cambium, cork cambium) Rings or clusters of meristematic cells that increase the width of stems and roots when the divide. (18)

Secondary Sex Characteristics Those characteristics other than the gonads and reproductive tract that develop in response to sex hormones. For example, breasts and pubic hair in women and a deep voice and pubic hair in men. (31)

Secondary Succession The development of a community in an area previously occupied by a community, but which was disturbed in some manner; for example, fire, development, or clear-cutting forests. (41)

Secondary Tissues Tissues produced to accommodate new cell production in plants with woody growth. Secondary tissues are produced from cambia, which produce vascular and cork tissues, leading to an increase in plant girth. (18)

Second Messenger Many hormones, such as glucagon and thyroid hormone, evoke a response by binding to the outer surface of a target cell and causing the release of another substance (which is the second messenger). The best-studied second messenger is cyclic AMP which is formed by an enzyme on the inner surface of the plasma membrane following the binding of a hormone to the outer surface of the membrane. The cyclic AMP diffuses into the cell and activates a protein kinase. (25)

Secretion The process of exporting materials produced by the cell. (5)

Seed A mature ovule consisting of the embryo, endosperm, and seed coat. (20)

Seed Dormancy Metabolic inactivity of seeds until favorable conditions promote seed germination. (20)

Secretin Hormone secreted by endocrine cells in the wall of the intestine that stimulates the release of digestive products from the pancreas. (25)

Segmentation A condition in which the body is constructed, at least in part, from a series of repeating parts. Segmentation occurs in annelids, arthropods, and vertebrates (as revealed during embryonic development). (39)

Selectively Permeable A term applied to the plasma membrane because membrane proteins control which molecules are transported. Enables a cell to import and accumulate the molecules essential for normal metabolism. (7)

Semen The fluid discharged during a male orgasm. (31)

Semiconsevative Replication The manner in which DNA replicates; half of the original DNA strand is conserved in each new double helix. (14)

Seminal Vesicles The organs which produce most of the ejaculatory fluid. (31)

Seminiferous Tubules Within the testes, highly coiled and compacted tubules, lined with a self-perpetuating layer of spermatogonia, which develop into sperm. (31)

Senescence Aging and eventual death of an organism, organ or tissue. (3, 18)

Sense Strand The one strand of a DNA double helix that contains the information that encodes the amino sequence of a polypeptide. This is the strand that is selectively transcribed by RNA polymerase forming an mRNA that can be properly translated. (14)

Sensory Neurons Neurons which relay impulses to the central nervous system. (23)

Sensory Receptors Structures that detect changes in the external and internal environment and transmit the information to the nervous system. (24)

Sepal The outermost whorl of a flower, enclosing the other flower parts as a flower bud; collectively called the calyx. (20)

Sessile Sedentary, incapable of independent movement. (39)

Sex Chromosomes The one chromosomal pair that is not identical in the karyotypes of males and females of the same animal species. (10, 13)

Sex Hormones Steroid hormones which influence the production of gametes and the development of male or female sex characteristics. (25)

Sexual Dimorphism Differences in the appearance of males and females in the same species. (33)

Sexual Reproduction The process by which haploid gametes are formed and fuse during fertilization to form a zygote. (31)

Sexual Selection The natural selection of adaptations that improve the chances for mating and reproducing. (33)

Shivering Involuntary muscular contraction for generating metabolic heat that raises body temperature. (28)

Shoot In angiosperms, the system consisting of stems, leaves, flowers and fruits. (18)

Shoot System The above-ground portion of an angiosperm plant consisting of stems with nodes, including branches, leaves, flowers and fruits. (18)

Short-Day Plants Plants that flower in late summer or fall when the length of daylight becomes shorter than some critical period. (21)

Shrubland A biome characterized by densely growing woody shrubs in mediterranean type climate; growth is so dense that understory plants are not usually present. (40)

Sickle Cell Anemia A genetic (recessive autosomal) disorder in which the beta globin genes of adult hemoglobin molecules contain an amino acid substitution which alters the ability of hemoglobin to transport oxygen. During times of oxygen stress, the red blood cells of these individuals may become sickle shaped, which interferes with the flow of the cells through small blood vessels. (4, 17)

Sieve Plate Found in phloem tissue in plants, the wall between sieve-tube members, containing perforated areas for passage of materials. (18)

Sieve-Tube Member A living, food-conducting cell found in phloem tissue of plants; associated with a companion cell. (18)

Sigmoid Growth Curve An S-shaped curve illustrating the lag phase, exponential growth, and eventual approach of a population to its carrying capacity. (43)

Sign Stimulus An object or action in the environment that triggers an innate behavior. (44)

Simple Fruits Fruits that develop from the ovary of one pistil. (20)

Simple Leaf A leaf that is undivided; only one blade attached to the petiole. (18)

Sinoatrial (SA) Node A collection of cells that generates an action potential regulating heart beat; the heart's pacemaker. (28)

Skeletal Muscles Separate bundles of parallel, striated muscle fibers anchored to the bone, which they can move in a coordinated fashion. They are under voluntary control. (26)

Skeleton A rigid form of support found in most animals either surrounding the body with a protective encasement or providing a living girder system within the animal. (26)

Skull The bones of the head, including the cranium. (26)

Slow-Twitch Fibers Skeletal muscle fibers that depend on aerobic metabolism for ATP production. These fibers are capable of undergoing contraction for extended periods of time without fatigue, but generate lesser forces than fast-twitch fibers. (9)

Small Intestine Portion of the intestine in which most of the digestion and absorption of nutrients takes place. It is so named because of its narrow diameter. There are three sections: duodenum, jejunum, and ilium. (27)

Smell Sense of the chemical composition of the environment. (24)

Smooth ER (SER) Membranes of the endoplasmic reticulum that have no ribosomes on their surface. SER is generally more tubular than the RER. Often acts to store calcium or synthesize steroids. (5)

Smooth Muscle The muscles of the internal organs (digestive tract, glands, etc.). Composed of spindle-shaped cells that interlace to form sheets of visceral muscle. (26)

Social Behavior Behavior among animals that live in groups composed of individuals that are dependent on one another and with whom they have evolved mechanisms of communication. (44)

Social Learning Learning of a behavior from other members of the species. (44)

Social Parasitism Parasites that use behavioral mechanisms of the host organism to the parasite's advantage, thereby harming the host. (42)

Solute A substance dissolved in a solvent. (3)

Solution The resulting mixture of a solvent and a solute. (3)

Solvent A substance in which another material dissolves by separating into individual molecules or ions. (3)

Somatic Cells Cells that do not have the potential to form reproductive cells (gametes). Includes all cells of the body except germ cells. (11)

Somatic Nervous System The nerves that carry messages to the muscles that move the skeleton either voluntarily or by reflex. (23)

Somatic Sensory Receptors Receptors that respond to chemicals, pressure, and temperature that are present in the skin, muscles, tendons, and joints. Provides a sense of the physiological state of the body. (24)

Somites In the vertebrate embryo, blocks of mesoderm on either side of the notochord that give rise to muscles, bones, and dermis. (32)

Speciation The formation of new species. Occurs when one population splits into separate populations that diverge genetically to the point where they become separate species. (33)

Species Taxonomic subdivisions of a genus. Each species has recognizable features that distinguish it from every other species. Members of one species generally will not interbreed with members of other species. (33)

Specific Epithet In taxonomy, the second term in an organism's scientific name identifying its species within a particular genus. (1)

Spermatid Male germ cell that has completed meiosis but has not yet differentiated into a sperm. (31)

Spermatogenesis The production of sperm. (31)

Spermatogonia Male germ cells that have not yet begun meiosis. (31)

Spermatozoa (Sperm) Male gametes. (31)

Sphinctors Circularly arranged muscles that close off the various tubes in the body.

Spinal Cord A centralized mass of neurons for processing neurological messages and linking the brain with that part of peripheral nervous system not reached by the cranial nerves. (23)

Spinal Nerves Paired nerves which emerge from the spinal cord and innervate the body. Humans have 31 pairs of spinal nerves. (23)

Spindle Apparatus In dividing eukaryotic cells, the complex rigging, made of microtubules, that aligns and separates duplicated chromosomes. (10)

Splash Zone In the intertidal zone, the uppermost region receiving splashes and sprays of water to the mean of high tides. (40)

Spleen One of the organs of the lymphatic system that produces lymphocytes and filters blood; also produces red blood cells in the human fetus. (28)

Splicing The step during RNA processing in which the coding segments of the primary transcript are covalently linked together to form the mRNA. (15)

Spongy Parenchyma In monocot and dicot leaves, loosely arranged cells functioning in photosynthesis. Found above the lower epidermis and beneath the palisade parenchyma in dicots, and between the upper and lower epidermis in monocots. (18)

Spontaneous Generation Disproven concept that living organisms can arise directly from inanimate materials. (2)

Sporangiospores Black, asexual spores of the zygomycete fungi. (37)

Sporangium A hollow structure in which spores are formed. (37)

Spores In plants, haploid cells that develop into the gametophyte generation. In fungi, an asexual or sexual reproductive cell that gives rise to a new mycelium. Spores are often lightweight for their dispersal and adapted for survival in adverse conditions. (37)

Sporophyte The diploid spore producing generation in plants. (38)

Stabilizing Selection Natural selection favoring an intermediate phenotype over the extremes. (33)

Starch Polysaccharides used by plants to store energy. (4)

Stamen The flower's male reproductive organ, consisting of the pollen-producing anther supported by a slender stalk, the filament. (20)

Stem In plants, the organ that supports the leaves, flowers, and fruits. (18)

Stem Cells Cells which are undifferentiated and capable of giving rise to a variety of different types of differentiated cells. For example, hematopoietic stem cells are capable of giving rise to both red and white blood cells. (17)

Steroids Compounds classified as lipids which have the basic four-ringed molecular skeleton as represented by cholesterol. Two examples of steroid hormones are the sex hormones; testosterone in males and estrogen in females. (4, 25)

Stigma The sticky area at the top of each pistil to which pollen adheres. (20)

Stimulus Any change in the internal or external environment to which an organism can respond. (24)

Stomach A muscular sac that is part of the digestive system where food received from the esophagus is stored and mixed, some breakdown of food occurs, and the chemical degradation of nutrients begins. (27)

Stomates (Pl. Stomata) Microscopic pores in the epidermis of the leaves and stems which allow gases to be exchanged between the plant and the external environment. (18)

Stratified Epithelia Multicellular layered epithelium. (22)

Stream Flowing body of surface fresh water; streams merge together into larger streams and rivers. (40)

Stretch Receptors Sensory receptors embedded in muscle tissue enabling muscles to respond reflexively when stretched. (23, 24)

Striated Referring to the striped appearance of skeletal and cardiac muscle fibers. (26)

Strobilus In lycopids, terminal, cone-like clusters of specialized leaves that produce sporangia.

Stroma The fluid interior of chloroplasts. (8)

Stromatolites Rocks formed from masses of dense bacteria and mineral deposits. Some of these rocky masses contain cells that date back over three billion years revealing the nature of early prokaryotic life forms. (35)

Structural Genes DNA segments in bacteria that direct the formation of enzymes or structural proteins. (15)

Style The portion of a pistil which joins the stigma to the ovary. (20)

Substrate-Level Phosphorylation The formation of ATP by direct transfer of a phosphate group from a substrate, such as a sugar phosphate, to ADP. ATP is formed without the involvement of an electron transport system. (9)

Substrates The reactants which bind to enzymes and are subsequently converted to products. (6)

Succession The orderly progression of communities leading to a climax community. It is one of two types: primary, which occurs in areas where no community existed before; and secondary, which occurs in disturbed habitats where some soil and perhaps some organisms remain after the disturbance. (41)

Succulents Plants having fleshy, water-storing stems or leaves. (40)

Suppressor T Cells A class of T cells that regulate immune responses by inhibiting the activation of other lymphocytes. (30)

Surface Area-to-Volume Ratio The ratio of the surface area of an organism to its volume, which determines the rate of exchange of materials between the organism and its environment. (22)

Surface Tension The resistance of a liquid's surface to being disrupted. In aqueous solutions, it is caused by the attraction between water molecules. (3)

Survivorship Curve Graph of life expectancy, plotted as the number of survivors versus age. (43)

Sweat Glands Exocrine glands of the skin that produce a dilute salt solution, whose evaporation cools in the body. (26)

Symbiosis A close, long-term relationship between two individuals of different species. (42)

Symmetry Referring to a body form that can be divided into mirror image halves by at least one plane through its body. (39)

Sympathetic Nervous System Part of the autonomic nervous system that tends to stimulate bodily activities, particularly those involved with coping with stressful situations. (23)

Sympatric Speciation Speciation that occurs in populations with overlapping distributions. It is common in plants when polyploidy arises within a population. (33)

Synapse Juncture of a neuron and its target cell (another neuron, muscle fiber, gland cell). (23)

Synapsis The pairing of homologous chromosomes during prophase of meiosis I. (11)

Synaptic Cleft   Small space between the synaptic knobs of a neuron and its target cell. (23)

Synaptic Knobs   The swellings that branch from the end of the axon. They deliver the neurological impulse to the target cell. (23)

Synaptonemal Complex   Ladderlike structure that holds homologous chromosomes together as a tetrad during crossing over in prophase I of meiosis. (11)

Synovial Cavities   Fluid-filled sacs around joints, the function of which is to lubricate and separate articulating bone surfaces. (26)

Systemic Circulation   Part of the circulatory system that delivers oxygenated blood to the tissues and routes deoxygenated blood back to the heart. (28)

Systolic Pressure   The first number of a blood pressure reading; the highest pressure attained in the arteries as blood is propelled out of the heart. (28)

◄ T ►

Taiga   A biome found south of tundra biomes; characterized by coniferous forests, abundant precipitation, and soils that thaw only in the summer. (40)

Tap Root System   Root system of plants having one main root and many smaller lateral roots. Typical of conifers and dicots. (18)

Taste   Sense of the chemical composition of food. (24)

Taxonomy   The science of classifying and grouping organisms based on their morphology and evolution. (1)

T Cell   Lymphocytes that carry out cell-mediated immunity. They respond to antigen stimulation by becoming helper cells, killer cells, and memory cells. (30)

Telophase   The final stage of mitosis which begins when the chromosomes reach their spindle poles and ends when cytokinesis is completed and two daughter cells are produced. (10)

Tendon   A dense connective tissue cord that connects a skeletal muscle to a bone. (26)

Teratogenic   Embryo deforming. Chemicals such as thalidomide or alcohol are teratogenic because they disturb embryonic development and lead to the formation of an abnormal embryo and fetus. (32)

Terminal Electron Acceptor   In aerobic respiration, the molecule of O_2 which removes the electron pair from the final cytochrome of the respiratory chain. (9)

Terrestrial   Living on land. (40)

Territory (Territoriality)   An area that an animal defends against intruders, generally in the protection of some resource. (42, 44)

Tertiary Consumer   Animals that feed on secondary consumers (plant or animal) or animals only. (41)

Test Cross   An experimental procedure in which an individual exhibiting a dominant trait is crossed to a homozygous recessive to determine whether the first individual is homozygous or heterozygous. (12)

Testis   In animals, the sperm-producing gonad of the male. (23)

Testosterone   The male sex hormone secreted by the testes when stimulated by pituitary gonadotropins. (31)

Tetrad   A unit of four chromatids formed by a synapsed pair of homologous chromosomes, each of which has two chromatids. (11)

Thallus   In liverworts, the flat, ground-hugging plant body that lacks roots, stems, leaves, and vascular tissues. (38)

Theory of Tolerance   Distribution, abundance and existence of species in an ecosystem are determined by the species' range of tolerance of chemical and physical factors. (41)

Thermoreceptors   Sensory receptors that respond to changes in temperature. (24)

Thermoregulation   The process of maintaining a constant internal body temperature in spite of fluctuations in external temperatures. (28)

Thigmotropism   Changes in plant growth stimulated by contact with another object, e.g., vines climbing on cement walls. (21)

Thoracic Cavity   The anterior portion of the body cavity in which the lungs are suspended. (39)

Thylakoids   Flattened membrane sacs within the chloroplast. Embedded in these membranes are the light-capturing pigments and other components that carry out the light-dependent reactions of photosynthesis. (8)

Thymus   Endocrine gland in the chest where T cells mature. (30)

Thyroid Gland   A butterfly-shaped gland that lies just in front of the human windpipe, producing two metabolism-regulating hormones, thyroxin and triodothyronine. (25)

Thyroid Hormone   A mixture of two iodinated amino acid hormones (thyroxin and triiodothyronine) secreted by the thyroid gland. (25)

Thyroid Stimulating Hormone (TSH)   An anterior pituitary hormone which stimulates secretion by the thyroid gland. (25)

Tissue   An organized group of cells with a similar structure and a common function. (22)

Tissue System   Continuous tissues organized to perform a specific function in plants. The three plant tissue systems are: dermal, vascular, and ground (fundamental). (18)

Tolerance Range   The range between the maximum and minimum limits for an environmental factor that is necessary for an organism's survival. (41)

Totipotent   The genetic potential for one type of cell from a multicellular organism to give rise to any of the organism's cell types, even to generate a whole new organism. (15)

Trachea   The windpipe; a portion of the respiratory tract between the larynx and bronchii. (29)

Tracheal Respiratory System   A network of tubes (tracheae) and tubules (tracheoles) that carry air from the outside environment directly to the cells of the body without involving the circulatory system. (29, 39)

Tracheid   A type of conducting cell found in xylem functioning when a cell is dead to transport water and dissolved minerals through its hollow interior. (18)

Tracheophytes   Vascular plants that contain fluid-conducting vessels. (38)

Transcription   The process by which a strand of RNA assembles along one of the DNA strands. (14)

Transcriptional-Level Control   Control of gene expression by regulating whether or not a specific gene is transcribed and how often. (15)

Transduction   A type of genetic recombination resulting from transfer of genes from one organism to another by a virus.

Transfer RNA (tRNA)   A type of RNA that decodes mRNA's codon message and translates it into amino acids. (14)

Transgenic Organism   An organism that possesses genes derived from a different species. For example, a sheep that carries a human gene and secretes the human protein in its milk is a transgenic animal. (16)

Translation   The cell process that converts a sequence of nucleotides in mRNA into a sequence of amino acids in a polypeptide. (14)

Translational-Level Control   Control of gene expression by regulating whether or not a specific mRNA is translated into a polypeptide. (15)

Translocation   The joining of segments of two nonhomologous chromosomes (13)

Transmission Electron Microscope (TEM)   A microscope that works by shooting electrons through very thinly sliced specimens. The result is an enormously magnified image, two-dimensional, of remarkable detail. (App.)

Transpiration   Water vapor loss from plant surfaces. (19)

Transpiration Pull   The principle means of water and mineral transport in plants, initiated by transpiration. (19)

Transposition The phenomenon in which certain DNA segments (mobile genetic elements, or jumping genes) tend to move from one part of the genome to another part. (15)

Transverse Fission The division pattern in ciliated protozoans where the plane of division is perpendicular to the cell's length.

Trimester Each of the three stages comprising the 266-day period between conception and birth in humans. (32)

Triploid Having three sets of chromosomes, abbreviated 3N. (11)

Trisomy Three copies of a particular chromosome per cell. (17)

Trophic Level Each step along a feeding pathway. (41)

Trophozoite The actively growing stage of polymorphic protozoa. (37)

Tropical Rain Forest Lush forests that occur near the equator; characterized by high annual rainfall and high average temperature. (40)

Tropical Thornwood A type of shrubland occurring in tropical regions with a short rainy season. Plants lose their small leaves during dry seasons, leaving sharp thorns. (40)

Tropic Hormones Hormones that act on endocrine glands to stimulate the production and release of other hormones. (25)

Tropisms Changes in the direction of plant growth in response to environmental stimuli, e.g., light, gravity, touch. (21)

True-Breeder Organisms that, when bred with themselves, always produce offspring identical to the parent for a given trait. (12)

Tubular Reabsorption The process by which substances are selectively returned from the fluid in the nephron to the bloodstream. (28)

Tubular Secretion The process by which substances are actively and selectively transported from the blood into the fluid of the nephron. (28)

Tumor-Infiltrating Lymphocytes (TILs) Cytotoxic T cells found within a tumor mass that have the capability to specifically destroy the tumor cells. (30)

Tumor-Suppressor Genes Genes whose products act to block the formation of cancers. Cancers form only when both copies of these genes (one on each homologue) are mutated. (13)

Tundra The marshy, unforested biome in the arctic and at high elevations. Frigid temperatures for most of the year prevent the subsoil from thawing, which produces marshes and ponds. Dominant vegetation includes low growing plants, lichens, and mosses. (40)

Turgor Pressure The internal pressure in a plant cell caused by the diffusion of water into the cell. Because of the rigid cell wall, pressure can increase to where it eventually stops the influx of more water. (7)

Turner Syndrome A person whose cells have only one X chromosome and no second sex chromosome (XO). These individuals develop as immature females. (17)

◄ **U** ►

Ultimate (Top) Consumer The final carnivore trophic level organism, or organisms that escaped predation; these consumers die and are eventually consumed by decomposers. (41)

Ultracentrifuge An instrument capable of spinning tubes at very high speeds, delivering centrifugal forces over 100,000 times the force of gravity. (9)

Unicellular The description of an organism where the cell is the organism. (35)

Uniform Pattern Distribution of individuals of a population in a uniform arrangement, such as individual plants of one species uniformly spaced across a region. (43)

Urethra In mammals, a tube that extends from the urinary bladder to the outside. (28)

Urinary Tract The structures that form and export urine: kidneys, ureters, urinary bladder, and urethra. (28)

Urine The excretory fluid consisting of urea, other nitrogenous substances, and salts dissolved in water. It is formed by the kidneys. (28)

Uterine (Menstrual) Cycle The repetitive monthly changes in the uterus that prepare the endometrium for receiving and sustaining an embryo. (31)

Uterus An organ in the female reproductive system in which an embryo implants and is maintained during development. (31)

◄ **V** ►

Vaccines Modified forms of disease-causing microbes which cannot cause disease but retain the same antigens of it. They permit the immune system to build memory cells without diseases developing during the primary immune response. (30)

Vacoconstriction Reduction in the diameter of blood vessels, particularly arterioles. (28)

Vacuole A large organelle found in mature plant cells, occupying most of the cell's volume, sometimes more than 90% of it. (5)

Vagina The female mammal's copulatory organ and birth canal. (31)

Variable (Experimental) A factor in an experiment that is subject to change, i.e., can occur in more than one state. (2)

Vascular Bundles Groups of vascular tissues (xylem and phloem) in the shoot of a plant. (19)

Vascular Cambrium In perennials, a secondary meristem that produces new vascular tissues. (18)

Vascular Cylinder Groups of vascular tissues in the central region of the root. (18)

Vascular Plants Plants having a specialized conducting system of vessels and tubes for transporting water, minerals, food, etc., from one region to another. (18)

Vascular Tissue System All the vascular tissues in a plant, including xylem, phloem, and the vascular cambium or procambium. (18)

Vasodilation Increase in the diameter of blood vessels, particularly arterioles. (28)

Veins In plants, vascular bundles in leaves. In animals, blood vessels that return blood to the heart. (28)

Venation The pattern of vein arrangement in leaf blades. (18)

Ventricle Lower chamber of the heart which pumps blood through arteries. There is one ventricle in the heart of lower vertebrates and two ventricles in the four-chambered heart of birds and mammals. (28)

Venules Small veins that collect blood from the capillaries. They empty into larger veins for return to the heart. (28)

Vertebrae The bones that form the backbone. In the human there are 33 bones arranged in a gracefully curved line along the bone, cushioned from one another by disks of cartilage. (26)

Vertebral Column The backbone, which encases and protects the spinal cord. (26)

Vertebrates Animals with a backbone. (39)

Vesicles Small membrane-enclosed sacs which form from the ER and Golgi complex. Some store chemicals in the cells; others move to the surface and fuse with the plasma membrane to secrete their contents to the outside. (5)

Vessel A tube or connecting duct containing or circulating body fluids. (18)

Vessel Member A type of conducting cell in xylem functioning when the cell is dead to transport water and dissolved minerals through its hollow interior; also called a vessel element. (18)

Vestibular Apparatus A portion of the inner ear of vertebrates that gathers information about the position and movement of the head for use in maintaining balance and equilibrium. (24)

Vestigial Structure Remains of ancestral structures or organs which were, at one time, useful. (34)

Villi Finger-like projections of the intestinal wall that increase the absorption surface of the intestine. (27)

Viroids are associated with certain diseases of plants. Each viroid consists solely of a small single-stranded circle of RNA unprotected by a protein coat. (36)

Virus Minute structures composed of only heredity information (DNA or RNA), surrounded by a protein or protein/lipid coat. After infection, the viral nucleic acid subverts the metabolism of the host cell, which then manufactures new virus particles. (36)

Visible Light The portion of the electromagnetic spectrum producing radiation from 380 nm to 750 nm detectable by the human eye.

Vitamins Any of a group of organic compounds essential in small quantities for normal metabolism. (27)

Vocal Cords Muscular folds located in the larynx that are responsible for sound production in mammals. (29)

Vulva The collective name for the external features of the human female's genitals. (31)

◀ W ▶

Water Vascular System A system for locomotion, respiration, etc., unique to echinoderms. (39)

Wavelength The distance separating successive crests of a wave. (8)

Waxes A waterproof material composed of a number of fatty acids linked to a long chain alcohol. (4)

White Matter Regions of the brain and spinal cord containing myelinated axons, which confer the white color. (23)

Wild Type The phenotype of the typical member of a species in the wild. The standard to which mutant phenotypes are compared. (13)

Wilting Drooping of stems or leaves of a plant caused by water loss. (7)

Wood Secondary xylem. (18)

◀ X ▶

X Chromosome The sex chromosome present in two doses in cells of a female, and in one dose in the cells of a male. (13)

X-Linked Traits Traits controlled by genes located on the X chromosome. These traits are much more common in males than females. (13)

Xylem The vascular tissue that transports water and minerals from the roots to the rest of the plant. Composed of tracheids and vessel members. (18)

◀ Y ▶

Y Chromosome The sex chromosome found in the cells of a male. When Y-carrying sperm unite with an egg, all of which carry a single X chromosome, a male is produced. (13)

Y-Linked Inheritance Genes carried only on Y chromosomes. There are relatively few Y-linked traits; maleness being the most important such trait in mammals. (13)

Yeast Unicellular fungus that forms colonies similar to those of bacteria. (37)

Yolk A deposit of lipids, proteins, and other nutrients that nourishes a developing embryo. (32)

Yolk Sac A sac formed by an extraembryonic membrane. In humans, it manufactures blood cells for the early embryo and later helps to form the umbilical cord. (32)

◀ Z ▶

Zero Population Growth In a population, the result when the combined positive growth factors (births and immigration). (43)

Zooplankton Protozoa, small crustaceans and other tiny animals that drift with ocean currents and feed on phytoplankton. (37, 40)

Zygospore The diploid spores of the zygomycete fungi, which include Rhizopus, a common bread mold. After a period of dormancy, the zygospore undergoes meiosis and germinates. (36)

Zygote A fertilized egg. The diploid cell that results from the union of a sperm and egg. (32)

Photo Credits

Part 1 Opener Norbert Wu. **Chapter 1** Fig. 1.1: Jeff Gnass. Fig. 1.2a: Frans Lanting/ Minden Pictures, Inc. Fig. 1.2b: David Muench. Fig. 1.3a: Courtesy Dr. Alan Cheetham, National Museum of National History, Smithsonian Institution. Fig. 1.3b: Manfred Kage/Peter Arnold. Fig. 1.3c: Stephen Dalton/NHPA. Fig. 1.3d: Charles Summers, Jr. Fig. 1.3e: Larry West/ Photo Researchers. Fig. 1.3f: Bianca Lavies. Fig. 1.3g: Steve Allen/Peter Arnold. Fig. 1.3h: George Grall. Fig. 1.3i: Michio Hoshino/Minden Pictures, Inc. Fig. 1.6a: CNRI/Science Photo Library/Photo Researchers. Fig. 1.6b: M. Abbey/ Visuals Unlimited. Fig. 1.6c: Steve Kaufman/ Peter Arnold. Fig. 1.6d: Willard Clay. Fig. 1.6e: Jim Bradenburg/Minden Pictures, Inc. Fig. 1.7: (pages 16-17) Charles A. Mauzy; (page 16, top) Anthony Mercieca/Natural Selection; (page 16, center) Wolfgang Bayer/Bruce Coleman, Inc.; (page 16, bottom) Dr. Jeremy Burgess/Science Photo Library/Photo Researchers; (page 17, top) Dr. Eckart Pott/Bruce Coleman, Ltd.; (page 17, bottom) Art Wolfe. Fig. 1..8: Paul Chesley/ Photographers Aspen. Fig. 1.9a: Rainbird/Robert Harding Picture Library. Fig. 1.11: Dick Luria/ FPG International. **Chapter 2** Fig. 2.2a: Couresy Institut Pasteur. Fig. 2.3a: Topham/The Image Works. Fig. 2.3b: Leonard Lessin/Peter Arnold. Fig. 2.3c: Bettmann Archive. Fig. 2.5a: Laurence Gould/Earth Scenes/Animals Animals. Fig. 2.5b: Ted Horowitz/The Stock Market. **Part 2 Opener:** Nancy Kedersha. **Chapter 3** Fig. 3.1: Franklin Viola. Fig. 3.2: Courtesy Pachyderm Scientific Industries. Bioline: Jacan-Yves Kerban/The Image Bank. Fig. 3.7: Courtesy Stephen Harrison, Harvard Biochemistry Department. Fig. 3.11: Gary Milburn/Tom Stack & Associates. Fig. 3.12: Courtesy R. S. Wilcox, Biology Department, SUNY Binghampton. **Chapter 4** Fig. 4.1: Alastair Black/Tony Stone World Wide. Fig. 4.5a: Don Fawcett/Visuals Unlimited. Fig. 4.5b: Jeremy Burgess/Photo Researchers. Fig. 4.5c: Cabisco/Visuals Unlimited. Fig. 4.6: Tony Stone World Wide. Fig. 4.7: Zig Leszczynski/Animals Animals. Fig. 4.8: Robert & Linda Mitchell. Human Perspective: (a) Photofest (b) Bill Davila/Retna. Fig. 4.12a: Frans Lanting/AllStock, Inc. Fig. 4.12b: Mantis Wildlife Films/Oxford Scientific Films/Animals Animals. Fig. 4.17: Stanley Flegler/Visuals Unlimited. Fig. 4.18a: Courtesy Nelson Max, Lawrence Livermore Laboratory. Fig. 4.18b: Tsuned Hayashida/The Stock Market. **Chapter 5** Fig. 5.1a: The Granger Collection. Fig. 5.1b: Bettmann Archive. Fig. 5.2a: Dr. Jeremey Burgess/ Photo Researchers. Fig. 5.2b: CNRI/Photo Researchers. Fig. 5.4: Omikron/Photo Researchers. Fig. 5.6a: Courtesy Richard Chao, California State University at Northridge. Fig. 5.6c: Courtesy Daniel Branton, University of Berkeley. Fig. 5.7: Courtesy G. F. Bohr. Fig. 5.8: Courtesy

Michael Mercer, Zoology Department, Arizona State University. Fig. 5.9: Courtesy D. W. Fawcett, Harvard Medical School. Fig. 5.10a: Courtesy U.S. Department of Agriculture, Fig. 5.11: Courtesy Dr. Birgit H. Satir, Albert Einstein College of Medicine. Fig. 5.13: Courtesy Lennart Nilsson, from *A Child Is Born*. Fig. 5.14a: K. R. Porter/Photo Researchers. Fig. 5.14c: Courtesy Lennart Nilsson, BonnierAlba. Fig. 5.14d: Courtesy Lennart Nilsson, From *A Child Is Born*. Fig. 5.15a: Courtesy J. Elliot Weier. Fig. 5.16a: Courtesy J.V. Small. Fig. 5.17a: Peter Parks/Animals Animals. Fig. 5.17b: Courtesy Dr. Manfred Hauser, RUHR-Universitat Rochum. Fig. 5.18: Courtesy Jean Paul Revel, Division of Biology, California Institute of Technology. Fig. 5.19a: Courtesy E. Vivier, from *Paramecium* by W.J. Wagtendonk, Elsevier North Holland Biomedical Press, 1974. Fig. 5.19b: David Phillips/Photo Researchers. Fig. 5.20: Courtesy C.J. Brokaw and T.F. Simonick, *Journal of Cell Biology*, 75:650 (1977). Reproduced with permission. Fig. 5.21: Courtesy L.G. Tilney and K. Fujiwara. Fig. 5.22a: Courtesy W. Gordon Whaley, University of Texas, Austin. Fig. 5.22c: Courtesy R.D. Preston, University of Leeds, London. **Chapter 6** Fig. 6.1a: Alex Kerstitch. Fig. 6.1b: Peter Parks/Earth Scenes. Fig. 6.2: Marty Stouffer/Animals Animals. Fig. 6.3: Kay Chernush/The Image Bank. Fig. 6.4b: Courtesy Computer Graphics Laboratory, University of California, San Francisco. Fig. 6.8: Courtesy Stan Koszelek, Ph.D., University of California, Riverside. Fig. 6.10: Swarthout/The Stock Market. **Chapter 7** Fig. 7.5: Ed Reschke. Fig. 7.10a: Lennart Nilsson, ©Boehringer Ingelheim, International Gmbh; from *The Incredible Machine*. Human Perspective: (Fig. 1a) Martin Rotker/Phototake; (Fig. 1b) Cabisco/Visuals Unlimited. Fig. 7.11a: Courtesy Dr. Ravi Pathak, Southwestern Medical Center, University of Texas. **Chapter 8** Fig. 8.1a: Carr Clifton. Fig. 8.1b: Shizuo Iijima/Tony Stone World Wide. Fig. 8.5: Joe Englander/ Viesti Associates, Inc. Fig. 8.6: Courtesy T. Elliot Weier. Fig. 8.10: Courtesy Lawrence Berkeley Laboratory, University of California. Fig. 8.13: Pete Winkel/Atlanta/Stock South. Bioline: Courtesy Woods Hole Oceanographic Institution. **Chapter 9** Fig. 9.1: Stephen Frink/ AllStock, Inc. Bioline: (Fig. 1) Frank Oberle/ Bruce Coleman, Inc.; (Fig. 2) Movie Stills Archive. Fig. 9.6: Topham/The Image Works. Fig. 9.9: Courtesy Gerald Karp. Human Perspective: (top and bottom insets) Courtesy MacDougal; (top) Jerry Cooke; (bottom) Richard Kane/Sportchrome East/West. Fig. 9.11: Peter Parks/NHPA. Fig. 9.12a: Grafton M. Smith/ The Image Bank. **Chapter 10** Fig. 10.1: Sipa Press. Fig. 10.2a: Institut Pasteur/CNRI/ Phototake. Fig. 10.3a: Dr. R. Vernon/Phototake.

Fig. 10.3b: CNRI/Science Photo Library/Photo Researchers. Fig. 10.6: Courtesy Dr. Andrew Bajer, University of Oregon, Fig. 10.7: CNRI/ Science Photo Library/Photo Researchers. Fig. 10.8: Dr. G. Shatten/Science Photo Library/ Photo Researchers. Fig. 10.9: Courtesy Professor R.G. Kessel. Fig. 10.10: David Phillips/ Photo Researchers. **Chapter 11** Fig. 11.3: Courtesy Science Software Systems. Fig. 11.5a: Courtesy Dr. A.J. Solari, from *Chromosoma*, vol. 81, p. 330 (1980), Human Perspective: Donna Zweig/Retna. Fig. 11.7: Cabisco/Visuals Unlimited. **Part 3 Opener:** David Scharf. **Chapter 12** Fig. 12.1: Courtesy Dr. Ing Jaroslav Krizenecky. Fig. 12.6a: Doc Pele/Retna. Fig. 12.6b: FPG International. Fig. 12.9: Sydney Freelance/Gamma Liaison. Fig. 12.11 Hans Reinhard/Bruce Coleman, Inc. **Chapter 13** Fig. 13.2a: Robert Noonan. Fig. 13.6: Biological Photo Service. Bioline: Norbert Wu. Fig. 13.9: Historical Pictures Service. Fig. 13.11a: Courtesy M.L. Barr. Fig. 13.11b: Jean Pragen/Tony Stone World Wide. Fig. 13.12: Courtesy Lawrence Livermore National Laboratory. **Chapter 14** Fig. 14.1b: Lee D. Simon/Science Photo Library/Photo Researchers. Fig. 14.4: David Leah/Science Photo Library/Photo Researchers. Fig. 14.5: Dr. Gopal Murti/Science Photo Library/Photo Researchers. Fig. 14.6b: Fawcett/ Olins/Photo Researchers. Fig. 14.7: Courtesy U.K. Laemmli. Fig. 14.10a: From M. Schnos and.R.B. Inman, *Journal of Molecular Biology*, 51:61-73 (1970), ©Academic Press. Fig. 14.10b: Courtesy Professor Joel Huberman, Roswell Park Memorial Institute. Human Perspective: (Fig. 2) Courtesy Skin Cancer Foundation; (Fig. 3) Mark Lewis/Gamma Liaison. Fig. 14.17: Courtesy Dr. O.L. Miller, Oak Ridge National Laboratory. **Chapter 15** Fig. 15.1: Courtesy Richard Goss, Brown University. Fig. 15.2a: (top left) Oxford Scientific Films/ Animals Animals; (top right) F. Stuart Westmorland/Tom Stack & Associates. Fig. 15.2b: Courtesy Dr. Cecilio Barrera, Department of Biology, New Mexico State University. Fig. 15.3b: Courtesy Michael Pique, Research Institute of Scripps Clinic. Fig. 15.7b: Courtesy Wen Su and Harrison Echols, University of California, Berkeley. Fig. 15.8a: Courtesy Stephen Case, University of Mississippi Medical Center. Fig. 15.9: Roy Morsch/The Stock Market. **Chapter 16** Fig. 16.1a: David M. Dennis/Tom Stack & Associates. Fig. 16.1b: Courtesy Lakshmi Bhatnagor, Ph.D., Michigan Biotechnology Institute. Fig. 16.2: Art Wolfe/All Stock, Inc. Fig. 16.3: Ken Graham. Fig. 16.4a: Courtesy R.L. Brinster, Laboratory for Reproductive Physiology, University of Pennsylvania. Fig. 16.4b: John Marmaras/Woodfin Camp & Associates. Fig. 16.5: Courtesy Robert Hammer, School of Veterinary Medicine, University of

Pennsylvania. Human Perspective: Courtesy Dr. James Asher, Michigan State University. Fig. 16.6b: Professor Stanley N. Cohen/Photo Researchers. Fig. 16.10: Ted Speigel/Black Star. Fig. 16.11b: Philippe Plailly/Science Photo Library/Photo Researchers. Bioline: Photograph by Anita Corbin and John O'Grady, courtesy the British Council. **Chapter 17** ig. 17.2d: Courtesy Howard Hughs Medical Institution. Fig. 17.3: Tom Raymond/Medichrome/The Stock Shop. Fig. 17.7a: Culver Pictures. Fig. 17.7b: Courtesy Roy Gumpel. Fig. 17.9a: Will & Deni McIntyre/Photo Researchers. Human Perspective: Gamma Liaison. **Part 4 Opener:** J. H. Carmichael, Jr./The Image Bank. **Chapter 18** Fig. 18.1a: Ralph Perry/Black Star. Fig. 18.1b: Jeffrey Hutcherson/DRK Photo. Fig. 18.4: Courtesy Professor Ray Evert, University of Wisconsin, Madison. Fig. 18.5: Walter H. Hodge/Peter Arnold; (inset) Courtesy Carolina Biological Supply Company. Bioline: (Fig. 1) Doug Wilson/West Light; (Fig. 2) Courtesy Gil Brum. Fig. 18.7: Luiz C. Marigo/Peter Arnold. Fig. 18.8: Dr. Jeremy Burgess/Science Photo Library/Photo Researchers. Fig. 18.9: Saul Mayer/The Stock Market; (inset) David Scharf/Peter Arnold. Fig. 18.10a: Carr Clifton. Fig. 18.10b: Ed Reschke. Fig. 18.11: Courtesy Thomas A. Kuster, Forest Products Laboratory/USDA. Fig. 18.12: A.J. Belling/Photo Researchers. Fig. 18.13: Courtesy DSIR Library Centre. Fig. 18.15: Biophoto Associates/Photo Researchers; (inset) Courtesy C.Y. Shih and R.G. Kessel, reproduced from *Living Images*, Science Books International, 1982. Fig. 18.16: Courtesy Industrial Research Ltd, New Zealand. Fig. 18.19: Robert & Linda Mitchell; (inset) From *Botany Principles* by Roy H. Saigo and Barbara Woodworth Saigo, ©1983 Prentice-Hall, Inc. Reproduced with permission. Fig. 18.22a: Jerome Wexler/Photo Researchers. Fig. 18.22b: Fritz Polking/Peter Arnold. Fig. 18.22c: Dwight R. Kuhn. Fig. 18.22d: Richard Kolar/Earth Scenes. Fig. 18.24: Biophoto Associates/Science Source/Photo Researchers. **Chapter 19** Fig. 19.1a: Courtesy of C.P. Reid, School of Natural Resources, University of Arizona. Fig. 19.3a: Peter Beck/The Stock Market. Fig. 19.3b: Dr. J. Burgess/Photo Researchers. Fig. 19.4: D. Cavagnaro/DRK Photo. Fig. 19.7a: Biophoto Associates/Photo Researchers. Fig. 19.7b: Alfred Pasieka/Peter Arnold. Fig. 19.7c: Courtesy DSIR Library Centre. Fig. 19.8: Martin Zimmerman. **Chapter 20** Fig. 20.1: G.I. Bernard/Earth Scenes/Animals Animals. Fig. 20.2a: Dwight R. Kuhn/DRK Photo. Bioline (Cases 1 and 3) Biological Photo Service; (Case 2) P.H. and S.L. Ward/Natural Science Photos; (Case 4): G.I. Bernard/Animals Animals. Fig. 20.3: E.R. Degginger. Fig. 20,4a: William E. Ferguson. Fig. 20.4b: Phil Degginger. Fig. 20.4c: E.R. Degginger. Fig. 20.5a: Runk/Shoenberger/Grant Heilman Photography. Fig. 20.5b: William E. Ferguson. Fig. 20.6a: Robert Harding Picture Library. Fig. 20,6b Richard Parker/Photo Researchers. Fig. 20.6c: Jack Wilburn/Earth Scenes/Animals Animals. Fig. 20.9a: Visuals Unlimted. Fig. 20.9b: Manfred Kage/

Peter Arnold. Human Perspective: Harold Sund/The Image Bank. Fig. 20.15: Courtesy Media Resources, California State Polytechnic University. Fig. 20.16a: John Fowler/Valan Photos. Fig. 20.16b: Richard Kolar/Earth Scenes/Animals Animals. Fig. 20.17: Breck Kent. **Chapter 21** Fig. 21.1a: Gary Milburn/Tom Stack & Associates. Fig. 21.1b: Schafer & Hill/Peter Arnold. Human Perspective: (Fig. 1) Enrico Ferorelli; (Fig. 2) Michael Nichols/Magnum Photos, Inc. Fig. 21.5: Runk/Schoenberger/Grant Heilman Photography. Fig. 21.6: Courtesy Dr. Harlan K. Pratt, University of California, Davis. Fig. 21.7: Scott Camazine/Photo Researchers. Fig. 21.9a: E.R. Degginger. Fig. 21.9b: Fritze Prenze/Earth Scenes/Animals Animals. Fig. 21.12a: Pam Hickman/Valan Photos. Fig. 21.12b: R.F. Head/Earth Scenes/Animals Animals. **Part 5 Opener:** Joe Devenney/The Image Bank. **Chapter 22** Fig. 22.2b: Dr. Mary Notter/Phototake. Fig. 22.3b: Professor P. Motta, Department of Anatomy, University of "La Sapienza", Rome/Photo Researchers. Fig. 22.6a: Fred Bavendam/Peter Arnold. Fig. 22.6b: Michael Fogden/DRK Photo. Fig. 22.6c: Peter Lamberti/Tony Stone World Wide. Fig. 22.6d: Gerry Ellis Nature Photography. Fig. 22.7a: Mike Severns/Tom Stack & Associates. Fig. 22.7b: Norbert Wu. **Chapter 23** ig 23.1:Fawcett/Coggeshall/Photo Researchers. Fig. 23.5: (top left) Vu/©T. Reese and D.W. Fawcett/Visuals Unlimited; (bottom left) Courtesy Lennart Nilsson, *Behold Man*, Little Brown & Co., Boston. Fig. 23.9: Courtesy Lennart Nilsson, *Behold Man*, Little, Brown and Co., Boston. Fig. 23.10b: Alan & Sandy Carey. Human Perspective: Science Vu/Visuals Unlimited. Fig. 23.14: Focus on Sports. Fig. 23.15: The Image Works. Fig. 23.20a: Doug Wechsler/Animals Animals. Fig. 23.20b: Robert F. Sisson/National Geographic Society. **Chapter 24** Fig. 24.1a: Anthony Bannister/NHPA. Fig. 24.1b: Giddings/The Image Bank. Fig. 24.1c: Michael Fogden/Bruce Coleman, Inc. Fig. 24.4: (left) Omikron/Photo Researchers. Fig. 24.4: (inset) Don Fawcett/K. Saito/K. Hama/Photo Researchers. Fig. 24.5a: (top) Courtesy Lennart Nilsson, *Behold Man*, Little Brown & Co., Boston; (center) Don Fawcett/K. Saito/K. Hama/Photo Researchers. Fig. 24.5b: Star File. Fig. 24.6: Philippe Petit/Sipa Press. Fig. 24.7b: Jerome Shaw. Fig. 24.10a: Oxford Scientific Films/Animals Animals. Fig. 24.10b: Kjell Sandved/Photo Researchers. Fig. 24.10c: George Shelley; (inset) Raymond A. Mendez/Animals Animals. **Chapter 25** Fig. 25.1: Robert & Linda Mitchell. Fig. 25.7: Courtesy Circus World Museum. Fig. 25.8: Courtesy A.I. Mindelhoff and D.E.. Smith, *American Journal of Medicine*, 20: 133 (1956). Fig. 25.13: Graphics by T. Hynes and A.M. de Vos using University of California at San Francisco MIDAS-plus software; photo courtesy of Genentech, Inc. **Chapter 26** Fig. 26.1: Ed Reschke/Peter Arnold. Fig. 26.2a: Joe Devenney/The Image Bank. Fig. 26.2b: John Cancalosi/DRK Photo. Fig. 26.3: Wallin/Taurus Photos. Fig. 26.4a: E.S. Ross/Phototake. Fig. 26.5a: D. Holden Bailey/Tom

Stack & Associates. Fig. 26.5b: Jany Sauvanet/Photo Researchers. Fig. 26.6a: (top) Courtesy Lennart Nilsson, *Behold Man*, Little Brown & Co., Boston; (center) Michael Abbey/Science Source/Photo Researchers. Fig. 26.6b: F. & A. Michler/Peter Arnold. Human Perspective: Courtesy Professor Philip Osdoby, Department of Biology, Washington University, St. Louis. Fig. 26.8: Duomo Photography, Inc. Fig. 26.12: A. M. Siegelman/FPG International. Fig. 26.19a: Alese & Mort Pechter/The Stock Market. Fig. 26.19b-c: Stephen Dalton/NHPA. **Chapter 27** Bioline, Fig. 27.3, and Fig. 27.5: Courtesy Lennart Nilsson, *Behold Man*, Little Brown & Co., Boston, Fig. 27.6: Micrograph by S.L. Palay, courtesy D.W. Fawcett, from *The Cell*, © W..B. Saunders Co. Fig. 27.8a: Courtesy Gregory Antipa, San Francisco State University. Fig. 27.8b: Gregory Ochocki/Photo Researchers. Human Perspective: Derik Muray Photography, Inc. Fig. 27.9a: Warren Garst/Tom Stack & Associates. Fig. 27.9b: Hervé Chaumeton/Jacana. Fig. 27.9c: D. Wrobel/Biological Photo Service. Fig. 27.10: Sari Levin. **Chapter 28** Fig. 28.2: Biophoto Associates/Photo Researchers. Fig. 28.4: Courtesy Lennart Nilsson, *Behold Man*, LIttle Brown & Co., Boston. Fig. 28.10: CNRI/Science Photo Library/Photo Researchers. Fig. 28.11a-b: Howard Sochurek. Fig. 28.11c: Dan McCoy/Rainbow. Fig. 28.12: Jean-Claude Revy/Phototake. Fig. 28.15: Dr. Tony Brain/Science Photo Library/Photo Researchers. Fig. 28.16: Alan Kearney/FPG International. Fig. 28.17a: Ed Reschke/Peter Arnold. Fig. 28.18: Manfred Kage/Peter Arnold. Fig. 28.19b: Runk/Schoenberger/Grant Heilman Photography. **Chapter 29** Fig. 29.1: Richard Kane/Sports Chrome, Inc. Fig. 29.2: Courtesy Ewald R. Weibel, Anatomisches Institut der Universität Bern, Switzerland, from "Morphological Basis of Alveolar-Capillary Gas Exchange", *Physiological Review*, Vol. 53, No. 2, April 1973, p. 425. Fig. 29.3: (top left and right insets) Courtesy Lennart Nilsson, *Behold Man*, Little Brown & Co., Boston; (center) CNRI/Science Photo Library/Photo Researchers. Fig. 29.7: Four By Five/SUPERSTOCK. Human Perspective: (Fig. 2) Vu/O. Auerbach/Visusals Unlimited; (Fig. 3) Photofest. Fig. 29.8: Kjell Sandved/Bruce Coleman, Inc. Fig. 29.10: Robert F. Sisson/National Geographic Society. Fig. 29.11a: Robert & Linda Mitchell. **Chapter 30** Fig. 30.1 and Fig. 30.5: Courtesy Lennart Nilsson, Boehringer Ingelheim International GmbH. Bioline: Courtesy Steven Rosenberg, National Cancer Institute. Fig. 30.6: G. Robert Bishop/AllStock, Inc. Fig. 30.7b: Courtesy A.J. Olson, TSRI, Scripps Institution of Oceanography. Fig. 30.10: Alan S. Rosenthal, from "Regulation of the Immune Response—Role of the Macrophage', *New England Journal of Medicine*, November 1980, Vol. 303, #20, p. 1154, courtesy Merck Institute for Therapeutic Research. Fig. 30.11: Nancy Kedersha. Human Perspective: (Fig. 1a) David Scharf/Peter Arnold; (Fig. 1b) Courtesy Acarology Laboratory, Museum of Biological Diversity, Ohio State University, Columbus; (Fig. 1c) Scott Camazine/

Index